P9-EDJ-886

Ancestors:
The Hard Evidence

Most of the participants in the "Ancestors" scientific sessions
outside the American Museum of Natural History, April 6, 1984.

Ancestors: *The Hard Evidence*

Proceedings of the Symposium held at the
American Museum of Natural History
April 6–10, 1984
to mark the opening of the exhibition
"Ancestors: Four Million Years of Humanity"

Editor

Eric Delson
American Museum of Natural History
New York, New York

Lehman College, City University of New York
Bronx, New York

Alan R. Liss, Inc., New York

To Bill Howells, who sparked my interest in paleoanthropology, and to my wife, Roberta, and my son, Bill, who have endured it.

Address all Inquiries to the Publisher
Alan R. Liss, Inc., 41 East 11th Street, New York, NY 10003

Printed in the United States of America

Second Printing, August 1986

Library of Congress Cataloging in Publication Data

Main entry under title:

Ancestors, the hard evidence.

Bibliography: p.
Includes index.
1. Human evolution—Congresses. 2. Fossil man—Congresses. I. Delson, Eric.
GN282.A53 1985 573.3 85-10260
ISBN 0-8451-0249-4

Contents

Contributors

Karl Dietrich Adam, Staatliches Museum für Naturkunde, 714 Ludwigsburg BRD, Arsenalplatz 3, West Germany [272]

Peter Andrews, Department of Palaeontology, British Museum (Natural History), Cromwell Road, London SW7 5BD, England [14]

Anna K. Behrensmeyer, Department of Paleobiology, National Museum of Natural History, NHB-E207 M.S. 121, Washington, DC 20560 U.S.A. [60]

Amilcare Bietti, Departimento di Biologia Animale e dell'Uomo, Sezione di Antropologia, Università di Roma "La Sapienza", Città Universitaria, 00100 Rome, Italy [277]

C.K. Brain, Transvaal Museum, P.O. Box 413, Pretoria, South Africa 0001 [72]

Francis H. Brown, Department of Geology and Geophysics, University of Utah, 1400 East 2nd South, Salt Lake City, UT 84112 U.S.A. [82]

Ronald J. Clarke, Department of Anatomy, Medical School, University of the Witwatersrand, York Road, Parktown, Johannesburg, 2193 South Africa [171,301]

H.B.S. Cooke, 2133 154th Street, White Rock (South Surrey), British Columbia V4A 4S5, Canada [60]

T. Davies, Research School of Earth Sciences, Australian National University, Canberra, A.C.T. Australia [82]

Michael H. Day, Department of Anatomy, St. Thomas' Hospital Medical School, Lambeth Palace Road, London SE1 7EH, England [91]

Eric Delson, Department of Anthropology, Herbert H. Lehman College, City University of New York, Bronx, NY 10468; and Department of Vertebrate Paleontology, American Museum of Natural History, New York, NY 10024 U.S.A. [xi,1,9,296]

John de Vos, Rijksmuseum van Natuurlijke Historie, Posbus 9517, 2300 RA Leiden, Netherlands [215]

John G. Fleagle, Department of Anatomical Sciences, State University of New York, Stony Brook, NY 11794 U.S.A. [23]

Jens Lorenz Franzen, Abteilung Paläoanthropologie, Forschungsinstitut Senckenberg, 6 Frankfurt-am-Main 1 BRD, Senckenberg-Anlage 25, West Germany [221]

Frederick E. Grine, Department of Anthropology, State University of New York, Stony Brook, NY 11794 U.S.A. [153]

John M. Harris, Department of Vertebrate Paleontology, Natural History Museum of Los Angeles County, 900 Exposition Boulevard, Los Angeles, CA 90007 U.S.A. [76]

Ralph L. Holloway, Department of Anthropology, Columbia University, New York, NY 10027 U.S.A. [319]

Jean-Jacques Hublin, Laboratoire de Paléontologie des Vertébrés et Paléontologie Humaine, Université de Paris VI, 75230 Paris Cedex 05, France [283]

Glynn Ll. Isaac, Department of Anthropology, Harvard University, Cambridge, MA 02138 U.S.A. [346]

Donald C. Johanson, Institute of Human Origins, 2453 Ridge Road, Berkeley, CA 94709 U.S.A. [120]

William L. Jungers, Department of Anatomical Sciences, State University of New York, Stony Brook, NY 11794 U.S.A. [184]

Richard F. Kay, Department of Anatomy, Duke University Medical Center, Durham, NC 27710 U.S.A. [23]

William H. Kimbel, Institute of Human Origins, 2453 Ridge Road, Berkeley, CA 94709; and Departments of Anthropology and Biology, Kent State University, Kent, OH 44242 U.S.A. [120]

Reinhart Kraatz, Geologisch-Paläontologisches Institut der Universität Heidelberg, Im Neuenheimer Feld 234, 69 Heidelberg 1 BRD, West Germany [268]

Jeffrey T. Laitman, Department of Anatomy, Mount Sinai School of Medicine, One Gustave Levy Place, New York, NY 10029 U.S.A. [265]

The number in brackets designates the first page number of the contributor's chapter.

R. Maier, Research School of Earth Sciences, Australian National University, Canberra, A.C.T. Australia [82]

Wolfgang O. Maier, Zentrum der Morphologie, D 6000 Frankfurt-am-Main 70 BRD, Theodor-Stern-Kai 7, West Germany [249]

Ian McDougall, Research School of Earth Sciences, Australian National University, Canberra, A.C.T. Australia [82]

Henry M. McHenry, Department of Anthropology, University of California, Davis, CA 95616 U.S.A. [178]

Abel T. Nkini, Zentrum der Morphologie, D 6000 Frankfurt-am-Main 70 BRD, Theodor-Stern-Kai 7, West Germany [202,249]

Todd R. Olson, Anatomical Sciences Division, City University of New York Medical School, Convent Ave & 138 St., New York, NY 10031 U.S.A. [102]

David Pilbeam, Department of Anthropology, Harvard University, Cambridge, MA 02138 U.S.A. [51]

Jakov Radovčić, Geološko-Paleontološki Muzej, Demetrova 18/II, 41000 Zagreb, Yugoslavia [310]

Yoel Rak, Department of Anatomy, Sackler Medical School, Tel-Aviv University, Ramat-Aviv, Tel-Aviv, Israel [168]

S. Mahmood Raza, 917 Gali 107, G-9/4 Islamabad, Pakistan [339]

G.P. Rightmire, Department of Anthropology, State University of New York, Binghamton, NY 13901 U.S.A. [255]

Brigitte Senut, Laboratoire d'Anthropologie, Musée de l'Homme, Place du Trocadéro, 75116 Paris, France [193]

S.M. Ibrahim Shah, Director, Geological Survey of Pakistan, Quetta, Pakistan [339]

Harry L. Shapiro, Department of Anthropology, American Museum of Natural History, New York, NY 10024 U.S.A. [6]

Elwyn L. Simons, Duke University Primate Center, 3705 Erwin Road, Durham, NC 27705 U.S.A. [37]

Fred H. Smith, Department of Anthropology, 252 South Stadium Hall, University of Tennessee, Knoxville, TN 37916 U.S.A. [325]

Arun Sonakia, Regional Palaeontological Laboratory, Geological Survey of India, Central Region, 46 Ramdaspeth, Nagpur 440 010, India [334]

Jack T. Stern, Jr., Department of Anatomical Sciences, State University of New York, Stony Brook, NY 11794 U.S.A. [184]

C.B. Stringer, Department of Palaeontology, British Museum (Natural History), Cromwell Road, London SW7 5BD, England [289]

Randall L. Susman, Department of Anatomical Sciences, State University of New York, Stony Brook, NY 11794 U.S.A. [184]

Christine Tardieu, Laboratoire d'Anatomie Comparée, 55 rue Buffon, 75005 Paris, France [193]

Ian Tattersall, Department of Anthropology, American Museum of Natural History, New York, NY 10024 U.S.A. [1]

Herbert Thomas, Institut de Paléontologie, 8 Rue de Buffon, 75005 Paris, France [42]

Phillip V. Tobias, Department of Anatomy, Medical School, University of the Witwatersrand, York Road, Parktown, Johannesburg, 2193 South Africa [94]

Erik Trinkaus, Department of Anthropology, University of New Mexico, Albuquerque, NM 87131 U.S.A. [325]

John A. Van Couvering, Micropaleontology Press, American Museum of Natural History, New York, NY 10024 U.S.A. [1]

Bernard Vandermeersch, Laboratoire d'Anthropologie, Université de Bordeaux I, Avenue des Facultés, 33405 Talence, France [306]

E.S. Vrba, Department of Vertebrate Palaeontology, Transvaal Museum, P.O. Box 413, Pretoria, South Africa 0001 [63]

Tim D. White, Department of Anthropology, University of California, Berkeley, CA 94720 U.S.A. [120,138]

Milford H. Wolpoff, Department of Anthropology, University of Michigan, Ann Arbor, MI 48104 U.S.A. [202]

Bernard Wood, Department of Anatomy and Biology as Applied to Medicine, The Middlesex Hospital Medical School, Cleveland Street, London W1P 6DB, England [206]

Wu Rukang, Institute of Vertebrate Paleontology and Paleoanthropology, Academia Sinica, P.O. Box 643, Beijing 28, People's Republic of China [245]

Preface

From early in its history, the "Ancestors" project was designed to be multifaceted. In addition to the exhibition of original human fossils, the idea of a symposium at which curators of these fossils could present the results of their current work was among the first aspects planned. Once approval for the project was received, it was decided that a symposium volume would be a logical addition. This book is thus the outcome of the Ancestors symposium, "Paleoanthropology: The Hard Evidence," held from April 6 to April 10, 1984.

Each curator bringing fossils to the exhibit was invited to present a paper. About half accepted; the others, not specialists in paleoanthropology, declined. On the basis of the titles offered by these curators, a second group, consisting of colleagues with major research interests complementing those of the curators, was invited to participate in the scientific aspect of "Ancestors" by presenting a symposium paper (or chairing a session) and taking part in the fossil-study sessions. Almost all of the contributors to the symposium submitted papers for publication—the few exceptions are noted in the introductory sections. Owing to financial exigencies, the majority of the participants thus invited were Americans, but a number of our most active colleagues from Eurasia and Africa were also included.

The symposium was planned to represent more of a stock-taking—a summary of the state of paleoanthropology in the mid-1980's—than a source of new directions, although some of those were included. The speakers were asked to concentrate on aspects or implications of the "hard evidence," the fossils themselves, in light of the direct association with the exhibition of original fossils. Again, in view of this focus, archeological presentations were not solicited, since within the time available broadening the range of topics would have meant reducing the depth of those covered.

The Wenner-Gren Foundation for Anthropological Research cosponsored the symposium with the American Museum of Natural History. In addition, Mrs. Lita Osmundsen, Director of Research at Wenner-Gren, joined Ian Tattersall, John Van Couvering, and myself on the Symposium Committee and gave us the benefit of her years of experience in organizing meetings of all sizes. The generous support provided by Wenner-Gren for the symposium and its publication was vital, and Mrs. Osmundsen's active participation was itself an indispensible element in the success of the entire scientific program.

Although the study session is described in the first article, it is appropriate to note two aspects here. First, most of the financial support for that period of comparative study and research (apart from expenses for the curators) came from the Anthropology Program of the National Science Foundation and the L.S.B. Leakey Foundation; without this aid, combined with that from Wenner-Gren for the symposium, the scientific sessions would simply have been limited to the participating curators and would therefore have had much less lasting value and impact.

Second, it was decided from the earliest discussions of the specimen study format that no photography by participants would be allowed. The prospect of a queue of scientists waiting to photograph the Taung skull or the Heidelberg jaw created visions of possible damage and disaster. Instead, I arranged with Chester Tarka, of the AMNH Department of Vertebrate Paleontology, to take a series of color slides comparing pairs of the available fossils. Once permission was obtained from all curators, Chester set up his equipment in the vault storage area where the fossils were kept, working in the evenings during the study session and on the following weekend, before the specimens were installed in the exhibition. Paul Beelitz sacrificed his time to accompany Chester during these off-hours in order that security arrangements could be fully observed. The resulting group of color slides has been distributed to the "Ancestors" participants, and thanks to support from the Wenner-Gren and L.S.B. Leakey Foundations it was possible to include a color insert of 16 pages of these photos in this book.

The organization of the papers in this volume closely corresponds to the plan of the symposium. Following two introductory essays, the remaining contributions are organized in seven topical sections reflecting the original half-day sessions. In the symposium, each group of three or four papers was followed by a discussion interval during which the session chair posed questions received in written form from the audience. The entire symposium was tape recorded, and transcripts of many papers and all the discussions were prepared by assistants to Ken Weaver of *National Geographic*, to whom we are most grateful. The discussion transcripts served as guides to the session chairs, who were asked to introduce each section of the book and to summarize the discussion they moderated. They were also encouraged to add comments of their own on the topics under consideration.

The task of putting this volume together has fallen mainly to me, but I was assisted with the editing by my wife, Dr. Roberta Marx Delson, and by my students Elizabeth Strasser, Robert Kluberdanz and, especially, David Dean. Ian Tattersall and John Van Couvering also read and commented upon a number of specific papers, and Barbara Werscheck and Lorraine Meeker of the Department of Vertebrate Paleontology provided help with German translation and illustration makeovers, respectively. Paulette Cohen of Alan R. Liss, Inc. first accepted this volume for publication and was most helpful throughout its gestation, as were her colleagues Doug McLaurine, Michael O'Connor, and especially Amy Kramer. As a result of their efforts, it will appear little more than a year after the opening of "Ancestors: Four Million Years of Humanity." I thank all of them for their much-appreciated assistance. Lastly, I take this final opportunity to thank once again Dr. Thomas D. Nicholson, Director of the American Museum of Natural History, for his foresight in supporting the "Ancestors" project and in financially underwriting almost all of it from Museum sources; the many colleagues who participated in all phases of the scientific sessions and submitted papers to this volume; and, especially, the curators who brought their fossils to New York and left them in our care so that the "Ancestors" exhibit could become a model of international cooperation in paleoanthropology.

Eric Delson
New York and Fort Lee
October, 1984

Ancestors: The Hard Evidence, pages 1–5

The "Ancestors" Project: An Expurgated History

Ian Tattersall, John A. Van Couvering, AND Eric Delson
Department of Anthropology (I.T.), Micropaleontology Press (J.A.V.C.), and Department of Vertebrate Paleontology (E.D.), American Museum of Natural History, New York, New York 10024.

One question often asked us is, "How did you think of 'Ancestors'?" In fact, it doesn't take a genius, let alone three geniuses, to conceive of putting the world's most famous human fossils on display, especially if you work at one of the great natural history museums and spend a lot of time thinking and writing about prehistoric primates. What it does take, perhaps, is chutzpah: John's initial memo to Ian, dated April 27, 1979, begins: "Conceivably the most grandiose idea I have ever aspired to entertain." We soon discovered, however, that grandiose as it may have seemed to us, we were by no means the first to have had the idea of staging an international exhibition of original human fossil specimens. But always before, in spite of the great popular and scientific interest such an event would arouse, the notion had been dismissed in the face of apparently insurmountable problems. So the question that people ought perhaps to ask us is, "What made you think 'Ancestors' *could* happen?"

Over the months that followed the initial floating of the idea it may simply have been the benign ambience of the old "Blarney Castle" on 72nd Street, an establishment where excellent ideas draw benefit from collegial encouragement and distance from care, that kept the project alive and gestating. However, even in less euphoric surroundings we soon began to sense that the time was ripe: public as well as scientific interest was at a new high due to a spate of well-publicized discoveries and to the popular books derived from them, as well as because of the escalation of the "creationist" assault on evolutionary biology, a matter of great and growing concern at the Museum. Nevertheless, when we presented the outline of the proposed exhibition to Dr. Thomas D. Nicholson, the Museum's Director, we suggested that an advisory committee first be assembled to help us determine if the proposal was even feasible. If they agreed with us that it was, we hoped (rightly) that the members would then become a steering committee that would advise us, from expert knowledge, on the complex and delicate task of getting the specimens needed from their various vaults and locked cases around the world.

Our proposal met with Dr. Nicholson's enthusiastic support, and the committee eventually convened for its first meeting on January 27, 1981, in the Portrait Room at the Museum. Every person we contacted agreed to come. Present in addition to Museum personnel were four eminent paleoanthropologists: David Pilbeam of Harvard, Elwyn Simons of Duke, and Clark Howell and Desmond Clark, both of Berkeley. With them sat the heads of four major private foundations that support the search for fossil man: Lita Osmundsen of the Wenner–Gren Foundation, Mary Pechanec of the L.S.B. Leakey Foundation, Melvin Payne of the National Geographic Society, and David Mash of the Foundation for Research into the Origins of Man. Wenner–Gren and the Leakey Foundation later became involved in the scientific activities that preceded the exhibition, as did the four academic members.

This January 1981 meeting was the first real trial for the "Ancestors" project. We began by outlining our concept of the exhibition, to justify asking the assistance of the committee. Our view was that the primary purpose of the exhibition was not, in any political sense, to make a statement about human evolution, as if to dignify the challenge of "creation science," but simply to make it possible for the public, lay and professional, to witness at first hand and for the first time a full sample of what had been found in the century or so since the search for prehistoric humans began. The special significance of the specimens as, in a broad sense, the ancestors of each person who came to the exhibition, would be acknowledged by presenting the fossils as objects of beauty and fascination in their own right, separate from the explanatory material needed to place them in the context of current scientific understanding.

The committee was unanimous in applauding the concept of the exhibition, but its members varied in the degree of their optimism that the goal could be achieved. The unmistakable conclusion, however, was that it was worth a try, if the Museum were prepared to invest the resources necessary to ensure the security of the fossils should the exhibition become a reality. For it was overwhelmingly clear that if their guardians could not be assured that every possible precaution was being taken, not a single specimen would come to New York, and rightly so. The committee returned more than once to the subject of physical security during the exhibition, and even showed a rather ghoulish fascination with ever more ingenious ways (which we duly noted) to smash delicate fossils to bits. Of course, we had not yet designed and built the display cases—a project which ultimately involved a lot of the New York experience gained from bank teller cages, armored car windows, and Transit Authority token booths. The Police Department supplied specialists on criminal and lunatic violence, and Chemical Bank offered the expertise of its security chief. As it turned out, of the entire supply of active maniacs in the Tri-State area, not one chose in the event to test the system that resulted, possibly because of the enthusiasm and vigilance of the Museum guards assigned to the exhibit. Also, in view of the fragility of many of the specimens, we proposed that no fossil would be touched, photographed, or viewed in the open except by personal arrangement with the responsible curator, who would retain absolute control of the specimen whenever it was out of the closed, high-security exhibit case.

With the most convincing of reasons and the best in security, could we assemble enough specimens to justify an exhibition? A draft "wish list" had been presented to the advisory committee, and names were suggested of those curators from around the world whose participation was considered most important to the success of the venture. Over the following several months we made both informal and formal approaches to these colleagues, and were delighted by the generally positive responses we received. We were particularly gratified by the immediate enthusiasm for the project shown by Elisabeth Vrba and Bob Brain of the Transvaal Museum; it may well be that their early commitment to participate with a large number of the important fossils in their care helped to sway the decisions of many other curators. For, frankly, we were surprised by the number of affirmative replies we received over these months, since our requests presented each individual we approached with a difficult curatorial decision. It cannot have been easy for them to justify the risks of transporting such delicate and irreplaceable specimens over vast distances, and of placing them on prolonged exhibit in the homeland of creationism. Some curators or institutional trustees, after longer or shorter periods of indecision, quite understandably declined to take those risks. Nevertheless, most of those approached at this initial stage shared our enthusiasm for this unique project, and agreed on principle to participate in "Ancestors" if adequate security for their specimens could be assured. Accordingly, at a second meeting of the Advisory Committee, held on June 18, 1982, it was formally decided to proceed with "Ancestors," and the date of April, 1984 was set for the opening. At this time the American Museum committed itself to underwriting the unmet costs of the exhibition (which in the end meant almost all of the costs, since the exhibition ultimately proved to be too "controversial" to attract either Federal or private funding (with the exception of a valuable indemnity from the Federal Council on the Arts and the Humanities), although small but useful grants were generously made by the New York State Council on the Arts), and to actively seek the participation of additional institutions holding important human fossils. In addition, since every fossil would be personally hand-carried to New York by its curator, it was agreed that the resulting assembly of scientists should be made the occasion of a scientific symposium in advance of the opening of the exhibition.

Following this meeting, we began planning details of the exhibition with Michael Blakeslee of the Museum's Department of

Exhibitions, who had been assigned to "Ancestors" as designer. Scientifically, however, the major turning point of the event occurred in October 1982, when Eric and Ian attended the First International Congress of Human Paleontology, held in Nice. Here, for the first time, it was possible for a substantial number of potential and committed participants to gather in person, and to discuss the concerns of those contributing fossils to the exhibit. Besides permitting a consensus on technical matters such as security and insurance, this meeting was memorable for an impassioned appeal by Phillip Tobias to exploit the scientific potential of "Ancestors" to the utmost. Phillip recalled the excitement of those three days in 1964 when, in Cambridge, he and Ralph von Koenigswald were able to compare the Olduvai *Homo habilis* fossils with Ralph's material from Java. The idea of staging a major scientific comparison session of the fossils assembled in New York struck a chord with all those present, and the idea was enthusiastically endorsed by the meeting. Many, Phillip among them, felt that the addition of this unique scientific aspect to the proceedings would be of great help in persuading their institutions or governments to permit the showing of their fossils in "Ancestors." We were ourselves delighted by this development; for although the exciting possibilities of scientific comparison of the assembled specimens had certainly not failed to occur to us, we had felt somewhat constrained from actually suggesting this because we had from the beginning assured potential participants that their fossils would remain under their personal control at all times, and would not be handled by third persons except with their express permission. Since the suggestion had now come from a curator, however, and had received the collective endorsement of a substantial number of those who had agreed to participate in "Ancestors," the way was now open to making the scientific sessions a reality.

The organization of the scientific sessions received detailed consideration at the final meeting of the Advisory Commitee, held on January 14, 1983. It was decided to seek funding for both a scientific comparison session and a symposium, each of several days' duration, in which not only transporting curators, but also invited paleoanthropologists from the U.S.A. and abroad would participate. A subcommittee consisting of Eric, John, Ian, and Lita Osmundsen was delegated to plan these scientific activities in detail. That the resulting plans were successfully implemented, as the existence of this volume testifies, is due to the generous support of the Wenner–Gren Foundation (which funded the attendance at the symposium of non-curatorial participants and helped to defray the expenses of this volume), and of the National Science Foundation and L.S.B. Leakey Foundation (both of which subsidized the attendance of non-curatorial invited participants at the comparison sessions and paid the expenses of comparison photography), as well as to that of the American Museum of Natural History, which provided facilities and underwrote the expenses of all "Ancestors" curators who attended the scientific sessions.

Arrangements for the comparison sessions posed something of a challenge. Obviously, we could not simply lay all of the available fossils on a table and invite a free-for-all. In the end we decided that a maximum of six scientists at a table making comparisons between a maximum of seven specimens would be as much as would be manageable; with four or five such sessions running concurrently, and ten such periods over the four days of comparison, we concluded that we would be able to offer each of the sixty-plus participating scientists a reasonable choice of comparisons to make, as well as to provide opportunities for other scientists and students who had obtained curatorial permission to examine specimens. Little suspecting what we were getting into, instead of setting up comparison sets and inviting applications for seats, we then circulated a list of included fossils and requested that each curator and invited scientist send us a choice of four comparisons among six fossils, plus a couple of additional choices. Some of the requests we received were idiosyncratic to say the least, multiplying the number of choices vastly beyond what we had expected; but fortunately Philippe Lampietti and Les Marcus came to our aid with a computer program which managed to match scholars and specimens more or less to everyone's satisfaction.

Time-consuming though its organization had been, and nerve-racking though the procedure was itself, with dozens of delicate fossils being repeatedly handled by large numbers of excited scientists, the comparison session turned out to be an extraordinary event that far surpassed even our optimistic expectations. Even those who had begun by believing that casting technology was by now able to provide us with replicas virtually as good as the originals had by the end to admit that technology still has a long way to go. There is nothing like an original fossil—and for an egotistical species, particularly a hominid fossil—to get the juices flowing; and in this respect two fossils, especially if they normally reside in institutions hundreds or thousands of miles apart, are much more powerful than one. The comparison session also attracted a good deal of attention in the popular media; of the various articles that resulted, the excitement of the event was probably best caught by John Pfeiffer's August contribution in *Smithsonian* magazine (Pfeiffer, 1984). When Paul Beelitz and Gary Sawyer returned the specimens to the Museum's vault at the end of the last day of comparisons we breathed a deep sigh of relief, but could not help noting that the fossils had withstood the proceedings at least as well as the scientists, most of whom were beginning to suffer from sensory overload.

The period between January 1983 and the opening of the exhibition on April 9, 1984 had been a frantic time of writing label copy, finalizing details of design, making transport, insurance, and other logistical arrangements, attempting to recruit additional fossils for display, and attending to a thousand other exhibition-related tasks, as well as of organizing the symposium and comparison sessions. A particular problem lay in the design of the free-standing exhibition cases in which the original fossils would be displayed; for while these had to be as secure and as shock-resistant as possible, they also had to be aesthetically pleasing. Michael Blakeslee's imaginative and highly successful solution consisted of a case within a case, in which the fossils stood on a free-standing internal pedestal; this was surrounded but not contacted by an angular armored structure supporting a subsquare bulletproof laminated Lexan cover with mullioned joints. Fifteen of these cases (see page 228), each containing between one and four specimens, were disposed, following a branching time line, along the central area of the Museum's new Gallery 1; explanatory materials, visuals, and the few casts considered necessary to round out the story were ranged along the walls. In the first place the exhibit was set up using casts; the originals were finally substituted for these during the four days of the symposium, the result of which is this volume.

Virtually up to the last minute, we expected the participation in "Ancestors" of 25 institutions in a dozen different countries. However, only two weeks before opening date the Museum of Victoria was obliged to withdraw the Kow Swamp 1 and Keilor crania because of Aboriginal opposition to their inclusion, while a week later the Dutch government rescinded its permission for the Rijksmuseum van Natuurlijke Historie to participate with Dubois' Trinil 2 calotte, the Trinil 3 femur, and the Wadjak skull. And finally, on the eve of the arrival of four curators and eight specimens, the Tanzanian government withdrew in the face of pressure exerted at the United Nations by a political group objecting to the inclusion of South African specimens in the exhibition. Nonetheless, on April 13, 1984, following the highly successful scientific program reported in this volume, "Ancestors: Four Million Years of Humanity" opened to the public, and to widespread media acclaim, with 40 of the world's most important human and prehuman fossil specimens on display. Eleventh-hour disappointments notwithstanding, "Ancestors" was the unique and highly gratifying result of an unprecedented act of international paleoanthropological cooperation between 21 institutions in nine countries. By the time the exhibition closed on September 9, 1984, almost half a million people, not only from this country but from all over the world, had had the opportunity to see and to appreciate for themselves a substantial proportion of the original, tangible evidence upon which our present understanding of human evolution is based.

ACKNOWLEDGMENTS

"Ancestors" could only have come to fruition as a result of the wholehearted cooperation both of the institutions listed below and of the individual curators of the specimens, most of whom have contributed to this volume. To all of these individuals and institutions we express our profound gratitude. Within the American Museum itself, the success of the venture was similarly due to the efforts of many people in several different departments. Apart from Dr. Thomas D. Nicholson and the staff of the Director's Office, we particularly wish to thank Michael Blakeslee and his team of preparators from the Department of Exhibitions who worked so willingly under difficult conditions (for example, during the rush at the last minute when we found that our precision specimen mounts, so carefully prepared on the basis of casts supplied in advance, in most cases failed to fit the original specimens—as dramatic an illustration as one could wish that casts are no substitute for originals); Charles Miles and Sankar Gokool and all of the Museum attendants who under their supervision maintained a high level of enthusiasm and vigilance for five months to ensure that the public exhibition passed without incident; Paul Beelitz, Gary Sawyer, and Barbara Conklin, who took responsibility for the original specimens during the harrowing days of the comparison sessions, and Chester Tarka who so expertly photographed them; and Priscilla Ward, Pat Bramwell, Clarissa Wilbur, and Barbara Werschek, who helped with the organization. Others outside the Museum also contributed: Joan Fellerman, Rocky Covino, and Roy Johnson made travel and arrival/departure arrangements, and Joe Maisano and his colleagues ensured swift and comfortable ground transportation, while the New York City Police and the Port Authority of New York and New Jersey Police most efficiently helped to ensure the safety of the fossils during transit. Officials of both the U.S. Customs Service and the Immigration and Naturalization Service did everything possible to speed the passage through formalities of the fossils and their transporting curators. And once again we would like to record our gratitude to the members of the "Ancestors" Steering Committee who guided the entire process from the beginning, and the foundations already mentioned whose support made the full scientific program possible. Finally, we should note that our families, and especially our wives, Andrea, Enid, and Bobbie, endured almost as much as we did: the 4:00 A.M. telephone calls, the constant string of crises, and of course the griping. Thank you all three.

LIST OF PARTICIPATING INSTITUTIONS AND SPECIMENS CONTRIBUTED TO "ANCESTORS: FOUR MILLION YEARS OF HUMANITY"

American Museum of Natural History, New York, U.S.A. (Cerro Sota 2)

Geological Museum, Cairo, Egypt (CGM 40237 *Aegyptopithecus* cranium)

Geological Survey of Pakistan, Quetta (GSP 15000 *Sivapithecus* face)

Geologisch-Paläontologisches Institut, Universität Heidelberg, Federal Republic of Germany [Mauer 1 mandible]

Geološko-Paleontološki Muzej, Zagreb, Yugoslavia (Krapina A/1 part calotte, C/3 part skull)

Institut de Paléontologie Humaine, Paris, France (Arago 2 mandible, 13 mandible, 21 face, 44 part pelvis, 47 parietal)

Israel Department of Antiquities and Museums, Jerusalem (Amud 1 skull, Zuttiyeh 1 frontal)

Musée de l'Homme, Laboratoire d'Anthropologie, Paris, France (Cro-Magnon 2 skull, La Ferrassie 1 skull, La Quina H5 skull)

Muséum National d'Histoire Naturelle, Institut de Paléontologie, Paris, France (Ternifine 3 mandible, 1857-1 *Dryopithecus* mandible)

National Museum, Bloemfontein, South Africa (Saldanha 1 calotte)

Natural History Museum of Los Angeles County, Los Angeles, California, U.S.A. (La Brea 1 skull)

Peabody Museum, Harvard University, Cambridge, Massachusetts, U.S.A. (Skhul 5 skull)

Rheinisches Landesmuseum, Bonn, Federal Republic of Germany (RLB 332 Neanderthal calotte, left femur)

Senckenberg Museum, Frankfurt am Main, Federal Republic of Germany (Sangiran 2 calotte, 4 palate)

South African Museum, Cape Town (Saldanha 1 calotte)

Staatliches Museum für Naturkunde, Stuttgart, Federal Republic of Germany (Steinheim 1 cranium)

Transvaal Museum, Pretoria, South Africa (Sts 5 cranium, 14 part skeleton, 52a&b maxilla/mandible, 71 cranium; Sk 23 mandible, 48 cranium, 847 part cranium)

Università di Roma "La Sapienza," Rome, Italy (Saccopastore 1 cranium)

Université de Bordeaux 1, Talence, France (Biache 1 part cranium, St. Césaire part skull)

Université de Paris VI, Paris, France (Salé 1 cranium)

University of the Witwatersrand, Johannesburg, South Africa (Taung 1 part skull, Border Cave 1 skull, Stw 53 part cranium)

LITERATURE CITED

Pfeiffer, J (1984) Early man stages a summit meeting in New York City. Smithsonian *15*(5):51–57.

Ancestors: The Hard Evidence, pages 6–8

The Role of the American Museum of Natural History in 20th Century Paleoanthropology

Harry L. Shapiro
Department of Anthropology, American Museum of Natural History, New York, New York 10024

One of the hazards, among a variety of others, on becoming an oldtimer in an institution such as this, is the expectation of one's colleagues that one has become an infallible reservoir of all that has transpired. Although I have been associated with the A.M.N.H. for 58 years, I make no such claims, yet many things and various events still survive vividly in my memories. Thus, in fulfilling the implications of the subject assigned to me, I could either rely on my memory or do a vast and tedious research program for which I had neither the inclination nor the time. I can, therefore, only hope that I do not leave out some significant event or allow time to distort my memory.

When the Museum was founded 116 years ago, there was no paleoanthropology. The two discoveries of fossil hominid relics, Neandertal and Gibraltar, had not yet been accepted as evidence of the evolutionary past of modern man. Virchow, for example, had brushed Neandertal man aside as a pathological specimen and Darwin's recently published exegesis of evolution had not yet established a knowledgeable and firmly based branch of paleontology dealing with man.

Vertebrate paleontology encouraged by 19th century geological research and study was, on the contrary, already well underway even though it had not progressed very far. This new science was incorporated into the Museum in 1891 with Henry Fairfield Osborn as curator. There already existed a number of collections of vertebrate fossils that were purchased by the Museum, and the initiation of active field work added to the Museum's resources, making it one of the great collections in the world.

I give this brief reference to the Museum's activity in Vertebrate Paleontology because Osborn, as a result of his research, made perhaps the first theoretical contribution by the A.M.N.H. to human paleontology. As a result of his study of vertebrate fossils and their distribution by taxa he had concluded that Asia was perhaps the major center of vertebrate evolution leading to the hominid line. This theory created considerable interest and discussion and was subsequently developed and elaborated by W.D. Matthew in a number of articles and monographs that received considerable attention. Although this is not the current view, it has, with some modifications, a relevance that has survived.

It was also these developments that led, at least in part, to Roy Chapman Andrews' famous expeditions in the 1920s to Mongolia. He specifically declared that one of his motives was to search for hominid origins. And to support this interest he took with him Nels Nelson, an archeologist in the A.M.N.H. Department of Anthropology, as an expert to supervise excavations revealing traces of human occupation and possibly hominid fossils. Unfortunately, traces of early man were not discovered, although Nelson carried out important archeological research that has not as yet been fully published.

However, despite Roy Chapman Andrews' failure to unearth any hominid fossils in Mongolia, his interest in this pursuit did have some positive but little known effect on the discovery of Peking Man in China. One of Andrews' expert assistants, a distinguished paleontologist, was Walter Granger, who had made several trips to Asia and

China in connection with Andrews' expeditions. He was one of the paleontologists who had become aware of the existence of "dragon bones" in Chinese pharmacies. And he, along with Anderson, had traced these fossils they found in Peking to their source at Chou-Kou-Tien. After Anderson had established a team to explore the chicken bone hills at Chou-Kou-Tien, he eagerly sought and used Granger's help and advice in exploring these sites. Thus Granger played a part in the discovery of "*Sinanthropus pekinensis.*"

The interest of the Department of Vertebrate Paleontology in hominid evolution was also reflected in its activity in primate anatomy and behavior. Harry Raven was occupied for many years with the former and, together with W.K. Gregory, a successor to Osborn as chairman of the Department of Vertebrate Paleontology, in the latter. In 1929 Gregory, together with MacGregor, a professor at Columbia, and Harry Raven embarked on one of the first field studies of chimpanzee behavior for insights it might provide into human evolution.

I have up to now stressed the contributions of our Department of Vertebrate Paleontology, with little or no reference to the Department of Anthropology, where one would perhaps expect a natural interest and possible activity in the study of human evolution. At the beginning of my talk I suggested very briefly one reason for this delay in the Department of Anthropology to become active in this field. This was in part due to the absence of any organized and active field of human paleontology at the time the Anthropology Department was established, in 1873. But besides this there were several other reasons. One was the fact that physical anthropology was not included in the initial composition of the department. In the latter half of the 19th century, when anthropology was becoming established as a science in the U.S., most of its concerns were focused on ethnology and archeology. The museums and universities that were incorporating this field of research stressed these aspects of anthropology, founding professorships or curatorships of ethnology and archeology but omitting physical anthropology. When such a department was established in 1873 at the A.M.N.H., there were no physical anthropologists on the staff at Harvard University or the Peabody Museum and none at the Smithsonian, nor at any institution where anthropology was represented.

It was not until the beginning of the present century that physical anthropology was recognized by the A.M.N.H. in its research coverage, when A. Hrdlička was employed for some field studies of American Indians, and not until 1916 that a curatorship was established with the appointment of Louis Sullivan. Until his death in 1923, he was largely absorbed in the study of racial morphology. But in those days, the opportunities for physical anthropologists to engage in human paleontological research or in active field work were limited. American physical anthropologists did not have the advantage of being near where the action was.

The Museum and the Departments of Anthropology and Vertebrate Paleontology did, however, enter into an active role in the area of human paleontology by means of their exhibition function. In the first decade of the 20th century they mounted the first major exhibit of human evolution in the U.S. One hall was devoted to the subject, with a series of murals reconstructing the appearance and environment of some of the fossil types known at that time. In 1962 this was replaced by a much more detailed and extensive exhibit made possible by the many newer discoveries.

In the history of human paleontology, two researchers stand out among the distinguished representatives of this field. These are Franz Weidenreich and Ralph von Koenigswald. It was my great pleasure while chairman of the Department of Anthropology to serve as host to their sojourn in New York. Weidenreich, on leaving China in 1940, joined us here and continued his research in one of the laboratories assigned to him. For eight years, until his death, he continued his monumental studies here, publishing extensively. His last work on Solo Man was still unfinished, but I edited and issued it as one of the monographs in our series of publications.

Von Koenigswald, on leaving Java at the end of World War II, joined us and spent a year and a half in one of our departmental laboratories. His visit was very fruitful and gave him the opportunity to renew his ties with Weidenreich.

And today we have two distinguished contributors to human paleontology: Ian Tattersall and Eric Delson. I shall not embarrass them with encomia familiar to all of you. Dr.

Tattersall has engaged in various projects relating to the diversification and relations of the primates around the world. His recent book *The Myths of Human Evolution* is an example of his approach to human paleontology. Eric Delson, who has been associated with the Department of Vertebrate Paleontology from his student days and, since 1975, a Research Associate in that department, has accumulated an impressive array of contributions to our knowledge of the primate relatives of man, exemplified by his book *Evolutionary History of the Primates*.

But of all the contributions I have mentioned and perhaps some I have overlooked, one stands out as literally unique and will not soon be forgotten. I hope it may become a precedent for an invaluable experience. This is the "Ancestors" project in which we are now participating. We owe a great debt for this to John Van Couvering, who initiated the idea, and to Eric Delson and Ian Tattersall, who labored for over three years to achieve it and suffered the inevitable disappointments and enjoyed their hard-earned successes, which we now share with them.

Ancestors: The Hard Evidence, pages 9–13

Catarrhine Evolution

Eric Delson
Department of Anthropology, Herbert H. Lehman College, City University of New York, Bronx, New York 10468; and Department of Vertebrate Paleontology, American Museum of Natural History, New York, New York 10024

The first scientific session included six presentations on a variety of topics dealing with the "prehuman" phases of paleoanthropology. Elwyn Simons began the program with a discussion of the Fayum primates of the Egyptian Oligocene, especially the *Aegyptopithecus–Propliopithecus* group, which he placed in the Hominoidea. One question asked of him was whether those taxa might also be potentially related or ancestral to the Old World monkeys and went on to ask what features made them hominoids, as opposed to members of yet a third major group of catarrhines. Simons replied that both the limb bones and teeth of cercopithecoids have a rather stereotyped morphology quite different from that of living hominoids or of *Aegyptopithecus*. Although the dentition of Old World monkeys *could* have been derived from something like *Aegyptopithecus*, as perhaps could have been the postcranium, the oldest known monkey postcrania suggest to Simons a rather different predecessor. In fact, he argued, features of the teeth in *Parapithecus* resemble those of cercopithecids in ways not seen in the Fayum hominoids, while the hind limb of the parapithecid *Apidium*, although "strangely adapted for leaping," similarly resembles monkeys. The best way to answer the question, however, would be to find fossils that are transitional between later cercopithecoids and one or another early catarrhine.

Asked how *Aegyptopithecus* is linked to the Early Miocene African hominoids such as *Proconsul*, Simons cited the close similarity of their talus, calcaneum, and tibia, which differ greatly from those of early Old World monkeys. He suggested that if "monkey-like creatures [were found] contemporary with *Proconsul*, it would be possible to suppose that an earlier form had given rise to both" lines. Simons noted that those who wish to use the features of Miocene to modern hominoids as a basis for a definition of Hominoidea would find *Aegyptopithecus* "an uncomfortable little creature," because it is a mosaic of hominoid molars, fused frontal and hind limb, and other features that are more primitive.

David Pilbeam reviewed his recent finds in the Miocene of Pakistan and their implications for human origins. In response to the same questions asked of Simons, he indicated that he would include *Proconsul* in Hominoidea without question, one reason being its hominoid-like vertebral column: that of *Proconsul africanus* is gibbon-like, while the vertebrae from the Middle Miocene site of Moroto (Uganda) are rather like those of chimpanzees. However, he added that it is possible that these are convergences, because they are not too different from vertebrae of spider monkeys. The oldest fossil that Pilbeam would accept as a cercopithecoid is a single, clearly bilophodont molar from Napak, close to 20 million years (m.y.) old. He said that he is "enough of a cladist to feel most comfortable with a definition of Hominoidea which links everything on the basis of shared-derived features," but that few *Proconsul* features outside the vertebrae are synapomorphic with Middle Miocene to living hominoids.

Herbert Thomas discussed the land connections and faunal migration patterns among Africa, Arabia, and Eurasia in the earlier Miocene, with special reference to the hominoids. He was asked about the relationship between woodlands and aridity after Africa connected to Eurasia. He replied that the bovids of the "Chinji" horizon in the Siwaliks revealed a fairly closed environment, while apparently contemporaneous environments in Arabia (Hofuf Formation) were very open, based on rodents such as the sciurids and ctenodactylids.

Alan Walker (Johns Hopkins University) spoke on functional morphology and Miocene hominoids. Although he was unable to prepare a manuscript, I will briefly summarize his major findings. He first reviewed recent finds of additional portions of the skeleton of *Proconsul africanus* and *P. nyanzae*, both in the field and in wrongly identified museum collections, much of which information has been published by Walker and Pickford (1983). In sum, the brachial index of *P. africanus* is gorilla-like, while the crural and intermembral indexes are more like those of macaques. "In robusticity, it's quite a chunky little animal, not a slim, gracile leaper of the sort that perhaps we've been led to believe from the forelimb analysis." Moreover, while the hand, forelimb and thigh are rather monkey-like, the foot, lower leg (with strong fibula), scapula, upper arm, and even elbow are more apelike. He then noted even more recent discoveries at the late Early Miocene (older than 17.5 m.y.) site of Buluk in northern Kenya. The primates include one or two species of monkey, like *Prohylobates*, being described by Meave Leakey; two bits of a small hominoid like *Micropithecus clarki*; and some jaw fragments and isolated teeth of larger hominoid, which, he said, is definitely not a *Proconsul*. The maxilla has a long, curved canine root and deep canine fossa, with the premaxillary suture curving into the nasal aperture. An M^2 in place has thick enamel and no cingulum. The reconstructed mandible showed close-set parallel toothrows, double mandibular torus, rotated stubby canines with a hollow behind the root, inflated corpus, and steep ramus. All of these features differed greatly from *Proconsul* specimens to which he compared them, but were similar to those of the younger Pakistan *Sivapithecus*. Walker suggested that the Buluk fossils might represent the morphology of the common ancestor of the great apes and humans, but not a special link to the Asian orangutans.

Neither John Fleagle nor Richard Kay was able to be present at the symposium, although their paper is included in this section. Instead, Jeffrey Schwartz (University of Pittsburgh), who had participated in the pre-symposium study session, was willing to fill in for them on short notice by presenting a summary of his paper that had just appeared in *Nature* and drawn much attention (Schwartz, 1984). Schwartz's analysis of the cladistic relationships of the living larger hominoids resulted in the surprising finding that the orangutan (*Pongo*) shares the greatest number of derived features with *Homo* and thus is the closest relative of humans. He considered that most of the features shared by humans and African apes are either primitive retentions or of low phyletic weight, while the synapomorphies joining *Homo* and *Pongo* are stronger. Reinterpretation of the previously widely accepted molecular biological analyses suggested to Schwartz that most of the protein and DNA similarities between humans and African apes were symplesiomorphies or that it was impossible to determine the polarity of transformation in these characters, thus rendering them useless for phylogeny reconstruction. The fossil *Sivapithecus* was found to share a number of the *Homo–Pongo* synapomorphies and thus was placed (along with *Gigantopithecus*) in a derived clade separate from the African great apes or gibbons. When asked about chromosomal and other features that seem to link humans and chimpanzees, specifically, Schwartz suggested that most were probably symplesiomorphies—that is, shared primitive retentions. He noted that a recent paper by Marks (1983) reviewed the hominoid cytological literature, concluding that many of the published conclusions were insufficiently supported by the underlying data.

Schwartz was also asked to comment on how his study of the palatine fenestrae and related morphology led to a diametrically opposing result from that reached by Ward (Ward and Pilbeam, 1983; Ward and Kimbel, 1983) on similar material. He replied that Ward and colleagues, as he read them, compared the floor of the nasal cavity in *Pan* and *Australopithecus afarensis* and found them similar in having large foramina on either side of the midline; from this, they generalized to all African apes and humans sharing this complex and thus distinct from the *Sivapithecus–Pongo* group with small foramina. In fact, Schwartz argued, both orangs and modern humans have small foramina; this did not suggest that they were specially related, but that the large foramina might be the primitive condition. On the other (oral cavity) side of the palate, as Schwartz had shown previously, humans and orangs share a single foramen as opposed to the double, primitive condition seen in other catarrhines. The confusion between these two aspects of the problem might have resulted in some uncertainty as to their interpretations.

Peter Andrews finished the program with a broad survey of the relationships among all major groups of catarrhines. He differed from most of the preceding speakers in his interpretation of the phyletic positions of *Aegyptopithecus, Proconsul, and Pongo*, supporting each decision with an analysis of the polarity of transformation at one of several major branching points on his catarrhine cladograms. One specific point on which he was questioned was the placement of the Early Miocene East African *Dendropithecus* outside of both the hominoid and cercopithecoid lineages, as defined by derived characters shared among their respective living members. He replied that the lack of an auditory region made it difficult to interpret this taxon: the parapithecoids and propliopithecids of the Fayum have a ring-like meatus, living hominoids and cercopithecoids a tube, and Miocene pliopithecids an intermediate condition. *Dendropithecus* shares all the features found in common among hominoids and cercopithecoids (for which fossils are known), but none of the derived characters distinguishing the two modern superfamilies. Therefore, it was placed *incertae sedis* within Andrews's "true" Catarrhini.

A question about the relationships of *Proconsul*, whether it is phyletically closer than the gibbons to the living large hominoids or more distant, was taken up by both Andrews and Walker. Walker indicated that gibbons were quite distinctive morphologically, with features he could not see in *Proconsul*; he thus did not consider them to have been derived from that taxon. *Proconsul*, Walker said, presents a number of quite ape-like characters that are not just primitive. Andrews replied that "you don't have to be able to see gibbon characters in *Proconsul* in order to say whether its position is 'above' or 'below' that of gibbons." Instead, Andrews argued, he finds that *Proconsul* is more derived than *Dendropithecus* in sharing features found otherwise only in both gibbons and great apes (including humans), but not in either of those two groups alone. Andrews asked Walker which characters of *Proconsul* might be shared with great apes but not by gibbons, thus linking the former two groups most closely. Walker, evading the question, responded that he understood this method of analysis but did not agree with it. Rather, he perferred to "pull it together and have a feel for the organism. As an old-fashioned biologist-paleontologist," Walker would "like to think of things as animals, not as traits."

The discussion during this symposium session revealed an apparent dichotomy among primate systematists that is common to most such disciplines: "cladists" on the one hand and "evolutionary systematists" on the other. In fact, however, I see it more as a morphocline or graded series than as a strict dichotomy. Walker and Simons argued for relationships based on the overall pattern of resemblance among the animals involved, looking for ways to incorporate the paleobiology of the fossils and their "total morphological pattern" into the analysis (see also Tobias, 1985). Pilbeam, like a number of current workers, has accepted part of the cladists' mode of approach, but not all; while Andrews and Schwartz (and I) emphasize that relationship depends on recency of common ancestry as reflected by the sharing of derived (or "advanced") characters distinct from those ancestral features of wider distribution. For Walker to say that he speaks of animals rather than traits (or collections of traits, as some phrase it) sidesteps the issue.

For most extinct mammals, we have only a collection of fragmentary fossils from which we can try to deduce two different aspects of the animal's natural history: its phyletic position and its paleobiology. The only way to approach the former with reasonable reliability is to search out the derived characters that link it to its closest genealogical relatives. Neither shared conservative retentions from an earlier ancestry nor convergent or analogous similarities tell anything about its phylogeny—that can be deduced only from a study of homologous traits shared with a small group of closest relatives. On that basis, the position of *Proconsul* is still unclear: if Andrews is correct, it is near the base of the hominoid radiation, but Walker appeared to interpret it as closer to great apes in certain features. If he were willing to elucidate those features, I would be most interested, not least because I argued (Szalay and Delson, 1979) that *Procon-*

sul did share certain postcranial features with the great ape and human clade not seen in gibbons. Both Andrews and I now agree that this clade can be termed Hominidae, as separate from the Hylobatidae, and the question thus resolves to: Is *Proconsul* a hominid?

The second aspect of paleontological reconstruction, paleobiology, is well treated in the paper by Fleagle and Kay. Although teeth surely give evidence about dietary adaptations, the limbs are usually dominant in these considerations. Thus, Walker's careful poring over the old collections and backdirt on Rusinga was important for both paleobiology and phylogeny. He ended his symposium discussion of *Proconsul* hoping someday to see the pelvis, sacrum, and tail of this animal. Incredibly, several months later, he and Kenyan colleagues recovered most of the skeletons of five individuals on Rusinga—the power of positive thinking!

Three other questions brought up in this session lead me to add some comments. Simons noted that the distal hindlimb of *Apidium* and the molars of *Parapithecus* are rather similar to those of Old World monkeys, supporting his long-held views of a special relationship between parapithecids and cercopithecids. But several recent papers by Simons and his co-workers have actually weighed heavily against these views. Fleagle and Simons (1979) indicated that parapithecid pelves show no trace of the ischial callosities expected not only in early cercopithecoids but in the common ancestor of all living catarrhines: the parapithecids must thus have predated such an ancestor and cannot be related to cercopithecoids specifically. Fleagle and Simons (1983) reported that the tibia and fibula of *Apidium phiomense* were partly fused or at least more strongly appressed distally than in any other known anthropoid. Although Simons argued in the discussion that this might have been the ancestral condition for Old World monkeys, there is no evidence for that supposition, and it seems more likely that *Apidium* (and perhaps other parapithecids) were uniquely derived in this region. Finally, Kay and Simons (1983) showed that the permanent lower incisors of *Parapithecus grangeri* were entirely lost, leaving the lower canines to meet in the symphyseal midline. This is also a unique derived condition (not seen in *Apidium*), which suggests that this animal was not likely ancestral to any later group. It also may be the only valid reason for resuscitating Gingerich's (1978) genus *Simonsius* for *P. grangeri*.

In sum, the two well-known genera of Parapithecidae show a number of highly derived features distinct from those of all later catarrhines. I continue to argue that parapithecids—or paracatarrhines, as I called them (Delson, 1977), based on the "para" in *Parapithecus*—represent the sister-taxon of all the remaining Old World higher primates, which I have termed eucatarrhines. Andrews argues here that the shared retentions used by Szalay and Delson (1979) to unite *Propliopithecus* (including *Aegyptopithecus*) and *Pliopithecus* in the Pliopithecidae are insufficient; he thus recognized two essentially monotypic superfamilies for thse genera. Fleagle and Kay (1983) used similar arguments to suggest two subfamilies for these taxa, as had Delson and Andrews (1975) earlier. I agree that proliferation of new categories and new higher taxa is the bane of a purely cladistic classification, and this would result from Andrews' scheme. The alternative is to accept the paraphyletic combination of the two families in the Pliopithecoidea, which I tentatively favor.

Turning to later hominoids, *Proconsul* has taken center stage once again. Andrews has argued persuasively that it has no hominid or hylobatid features and thus has placed it in a separate hominoid family, Proconsulidae. I would accept this taxonomy if Andrews' interpretation of the branching sequence were further supported, but I would place *Proconsul* within Hominidae (perhaps as a third subfamily or within Dryopithecinae) if Walker's *gestalt* interpretation is confirmed by new data. Andrews and I agree (at present) on the inclusion of most larger African Early Miocene hominoids within *Proconsul*.

On the other hand, two groups of African earlier Miocene fossils clearly do not belong in that genus: *Kenyapithecus* and specimens from Buluk (and perhaps Moroto). Walker argued that the Buluk fossils would represent *Silvapithecus*, distinct from *Proconsul* in a number of gnathic features. Instead, having seen casts through Walker's courtesy, I would suggest that these specimens are best placed in *Dryopithecus*, known from Europe and less widely from Asia ("*Sivapithecus*" *simonsi* and a new jaw from Wudu, Gansu province, China). The Buluk mandible fragment shows double symphyseal tori as do other jaws I place in *Dryopithecus*, and the maxilla does not present the diagnostic *Sivapithecus* morphology described by Ward and Pilbeam (1983). Enamel thickness and microstructure may yield further information. For example, Martin (1983) has said that "*S.*" *simonsi* is less derived dentally (as far as can be told from a long-lost holotype) than *Kenyapithecus* and suggested that the Moroto fossils are not *Proconsul*, but that the new Saudi Arabian hominoid is *Dryopithecus*. If *Dryopithecus* is thus redefined, it falls between *Proconsul* and *Sivapithecus* along a hominoid morphocline. It is still unclear where *Dryopithecus* falls with respect to the hominine-pongine divergence, but neither Andrews nor I think it shares derived features with members of either extant lineage. We further agree that Schwartz's arguments, while provocative, are not yet sufficient to rebut the bulk of the molecular evidence for a *Homo-Pan* clade.

I disagree with Andrews, however, on the relative placement of *Dryopithecus* and *Kenyapithecus*. He argues that the latter taxon shares more derived conditions with the modern hominid clades than does the former. In fact, his article lists four synapomorphies linking these fossils and younger hominids and five features with respect to which the two fossils differ. Of these five, *Dryopithecus* shares three with hominids: an enlarged maxillary sinus, spatulate I^2, and keeled humeral trochlea; *Kenyapithecus* presents the primitive hominoid condition for all of these. On the other hand, *Kenyapithecus* shares thick enamel with the pongines and *Australopithecus*, but more data on its formation rate is needed to understand its phyletic position; Martin (1983) has shown that thickness alone can be convergent and confusing. In addition, *Kenyapithecus* has mesiodistally elongated upper premolars, found in living hominids but no Miocene fossils. I tend to see both these characters as having low phyletic weight, because convergence is likely, and thus I consider Andrews' interpretation the weaker of the two alternatives, on available data. Analysis of the new specimens attributed to *Kenyapithecus* from Baragoi (Pickford, 1983) will be important in assessing the affinities of this group.

The original charge to the participants in this session was to concentrate on the content and relationships of the higher taxa of catarrhines (families and superfamilies). Further analysis of *Dendropithecus*, the Late Miocene Italian *Oreopithecus*, and other, lesser-known taxa should lead to a reevaluation of the basal characters of the Cercopithecoidea and Hominoidea and, one hopes, greater understanding of their original divergence.

ACKNOWLEDGMENTS

I thank Drs. Frederick S. Szalay and Ian Tattersall for helping to clarify earlier versions of this report.

LITERATURE CITED

Delson, E (1977) Catarrhine phylogeny and classification: principles, methods and comments. J. Human Evol. *6*:433–459.

Delson, E, and Andrews, P (1975) Evolution and interrelationships of the catarrhine primates. In WP Luckett and FS Szalay (eds): Phylogeny of the Primates: A Multidisciplinary Approach. New York: Plenum, pp. 405–446.

Fleagle, JG, and Kay, RF (1983) New interpretations of the phyletic position of Oligocene hominoids. In RL Ciochon and RS Corruccini (eds): New Interpretations of Ape and Human Ancestry. New York: Plenum, pp. 181–210.

Fleagle, JG, and Simons, EL (1979) Anatomy of the bony pelvis of parapithecid primates. Folia Primatol. *31*:176–186.

Fleagle, JG, and Simons, EL (1983) The tibio-fibular articulation in *Apidium phiomense*, an Oligocene anthropoid. Nature *301*:238–239.

Gingerich, PD (1978) The Stuttgart collection of Oligocene primates from the Fayum province of Egypt. Paläont. Zeits. *52*:82–92.

Kay, RF, and Simons, EL (1983) Dental formulae and dental eruption patterns in parapithecids (Primates, Anthropoidea). Am. J. Phys. Anthropol. *62*:363–375.

Marks, J (1983) Hominoid cytogenetics and evolution. Yrbk. Phys. Anthropol. *26*:131–159.

Martin, LB (1983) The relationships of the later Miocene Hominoidea. Unpublished Ph.D. thesis, University of London.

Pickford, M (1983) An account of the new Kenyan fossil discoveries. Interim Evidence *5*(2):1–6.

Schwartz, JH (1984) The evolutionary relationships of man and orang-utans. Nature *308*:501–505.

Szalay, FS, and Delson, E (1979) Evolutionary History of the Primates. New York: Academic Press.

Tobias, PV (1985) Single characters and the total morphological pattern re-defined: The sorting effected by a selection of morphological features of the early hominids. In E Delson (ed): Ancestors: The Hard Evidence. New York: Alan R. Liss, Inc., pp. 94–101.

Walker, AC, and Pickford, M (1983) New postcranial fossils of *Proconsul africanus* and *Proconsul nyanzae*. In RL Ciochon and RS Corruccini (eds): New Interpretations of Ape and Human Ancestry. New York: Plenum, pp. 325–351.

Ward, SC, and Kimbel, WH (1983) Subnasal alveolar morphology and the systematic position of *Sivapithecus*. Am. J. Phys. Anthropol. *61*:157–171.

Ward, SC, and Pilbeam, DR (1983) Maxillofacial morphology of Miocene hominoids from Africa and Indo-Pakistan. In RL Ciochon and RS Corruccini (eds): New Interpretations of Ape and Human Ancestry. New York: Plenum, pp. 211–238.

Ancestors: The Hard Evidence, pages 14–22

Family Group Systematics and Evolution Among Catarrhine Primates

Peter Andrews
British Museum (Natural History), London SW7 5BD, England

ABSTRACT The ancestral catarrhine morphotype is reconstructed on the basis of shared homologous characters present in the extant catarrhines. An attempt is made to distinguish between primitive and derived characters at this level, and the latter are held to be the defining characters of the Catarrhini. Comparisons of the fossil primates usually referred to as catarrhines shows that the Oligocene anthropoids *Parapithecus* and *Propliopithecus* and the Miocene *Pliopithecus* share some but not all of the extant catarrhine derived characters. They are therefore referred to extinct superfamilies as sister groups to the extant superfamilies.

The Early Miocene *Dendropithecus* shares all extant catarrhine characters for which the condition is known in the fossil, but no cercopithecoid or hominoid apomorphies; it thus cannot be assigned to either of these extant superfamilies. The Early Miocene *Proconsul*, on the other hand, shares all known catarrhine characters and a number of derived characters specifically with the Hominoidea. It is therefore considered to be hominoid, but since it shares no character exclusively with any one part of the Hominoidea it is considered the sister group of the extant hominoid families and therefore classified at the family level. *Dryopithecus* also shares hominoid characters, and in addition it shares some of the characters of the great ape and man clade (Hominidae), and so it is linked with that group, but in its own family. Similarly, *Sivapithecus* is also a member of the great ape and man clade, and since it shares characters exclusively with *Pongo* it is linked with it in the Ponginae.

My aim in this paper is to clarify the relationships among catarrhine primates. This will be done by defining the characters present at the various branching points for the clades that make up the Catarrhini by reference to the two extant superfamilies, the Cercopithecoidea and Hominoidea. Further reference will be made where necessary to the Platyrrhini as the nearest outgroup to the Catarrhini.

The sharing of any homologous characters by the cercopithecoid monkeys and the hominoids will be taken to indicate that these characters were present in the common ancestor of the two groups. They therefore describe the ancestral catarrhine condition, and they will consist of both primitive retentions from earlier branching points and derived characters or synapomorphies of the Catarrhini. In order to distinguish between these two situations, and also to provide additional information when the monkeys and apes do not share the same character states, reference will be made to the platyrrhine monkeys as the nearest outgroup, so that when two of the three groups share the same state, but the third is different, it will be taken that the shared condition is primitive for the Catarrhini and the different condition is derived. A simple example of this procedure can be given for the presence or absence of a tail: Old World monkeys have one and apes do not, so that the ancestral condition for the catarrhines cannot be determined from this evidence. But the platyrrhine outgroup is similar to the Old World monkeys in the presence of the tail, and the most parsimonious explanation for the distribution of this character is that it is plesiomorphic for the catarrhines and being primitively retained in the monkeys, and that its loss in the apes is a derived character.

This method draws on a wide range of taxonomic groups to determine the ancestral morphotype at any given branching point. It does not depend on any one group that is intuitively thought to be "primitive," such as the platyrrhines, and equally it does not depend on assumptions of "primitiveness" of a fossil primate such as *Propliopithecus*. Both of these may well retain more plesiomorphies than many of the Old World monkeys and apes, but they would also be expected to have some derived characters of their own, and the problem would then be to distinguish between them without any means to do so.

The results of the character analysis based on the morphology of the extant catarrhines and platyrrhines are given in Figure 1. Only one branching point has been defined in this cladogram, that of the Old World monkeys and apes at node 1, and the characters that describe the hypothetical common ancestor at this point are listed in Table 1. Many of these are primitive retentions from an earlier but undefined node, but some, which are marked with an asterisk, appear to be synapomorphies that define the catarrhine clade, as opposed to the others that merely describe it, and these are the characters to look for in the fossil record. Any fossil exhibiting all or most of these catarrhine synapomorphies must by definition be included in the infra-order. The rest of this paper will consider a number of fossil groups to determine their status with or in the Catarrhini.

PARAPITHECOIDEA

It has been argued by Delson (1977; Delson and Andrews, 1975) that *Parapithecus* and *Apidium* represent a major subdivision of the Catarrhini. This is based on the retention of three premolars, where other catarrhines have only two, and on uniquely derived characters such as the molar waisting and absence of interconnecting ridges on the molar cusps (Szalay and Delson, 1979). The parapithecids are shown here (Table 1) to share at least some of the catarrhine synapomorphies, although they lack the great majority of them, and so if they were included in the infra-order its definition would have to be changed to accomodate them (as indeed Szalay and Delson (1979) have done). They are therefore considered here to belong to a separate superfamily that is the sister group to the extant catarrhines but is of uncertain status with respect to the infra-order.

PROPLIOPITHECOIDEA AND PLIOPITHECOIDEA

Propliopithecus (including *Aegyptopithecus*) from Oligocene deposits in Egypt is placed in the Propliopithecoidea, and *Pliopithecus* and *Crouzelia* from Middle Miocene deposits of Europe are linked in the Pliopithecoidea. The character states of the two superfamilies are shown in Table 1, where it can be seen that the former shares seven synapomorphies with the extant catarrhines and the latter 11 synapomorphies, compared with five for the Parapithecoidea. Both share the loss of P_2 with the extant catarrhines, indicating that they are more closely related to this group that are the Parapithecoidea, but they still lack many of the other synapomorphies of the group. If they were included in the Catarrhini, the common

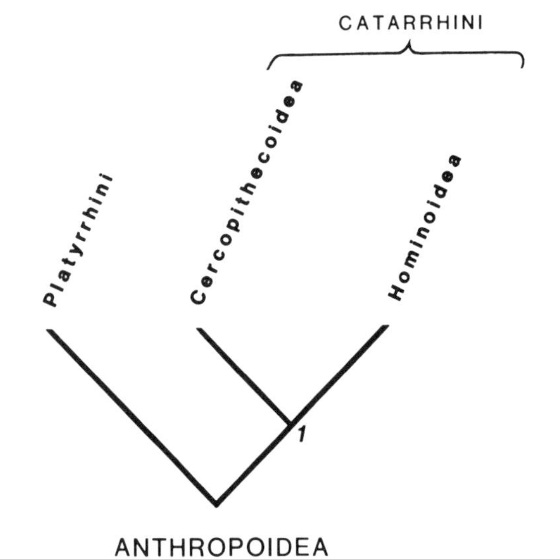

Fig. 1. Cladogram of relationships among the higher primates. The characters defining the catarrhine morphotype at node 1 are listed in Table 1.

TABLE 1. Characters of the ancestral (extant) catarrhine morphotype at node 1 of Figure 1, including both plesiomorphic and apomorphic states for this node.[1]

Ancestral catarrhine morphotype	Parapithe-coidea	Pro-pliopithecoidea	Pliopithe-coidea	*Dendro-pithecus*
1 thin cranial vault bones	x[2]	x	x	—
2 skull breadth greatest at supramastoid crests	—[3]	x	—	—
3 moderate postorbital constriction	x	p[4]	x	—
4 frontal narrows posteriorly to bregma	—	x	—	—
5 low rounded frontal	x	d[5]	x	—
6 frontal long relative to parietal	—	—	—	—
7 parietal flat or gently rounded	—	x	x	—
8 sagittal crest may extend to frontal	x	x	—	—
9 flattened zygomatic, with inferior inclination	—	x	x	x
10 rounded occipital	—	p	—	—
11 inion high relative to orbits, large nuchal area	—	x	—	—
12 upper scale of occipital shorter than lower	—	x	—	—
13 foramen magnum on base of skull, but posterior	—	x	—	—
14 foramen magnum as broad as long	—	x	—	—
15 low temporal, top edge with posterior slope	—	x	—	—
16 glenoid fossa flat, no articular process	—	x	—	—
17 no mastoid process	—	x	—	—
18 no digastric groove	—	x	—	—
19 large supramastoid crest	—	x	—	—
20 occipital crest often present	—	x	—	—
21 lateral pterygoid much larger than medial	—	x	—	—
22 sinuses not enlarged	x	x	x	x
23 prominent glabella, separate from tori*	x	p	x	—
24 broad interorbital pillar*	x	x	x	—
25 deep, guttered temporal fossa, elongated	—	x	—	—
26 small supraorbital tori, divided at glabella	x	x	x	—
27 no post-toral sulcus	x	x	x	—
28 lateral margins of face vertical	x	x	x	—
29 orbits as broad as high	x	x	p	—
30 nasal aperture overlaps lower rim of orbits	x	x	x	—
31 nasal aperture high, oval shaped	x	x	x	x
32 moderate facial prognathism	x	d	x	x
33 recession of snout and shortened nasals	x	d	x	x
34 broad parallel-sided nasal bones	x	d	x	—
35 premaxilla short but broad	x	x	x	x
36 premaxilla overlaps maxilla, resulting in stepped floor of nasal chamber	—	x	x	x
37 large incisive foramina	x	x	—	x
38 palatine makes up at least ¼ hard palate	—	—	—	p
39 tubular external auditory meatus*	pp[6]	pp	pp	—
40 spatulate I^1 higher than broad	x	x	x	x
41 I_1 may have lingual pillar	—	x	x	x
42 pointed caniniform I^2	x	x	x	x
43 upper incisors similar size	x	x	x	x
44 slender canines, high crowned relative to length	x	x	x	x
45 canines sexually dimorphic, mesial sulcus on crown only	x	x	x	x
46 P_3 single-cusped, bilaterally compressed*	p	x	x	x
47 P_3 high crowned relative to length	p	x	x	x
48 P_4 bicuspid, slightly longer than broad*	p	p	x	x
49 P_{3-4} buccolingually broad	x	x	x	x
50 P^{3-4} with heteromorphic cusps*	x	x	x	x
51 P^4 smaller than P^3	p	x	x	x
52 lower molars: 5 cusps, small hypoconulid*	x	x	x	x
53 cingulum developed	x	d	—	—
54 broad talonid	d	x	x	x
55 $M_1 < M_2 < M_3$*	d	x	x	x
56 M_3 only slightly larger than M_2, small hypoconulid*	d	d	x	x
57 upper molars: 4 cusps, hypocone small	x	x	x	x
58 cingulum developed	d	—	—	—
59 $M^1 < M^2 > M^3$*	x	x	x	x
60 buccolingually broad	x	x	x	x
61 crista obliqua well developed	d	x	x	x
62 molar enamel thin	x	x	x	x
63 dental formula 2 – 1 – 2 – 3*	p	x	x	x
64 mandible relatively deep and gracile	x	x	x	x
65 tooth rows V-shaped	x	x	x	x
66 inferior and superior symphyseal tori present	—	x	x	x

(continued)

TABLE 1. Characters of the ancestral (extant) catarrhine morphotype at node 1 of Figure 1, including both plesiomorphic and apomorphic states for this node.[1] *(continued)*

Ancestral catarrhine morphotype	Parapithecoidea	Propliopithecoidea	Pliopithecoidea	*Dendropithecus*
67 maxillary body lightly built	x	x	x	x
68 maxillary jugum small or absent	x	x	x	x
69 palate shallow	x	x	x	x
70 palate longer than broad	x	x	x	x
71 ischial tuberosities present	—	—	—	—
72 clavicle short and stout	—	—	x	—
73 scapula laterally placed	—	—	—	—
74 humeral head oval	—	x	x	x
75 humeral head directed posteriorly	—	x	x	x
76 flat deltoid plane on humerus*	—	p	p	x
77 narrow bicipital groove*	—	p	p	x
78 entepicondylar foramen absent*	p	p	p	x
79 trochlea width similar to capitulum width	x	x	x	x
80 low to moderate trochlear keels	x	x	x	x
81 capitulum rounded, but not on distal surface	d	x	x	x
82 low ridge separates capitulum from trochlea	x	x	x	x
83 olecranon fossa deep and well defined with a well developed lateral ridge*	p	p	p	p
84 large medial epicondyle	x	x	x	x
85 brachialis flange moderately developed	x	x	x	x
86 olecranon process of ulna long and high	x	x	x	x
87 long ulnar styloid process, contacts carpals	x	x	x	—
88 sigmoid notch long and narrow with rounded keel	x	x	x	x
89 radius and ulna straight shafted, stout	x	x	x	x
90 ulna coronoid process = olecranon process	x	x	x	x
91 radial head oval	—	—	x	x
92 femoral head small, rounded	—	—	x	—
93 femoral neck long and at a high angle	—	—	x	—
94 broad patellar groove	—	—	—	—
95 femoral condyles subequal	—	—	x	—
96 ilium long and narrow	—	—	x	—
97 talus longer than broad, long neck*	x	x	x	x
98 calcaneus long and narrow	x	x	x	x
99 astragalocalcaneal joint helical	—	—	—	x
100 os centrale separate	—	—	x	—
101 scaphoid lacks beak like process	—	—	x	x
102 hallux long and abductable	—	—	x	—
103 hallux with modified saddle joint*	—	—	x	—
104 prehallux absent*	—	p	p	—
105 metacarpals and phalanges short and curved	—	x	x	x
106 metacarpal distal extremities narrow dorsally	—	—	x	x
107 vertebrae: thoracic 13	—	—	—	—
108 lumbar 6–7	—	—	x	—
109 sacral 3	—	—	x	—
110 caudal 20–30	—	—	—	—
111 sternum long and narrow	—	—	x	—
112 tail present	—	—	—	—
113 intermembral index variable	—	—	—	—
114 origin of pectoralis major long	—	—	—	—
115 origin of pectoralis minor short	—	—	—	—
116 small pectoralis abdominis	—	—	—	—
117 deltoideus thin and aponeurotic	—	—	—	—
118 deltoid insertion top third of humerus	—	—	—	—
119 trapezius insertion not on clavicle	—	—	—	—
120 trapezius origin extensive	—	—	—	—
121 dorsoepitrochlearis insertion not on medial epicondyle	—	—	—	—
122 fleshy origin of teres major	—	—	—	—

[1]The latter, which consist of characters derived with respect to the ancestral anthropoid morphotype, are marked*. It is assumed here that these are characters of common inheritance and have not arisen in parallel in the two extant superfamilies. All characters are determined by reference only to living anthropoids. To the right is indicated the condition for each character for the four extinct anthropoid taxa shown in Figure 2. (Data were compiled from the following references: Aiello, 1981; Andrews, 1978; Andrews and Cronin, 1982; Andrews and Groves, 1975; Delson and Andrews, 1975; Fleagle, 1983; Harrison, 1982; Napier and Napier, 1967; Rak, 1983; Rose, 1983; Szalay and Delson, 1979; Walker and Pickford, 1983; Ward and Pilbeam, 1983; Zapfe, 1960.)
[2]x: Character present as described.
[3]—: Condition unknown in fossil.
[4]p: Primitive retention for this character.
[5]d: Uniquely derived condition for this character (autapomorphy).
[6]pp: Retention of relatively more primitive condition for this character than present in extant catarrhine morphotype.

ancestor of the enlarged group would retain the primitive condition for the characters for which the extant catarrhines apparently share the derived condition. It is simplest, therefore, to consider both the Propliopithecoidea and the Pliopithecoidea in the same way as the Parapithecoidea; namely, as successive sister groups of the extant catarrhines but of uncertain status with respect to the infra-order. They are considered separate, rather than as a single taxon as suggested by Szalay and Delson (1979), because of both the lack of synapomorphies linking them and the change in the ectotympanic between *Propliopithecus* and *Pliopithecus*: in the former there is no development of a tubular ectotympanic, whereas in the latter there is a slight development of this character that typifies the extant catarrhines.

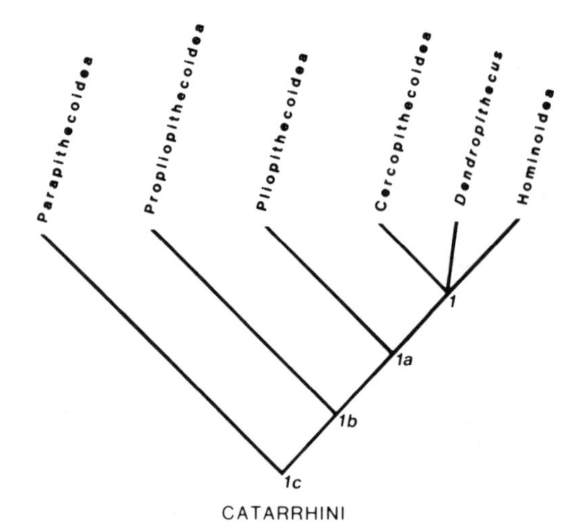

Fig. 2. Cladogram of relationships among catarrhine superfamilies. Characters for nodes 1c, 1b, and 1a are listed in Table 1, which reveals the progressive addition of catarrhine synapomorphies in the Parapithecoidea, Propliopithecoidea, and Pliopithecoidea. Tables 1 and 2 describe the trichotomy at node 1, where *Dendropithecus* shares all of the catarrhine synapomorphies for which the condition is known in this fossil, but lacks any of the synapomorphies defining either the Cercopithecoidea or Hominoidea.

DENDROPITHECUS

It has been suggested recently (Harrison, 1982) that two genera from the Early Miocene of East Africa, *Proconsul* and *Dendropithecus*, are best placed in a clade separate from the Hominoidea. *Dendropithecus* possesses all of the catarrhine characters listed in Table 1, but it lacks either hominoid or cercopithecoid synapomorphies (Table 2), whereas *Proconsul* shares many of them, and on this evidence there appears to be little reason for grouping them in one clade. Instead, *Dendropithecus* is shown in Figures 2 and 3 as equally related to both extant superfamilies in a trichotomy.

PROCONSULIDAE

It has also been questioned recently whether *Proconsul* can be included in the Hominoidea either (Harrison, 1982). Traditionally, it has been accepted without question as a "pongid" (Le Gros Clark and Leakey, 1951; Pilbeam, 1969; Andrews, 1978) or as a "dryopithecine" (Simons and Pilbeam, 1965), but the reasons for doing so are rarely made explicit. In order to demonstrate its relationship with the extant hominoids, it is first necessary to define the superfamily, and this is done in Table 2, which lists the synapomorphies of the Hominoidea at node 2, with the character states of *Proconsul* indicated on the right. It can be seen to share eight hominoid synapomorphies covering the canine/premolar complex, the mobility of the shoulder joint, and the stability of the elbow joint (Walker and Pickford, 1983; Fleagle, 1983). These provide strong evidence for the relationship of *Proconsul* with the hominoids, but many of the

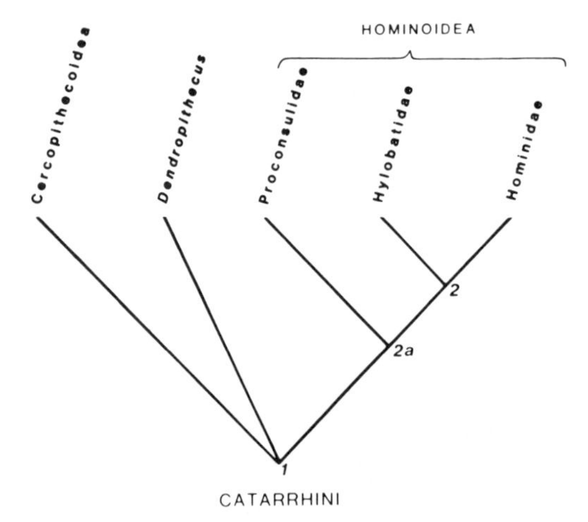

Fig. 3. Cladogram of relationships within the Hominoidea, with neighboring outgroups. See Tables 1 and 2 for defining characters at nodes 1, 2a, and 2 and indication of eight hominoid synapomorphies in *Proconsul* but none in *Dendropithecus*.

other synapomorphies of the rest of the body are not present in the fossil. It is therefore shown in Figure 3 as the sister group to the later Hominoidea. Further investigation requires the definition of the next branching point, as in Table 3, where node 3 is defined by the synapomorphies of the Hominidae, which distinguishes it from the Hylobatidae. Not one of these characters is found in *Proconsul*, showing that there is no evidence supporting hominid (pongine or hominine) affinities of this fossil genus.

DRYOPITHECIDAE

Dryopithecus from the Middle Miocene of Europe, on the other hand, shares a number of the hominid synapomorphies listed in Table 3. These are mainly dental characters, but the distal humerus is also more like that of the great apes and man in the morphology of the trochlear ridges. For these reasons it is shown in Figure 4 as a sister group of the hominids and separated from *Proconsul* (formerly considered a subgenus of *Dryopithecus*) by the Hylobatidae at node 2.

The positions of nodes 3b and 3a are somewhat arbitrary in Figure 4 because of the uncertain relationships of the African Middle Miocene genus *Kenyapithecus*. It shares about the same number of synapomorphies with the great ape and man clade as does *Dryopithecus*, but they are different ones and therefore difficult to compare. The two genera are put in the order shown following Martin (1983; Martin and Andrews, 1982), based on the probable polarity of enamel thickness evolution, with thick enamel being thought to be a hominid synapomorphy. The exact order is not important, however, but

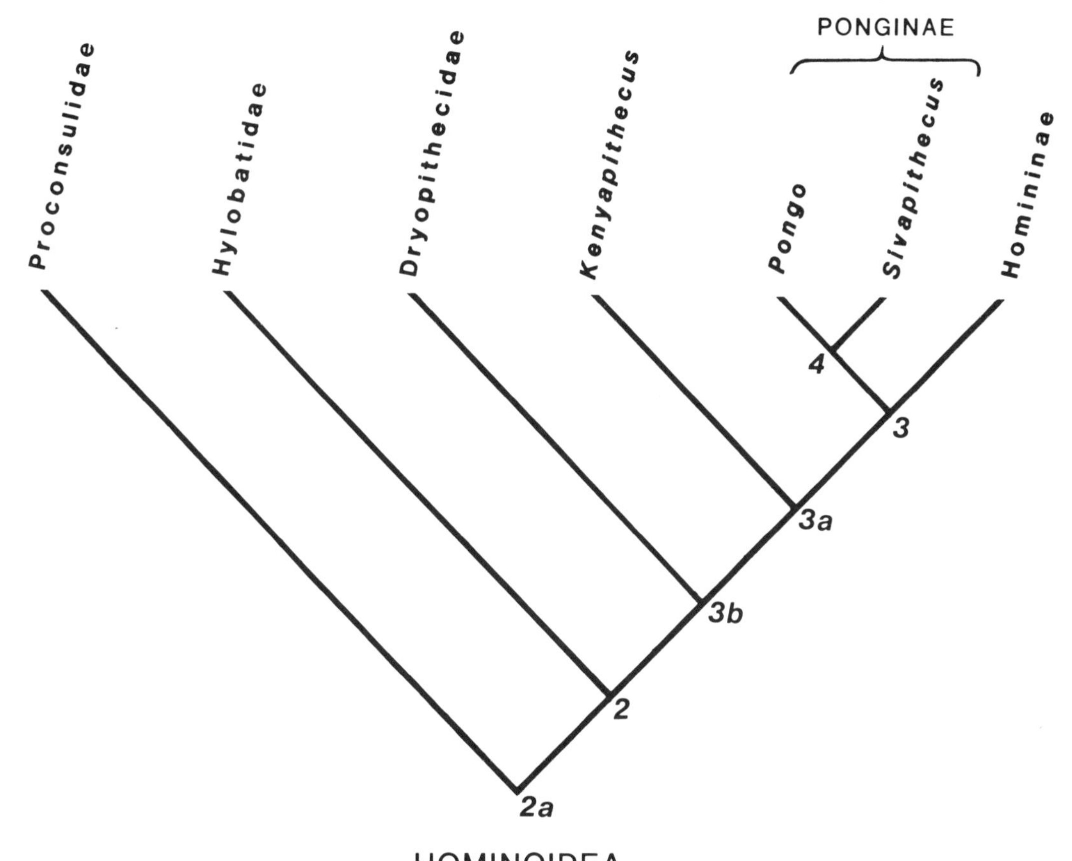

Fig. 4. Detailed cladogram of intra-Hominoidea relationships. Characters for nodes 3, 3a, and 3b are listed in Table 3, while those for node 4 are shown in Table 4. *Proconsul* shares no synapomorphies with individual members of this clade and thus is shown as the most extreme outgroup.

TABLE 2. Synapomorphies of the ancestral morphotypes of the Hominoidea and Cercopithecoidea (nodes 2 and 1 in Figure 3), with comparison of character state distribution for Proconsul *and* Dendropithecus

	Ancestral hominoid morphotype	*Proconsul*	*Dendropithecus*
4	frontal bone wide at bregma narrows anteriorly	x[1]	—[3]
22	enlarged sinuses	p[2]	p
40	broad I^1, as broad as high	p	p
41	no lingual pillar	p	p
47	low crowned P_3	x	p
50	cusp heteromorphy on upper premolars reduced	x	p
54	lower molars broad	p	p
68	maxillary jugum developed	x	p
69	palate deep	p	p
72	clavicle elongated	—	—
73	scapula with elongated vertebral border, robust acromion	x	—
74	humeral head rounded, larger than femoral head	x	—
75	humeral head more medially oriented	x	—
80	strong medial and lateral keels of the trochlea	x	p
86	short olecranon process of ulna	p	p
87	reduced styloid process of ulna; meniscus developed between the styloid process and carpals	p	—
88	sigmoid notch broad	p	p
89	ulna shaft bowed	p	p
90	ulna coronoid process more projecting than olecranon process	p	p
91	radial head rounded	p	p
95	femur with asymmetrical condyles	p	—
96	iliac blade broadened	p	—
97	talus neck short, broad	p	p
98	calcaneus short and broad	p	p
106	metacarpals with broad distal ends	p	—
107	vertebrae: thoracics protrude into thoracic cavity	—	—
108	lumbar 5, loss of accessory articular processes	—	—
109	sacral 4–5	—	—
110	caudal 6	—	—
111	sternum broad	—	—
112	no tail	—	—
118	low deltoid insertion	p	p
119	trapezius inserts on clavicle	—	—
121	dorsoepitrochlearis insertion on medial epicondyle	—	—
122	tendinous origin of teres major	—	—
	Ancestral cercopithecoid morphotype		
31	nasal aperture very high and narrow		
45	upper canines with mesial groove extending on to roots		
46	P_3 sectorial, expanded mesiobuccal flange		
52	lower molars: bilophodont; 4 cusps, no hypoconulid but large posterior fovea on M_{1-2}		
53	cingulum lost, great crown height and relief		
54	trigonid and talonid widths subequal		
56	$M_3 \gg M_2$, large hypoconulid		
57	upper molars: bilophodont; "mirror image" of lowers		
58	cingulum lost, great crown height and relief		
60	strongly elongate		
61	crista obliqua reduced or absent		
80	large lateral trochlear keel		
81	flattened capitulum		
82	no ridge between capitulum and trochlea		
84	reduced medial epicondyle, with slight posterior angulation		
92	head of femur is continuous with the neck posteriorly		
99	astragalocalcaneal joint rotational		
101	scaphoid has beak-like process		

Characters numbered as in Table 1.
[1]x: Character present as described.
[2]p: Primitive retention for this character.
[3]—: Condition unknown in fossil.

TABLE 3. Characters of the ancestral hominid morphotype (Fig. 4, node 3: synapomorphies of the Ponginae [Pongo] *and the Homininae* [Homo, Pan, *and* Gorilla]), *compared with states in three extinct hominoids*

Ancestral hominid morphotype	*Sivapithecus*	*Dryopithecus*	*Kenyapithecus*
17 mastoid processes distinct	—[1]	—	—
21 medial pterygoid as big as lateral	—	—	—
22 maxillary sinus enlarged	x[2]	x	p[3]
29 orbits higher than broad	x	—	—
32 increased alveolar prognathism	x	p	p
35 lengthened premaxilla	x	p	p
37 incisive foramen small	x	p	—
42 I^2 spatulate	x	x	p
44 canines robust, long relative to height	x	x	x
46 P_3 not bilaterally compressed, more robust	x	x	x
49 P_{3-4} more elongated mesiodistally	p	p	x
62 molars with thick enamel	x	p	x
65 toothrows wider apart, more parallel	x	x	x
66 deep symphysis with inferior torus larger than superior	x	x	x
71 ischial tuberosities absent	—	—	—
80 prominent trochlear keels, with deep sulci on either side of the lateral trochlear keel	x	x	p
88 sigmoid notch with strong keel	—	—	—
87 reduced ulnar styloid process, does not contact carpals	—	—	—
90 ulnar coronoid process much higher than olecranon process	—	—	—
108 vertebrae: lumbar 4–5	—	—	—
113 hindlimbs reduced in length, so high intermembral index	—	—	—
114 origin of pectoralis major short	—	—	—
116 pectoralis abdominis not developed	—	—	—
117 enlarged deltoid muscle	—	—	—
120 trapezius thickest fibres cranial	—	—	—

Characters numbered as in Table 1.
[1]—: Condition unknown in fossil.
[2]x: Character present as described.
[3]p: Primitive retention for this character.

TABLE 4. Characters of the ancestral pongine morphotype (Fig. 4, node 4).[1]

9	zygomatic flattened, facing anteriorly or even slightly superiorly
23	no glabellar thickening
24	interorbital distance narrow
36	smooth transition from premaxilla to maxilla, so that floor of nose is not stepped
37	extremely small incisive foramen
43	great size discrepancy between the I^1 and the I^2

[1]Following Andrews and Cronin (1982) and Ward and Pilbeam (1983), these characters are shared by *Pongo* and *Sivapithecus*, but there are many additional ones for *Pongo* for which the condition in the fossil genus is unknown.

TABLE 5. Catarrhine classification

Catarrhini	
Cercopithecoidea	Cercopithecidae
Hominoidea	
Proconsulidae	*Proconsul*
Hylobatidae	*Hylobates*
Dryopithecidae	*Dryopithecus*
Hominidae	Great ape and man clade
Ponginae	*Pongo, Sivapithecus*
Homininae	*Pan, Gorilla, Homo*
Uncertain family	*Kenyapithecus*
Uncertain affinities	
Parapithecoidea	*Parapithecus, Apidium*
Propliopithecoidea	*Propliopithecus*
Pliopithecoidea	*Pliopithecus, Crouzelia*

what matters in the context is the position of both groups as sister taxa of the hominid or great ape and man clade.

Finally, brief mention can be made of *Sivapithecus* (including *Ramapithecus, Ankarapithecus*, and *Ouranopithecus*), which is included with *Pongo* in the pongine clade at node 4. This relationship has been much publicized recently (Andrews and Cronin, 1982; Pilbeam, 1982; Ward and Pilbeam, 1983), and a suite of characters has been demonstrated in its support (see Table 4).

CLASSIFICATION

The relationships proposed here are shown in Table 5. Two extant families are recognized in the Hominoidea, along with two extinct ones: Dryopithecidae and Proconsulidae. To show the probable grouping of these four families would require a plethora of names that would be both ungainly and unnecessary, and so they are all shown here as equal in rank, but in the order of the successive branching points. The Miocene catarrhine taxa *Oreopithecus, Micropithecus, Dionysopithecus*, and *Platodontopithecus* are so poorly known or insufficiently studied as to be impossible to place realistically in this classification.

LITERATURE CITED

Aiello, LC (1981) Locomotion in the Miocene Hominoidea. In CB Stringer (ed): Aspects of Human Evolution. London: Taylor & Francis, pp. 63–97.

Andrews, P (1978) A revision of the Miocene Hominoidea of East Africa. Bull. Br. Mus. Nat. Hist. (Geol.) *30:*85–224.

Andrews, P, and Cronin, J (1982). The relationship of *Sivapithecus* and *Ramapithecus* and the evolution of the orang-utan. Nature *297:*541–546.

Andrews, P, and Groves, C (1975) Gibbons and brachiation. In DM Rumbaugh (ed): Gibbon and Siamang. Basel: Karger, pp. 167–218.

Delson, E (1977) Catarrhine phylogeny and classification: Principles, methods and comments. J. Hum. Evol. *6:*433–459.

Delson, E, and Andrews, P (1975) Evolution and interrelationships of the catarrhine primates. In WP Luckett and FS Szalay (eds): Phylogeny of the Primates: An Interdisciplinary Approach. New York: Plenum, pp. 405–446.

Fleagle, JG (1983) Locomotor adaptations of Oligocene and Miocene hominoids and their phyletic implications. In RL Ciochon and RS Corruccini (eds): New Interpretations of Ape and Human Ancestry. New York: Plenum, pp. 301–324.

Harrison, T (1982) Small-bodied Apes from the Miocene of East Africa. Ph.D. thesis, University College, London.

Le Gros Clark, WE, and Leakey, LSB (1951) The Miocene Hominoidea of East Africa. Fossil Mammals of Africa *1:*1–117. London: British Museum (Natural History).

Martin, L (1983) The Relationships of the Late Miocene Hominoidea. Ph.D. thesis, University of London.

Martin, L, and Andrews, P (1982) New ideas on the relationships of the Miocene hominoids. Primate Eye *18:*4–17.

Napier, JR, and Napier, PH (1967) A Handbook of Living Primates, London: Academic Press.

Pilbeam, DR (1969) Tertiary Pongidae of East Africa: Evolutionary relationships and taxonomy. Bull. Peabody Mus. Nat. Hist. *31:*1–185.

Pilbeam, DR (1982) New hominoid skull material from the Miocene of Pakistan. Nature *295:*232–234.

Rak, Y (1983) The Australopithecine Face. New York: Academic Press.

Rose, MD (1983) Miocene hominoid postcranial morphology: Monkey-like, ape-like, neither, or both. In RL Ciochon and RS Corruccini (eds): New Interpretations of Ape and Human Ancestry. New York: Plenum, pp. 405–417.

Simons, EL, and Pilbeam D (1965) Preliminary revision of the Dryopithecinae (Pongidae, Anthropoidea). Folia Primatol. *3:*81–152.

Szalay, FS, and Delson, E (1979) Evolutionary History of the Primates. New York: Academic Press.

Walker, AC, and Pickford, M (1983) New postcranial fossils of *Proconsul africanus* and *Proconsul nyanzae.* In RL Ciochon and RS Corruccini (eds): New Interpretations of Ape and Human Ancestry. New York: Plenum, pp. 325–351.

Ward, SC, and Pilbeam, D (1983) Maxillofacial morphology of Miocene hominoids from Africa and Indo-Pakistan. In RL Ciochon and RS Corruccini (eds): New Interpretations of Ape and Human Ancestry. New York: Plenum, pp. 211–238.

Zapfe, H (1960) Die primatenfunde aus der miozänen spaltenfüllung von Neudorf an der March (Devinska Nova Ves), Tschechoslowakei. Schweiz. Palaeont. Abh. *78:*1–293.

Ancestors: The Hard Evidence, pages 23–36

The Paleobiology of Catarrhines

John G. Fleagle and Richard F. Kay

Department of Anatomical Sciences, State University of New York, Stony Brook, New York 11794 (J.G.F.), and Department of Anatomy, Duke University Medical Center, Durham, North Carolina 27710 (R.F.K.)

ABSTRACT During the past 35 million years, catarrhine primates have undergone a series of adaptive radiations, parts of which are preserved in sediments of the early Oligocene, earlier Miocene, and later Miocene ages. An examination of the adaptive diversity of these successive radiations compared with modern catarrhines provides an overview of catarrhine paleobiology. Oligocene parapithecids and eucatarrhines were distinctly different from later radiations in their known size distribution as well as locomotor and dietary adaptations. In Miocene through Recent catarrhine faunas the overall range of adaptive zones occupied by catarrhines has remained similar despite changes in the composition of the faunas.

INTRODUCTION

The Anthropoidea are the most successful group of living primates. They include at least 36 genera and over 150 species, and their geographic range covers much of South and Central America, Africa, and tropical Asia, with a token foothold in Europe. Anthropoids of the Old World, the catarrhines, include Old World monkeys, apes, and humans. They are more abundant than platyrrhine anthropoids of the New World and had the distinction of giving rise to our own species, hence their significance for this volume.

The fossil record of catarrhine evolution is spotty in both time and space, and provides us with only occasional glimpses of these creatures during the known 35 million years of their evolution (Figure 1). We have virtually no record of fossil catarrhines from the two areas in which the group is most abundant today—the rainforests of Subsaharan Africa and Southeast Asia—whereas many of the most important early catarrhine fossils come from North Africa, western Europe, and the Himalayan foothills of India and Pakistan, which today have few if any nonhuman catarrhine residents. Therefore, we cannot discount the possibility that many of the supposed ecological "trends" in catarrhine evolution may be biased by the nature of the fossil record.

Despite this patchy record, we have an improving understanding of the broader aspects of catarrhine evolution. Most current workers agree that the fossil record of catarrhines constitutes a series of successive adaptive radiations by increasingly modern-looking groups, rather than a number of ancient lineages evolving in parallel through long stretches of time (e.g., Fleagle and Rosenberger, 1983). Most current debate involves the phyletic placement of these radiations *vis à vis* living catarrhine groups (e.g., Andrews, 1985; Ciochon, 1983; Harrison, 1982). The purpose of this contribution is to examine these successive radiations from an ecological or adaptive perspective to see how the adaptive spectrum of catarrhines has changed over the past 35 million years.

METHODS

We have focused on four major and several minor aspects of the catarrhine life style and comparative ecology that can be reasonably reconstructed from the fossil remains. Our

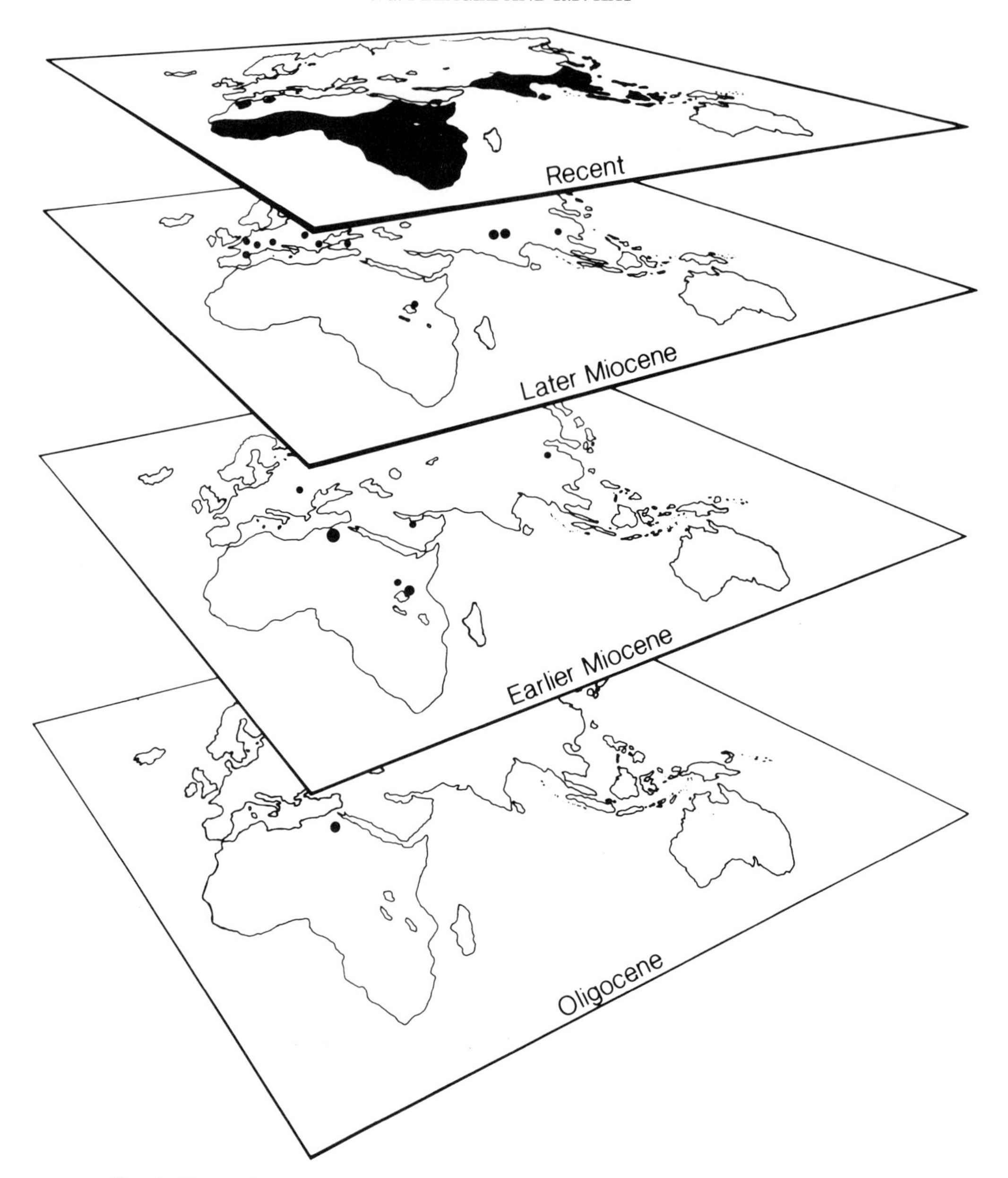

Fig. 1. Known distribution of catarrhine primates during the Oligocene, earlier Miocene, later Miocene, and present.

philosophy and methods, and the comparative studies underlying these reconstructions, as well as many of the results summarized here, have been discussed in much greater detail elsewhere (e.g., Fleagle, 1978, 1983; Fleagle et al., 1980; Kay, 1984; Kay and Cartmill, 1977; Kay and Hylander, 1978). In keeping with the "survey" nature of this paper, we have found it useful to apply several colloquial terms to describe a number of supergeneric taxa. A full list of these is provided in the caption to Figure 6.

TABLE I. Summary of inferred behavioral characteristics of extinct catarrhine genera

Taxon (Number of Species)	Weight Range	Substratum	Locomotion	Diet
Oligocene				
Qatrania (1)	300	?	?	F
Apidium (2)	900–1,600	A	Q	H
Parapithecus (1)	1,700	?	?	F
Simonsius (1)	3,000	?	?	F/L
Oligopithecus (1)	1,500	?	?	F
Propliopithecus (3)	3,700–4,000	A	Q	F
Aegyptopithecus (1)	5,900	A	Q	F
Subtotals (%)		100:00 (A:T)	100:00 (Q:S)	79:14:7 (F:H:L)
Earlier Miocene (20–15 MY)				
Pliopithecus (2)	5,900	A	S/Q	L/F
Proconsul (3)	15,700–41,400	T/A	Q	F
Rangwapithecus (2)	10,600–12,500	?	?	L/F
Dendropithecus (2)	7,500	A	S/Q	F
Limnopithecus (2)	5,000–5,500	A	Q	F
Micropithecus (2)	4,500–6,400	A	?S	F
Dionysopithecus (1)	3,500	?	?	F
cf. Sivapithecus (2)	34,500–58,100	?	?	H
Victoriapithecus (1)	8,600	A/T	Q	L/F
Prohylobates (2)	6,900–19,600	?A/T	?	L/F
Subtotals (%)		78:22 (A:T)	67:33 (Q:S)	70:10:20 (F:H:L)
Later Miocene (14–8 MY)				
Pliopithecus (3)	5,900–8,500	A	S/Q	L/F
Proconsul (1)	41,300	?	?	F
Limnopithecus (1)	4,100	?A	?	F
Oreopithecus (1)	14,800	A	S	L
Dryopithecus (2)	16,600–31,500	?A/T	?Q/S	F
Sivapithecus (6)	20,200–90,900	A/T	Q/S	H
Gigantopithecus (1)	142,100	T	Q	H
Victoriapithecus (1)	5,000	A/T	Q	L/F
Mesopithecus (1)	8,500	T	Q	L
?Presbytis (1)	5,600	?	Q	L/F
Subtotals (%)		57:43 (A:T)	69:31 (Q:S)	45:20:35 (F:H:L)

The number in parentheses following the genus is the number of species recognized. Range of weights is given when there is more than one species in the genus. Otherwise, the value is the mean weight in grams for the species based on the M_1 or M_2 tooth area following the formulae of Gingerich et al. (1982). Symbols: A, arboreal; T, terrestrial; Q, quadrupedal; S, suspensory; F, fruit-eating; H, hard-nut as well as fruit-eating; L, leaf-eating. When two symbols are presented, separated by a slash, this means either that the various species of a genus differ in habits or that the single species in question may have had intermediate habits.

Body size

How big were the fossil species? It is useful to have estimates of the body size of extinct species inasmuch as size is closely linked with so many aspects of ecology and behavior (e.g., Fleagle, 1978; Jungers, 1984, Kay, 1984; Kay and Hylander, 1978). We have estimated the body weights of fossil catarrhines on the basis of molar tooth size using the regression constants for living primate species provided by Gingerich et al. (1982), recognizing that any such estimates are more accurate as indications of relative differences than as absolute dimensions. Table 1 gives the data on which our estimates are based. Figure 2 summarizes this data in the fashion of histograms.

Locomotor behavior

How did they move? We have attempted to reconstruct two aspects of the locomotor habits of fossil catarrhines—substrate preferences (i.e., whether they moved primarily on the ground or in the trees) and locomotor

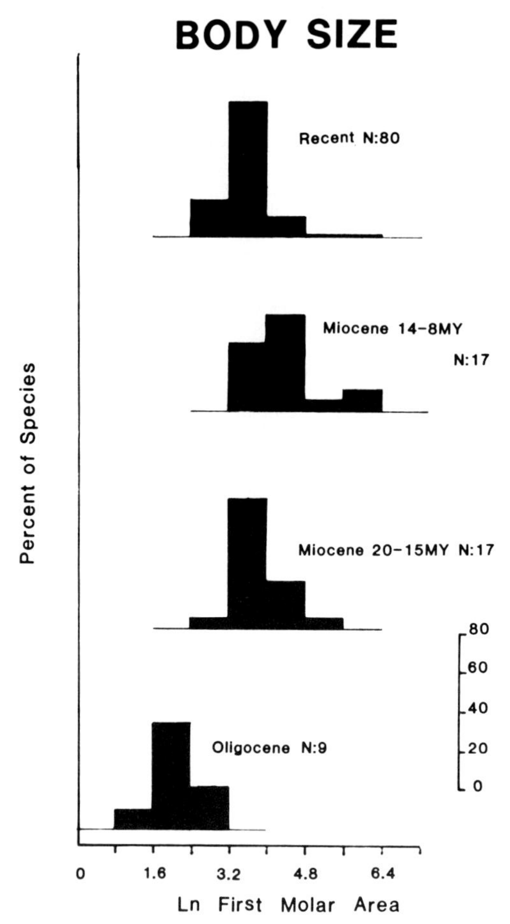

Fig. 2. Catarrhine body size. Percentage histograms of body size in living and extinct catarrhine species. Size axis is graduated in intervals of 0.80 Ln first molar area (initially in square mm).

patterns (whether they moved primarily by quadrupedal walking and running, or leaping, or whether they were suspensory). Indications of locomotor habits are derived from analyses of mostly fragmentary limb bones (e.g., Fleagle, 1983; Morbeck, 1983; Rose, 1983; Walker and Pickford, 1983). Figures 3 and 4 summarize the locomotor behavior of extant catarrhines from behavioral studies.

Social structure

What kind of groups did they live in? Living primates are social animals that live in a variety of different social groupings. One simple way of categorizing the social groupings of anthropoids is into monogamous groups conmposed of one adult male, one adult female, and their offspring; and polygynous ones composed of several adult females and either one adult male (harems) or several adult males (polygynous groups) (Hrdy, 1981). Among living anthropoids, the monogamous/polygynous dichotomy is tightly associated with the absence or presence, respectively, of sexual dimorphism in the canine teeth and the size of the lower jaw (Fleagle et al., 1980; Harvey et al., 1978). Approaches and techniques for identifying sexual dimorphism in fossil species are discussed by Gingerich (1981).

Diet and feeding behavior

What did they eat and how did they eat it? Diet and feeding behavior are among the best documented aspects of behavior in living species and seem to be related to many other aspects of their ecology, such as habitat preferences and ranging behavior. Figure 5 summarizes the diets of living catarrhines based on behavioral studies.

Dietary adaptations are clearly related to dental morphology among living species, so that the dentitions of fossil catarrhines can be used to reconstruct their likely dietary specializations (e.g., Kay, 1984). It is important to distinguish what is meant by ingestion and mastication, each of which are associated with distinctive dental adaptations. By ingestion is meant any process by which food is separated from its matrix as a "bite." Mastication is the process of mechanical breakup of the food once in the mouth. Teeth in the front of the mouth—incisors, canines, and sometimes premolars—are normally responsible for ingestion, whereas the molars are the principal agents for mastication. Incisors and canines often are adapted for many activities in addition to ingestion (i.e., grooming, defense, sexual displays). Moreover, some kinds of incisor designs are well-suited for ingesting *many* sorts of foods. For this reasons, reliable inferences about what kinds of foods are eaten cannot be based on incisor or canine structures alone, as is often done. Incisor structure can tell us about *feeding* behavior but not necessarily about diet. By contrast, molars are used for mastication and little else, so they tend to be much more reliable guides to a species' diet (Kay and Hylander, 1978). Among the most distinctive dietary specializations identifiable from dental structure are those for fruit- and nut-, leaf-, or insect-eating. Such

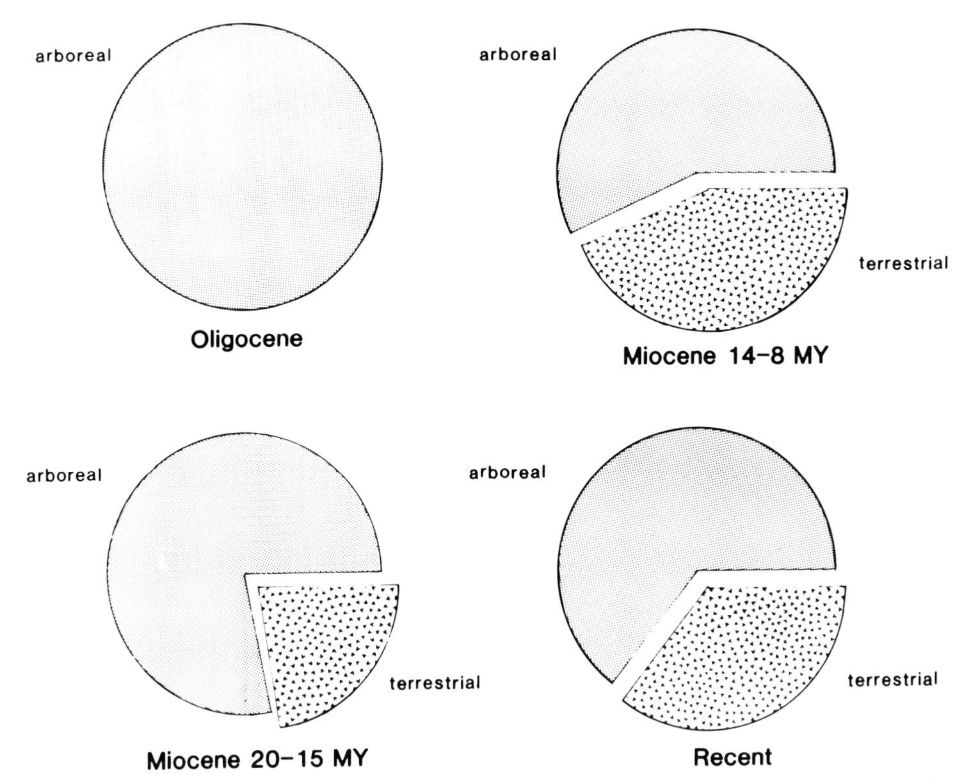

Fig. 3. Substratum preference of catarrhine genera through time. Each time interval includes all available taxa worldwide. If a particular genus occurs in two time intervals it is included twice. Data for Recent taxa are based on behavioral studies. Those for extinct animals are based on an assessment of anatomy of the best known forms (see Table 1). Extinct species for which there is no postcranial material are eliminated from consideration.

groups differ in body size and the relative development of features on the molars, like the cutting edges, crushing surfaces, crown heights, and enamel thickness.

Activity pattern

Although all living catarrhines are diurnal, it is well to remember that many prosimian primates are nocturnal. There is no reason to believe *a priori* that catarrhines were always diurnal (in evolutionary time). Fortunately, it is possible to judge reliably whether a primate is nocturnal or diurnal based on orbit size: nocturnal species tend to have relatively larger orbits than diurnal species of comparable body size (e.g., Kay and Cartmill, 1977; Kay, 1984).

LIVING CATARRHINES

Extant catarrhines are the dominant primates throughout the Old World, except for Madagascar. Their range extends from Gibraltar and Senegal in the west to Japan and the Philippines in the east; and from South Africa as far north as Japan and Pakistan. Living catarrhine faunas normally contain elements of two phyletically distinct groups—Old World monkeys (Cercopithecidae) and apes (Hominoidea). Figure 5 provides a breakdown of their relative contributions to the diversity of catarrhines as a whole.

The Old World monkeys are the more abundant and successful group, with about 15 genera and over 75 living species. There are two subfamilies with somewhat different biogeography. The Cercopithecinae are the common monkeys of Africa with only a single genus *(Macaca)* in Asia, while the leaf-eating Colobinae have only a few (numerically abundant) genera in Africa and a more diverse radiation in Asia.

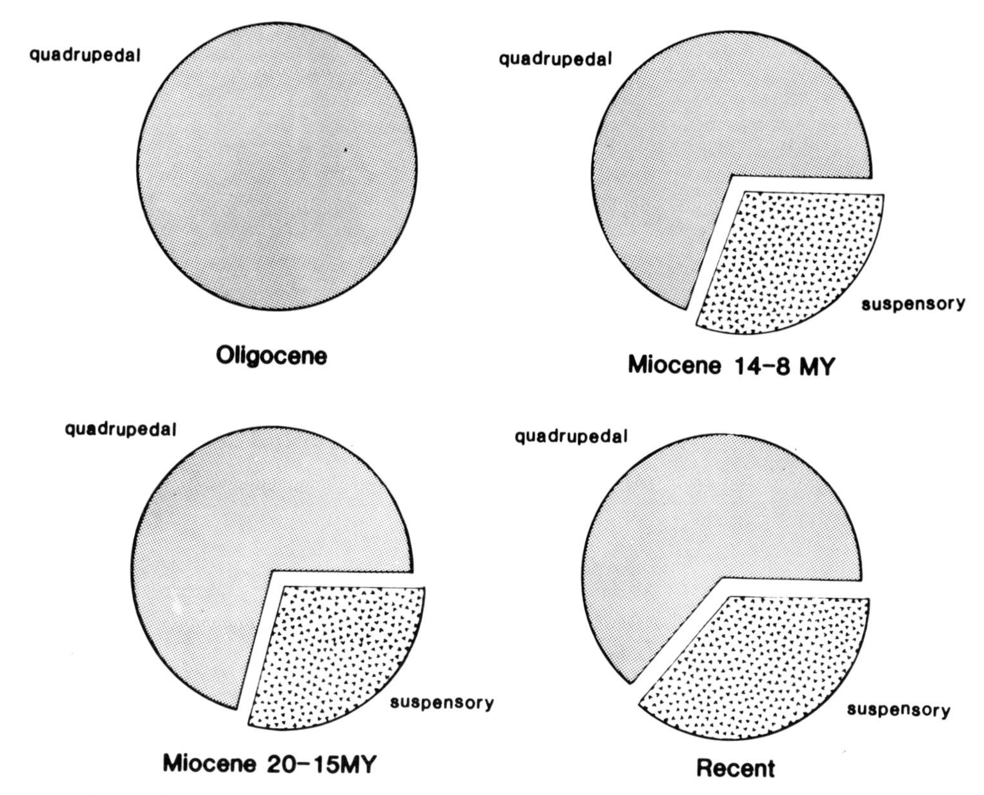

Fig. 4. Locomotor preference of catarrhine genera through time. Data base and conventions are the same as for substratum preference.

Among living primates, Old World monkeys are medium to large animals, ranging in body weight from the 2-kg talapoin monkey to baboons and proboscis monkeys that weigh up to 40 kg. Most cercopithecid species are arboreal monkeys, but there is considerable range in their substrate preferences. Some taxa are almost totally terrestrial *(Theropithecus, Erythrocebus)*, and many others move and feed both on the ground or in the trees (e.g., *Papio* species, *Cercopithecus aethiops, Macaca nemestrina, Presbytis entellus)*. All cercopithecids move on the ground or in the trees using all four limbs; i.e., as quadrupedal monkeys (Rose, 1974; Fleagle, 1980). Some are excellent leapers, but there are no species that spend a large part of their time in suspensory activities (i.e., hanging under the source of support either with arms, legs, or both).

Cercopithecid species primarily eat ripe and unripe fruit (most cercopithecines) and/ or leaves (most colobines). The two extremes of this dietary spectrum are readily distinguished anatomically: the former have short, rounded molar crests whereas the latter have long, trenchant ones (Kay, 1978). At least one genus, *Cercocebus*, eats a large amount of extremely hard nuts, an activity for which it is adapted by having an extremely thick covering of enamel on the teeth (e.g., Kay, 1981). There are a few species that specialize on seeds of various types *(Colobus satanas, Theropithecus)*, but there are no predominantly insect-eating or gum-eating species as are found among some other primate groups.

Most cercopithecids live in polygynous social groups (one-male harems or larger multimale troops where adult females outnumber males). Associated with this is the tendency for males to be much larger than females and to have relatively much larger canines. Strict monogamy is rare in Old World monkeys.

The Hominoidea, or apes, are the less abundant and less species-rich group of living catarrhines, with only four genera and no more than a dozen species in two families—the Hylobatidae or lesser apes, and the

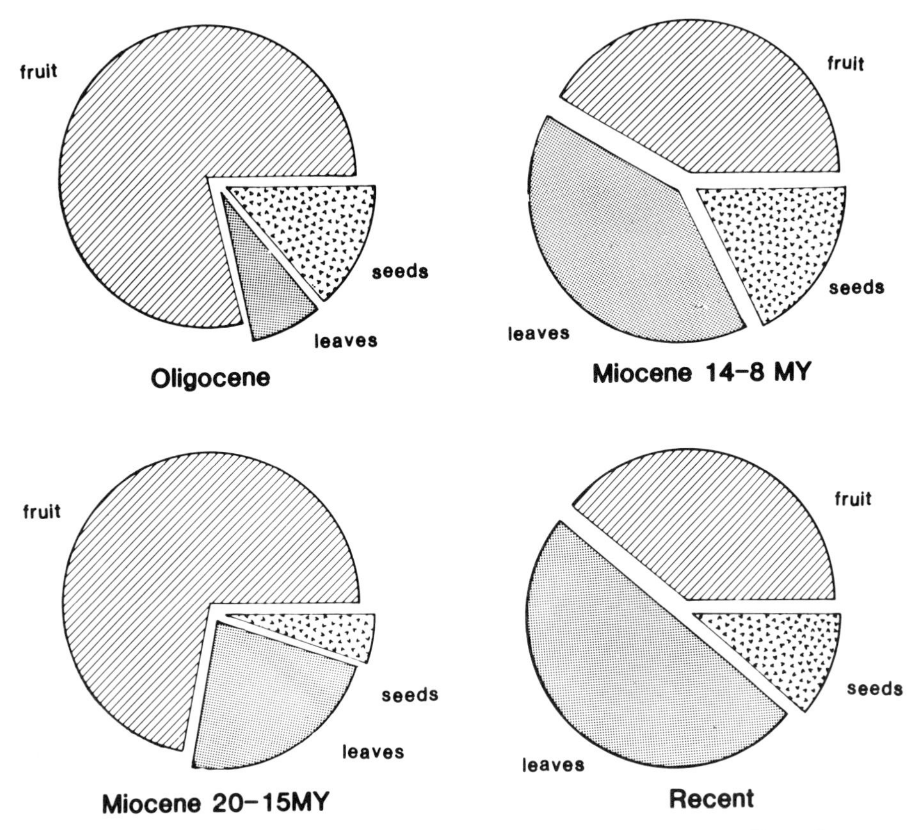

Fig. 5. Diets of catarrhine genera through time. Data are treated as in Figure 3.

Pongidae or great apes. Both the African *(Pan* and *Gorilla)* and Asian *(Hylobates* and *Pongo)* apes live sympatrically with cercopithecoids in forested habitats. The lesser apes *(Hylobates)* are medium-sized primates (4–12 kg), while the great apes (exemplified by *Gorilla)* are the largest of all living primates.

Living apes vary considerably in their substrate preferences. Hylobatids are completely arboreal, while great apes include species that are occasionally terrestrial *(Pongo)* and others (i.e., Mountain *Gorilla)* that are predominantly terrestrial. Irrespective of their habitat preferences, apes seem to be adept at various types of suspensory behavior and climbing (Fleagle, 1976). Of this group, only the African apes are regularly quadrupedal.

Most living ape species specialize in ripe fruit, but a few, like *H. (Symphalangus)* and *Gorilla* eat a considerable portion of leaves and herbaceous material. *Pongo* in some circumstances eats extremely hard nuts, which could explain its having particularly thick molar enamel. None specialize in gum- or insect-eating.

All hylobatids are monogamous; males and females are similar in body size and canine size, which goes along with this. Pongid social structures are difficult to categorize, but their reproduction activities involve males having reproductive access to more than one female; not surprisingly, this enhances male–male competition and selects for larger male body size and canine size.

In Figures 3–5 are plotted the ecological features of extant Old World monkeys and apes. Against this background we can now examine the catarrhines from earlier epochs (also summarized on those figures).

OLIGOCENE ANTHROPOIDS

The earliest well-documented Old World anthropoids are from the early Oligocene of Egypt. They include two very different phyletic groups—Parapithecidae and Propliopithecinae—which sympatrically and synchronously inhabited a riverine rainforest environment (Bown et al., 1982). Figure 4 gives a breakdown of the diversity of those forms at the species level. Interestingly, unlike Miocene to Recent catarrhines, these early anthropoids did not live sympatrically with any known lorisoids. This may explain in part why the adaptive roles filled by Oligocene catarrhines were so different from those of more recent forms.

The more primitive group of Oligocene anthropoids is the Parapithecidae, whose affinity with extant catarrhines is probably rather distant. In many features parapithecids retained primitive features found among platyrrhines or tarsioid prosimians (including three premolars), rather than catarrhine synapomorphies (Andrews, 1985; Kay and Simons, 1983b). They were very small monkeys, scarcely overlapping living catarrhines in body size. The smallest was *Qatrania wingi*, at 300 g the smallest known Old World anthropoid. The largest was *Simonsius grangeri* at about 3000 g. This size range fills in nicely for lorisoids, which today occupy the smaller arboreal primate niches of the Old World.

Available fragmentary remains of the faces of *Apidium* and *Simonsius* as well as of the propliopithecine *Aegyptopithecus* (see below) suggest that all had relatively small orbits, with the range of relative sizes found among extant diurnal anthropoids and strepsirhines. There is no hint of a nocturnal mode of activity among these or any other extinct catarrhine (Kay and Simons, 1980).

The limb skeleton is well-known for only one parapithecid, *Apidium phiomense*. This talapoin-sized monkey (ca. 1300 g) was an arboreal, quadrupedal leaper with numerous adaptations for saltatory behavior. For the other species, we can only postulate from the nature of the riverine environment and their relatively small size that they were probably arboreal and unlikely to have been suspensory.

The dental morphology of most parapithecids is quite similar to that found among any living anthropoids. *Apidium* had front teeth very similar to those of extant forms such as *Miopithecus* or *Saimiri*. However, *Simonsius grangeri* apparently lacked lower permanent incisors (Kay and Simons, 1983b), so at least some of these early catarrhines acquired their foods in a manner unlike that of extant ones. On the other hand, the front teeth were used for incising a bite of food, not in the role of "tooth-comb" as in extant lorises and lemurs. The molars are most comparable to those of living fruit-eating anthropoids, with low, rounded cusps and little molar shearing. Recent study of *Apidium phiomense* suggests it had very thick molar enamel, which would suggest, by analogy with the living platyrrhine *Cebus*, that *Apidium* may have been able to feed on small, hard nuts. *Simonsius* has enhanced molar shearing, suggesting a component of leaf-eating. The smallest species, *Qatrania wingi*, had rounded molar cusps and could not have been insectivorous; it was more probably a fruit-eater or gum-eater, the latter dietary guild not found among extant catarrhines. Even at this early stage in their career, the catarrhine diet probably contained little or no insects. No known catarrhine exhibits a combination of well-developed molar shearing and very small body size, hallmarks of such a feeding strategy. A recently discovered African Oligocene tarsioid hints that such adaptive niches were present and filled at least in part by primates, although not by catarrhines.

Two parapithecid species *(A. moustafai* and *A. phiomense)* show evidence of canine sexual dimorphism suggesting a polygynous social structure (Fleagle et al., 1980).

The other group of Fayum anthropids consists of *Propliopithecus* and *Aegyptopithecus*, early catarrhines (anthropoids with two premolars) whose phyletic descendants included both later apes and cercopithecids (e.g., Fleagle and Kay, 1983). Here we refer to them as "eucatarrhines." These were large animals (3–6 kg), similar in size (and in much of their morphology) to the larger living platyrrhines.

The limb skeletons of eucatarrhines indicate that they were arboreal quadrupeds whose best living analog is the Neotropical howling monkey *Alouatta*. There is no reason to suspect that they had a prehensile tail, but the foot structure does suggest the possibility of some suspensory abilities (Fleagle and Simons, 1983).

Eucatarrhines were predominantly frugivorous animals. Both well-known species had sexually dimorphic canines, suggesting a polygynous social structure (Fleagle et al., 1980).

Figures 3–5 summarize the behavior and ecology of Oligocene anthropoids deduced from their anatomy.

EARLIER MIOCENE (25-15 MY BP) PROTOAPES AND EARLY MONKEYS FROM EAST AFRICA

A temporal gap of over ten million years separates the Oligocene anthropoids from Egypt (31 my+) from the numerous fossil apes and (rare) Old World monkeys from the Early Miocene (17–19 my). In Kenya and Uganda, where these animals are best known, there are numerous localities representing a variety of habitats ranging from rainforest to woodland, each of which contains a somewhat different mixture of species (Harrison, 1982; Pickford, 1983; Walker and Leakey, 1984). Similar genera of fossil apes are known from both Saudi Arabia (Andrews et al., 1978) and China (Li, 1978; Gu and Lin, 1983), although their exact temporal and phyletic relationships to the African taxa are poorly understood. Fossil monkeys from this time were restricted to Africa, coming from Egypt (Simons, 1969) and Libya (Delson, 1979) as well as east Africa.

The most abundant of the Early Miocene catarrhines from East Africa include the somewhat ape-like *Proconsul* and its relatives. These numerous genera and species are here referred to collectively as protoapes. They were synchronous and partly sympatric with early Old World monkeys that appear to have been very much like modern forms. Collectively, protoapes and Old World monkeys were very diverse, with a size range comparable to that found among living catarrhines (Fleagle, 1978) (see Figure 2).

Relatively complete skeletal material is known from some species of protoapes, while others are only known from isolated elements attributed to dental taxa largely on the basis of size (e.g., Fleagle, 1983; Walker and Pickford, 1983; Harrison, 1982 and references therein). Nevertheless, the limb bones indicate a considerable diversity of locomotor abilities; some species were suspensory *(Dendropithecus macinnessi)*, others were arboreal quadrupeds *(Proconsul africanus)*, and two *(P. nyanzae* and *P. major)* were probably somewhat terrestrial.

There is considerable variation in the relative size of the incisors of the protoapes, suggesting considerable diversity in the importance of these structures for ingestion (although not necessarily indicative of dietary diversity, as Harrison (1982) claimed). In *Rangwapithecus* and *Dendropithecus* the incisors were relatively small, suggesting only limited use of these structures for food preparation. Other taxa like *Micropithecus* and *Limnopithecus* apparently more fully utilized their front teeth for incision of bites of food. Nearly all of the species studied have molars comparable to living catarrhines with a relatively frugivorous diet (Kay, 1977); certainly there were no extremely folivorous taxa. No convincing evidence has yet been advanced that any species of protoape had thick enamel and might have been a nut-eater, although such a possibility has been alluded to by Gantt (1983).

Because of the numerous sympatric taxa at most localities, sexual differences in dental size are difficult to dissociate from those attributable to different-sized species. Nevertheless, most taxa seem to have been sexually dimorphic in canine size (Greenfield, 1972; Harrison, 1982), suggesting polygnous social groups. *Dendropithecus macinnesi* is striking, in that—although there is considerable dimorphism in canine size—both sexes have relatively large canines. Whether this is attributable to the canines of both sexes having a role in defense, exclusion of conspecifics of the same sex from defended territories, or food acquisition cannot be determined from present evidence.

Although the Early Miocene primate faunas in East Africa are dominated by the protoapes, Old World monkeys are known from two localities (Napak and Buluk), as well as roughly contemporary localities in North Africa (Jebel Zelten and el Moghara). It is not clear at present why monkeys are at some localities but not others during this earliest time period. Cercopithecids are much more common at the younger, Middle Miocene locality of Maboko Island, and many authors have suggested on this basis that they intitially evolved in open woodland habitats (Pickford, 1983; Andrews, 1981).

There are probably two genera of Old World monkeys from the earlier Miocene, *Prohylobates* and *Victoriapithecus*, each with several species. They are not readily allocated to either living subfamily. The isolated teeth and fragmentary jaws show that they were small- to medium-sized monkeys by the standards of extant Old World monkeys. The few limb elements (from Maboko) are most comparable to those of the living vervet monkey, *Cercopithecus aethiops*, suggesting neither strict arboreal nor terrestrial habit and are concordant with the suggestion that they inhabited open woodland environment, as do vervet monkeys.

Reconstructing the dietary habits of these early Old World monkeys is a complicated problem, because there are several possible frames of reference (Kay, 1984). Compared with extant cercopithecids, *Victoriapithecus* had a molar morphology very similar to some macaques, suggesting a frugivorous diet. However, *Victoriapithecus* had greater molar shearing abilities (suggesting folivory) than any of the Miocene protoapes. The extremely high crowns of the molar teeth suggest the diet contained considerable grit, which would go along with other anatomical features pointing to terrestrial habits.

There was canine dimorphism in at least one taxon *(Victoriapithecus)*, indicating these early Old World monkeys may have had polygynous social groups.

Figures 3–5 summarize our findings about the ecology and behavior of earlier Miocene taxa.

LATER MIOCENE (14–8 MY BP) CATARRHINES FROM EURASIA

In contrast with the abundance of Early Miocene fossil catarrhines in Africa, there is a relative paucity of these forms in the Middle and Late Miocene record of the continent. The fossil record of catarrhines from the Middle and Late Miocene comes mainly from Europe and Asia, where numerous localities representing a variety of environments have yielded fossil catarrhines. The phyletic and biogeographical relationships of European and Asian apes are obviously more complicated than earlier authorities had realized. There are at least five distinct phyletic groups. The pliopithecids *(Pliopithecus* and *Crouzelia)* lack many features that characterize extant catarrhines and seem to be most closely allied with the Fayum eucatarrhines (Kay et al., 1981; Fleagle and Kay, 1983). A second group, exemplified by *Dropithecus,* were more modern-looking hominoids, as far as we can tell mainly from dental remains. *Sivapithecus* and *Gigantopithecus* appear to be a third group of specialized taxa with special phyletic relationships to some (Andrews and Cronin, 1982; Lipson and Pilbeam, 1982) or possibly all extant apes (Wolpoff, 1982; Walker and Leakey, 1984) or humans (Kay, 1982a; Kay and Simons, 1983a). Perhaps *Dryopithecus* is a close relative of the *Sivapithecus* clade. To these may be added a fourth enigmatic group known only by *Oreopithecus* and a fifth, the Cercopithecidae, which were late-comers to Eurasia, appearing after these other groups had become locally or regionally extinct in Late Miocene times. Figure 6 summarizes this taxonomic diversity.

Like the Early Miocene protoapes, the Middle and Late Miocene apes, here referred to collectively as pongids, had a range of body sizes comparable to that of extant catarrhines, from small gibbon-sized apes to gorilla-sized ones like *Sivapithecus* and *Gigantopithecus* (Figure 2).

The locomotor abilities of the Middle and Late Miocene apes are difficult to reconstruct with confidence because of the lack of associated limb bones for most taxa. *Pliopithecus* is the only eucatarrhine for which most of the skeleton is known. This siamang-sized ape was an arboreal quadrupedal/suspensory animal most comparable to a living platyrrhine, the spider monkey, in many of its limb elements. The skeletal elements of Middle to Late Miocene pongids from Austria (Ehrenberg, 1938) and Hungary demonstrate considerable diversity in locomotor abilities. Much of this material is similar to living hylobatids, chimpanzees or orangutans, suggesting suspensory abilities (Morbeck, 1983). For *Sivapithecus* species from the Late Miocene of Pakistan, the hand bones and large grasping hallux suggest arboreal climbing abilities, while many morphological details of the elbow suggest suspensory abilities (Morbeck, 1983; Raza et al., 1983). A large pongid from Lufeng in China had long curved digits, reinforcing this impression. Some of the *Sivapithecus* group were very large and probably partly terrestrial, at least as adults.

There have been no systematic attempts to reconstruct the diet of all the Middle and Late Miocene apes of Europe. The relatively trenchant crests on the molars of eucatarrhines like *Pliopithecus vindobonensis* suggest a somewhat folivorous diet; *P. antiquus* had more rounded molar crests, suggesting it was more frugivorous. Pongids like *Dryopithecus* from Europe and Asia were probably relatively frugivorous, while *Sivpithecus* with its thick enamel seems to have specialized on hard objects such as nuts (Kay, 1981; but see Teaford and Walker, 1984, for a slightly different view of *Sivapithecus* adaptations).

As among the Early Miocene protoapes from East Africa, there is considerable difficulty in sorting out species and sexes among

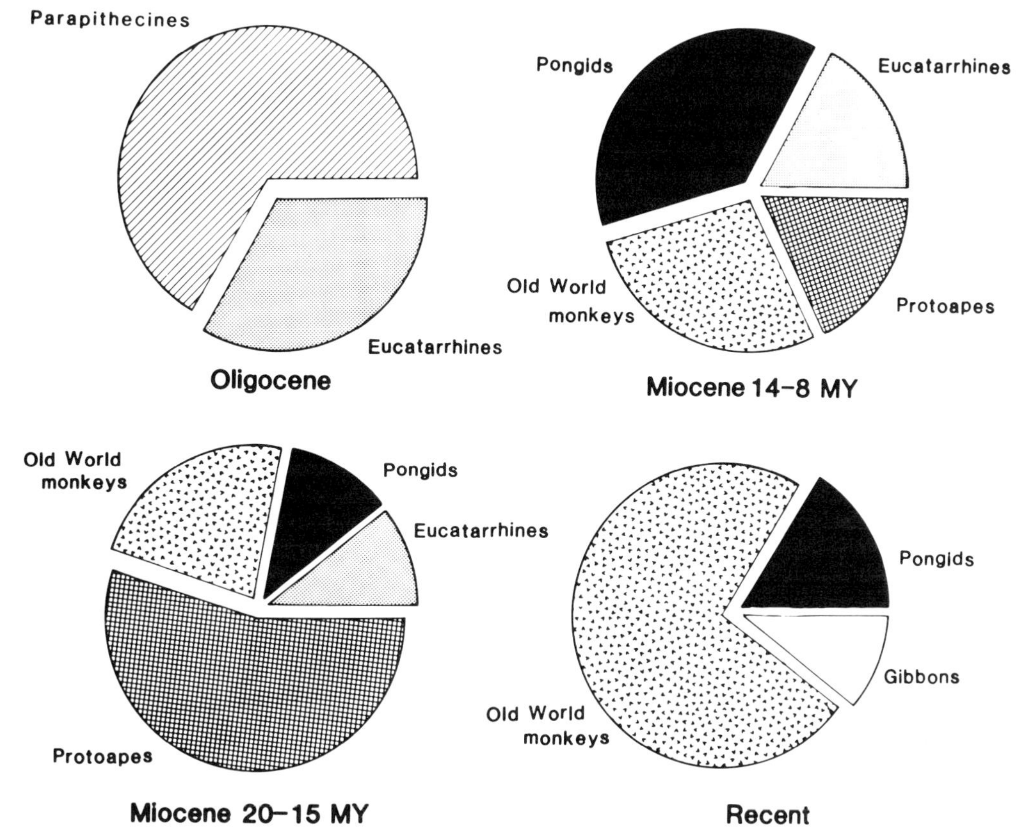

Fig. 6. Percentage contributions of informal catarrhine subunits to the overall pattern of catarrhine diversity at the generic level. Human beings are excluded. Key to the informal taxonomic terms in Figure 6 and the text: Parapithecines = Parapithecidae; Eucatarrhines = Pliopithecidae, Propliopithecinae, Pliopithecinae; Protoapes = Proconsulinae; Pongids = Modern great apes of Africa and Asia and their anatomically similar antecedents (Dryopithecinae, Ramapithecinae) and *Oreopithecus*; Gibbons = Hylobatidae; Old World monkeys = Cercopithecidae (excludes parapithecids).

the Middle and Late Miocene taxa. *Pliopithecus* has considerable sexual dimorphism, suggesting a polygynous social organization, and the same seems possible for *Dryopithecus*. In *Sivapithecus*, however, the variability in canine size among all known taxa is strikingly small, suggesting a lack of canine dimorphism and possibly a monogamous social structure (Kay, 1982a,b).

Despite a century of study, the suprageneric affinities of *Oreopithecus* are still the subject of extensive debate among those who feel it is an unusual cercopithecid, those who feel it is an unusual ape, and those who feel equally strongly both ways (see Harrison, 1982; Szalay and Delson, 1979 for reviews). It is almost certainly not a hominid or a parapithecid. In the taxonomic summary in Figure 6, *Oreopithecus* is lumped with pongids. In contrast, all agree on the assessment of the behavioral abilities of this very specialized Late Miocene catarrhine from Italy. *Oreopithecus bambolii* is known mainly from lignite deposits and apparently lived in a riverine or lacustrine swamp forest. It apparently weighed around 15 kg and was unquestionably an arboreal, suspensory species, as indicated by the very long slender arms, relatively short hindlimbs, and limb joints that allowed extraordinary ranges of mobility (Jungers, 1984). It was folivorous with its particularly long sharp-edged molar crests. There is considerable dimorphism in canine size among individuals of *Oreopithecus*, suggesting a polygynous social structure.

Figures 3–5 summarize our behavioral assessments of later Miocene catarrhines.

EVOLUTION AND RADIATION OF OLD WORLD MONKEYS

Because the evolution of Old World monkeys is collateral to the evolution of humans, we have not dealt with it in any great detail here. However, a few striking features of the evolutionary history of that group are useful for the contrast they provide to our knowledge of hominoid evolution.

The fossil record of Old World monkeys is dramatically different from that of hominoids in that the living apes are a group whose diversity in the fossil record vastly exceeds that in the present while cercopithecids have a meagre fossil record compared with their current diversity and abundance. This discrepancy is still present if we eliminate many of the Oligocene and Early Miocene species from consideration as either apes or monkeys. One possible explanation for this difference is that the evolution of cercopithecids has taken place primarily in those areas that lack a good fossil record—West Africa, the Zaire basin, and Southeast Asia—while ape evolution has taken place (or at least has been documented) in the areas from which we have a fairly good fossil record—East Africa, Europe, and Northern India and Pakistan, areas that are relatively depauperate of living primates today. This is the reverse of the scenario for Old World monkey evolution proposed by Napier (1970).

More likely however, the differences in the quality of the fossil record of ape and monkey evolution seem to be largely the result of differences in the timing of their adaptive radiations (e.g., Andrews, 1981). Living cercopithecids seem to be the result of a major radiation in the latest Miocene and Pliocene—times from which the fossil record of apes is relatively poor compared with earlier faunas.

A notable feature of fossil Old World monkeys is the abundance of terrestrial forms. The earliest known fossil colobines were much more terrestrial than the vast majority of their living relatives, and the arboreal *Cercopithecus* monkeys have no significant fossil record. It is very difficult to distinguish whether the initial change from ape-dominated faunas of the Miocene to more recent monkey-dominated ones should be viewed as 1) an ecological replacement within similar environments or 2) the result of an environmental change from forested environments to more open woodland habitats that seems to have occurred in the latest Miocene. Obviously monkeys dominate both sorts of environments today. However, it is surely significant that the earliest undoubted hominids appear in conjunction with the Pliocene and Pleistocene monkey faunas when there is no evidence of other apes.

DISCUSSION

When we examine the adaptations of catarrhines and especially hominoids over the past 35 million years, several surprising patterns emerge. Initially, the "adaptive breakdown" of catarrhine faunas in the Oligocene was very different from that of present-day catarrhines. However, since the Early Miocene there has been little change, despite considerable turnoever of the taxa involved, not only at the level of species and genera, but of families. Catarrhine evolution since the Early Miocene seems to be like a play in which the cast changes from production to production but the roles remain similar. At first glance it might appear that such a finding is mandated by the great breadth and limited resolution of the ecological parameters we have selected as well as by the presumption that some primate has to fill all these niches anyway. A consideration of other primate radiations shows, however, that this is not the case. If, for example, we tried to plot platyrrhine monkeys on our charts, we would find that the terrestrial "niche" remains unfilled and the folivorous niche is rather poorly represented. Along these lines it is intriguing that the Oligocene anthropoids from Egypt, frequently described as platyrrhine-like in their morphology and body size, are notably different from the Miocene and later catarrhines in the same ways platyrrhines differ from catarrhines. Of course, the true features of the Oligocene radiation may be obscured somewhat because the known fossils come from a few localities preserving a riverine, forested habitat. Nonetheless, it would appear that Old World primates only evolved (or invented) large size, folivory, and terrestriality in the Miocene at least 10 million years after the evolutionary origins of the group.

A last generalization is that several of the ecological parameters we have chosen to examine here tend to be relatively independent of one another from time frame to time frame (see Table 1). Thus the most folivorous species may turn out to be a terrestrial quadruped in one fauna and an arboreal quadruped in another. Or the hard-object feeder niche may be occupied by a small leaping form in one time frame and a large suspensory animal in another. This is simply an extension in the time dimension of the phenomenon reported by Fleagle and Mittermeier (1980), whose study of ecological partitioning of tropical-forest-dwelling primates identifies the absence of any consistent correlation between habitual mode of locomotion and dietary preference.

ACKNOWLEDGMENTS

We thank the organizers of the Ancestors Symposium for the invitation to participate in this venture. Figure 1 was drafted by Luci Betti. This work was funded in part by a fellowship from the John Simon Guggenheim Memorial Foundation and research grants 8210949 and 8209937 from the National Science Foundation. It profited greatly from comments and suggestions by William Jungers and Eric Delson.

LITERATURE CITED

Andrews, P (1981) Species diversity and diet in monkeys and apes during the Miocene. In CB Stringer (ed): Aspects of Human Evolution. London: Taylor and Francis, pp. 25–41.

Andrews, P (1985) Family group systematics and evolution among catarrhine primates. In E Delson (ed): Ancestors: The Hard Evidence. New York: Alan R. Liss, Inc., pp 14–22.

Andrews, P, and Cronin, JE (1982) The relationships of *Sivapithecus* and *Ramapithecus* and the evolution of the orangutan. Nature *297:*541–546.

Andrews, P, Hamilton, WR, and Whybrow, PJ (1978) Dryopithecines from the Miocene of Saudi Arabia. Nature *274:*249–250.

Ankel, F (1965) Der canalis sacralis als indikator für die Lange der caudalregion der primaten. Folia Primatol. *3:*263–276.

Bown, TM, Kraus, MJ, Wing, SL, Fleagle, JG, Tiffney, BH, Simons, EL, and Vondra, CF (1982) The Fayum primate forest revisited. Jour. Hum. Evol. *11:*603–632.

Ciochon, R (1983) Hominoid cladistics and the ancestry of modern apes and humans: A summary statement. In RL Ciochon and RS Corruccini (eds.): New Interpretations of Ape and Human Ancestry. New York: Plenum, pp. 781–844.

Delson, E (1979) *Prohylobates* (Primates) from the early Miocene of Libya: A new species and its implications for cercopithecid origins. Geobios (Lyon) *12:*725–733.

Ehrenberg, K (1938) *Austriacopithecus*, ein neuer menschenaffenartiger Primate aus dem Miozän von Klein-Hadersdorf bei Poysdorf in Niederösterreich (Nieder Donau). S.-Ber. Akad. Wiss. Wien, Math.-Nat. Kl. Abstr. 1, *147:*71–110.

Fleagle, JG (1976) Locomotion and posture of the Malayan Siamang and implications for human evolution. Folia Primatol. *26:*245–269.

Fleagle, JG (1978) Size distributions of living and fossil primate faunas. Paleobiol. *4:*67–76.

Fleagle, JG (1980) Locomotor behavior of the earliest anthropoids. A review of the current evidence. Z. Morphol. Anthropol. *71:*149–156.

Fleagle, JG (1983) Locomotor adaptations of Oligocene and Miocene hominoids and their phyletic implications. In RL Ciochon and RS Corruccini (eds): New Interpretations of Ape and Human Ancestry. New York: Plenum, pp. 301–324.

Fleagle, JG (1984) Primate locomotion and diet. In DJ Chivers, BA Wood, and A Bilsborough (eds): Food Acquisition and Processing in Primates. New York: Plenum, pp. 105–118.

Fleagle, JG, and Kay, RF (1983) New interpretations of the phyletic position of Oligocene hominoids, In RL Ciochon and RS Corruccini (eds): New Interpretations of Ape and Human Ancestry. New York: Plenum, pp. 181–210.

Fleagle, JG, Kay, RF, and Simons, EL (1980) Sexual dimorphism in early anthropoids. Nature *287:*328–330.

Fleagle, JG, and Mittermeier, R (1980) Locomotor behavior, body size and comparative ecology of Suriname monkeys. Am. J. Phys. Anthropol. *52:*301–314.

Fleagle, JG, and Rosenberger, AL (1983) Cranial morphology of the earliest anthropoids. In M. Sakka (ed): Morphologie Evolutive, Morphogenese du Crane et Origine de l'Homme. Paris: C.N.R.S., pp. 141–153.

Gantt, DG (1983) The enamel of Neogene hominoids. In RL Ciochon and RS Corruccini (eds): New Interpretations of Ape and Human Ancestry. New York: Plenum, pp. 249–298.

Gingerich, PD (1981) Cranial morphology and adaptations in Eocene Adapidae. I. Sexual dimorphism in *Adapis magnus* and *Adapis parisiensis.* Am. J. Phys. Anthropol. *56:*117–124.

Gingerich, PD, Smith, BH, and Rosenberg, K (1982) Allometric scaling in the dentition of primates and prediction of body weight from tooth size in fossils. Am. J. Phys. Anthropol. *58:*81–100.

Greenfield, L (1972) Sexual dimorphism in *Dryopithecus africanus.* Primates *13:*395–410.

Gu, Y, and Lin, Y (1983) First discovery of *Dryopithecus* in East China. Acta Anthropol. Sinica *2:*313–318.

Harrison, T (1982) Small-bodied apes from the Miocene of East Africa. Ph.D. Dissertation, University College, London.

Harvey, PH, Kavanagh, M, and Clutton-Brock, TH (1978) Sexual dimorphism in primate teeth. J. Zool. *186:*475–485.

Hrdy, S (1981) The Woman that Never Evolved. Cambridge, Massachusetts: Harvard University Press.

Jungers, W (1984) Aspects of size and scaling in primate biology with special reference to the locomotor skeleton. Yrbk Phys. Anthropol. *27:*73–97

Kay, RF (1977) Diets of early Miocene African hominoids. Nature *268:*628–630.

Kay, RF (1978) Molar structure and diet in extant Cercopithecoidea. In PM Butler and K Joysey (eds): Development, Function and Evolution of Teeth. London: Academic Press, pp. 309–339.

Kay, RF (1981) The nut-crackers—A new theory of the adaptations of the Ramapithecinae. Am. J. Phys. Anthropol. *55:*141–151.

Kay, RF (1982a) *Sivapithecus simonsi,* a new species of Miocene hominoid with comments on the phylogenetic status of the Ramapithecinae. Int. J. Primatol. *3:*113–174.

Kay, RF (1982b). Sexual dimorphism in Ramapithecinae. Proc. Nat. Acad. Sci. (USA)*79:*209–212.

Kay, RF (1984) On the use of anatomical features to infer foraging behavior in extinct primates. In P Rodman and JG Cant (eds): Adaptation for Foraging in Nonhuman Primates. New York: Columbia University Press, pp.21–53.

Kay, RF, and Cartmill, M (1977) Cranial morphology and adaptations of *Palaechthon nacimienti* and other Paromomyidae (Plesiadapoidea, ?Primates) with a description of a new genus and species. J. Hum. Evol. *6:*19–53.

Kay, RF, and Hylander, WL (1978) The dental structure of mammalian folivores with special reference to Primates and Phalangeroidea (Marsupialia). In GG Montogomery (ed): The Biology of Arboreal Folivores. Washington, D.C.: Smithsonian Inst. Press, pp. 173–191.

Kay, RF, and Simons, EL (1980) Ecology of Oligocene African Anthropoidea. Int. J. Primatol. *1:*21–37.

Kay, RF, and Simons, EL (1983a) A reassessment of the relationships between later Miocene and subsequent Hominoidea. In RL Ciochon and RS Corruccini: (eds.): New Interpretations of Ape and Human Ancestry, New York: Plenum, pp. 577–624.

Kay, RF, and Simons, EL (1983b) Dental formulae and dental eruption patterns in Parapithecidae (Primates, Anthropoidea). Am. J. Phys. Anthropol. *62:*363–375.

Li, C (1978) A Miocene gibbon-like primate from Shihhung, Kiangsu Province. Vert. Palasiat. *16:*187–192.

Lipson, S, and Pilbeam,DR (1982) *Ramapithecus* and hominoid evolution. J. Hum. Evol. *11:*545–548.

Morbeck, ME (1983) Miocene hominoid discoveries from Rudabanya: Implications from the postcranial skeleton. In RL Ciochon, and RS Corruccini (eds): New Interpretations of Ape and Human Ancestry. New York: Plenum, pp. 369–404.

Napier, JR (1970) Paleoecology and catarrhine evolution. In JR Napier and PH Napier (eds): Old World Monkeys. London: Academic Press, p. 55–95.

Pickford, M (1983) Sequence and environments of the Lower and Middle Miocene hominoids of Western Kenya. In RL Ciochon and RS Corruccini (eds): New Interpretations of Ape and Human Ancestry. New York: Plenum, pp. 421–439.

Raza, SM, Barry, JC, Pilbeam, DR, Rose, MD, Ibrahim Shah, SM, and Ward, S (1983) New hominoid primates from the Middle Miocene Chinji Formation, Potwar Plateau, Pakistan. Nature *306:*52–54.

Rose, MD (1974) Quadrupedalism in Primates. Primates *14:*337–357.

Rose, MD (1983) Miocene hominoid postcranial morphology: Monkey-like, ape-like, neither, or both? In RL Ciochon and RS Corruccini (eds): New Interpretations of Ape and Human Ancestry. New York: Plenum, pp. 405–420.

Simons, EL (1969) Miocene monkey *(Prohylobates)* from northern Egypt. Nature *223:*687–689.

Szalay, FS, and Delson, E (1979) Evolutionary History of the Primates. New York: Academic Press.

Teaford, MF, and Walker, AC (1984) Quantitative Differences in Dental Microwear Between Primate Species With Different Diets and a Comment on the Presumed Diet of *Sivapithecus.* Am. J. Phys. Anthropol. *62:*363–375.

Walker, AC, and Leakey, RE (1984) New fossil primates from the Lower Miocene site of Buluk, northern Kenya. Am. J. Phys. Anthropol. *63:*232.

Walker, AC, and Pickford, M (1983) New postcranial fossils of *Proconsul africanus* and *Proconsul nyanzae.* In RL Ciochon and RS Corruccini (eds): New Interpretations of Ape and Human Ancestry. New York: Plenum, pp. 325–352.

Wolpoff, M (1982) *Ramapithecus* and hominid origins. Curr. Anthropol. *23:*501–522.

Ancestors: The Hard Evidence, pages 37–41

Origins and Characteristics of the First Hominoids

Elwyn L. Simons
Duke University Primate Center, Durham, North Carolina 27705

ABSTRACT The earliest hominoids, *Aegyptopithecus* and *Propliopithecus* (family Propliopithecidae), occur in the Oligocene deposits of the Fayum depression, Egypt. Fifty years after the first Fayum primates had been described, I returned to the region to collect fossils, and in several field seasons greatly increased the number of both specimens and known taxa. Unfortunately, these expeditions recovered few postcranial elements and only one good cranium. Since 1977, seven additional field seasons have yielded three new faces of *Aegyptopithecus*, various other cranial parts, and numerous isolated postcrania, which provide a wealth of information about the lifeways of early catarrhines. The majority of the limb bones are allocated to *A. zeuxis* on the basis of size, while a few smaller but nearly identical elements can be assigned to *Propliopithecus*. Despite earlier suggestions, these two genera are dentally quite distinct, although closely related. The propliopithecids are clearly linked to later hominoids, such as the Miocene *Proconsul*, by shared-derived details of dental, postcranial, and cranial anatomy. They do not appear to be specially related to the Miocene "aberrant" pliopithecids, nor to any monkeys. Although recent finds in South Africa reveal that the oldest platyrrhines had teeth similar to those of propliopithecids, and the latter are postcranially most similar to certain New World monkeys (*Alouatta*), these similarities are likely to be shared-primitive features. Although it has been fashionable to suggest that propliopithecids might have been ancestral to cercopithecoid monkeys, that is doubtful on the basis of the morphological differences between Fayum forms and the earliest Old World monkeys.

The first hominoids belong to a family group, the Propliopithecidae, that occurs only in the Fayum Oligocene deposits of Egypt. Two genera are valid. One of these, *Propliopithecus*, has three species: *P. haeckeli*, *P. markgrafi*, and *P. chirobates*; the other, *Aegyptopithecus*, has but one species, *A. zeuxis*. These Oligocene apes have been recently reviewed in Kay et al. (1981). Another genus and species, *Oligopithecus savagei*, was named by Simons (1962), but it has not generally been associated closely with species of the aforementioned two genera. The possibility exists that *O. savagei* is more closely related to the two propliopithecid genera than it is to any other African Oligocene primates.

Fossils were collected in the Fayum Badlands of Egypt, perhaps in the spring of 1907, by the professional collector Richard Markgraf (Gingerich, 1978). Some of these, thought to be a marsupial and two primates, were sold by him to the Stuttgart "Naturalkabinett." In 1910 and 1911 Dr. Max Schlosser of Munich described these earliest African primates. The "marsupial" he described was the type of *Parapithecus fraasi*, a species we now include in the monkey family, Parapithecidae. The other two specimens became the types of *Propliopithecus haeckeli* and

Moeripithecus markgrafi. These taxa Schlosser compared favorably with *Pliopithecus* and even *Pithecanthropus*. In the 1920s, 1930s, 1940s, and 1950s, the better preserved of these two specimens, the type of *Propliopithecus haeckeli*, was often regarded as being related to *Hylobates* and/or equally to the ancestors of apes and man: see Gregory (1922), Piveteau (1957), or Le Gros Clark (1959). This type consists of a nearly complete lower jaw with incisor sockets, preserving as well all the posterior teeth save one canine. The lower premolars and molars of *Propliopithecus haeckeli* look very much like those of apes and hominids and do not at all resemble such teeth in prosimians or monkeys. Since this specimen shows a small canine, has premolars of subequal size, and has a jaw that is shallow anteriorly, it was even suggested by Pilbeam (1967) as representing a species especially related to the ancestry of hominids.

After a fifty year lapse in field work in Egypt, my expeditions beginning in 1961 started to produce new information about these earliest known higher primates of the Egyptian Fayum badlands. At first, only isolated teeth and fragmentary lower jaws were found by our successive expeditions. Specimens recovered in 1962 and 1963 made it possible to conclude (Simons, 1967b) that *Moeripithecus markgrafi* was nothing more than a distinct species of *Propliopithecus*, while other finds showed that a new genus and species of Fayum ape existed. This I named *Aegyptopithecus zeuxis* (Simons, 1965).

All the new teeth and jaws of these apes confirmed what had been indicated long before. Dentally, these creatures showed a whole range of similarities to modern apes and humans and to the many Miocene apes of Kenya and Uganda (East Africa) that had been discovered and named between 1911 and 1961. A series of expeditions in the 1960s recovered several new jaws of *Aegyptopithecus*, isolated teeth of *Propliopithecus*, and an extraordinarily large number of specimens of the monkeys *Parapithecus* and *Apidium*, including about two dozen isolated postcranial bones of these monkeys (Conroy, 1967).

Toward the end of the sixties, my field supervisor Mr. Grant E. Meyer found a nearly complete skull of *Aegyptopithecus*, reported in Simons (1967a). About the same time, an ulna of this ape was also found, but it was not described until eight years later by Fleagle et al. (1975). Thus, the "collecting" expeditions of the 1960s revealed little about the postcranial skeleton of these earliest apes, which, in remembrance of the "Dawn Horse" eohippus, Simons (1968) had termed the "Dawn Apes."

Between 1977 and 1983, I have directed a further seven field seasons in the Fayum badlands, each of two months duration. This second phase of our fieldwork there has produced a mounting series of isolated postcranial bones of *Aegyptopithecus*, as well as parts of several more skulls, including three well preserved faces, which were illustrated in connection with the "Ancestors" exhibit in Simons (1984) and are also shown here in Figure 1. The postcranial bones can be referred almost entirely to one species, *A. zeuxis*, because it is much the largest of the Fayum primates, and they are in its expected size range. These postcranials are also internally consistent with belonging to one large ape species. The distal humeral articular facets, for instance, fit exactly with those of the ulna. A few somewhat smaller but anatomically similar primate bones are thought to belong to *Propliopithecus*. Many more extensive remains of the postcranial bones of parapithecid monkeys have also come from the Fayum in the last seven years, but these are much smaller—only half the size of those of the Fayum apes.

All these many primate fossils from the Fayum, but especially those of the last four years, allow us to conclude several things about the origins and characteristics of the first hominoid primates. Propliopithecids are the oldest primates that show undoubted ties to the ancestral tree of man. First, it should be stressed that both *Aegyptopithecus zeuxis* and (a little less convincingly) the less-well-known species of *Propliopithecus* are anatomically, geographically, and temporally well situated to be in the direct ancestry of the apes and man. They could also be in or near the ancestry of monkeys, but the oldest known monkeys have not been convincingly demonstrated to resemble the Fayum apes. Hence, the advanced features of the Fayum apes resemble dryopithecine apes of the Miocene Epoch and do not resemble monkeys.

The question of the origin of the higher primates or Anthropoidea will eventually be resolved on the basis of Egyptian Oligocene fossils. Although dental resemblances between the Fayum parapithecids and the propliopithecids exist, they are not too close.

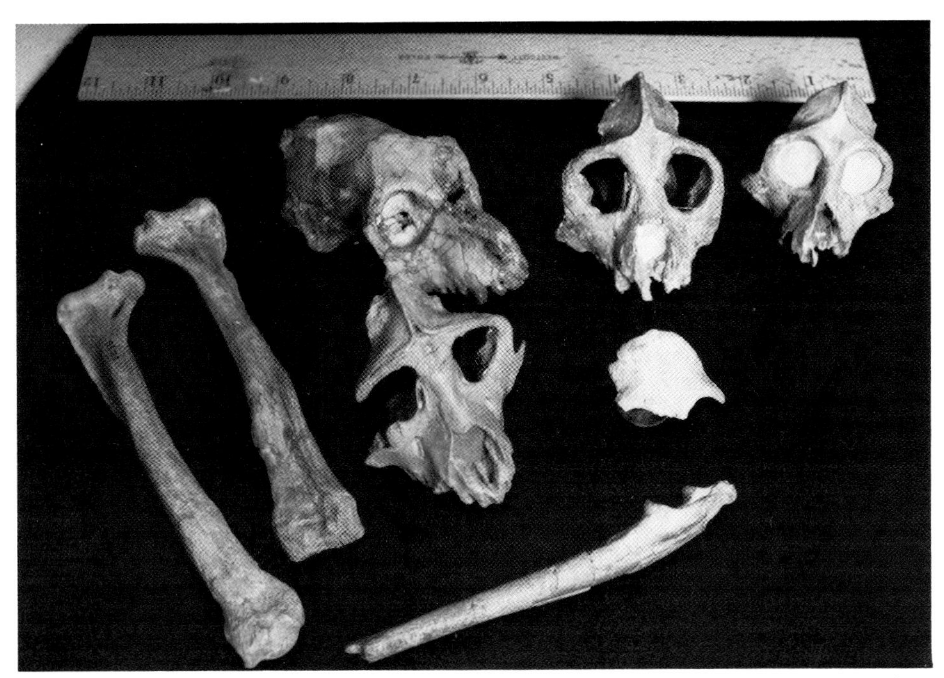

Fig. 1. Specimens of *Aegyptopithecus*. Left: complete humeri found in 1980 and 1981. Bottom: ulna found in 1967. Center: skull found by G.E. Meyer above face of 1981. Upper right: two faces found in 1982 above 1983 frontal. All from Quarry M, Fayum Oligocene.

The differences, such as retention of the P^2_2 in *Apidium* and *Parapithecus* together with their more quadrate molars, imply an ancestry separate for some considerable time. The long distal contact facet between tibia and fibula reported by Fleagle and Simons (1983) for *Apidium* is seen in several South-American monkeys and even in *Microcebus*. The propliopithecid tibia does not have this facet and resembles that of *Proconsul* (see Fleagle and Simons, 1982). *Oligopithecus savagei* shares the loss of P^2_2 with the propliopithecids and has a lower canine that is worn in a manner similar to some specimens of *Propliopithecus*. After surveying most early primates, P.M. Butler (personal communication) concluded that the unique type of *O. savagei* was not too different from these two genera and several species of propliopithecids and could perhaps be considered just another early hominoid. Even so, *Oligopithecus* might also represent a different group at least as distinct as the parapithecids. Finally, in 1983 a mandible closely resembling that of *Tarsius* was found in the Fayum. This diversity of primates of four different sorts, all variously resembling Haplorhini, strongly emphasizes the probability that Higher Primates arose in Africa. Although the sole mandible of *Oligopithecus savagei* has been said to suggest affinities with omomyids (Simons, 1962), the skull of *Aegyptopithecus* provides a much more extensive suite of anatomical characters, some of which demonstrate ties with Eocene groups. In overall aspect, it looks somewhat like the skull of a notharctid or adapid. Like members of these groups, the incisors and premolars of *Aegyptopithecus* are relatively generalized. The enlarged wing of the premaxilla, bordering the nasal aperture in *Aegyptopithecus*, resembles that of *Notharctus*. Such similarities led Gingerich (1973) to stress ties between the Adapidae, *sensu lato*, and the Anthropoidea. Nevertheless, a broad ascending wing of the premaxilla also exists in the non-adapid *Ourayia* as well as in other Eocene "tarsioids."

Although *Aegyptopithecus* and *Propliopithecus* have broadly similar premolars and molars, they are clearly different genera, *contra* Szalay and Delson (1979). As summarized in Kay et al. (1981), these two genera differ in cheek tooth crown anatomy at least as much as do *Pan* and *Gorilla*. *Aegyptopithecus* differs from *Propliopithecus* in having consistently larger canines in both sexes and an anteriorly much deeper mandible. In *Propliopithecus* M_{1-2} are the same size and P_4 relatively small in relation to $M\frac{-}{1}$. In contrast, *Aegyptopithecus* shows $M\frac{2}{2}$ distinctly larger than $M\frac{1}{1}$, with M_3 usually much larger still. This progressive molar size increase posteriorly, as well as the development of pronounced "beaded" lingual upper molar cingula, is a shared-derived resemblance to *Proconsul* not seen in *Propliopithecus*. On the whole then, *Aegyptopithecus* is more like members of the *Proconsul* group than is its contemporary. Apart from a series of details of upper and lower molar wrinkling, the beaded cingulum, and the Y-5 *Dryopithecus*-pattern, other shared-derived characters linking *Aegyptopithecus* with *Proconsul* include precise details of groove arrangment in the milk molars of both and a general overall (non-hylobatid-like) facial resemblance. Postcranially, the distal humerus, with its broadened trochlea, and the proximal ulnar articulation are quite reminiscent of those of *Proconsul*. Even more striking are the resemblances to corresponding bones of *Proconsul* seen in the calcaneum, astragalus, and hallucial phalanx of *Aegyptopithecus*. *Proconsul* itself, according to Walker and Pickford (1983), shows special resemblance to *Pan* in the hind limbs and feet.

By positing a whole range of assumed characteristics of the earliest catarrhines, Harrison (1982) recently attempted to disqualify *Aegyptopithecus* from the vicinity of the ancestry of the Anthropoidea. He argued that the rostrum or face was too long. Since modern cercopithecoids and hylobatids exhibit short snouts, their common ancestor must have been short snouted. This sort of reasoning held only when one face of *Aegyptopithecus* was known. The 3 new faces show that *A. zeuxis* sometimes had a shorter face than the single original find suggested. Szalay and Delson (1979) attempted to associate the propliopithecids with the seemingly aberrant European Miocene ape *Pliopithecus* as belonging to a group of apes unrelated to the ancestry of the higher Hominoidea. Such an association has little to recommend it. Besides the great geographic separation, the temporal separation between the two groups Propliopithecidae and Pliopithecidae is extreme, presumably in excess of 17 million years. Since the skull and face are known in both groups, it is easy to see that the skull of *Aegyptopithecus* bears no special facial or cranial resemblance to that of *Pliopithecus (Epipliopithecus) vindobonensis*. The latter, on the other hand, is very much more like modern *Hylobates*, with its flaring low-set zygomata, orbit-encircling flanges, and interorbital pit in the frontal just above and between the orbits, all shared-derived resemblances to *Hylobates*.

Finally, the question of whether or not *Aegyptopithecus* and *Propliopithecus* could be ancestral to monkeys, as well as apes, posed by Delson and Andrews (1975) must be addressed. In a series of papers, Fleagle and Simons (1978, 1982) and Fleagle (1980) have stressed the interpretation of the postcranial anatomy of *Aegyptopithecus* and *Propliopithecus* as having resembled that of a generalized arboreal quadruped not unlike *Alouatta*. The general assumption has been that these are grade, not clade, resemblances. Believing that South Atlantic rafting would have been impossible in the late Eocene, it seemed unlikely to me (Simons, 1976) that Egyptian Oligocene primates had anything directly to do with the ancestry of platyrrhines. It is difficult to determine whether shared-"primitive" features such as the distal articular facet between tibia and fibula of parapithecids and some platyrrhines are truly primitive, are shared-derived, or are independently acquired features. Better finds of archaic platyrrhines from the Oligocene of Argentina may resolve this problem. Fleagle and Bown (1983) report definite resemblances between molars from Gaiman, Argentina, and those of *Aegyptopithecus*, but these could all be shared-primitive features.

Although a strong consideration against linking Fayum primates with platyrrhines is geographic and not anatomical, the strongest disjunctions between Fayum apes and the cercopithecoid monkeys are anatomical. Teeth and limb-bones of the oldest African monkeys do not particularly resemble those of *Proconsul* or propliopithecids. Nevertheless, it has become fashionable to speculate that at least the Old World monkeys are derived from Fayum apes; e.g., Delson (1975) and Fleagle and Kay (1983).

The essence of this problem is briefly summarized by Simons (1985):

> Fleagle and Simons (1982b) conclude that in the skeleton Fayum apes much more closely resemble Miocene and living apes than cercopithecoid monkeys. My view has been that monkey and ape had already separated in Fayumian times (Simons, 1972). For this to be so, only a few skeletal features now seen in both cercopithecoids and hominoids, but not in the dawn apes, need have evolved in parallel since then, and these are all features which could easily have done so. Such easy-to-modify features include further lengthening of the ectotympanic of *Aegyptopithecus* (parapithecids could already have had a similar ectotympanic then), together with loss of entepicondylar foramen and facial foreshortening. Fleagle and Kay (1983) present the case that those dental and skeletal features which Oligocene apes share with Miocene-Recent apes may all be primitive features and therefore could have typified early ancestral cercopithecoids as well. Solution of the problem will not come until scientists have better skeletal material of undoubted early Old World monkeys.

LITERATURE CITED

Conroy, GC (1967) Primate postcranial remains from the Oligocene of Egypt. Contrib. Primatol. *8:*1–134.

Delson, E (1975) Toward the origin of the Old World monkeys. In: Evolution des Vertébrés—Problèmes Actuels de Paléontologie. Coll. Int. Cent. Nat. Rech. Sci., Paris *218:*839–850.

Delson, E and Andrews, P (1975) Evolution and interrelationships of the catarrhine primates. In WP Luckett and FS Szalay (eds): Phylogeny of the Primates. New York: Plenum, pp. 405–446.

Fleagle, JG (1980) Locomotor behavior of the earliest anthropoids: A review of the current evidence. Z. Morphol. Anthropol. *71:*149–156.

Fleagle, JG, and Bown, TM (1983) New primate fossils from late Oligocene (Colhuehaupian) localities of Chubut Province, Argentina. Folia Primatol. *41:*240–266.

Fleagle, JG, and Simons, EL (1978) Humeral morphology of the earliest apes. Nature *273:*705–707.

Fleagle, JG, and Simons, EL (1982) Skeletal remains of *Propliopithecus chirobates* from the Egyptian Oligocene. Folia Primatol. *39:*161–177.

Fleagle, JG, and Simons, EL (1983) The tibio-fibular articulation in *Apidium phiomense*, an Oligocene anthropoid. Nature *301:*238–239.

Fleagle, JG, Simons, EL, and Conroy, GC (1975) Ape limb-bone from the Oligocene of Egypt. Science *189:*135–137.

Gingerich, PD (1973) Anatomy of the temporal bone in the Oligocene anthropoid *Apidium* and the origin of the Anthropoidea. Folia Primatol. *19:*329–337.

Gingerich, PD (1978) The Stuttgart collection of Oligocene primates from the Fayum Province of Egypt. Paläontol. Z. *52:*82–92.

Gregory, WK (1922) The Origin and Evolution of the Human Dentition. Baltimore: Williams and Wilkins.

Harrison, T (1982) Small bodied apes from the Miocene of East Africa. (Unpublished doctoral dissertation, University College, London).

Kay, RF, Fleagle, JG, and Simons, EL (1981) A revision of the Oligocene Apes of the Fayum Province, Egypt. Am. J. Phys. Anthropol. *55:*293–322.

Le Gros Clark, WE (1959) The Antecedents of Man. Edinburgh: University Press.

Pilbeam, DR (1967) Man's earliest ancestors. Sci. Journ. (London) Feb. 1967:47–53.

Piveteau, J (1957) Traité de Paléontologie. Vol. 7. Paris: Masson.

Schlosser, M (1911) Beiträge zur Kenntnis der Oligozänen Landsäugetiere aus dem Fayum, Agypten. Beitr. Palaeontol. Geol. Öst.-Ung. *6:*1–227.

Simons, EL (1962) Two new primate species from the African Oligocene. Postilla *64:*1–12.

Simons, EL (1965) New fossil apes from Egypt and the initial differentiation of the Hominoidea. Nature *205:*135–139.

Simons, EL (1967a) The earliest apes. Sci. Am. *217:*28–35.

Simons, EL (1967b) Review of the phyletic interrelationships of Oligocene and Miocene Old World Anthropoidea. Problèmes Actuel de Paléontologie. Coll. Int. Cent. Nat. Rech. Sci., Paris *163:*597–602.

Simons, EL (1968) Hunting the "Dawn Apes" of Africa. Discovery (Peabody Museum, Yale University) *4:*19–32.

Simons, EL (1972) Primate Evolution. New York: Macmillan.

Simons, EL (1976) Primate radiations and the origin of hominoids. In RB Masterson (ed): Evolution of the Nervous System and Behavior. Washington, D.C.: V.H. Winston and Sons, Inc., pp. 383–391.

Simons, EL (1984) Dawn ape of the Fayum. Nat. Hist. *93(5):*18–20.

Simons, EL (1985) African Oligocene Primates: A Review. Ann. Geol. Surv. Egypt. (in press).

Szalay, FS, and Delson, E (1979) Evolutionary History of the Primates. New York: Academic Press.

Walker, AC, and Pickford, M (1983) New postcranial fossils of *Proconsul africanus* and *Proconsul nyanzae.* In RL Ciochon and RS Corruccini (eds) New Interpretations of Ape and Human Ancestry. New York: Plenum Press, pp. 325–352.

Ancestors: The Hard Evidence, pages 42–50

The Early and Middle Miocene Land Connection of the Afro-Arabian Plate and Asia: A Major Event for Hominoid Dispersal?

Herbert Thomas
Institut de Paléontologie, 75005 Paris, and Musée de l'Homme (L.A. 49) 75116 Paris, France

ABSTRACT New and current data on mammalian exchanges between Eurasia and Africa during the Early and Middle Miocene are used here to date the land connection of the Afro-Arabian plate and Asia. These data are consistent with the occurence of two main dispersal events and interchange phases, for which the names NDP 1 and NDP 2 (Neogene Dispersal Phases) are proposed. The earlier NDP 1 occurs during mammal units MN 3–4 (18 ± 1 m.y.) with NDP 2 approximately 15 ± 1 m.y. Finally, these data are used to test the different timing hypotheses of the Tethyan closure as deduced from marine organisms. It appears that terrestrial faunal data corroborate Rögl and Steininger's hypothesis that the Langhian marine peak reestablished the Mediterranean Indo-Pacific connection at 16.8 m.y. followed by a short land connection at 15–14 m.y. *Sivapithecus* participated in NDP 2, while *Pliopithecus* appeared slightly earlier; it is possible that a pliopithecid entered Asia even before NDP 1, as did anthracotheres.

INTRODUCTION

The disconnection of the proto-Mediterranean Sea and the Paratethys from the Indian Ocean during Miocene time is of great paleobiogeographical importance to understand the distribution and dispersal of the hominoid primates. If plate interaction and eustatic movements played a role in the branching of large hominoids into an "Asian" lineage leading to *Pongo*, and "African" lineages leading to *Pan*, *Gorilla*, and Hominidae, these geological events (in addition to climatic and ecological changes) would have played a major role for other mammal groups too.

Increasing knowledge of the phylogenetic relationships of many mammalian groups may now support the timing of the collision of the Afro-Arabian plate with the Asian mainland, in documenting an age for dispersal tracks. In some cases, however, it may not be possible to choose between two competing biogeographical processes: vicariance versus dispersal. Although it is not my subject here, I do not undervalue the role of climatic and ecological changes emphasized by many workers in the distribution of Miocene mammals. My aim will be to present a tentative hypothesis, using current and new data on mammalian exchanges between Eurasia and Africa. Finally, these data are used to test hypotheses on timing of Tethyan closure as deduced from marine organisms.

The existence of continental faunal exchanges between Eurasia and Africa through the eastern track during Early Miocene time has long been known. At the beginning of this century, Haug (1908–1911, p. 1733) assumed that the separation between the Indo-Pacific and the Mediterranean area was completed from the Pontian onwards. Meanwhile, Douvillé (1911) and Martin (1914) were still thinking that the Indo-Pacific had been detached from the Tethys since the Late Eocene. It was Umbgrove (1929) who finally demonstrated that the separation took place only during the Neogene. Continental faunal exchanges between Eurasia and Africa during the Early Miocene are nowadays admitted by most paleontologists (Van Couvering and Van Couvering, 1976). Some of them think that these exchanges may have taken place during the

Middle Burdigalian, or during the Langhian and possibly after the Langhian transgression. These land connections could be explained by Tethyan subduction under the Eurasian bloc, thus uniting the Afro-Arabian plate with the Asian mainland. According to Buchbinder and Gvirtzman (1976) and Bizon and Müller (1977), the elevation and emergence of the Afro-Arabian plate probably resulted in a separation of the Mediterranean from the Indo-Pacific at about 14 m.y. This separation would therefore have happened during the Serravallian, at the boundary of zones NN6 and NN7, and within zone N12, However, the absolute age of the tectonic movements that caused this event is thought to correspond more or less to the first appearance of *Orbulina*. This "*Orbulina* datum" had been dated at 14 m.y., but Gvirtzman and Buchbinder (1977) later finally accepted an age of 16 m.y. On the other hand, following investigations on nannoplankton and foraminifera, Rögl et al. (1978) estimated that a seaway splitting from the Mesopotamian trough reaching across northwestern Syria and the Taurid area into the Levantine Sea (eastern Mediterranean) existed from the Late Oligocene until the Middle Miocene (25 to 14 m.y.).

More recently, Rögl and Steininger (1983), modifying their previous point of view, concluded that there had been a temporary closure of the eastern Mediterranean at about 19 m.y., which provoked Eurasian/African mammal exchanges (ca. 19 m.y. and later on around 17.5–16.8 m.y.). According to them, the Langhian marine peak reestablished the Mediterranean/Indo-Pacific connection at 15–14.5 m.y., with the final closure of the Indo-Pacific seaway estimated at 12 m.y. At the other extreme, Drooger (1979), drawing evidence from two important larger foramifera groups, regarded the marine disconnection as having occured as early as the end of the Chattian! However, Adams et al. (1983), in an excellent review article using the geographical distribution of several age-diagnostic larger foraminifera, demonstrated that the Mediterranean and the Indian Ocean started to separate only by the mid-Burdigalian. According to Adams et al. (1983) (and Adams in litt.), no evidence seems to exist for a connection between the Mediterranean and Indian Ocean during the Langhian. This series of data leads us to question the classical paleontological conception of terrestrial communication between the Arabian peninsula and Eurasia during the Early Miocene and early Middle Miocene.

THE IMPLICATIONS OF RECENT DISCOVERIES IN SAUDI ARABIA

Our own discoveries and those of British paleontologists in Arabia are not without implications for these different paleogeographical hypotheses. In the eastern coastal zone of Saudi Arabia, at least two fossiliferous levels located in the Dam and Hofuf Formations are indeed particularly interesting in this respect, for they have yielded abundant and relatively varied faunas (Andrews et al., 1978; Hamilton et al., 1978; Thomas et al, 1978; Thomas et al., 1982).

Based on stratigraphical and paleontological results, it appears that these two levels should not be far apart chronologically. At the most, one to three million years lie between them. The most noticeable fact, in my opinion, is the difference in faunal composition between these two levels. In the most recent fossil locality (Al Jadidah, Hofuf Formation), one observes an important faunal renewal, quite comparable to the one that has been noticed in East Africa between the supposed unit fauna named "Rusingan"—minus the Maboko fossil locality—and the Fort Ternan one (Van Couvering and Van Couvering 1976; Pickford 1983).

The simultaneity of the faunal renewal both in the Arabian peninsula and in East Africa is certainly not fortuitous; besides, a local evolution could not account for this phenomenon. It is reasonable to think that the faunal renewal is due to an important immigration wave coming from Asia or Eurasia shortly before Fort Ternan or Al Jadidah times, and posteriorly to As Sarrar (Dam Formation); this immigration wave can therefore be dated 15 ± 1 m.y. according to the radiometric date (14 m.y.) of Fort Ternan.

Thus, the dispersal towards Eurasia of the *Ramapithecus–Sivapithecus* group can most probably be attributed to that exchange phase. On the other hand, the earlier dispersal of another group of Primates, the Pliopithecidae, remains uncertain (Bernor, 1983).

THE EARLIEST OCCURRENCES OF THE PLIOPITHECIDS AND THE *RAMAPITHECUS/SIVAPITHECUS* GROUP IN EURASIA

The *Pliopithecus* Group

The earliest European representatives appeared in zone MN 5 (Ginsburg and Mein, 1980) in the marine "Faluns de la Touraine" on the one hand, and in Neudorf-Spalte (central Paratethys) on the other. Based on the occurrence of *Pliopithecus* at Neudorf-Spalte, Rögl and Steininger (1983) consider that their penetration in Eurasia, which they date from the Karpatian (17.5–16.8 m.y.)—that is to say, before the Langhian transgression—precedes that of the *Ramapithecus–Sivapithecus* group. The Neudorf-Spalte fossil locality, though initially situated in zone MN 6, has been re-evaluated to be more ancient and positioned slightly above the Pontlevoy-Thenay localities of zone MN 5 (Mein 1979; Ginsburg and Mein, 1980). Furthermore, Ginsburg and Mein (1980) do not exclude an even earlier eastern immigration of *Pliopithecus*—at the same time as the first proboscideans—although they present no evidence to substantiate such an assumption.

The great antiquity of that immigration however could be indicated by the existence in Asia of ?Pliopithecidae *incertae sedis* (Szalay and Delson, 1979, p. 460) from the Taben Buluk badlands (Gansu Province, China; Bohlin, 1946). These very fragmentary remains (a molar fragment and a symphysis), named "Kansupithecus," were found in two different localities: Yindirte and a site near Hsi Shui, respectively. Although the Taben Buluk fauna had been initially considered as Late Oligocene in age, Conroy and Bown (1974) reassessed the supposed Oligocene "Kansupithecus" and, from the faunal association, regarded them to be "Early Middle Miocene, or very close to this age." However Russell and Zhai (in press), comparing the Taben Buluk assemblage (Yindirte) with the Early Miocene fauna from the Xieja Formation, Xiniang Basin (Qinghai Province), consider the Yindirte fauna (divided by Bohlin into three levels) as Late Oligocene in age and the Hsi Shui/Tieh Chiang Ku faunas as Early Miocene. It must be noted that only the small molar fragment (TB 557), which cannot be definitely identified as primate, was collected in the northern bed of Yindirte, the oldest level of the Taben Buluk area. Be that as it may, an early migration of the pliopithecids into Asia cannot be excluded.

The *Ramapithecus/Sivapithecus* group and *Dryopithecus*

The earliest occurrence of these hominoids in the Indian subcontinent is set at about 14.5 m.y. ago, or even slightly before (Raza et al., 1983). There precisely, we observe perennial closed environments, at least in the Lower Siwaliks (Thomas, in press). On the Anatolian platform, where this group appeared slightly earlier (14–15 m.y.) (Andrews and Tobien, 1977), more open woodland conditions are observed (Andrews, 1983). As discussed by Szalay and Delson (1979) and, in more detail, by Bernor (1983), this group entered the Paratethys region of Europe by 13–14 m.y., while *Dryopithecus* may have utilized the sub-Alpine arch to reach Spain and southern France at about the same time or slightly later (MN 7 or 8).

THE EARLY MIDDLE MIOCENE DESICCATION PHASE

Concurrently, a conspicuous climatic deterioration (drought) is suggested by the faunas during the beginning of the Middle Miocene in the Arabian peninsula, Western Asia, and low latitude Africa. Moreover, hydrodynamic data document that climatic evolution, from 15–16 m.y. on: during Langhian times (zones N8 to N10 of Blow) evaporitic sedimentation would have begun in the Red Sea, the Suez Gulf, and in many basins on both sides of the Zagros Range, and particularly in the most important of them, the Mesopotamian Basin (Lower Fars) (Heybroek, 1965; Kerdany, 1968; Stocklin, 1968; Gvirtzman and Buchbinder, 1977).

Evaporitic sedimentation demands a certain "climatic aridity," at least intermittently, without, however, requiring desert conditions; a drier climate simultaneously provokes an increase of evaporitic conditions and an inversion of the hydrological balance of the basins. In this respect, Rouchy (1982, pp. 217, 238) pointed out that it would not be necessary to suppose conditions implying extreme aridity, as the characteristics of present-day Mediterranean climate would be enough to develop even the saliferous phase of evaporitic deposition, should geographical isolation be sufficient. Thus, it seems clear that dispersal at about 15 m.y. was possible in the eastern Mediterranean (see Fig. 1).
15 m.y. was possible in the eastern Mediterranean (see Fig. 1).

Note, however, that the relative aridity of the basins on both sides of the Zagros, in the Red Sea and in the Suez Gulf, though leading one to imagine the existence of large discontinuities in the woodland tree canopy, obviously did not prevent the passage of several groups of primates. In this respect, it is interesting to note that according to Andrews (1983), *Sivapithecus* species were terrestrial at least in part.

THE EARLY MIOCENE MAMMALIAN DISPERSAL PHASE BETWEEN EURASIA AND THE AFRO-ARABIAN PLATE

If, on the one hand, that 15 m.y. exchange phase seems to be reliably established, many authors, on the other hand, admit the exis-

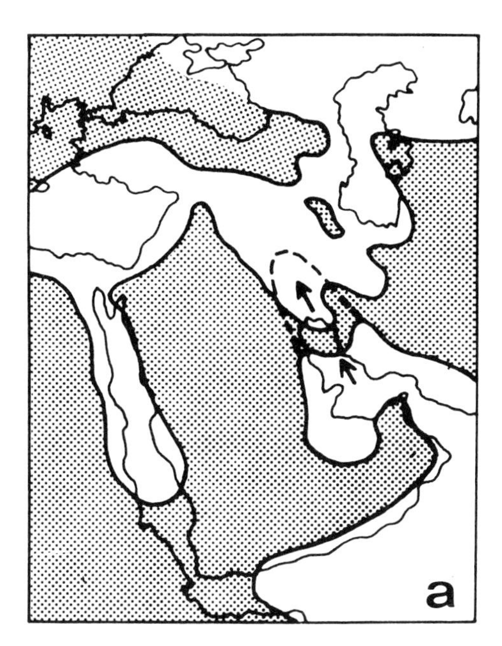

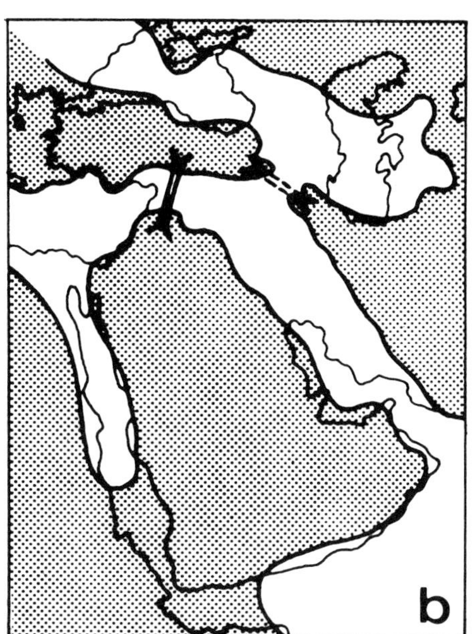

Fig. 1. Paleogeographic reconstruction during Langhian times (16.8–15.8 m.y.). a) After Adams et al. (1983): During the Burdigalian, an arm of the Indian Ocean extended into northern Iraq (dashed line). b) After Rögl and Steininger (1983): Shortly after the Langhian, during the Middle Serravalian (15.0–14.5 m.y.), the possibility of land bridges is indicated by inverted brackets.

tence of another exchange phase earlier in the Miocene. The proboscideans would represent the most prominent group involved in this early phase, consequently named the "Proboscidean Datum Event" (Madden and Van Couvering, 1976), owing to the many taxa of that group that would have migrated from Africa into Eurasia during the Burdigalian. Following the first detailed analysis of faunal movements and chronology by Van Couvering (1972), Berggren and Van Couvering (1974) dated that event at 17.5 m.y. (MN 3b).

This event is still difficult to date precisely, because the scanty information coming from only one paleontological source, does not allow us to avoid circular reasoning. Whereas the use of biostratigraphical subdivisions defined from mammalian units is satisfactory with regard to the Mediterranean Neogene (Mein's zones, 1975 and 1979), the Early Miocene faunal complexes of the Indian subcontinent (Bugti Beds, Murree Formation, Gaj/Manchar Formation, Kamlial Formation) and of North Africa (Moghara Formation, Djebel Zelten) cannot be easily situated in Mein's biozonation. For the time being, we have to cope with obvious correlation difficulties.

In spite of those uncertainties, it appears that several taxa of the Saudi Arabian As Sarrar fauna (Dam Formation), while presenting undoubted African affinities (Thomas et al., 1982), nevertheless provide evidence of a terrestrial communication between the Arabo-African platform and Asia before the deposition of the continental Dam Formation, the age of which is estimated at 16 ± 1 m.y. These forms include gomphotheriids and perhaps amebelodontines, the small felid *Pseudaelurus*, one amphicyonid, one giant suid of the *Bunolistriodon* type, and other mainly small and medium-sized vertebrates (lagomorphs, rodents, snakes, and turtles).

Other taxa—which have not been recovered from the Dam Formation—could be added to that list, which is not exhaustive and still subject to modification (see Table 1). In any case, the first vertebrates that emigrated from Africa involved two major groups: proboscideans and anthracotheres (as well as perhaps pliopithecids).

The Proboscidean Datum Event

As far as the proboscideans are concerned, Bernor (1983) indicated that Ginsburg's (1974) report of the possible presence of a proboscidean tooth from Condom (France) in zone MN 3a of Mein (1979) might imply an earlier migration date. In fact, that tooth was discovered while laying the foundations of the small Trianon near Condom (Gers), and it would come, according to the investigation carried out by F. Crouzel (Ginsburg, 1974), from a higher level than Estrepouy. So far no proboscideans have been recorded in Estrepouy.

A certain confusion remains regarding zones MN 3 and MN 4, depending on whether authors refer to Mein (1975) or to Mein (1979). Indeed, during the Athens Congress, Mein (1979) proposed to redefine zone MN 3 by distinguishing zone MN 3a (formerly, zone MN 3) and zone MN 3b, equivalent to the former zone MN 4a, each of them being characterized by an arrival of migrants (see Fig. 2). According to Mein (1979), the new zone MN 3 had the advantage of keeping the association *Brachyodus* + *Anchitherium*. Whichever position may be retained, no proboscidean has been found in Europe in levels anterior to MN 4 (Mein, 1975; i.e. MN 3b, Mein, 1979), and, in particular, none has been recovered from the Early Orleanian fossil localities of the Loire Basin (Chitenay, Chilleurs), of the Garonne Basin (Estrepouy), from Lisboa I, or from Wintershof-West. In Europe, the first proboscideans appeared at the same time as the first *Eotragus*, in Artenay, situated at the base of zone MN 4 (Mein, 1975) or in zone MN 3b (Mein, 1979). Besides, it is interesting to notice that in Northern Hungary, at Ipolytarnoc, footprints of Proboscideans have been recorded in the "Lower Rhyolite Tuff" (Rögl and Steininger 1983). This tuff, overlying terrestrial sediments with nannoplankton of zone NN 3, has been dated at 19. 6 ± 1.4 m.y. (Hámor and Ravasz-Baranyai, 1979).

In Asia, the situation regarding the oldest occurrence of proboscideans is more problematical. In the Eastern Paratethys, Gabunia (1979) indicates *Gomphotherium* aff. *cooperi* from Agenian levels (zones MN 1 and 2) of Nakhichevan, and in the Central Paratethys, *Zygolophodon* aff. *gromovae* is known in the Early Orleanian of the Tuchorice complex, placed in MN 3b by Mein (1979).

A much older locality is sometimes mentioned to have yielded proboscideans. In the Late Oligocene *Indricotherium* (= *Baluchitherium*) beds of St. Jacques (San-tao-Ho), which extend along the east bank of the Huanghe (Inner Mongolia), Teilhard de Chardin (1926) had indeed mentioned ?*Ser-*

TABLE 1. Mammalian taxa involved in Neogene Dispersal Phases 1 and 2

<table>
<tr><th></th><th>Dispersal towards Eurasia</th><th>Dispersal towards the Afro-Arabian plate</th></tr>
<tr><td>NDP 2</td><td>Catarrhini (Ramapithecus/
Sivapithecus)
Creodonta (Dissopsalis)
Hyracoidea
Tubulidentata
?Choerolophodontinae
Listriodontinae
Palaeotraginae
Bovidae (Caprotragoides)
Phiomyidae
Pedetidae
Thryonomyidae</td><td>Hyaenidae (Percrocuta)
Ischyrictini
Bovidae (Pachytragus/Protoryx)
Helicoportacina
Gliridae</td></tr>
<tr><td>NDP 1</td><td>Deinotheriidae
Gomphotheriinae
?Amebelodontinae
Mammutidae
Creodonta (Hyainailouros,
Megistotherium)</td><td>Felidae (Pseudaelurus)
Amphicyonidae
Herpestinae
Chalicotheriidae
Bovidae (?Eotragus, ?"Oioceros")
Tragulidae
Suidae (Bunolistriodon), Hyotheriinae,
?Sanitheriinae
Ochotonidae
Macroscelididae
Erinaceidae
Zapodidae
Sciuridae
Dipodidae
Cricetodontinae
Ctenodactylidae (Sayimys)</td></tr>
<tr><td></td><td colspan="2">Dicerorhininae
Aceratheriinae</td></tr>
<tr><td>Earlier?</td><td>Anthracotheriidae
?Pliopithecidae</td><td></td></tr>
</table>

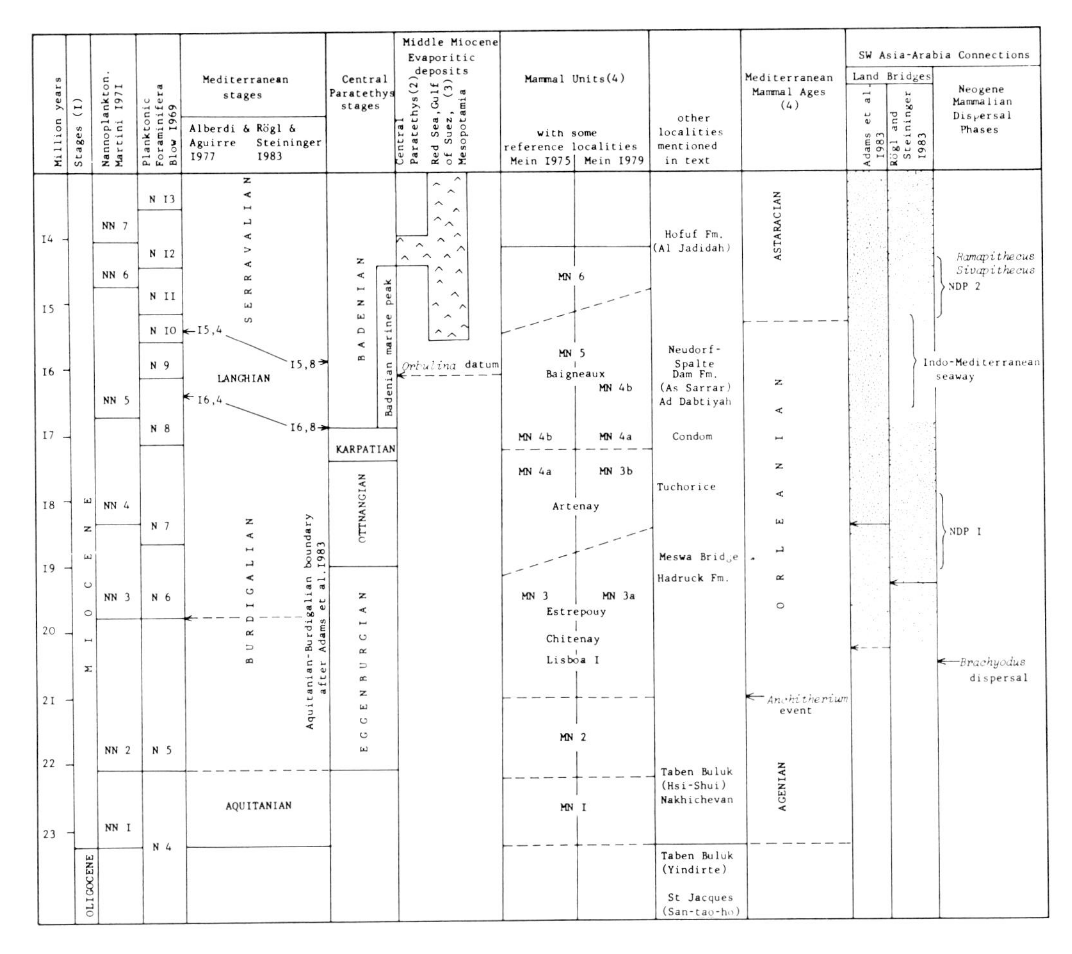

Fig. 2. Early and Middle Miocene biostratigraphy and tentative correlations of the mammalian dispersal phases between the Afro-Arabian plate and Asian mainland. (1) The Paleogene/Neogene boundary in Europe and the Indian Ocean is taken from Bizon and Müller (1979). (2) After Rögl et al. (1978). (3) After Gvirtzman and Buchbinder (1977). (4) From Alberdi and Aguirre (1977).

ridentinus from a fragment of ivory lamina and a fragment of molar. In fact, the tooth fragment, due to its enamel structures, cannot belong to a proboscidean but may be best referred to a hyracodontid. However the ivory fragment, with its structure "*en chevrons*," belongs definitely to a proboscidean, but its light fossilization suggests that it is probably intrusive. In fact, the very large diameter of the tusk, which can be inferred from the small piece, leaves little doubt that it belongs to a modern elephant.

Finally, it must be emphasized that the most primitive Neogene elephantoid, *Hemimastodon crepusculi*, known from two M^3, does not come from Africa, but from the Pakistani Bugti Beds (Tassy, 1982). We have then to face the following alternative: either the amebelodontines, the mammutids, the gomphotheres, and perhaps the choerolophodontines migrated from Africa into Eurasia (Asia at least) during zone MN 4 (Mein, 1975) or MN 3b (Mein, 1979) or even earlier, a position that Tassy (in press) is now somewhat reluctant to support; or the differentiation of the mammutids and gomphotheres did not take place in Africa, thus supporting a vicariant pattern. Contrary to this latter view, one has to mention the presence of the most primitive mammutid (*Zygolophodon morotoensis*) in the oldest fossiliferous levels of the African Neogene, at Meswa Bridge (20–23 m.y.?, Pickford and Tassy, 1980).

The *Brachyodus* Event

According to Dineur (1981), the first European representatives of *Brachyodus onoideus*, which belong to a relatively primitive form, appeared at Lisboa I, in the Early Orleanian of the Loire Basin (Chitenay, Chilleurs-aux-Bois, Neuville-aux-Bois, Les Beilleaux) and the Garonne Basin (Estrepouy) and also in Eggenburg (Austria). All these fossil localities are situated in zone MN 3a (Mein, 1979). Dineur (1981), though not categorically excluding an Asian origin of *Brachyodus*, rather favors an African origin, because of the very strong similarities between the postcranial skeletons of *Bothriogenys* from the Fayum and *B. onoideus* and because of the absence of *Brachyodus* in the Early Miocene of Asia (Bugti Beds).

Thus, it appears that the penetration into Europe of the African anthracotheres preceded that of the proboscideans. It is worth noticing that the most recent anthracotheres (e.g., *Merycopotamus*) were comparable to the modern hippopotamus in their general structure and probably also in their mode of life. It is therefore conceivable that *Brachyodus* could already have been a good swimmer, and consequently that it had not necessarily been compelled to use the "sub-Alpine arch migration route" (Antunes, 1979). As for the proboscideans, we must insist on the fact that the present-day proboscideans are excellent swimmers, since they can easily swim across large rivers (Zambezi, Nile) and lakes, using their trunks as snorkels. There is the famous case (Williams, 1951) of an elephant introduced into the Andaman Islands and found twelve years later on the largest of the northern islands, which implied three crossings of open sea at least a mile wide. Moreover, Sondaar (1977) noticed that elephants are often present in island faunas alongside deer and hippopotami, all of them extraordinary swimmers. However, Azzaroli (1982), on biological grounds, regarded the possibility of migrations by swimming as an unsatisfactory explanation.

Whatever one may think of those hypotheses, the proboscideans are in any case not the only mammals involved in this early faunal exchange (see Table 1). Coryndon and Savage (1973, p. 127) mentioned the appearance of 23 African immigrant families during the Early Miocene (without specifying them).

CONCLUSIONS

In conclusion, caution is required, due to the lack of a fossil record in Africa between ca. 32 m.y. (the base of the capping basalt of the Fayum has recently been dated to 31 m.y. by Obradovitch-Simons (Personal communication, 1984) and 22–23 m.y. The record in Asia is also poor (except for the Bugti Beds). It now appears that there is no real way to choose, at least for some of the mammals mentioned above, between the two biogeographic alternatives: a dispersal pattern or a vicariant pattern. The proboscideans, for example, were always considered as among the best evidence of a dispersal pattern, having arisen and differentiated in Africa. It appears nevertheless that the alternative hypothesis cannot be excluded for the differentiation of the Elephantoidea. It is clear also that the direction of the dispersal, inferred only from the earliest known record, has but little reliability. On the other hand, McKenna (1973) had stressed the fact that mammal dispersal via plate interaction "does not necessarily coincide with the time of plate collision but may occur significantly before or after such an event." Thus the collision of the Afro-Arabian plate with the Asian mainland may certainly have been preceded by epicontinental-sea conditions. Therefore, a sequence of sweepstake dispersal, natural raft, filter, and corridor conditions may have occurred before the oceanic strait became a collisional suture. Finally, I would emphasize the fact that the time necessary to connect the two land masses is within the time span of any detectable faunal differentiation (ca. 2 m.y.), although the evolution of many Early Miocene vertebrates can be traced down to a low taxonomic level.

Despite some vicariist reservation, the analyses of our paleontological data seem to be consistent with the occurrence of two main dispersal events and interchange phases (Fig. 2). I propose to formally name these two phases NDP 1 and 2 (Neogene Dispersal Phases). The earliest (NDP 1), occurring during mammal units MN 3–4 (18 ± 1 m.y.), is obviously related to the eastern occlusion of the Tethys. The latter event appears now to have been best dated by Adams et al. (1983), based largely on the distribution of age-diagnostic larger foraminifera but also on echinoids and molluscs. According to them, by mid-Burdigalian times at the latest, the eastern seaway was interrupted (Fig. 2).

As to the second dispersal wave, for which I propose the approximate age of 15 ± 1 m.y., Adams et al. (1983) assume that it was not directly relevant to any closure of the Tethys. Indeed, they found no evidence of the existence of any deep-water connection between the Mediterranean and the Indian Ocean that could have appeared then (during the Langhian). In fact, it must be noticed that the age proposed for the second wave is broadly consistent with the positioning of the land bridge between Southwest Asia and Arabia in Langhian times, as postulated by Adams et al. (1983). Regarding the number of the taxa involved in this second major wave, the question remains about the nature of the barrier that prevented their earlier dispersal. Although a marine barrier seems the most probable explanation, no definite statement can be made about it. However, the general scheme given by Rögl and Steininger (1983) is more consistent with the terrestrial faunal data. In fact, they assume that, contrary to Adams et al. (1983), the Langhian marine peak reestablishing the Mediterranean-Indo-Pacific connection at 16.8 m.y. was followed by a short land connection at 15–14 m.y. (Fig. 2).

In conclusion, apparently no catarrhine was part of the initial exchange phase NDP 1. However, the first ?pliopithecid may have entered Asia even earlier than NDP 1, if the age of the Hsi-Shui fauna in the Taben Buluk badlands has been correctly established and provided the specimen is really a primate. On the other hand, it was only during Langhian times (or shortly after the land connection was again restored—if we accept Rögl and Steininger's hypothesis) that the *Sivapithecus/Ramapithecus* group moved to Eurasia.

LITERATURE CITED

Adams, CG, Gentry, AW and Whybrow, PJ (1983) Dating the terminal Tethyan event. Utrecht Micropal. Bull. *30:*273–298.

Alberdi, MT and Aguirre, E (1977) Colloque sur stratigraphie mammalienne du Néogène méditerranéen. In Round-Table on Mastostratigraphy of the W. Mediterranean Neogene. Trabajos sobre Neogeno-Cuaternario, Secc. Paleont. Vert. y Humana, Madrid *7:*17–21.

Andrews, PJ (1983) The natural history of *Sivapithecus*. In RL Ciochon and RS Corruccini (eds): New Interpretations of Ape and Human Ancestry. New York: Plenum pp. 441–464.

Andrews, PJ and Tobien, H (1977) New Miocene locality in Turkey with evidence on the origin of *Ramapithecus* and *Sivapithecus*. Nature *268:*699–701.

Andrews, PJ, Hamilton, WR, and Whybrow, PJ (1978) Dryopithecines from the Miocene of Saudi Arabia. Nature *274:*249–250.

Antunes, MT (1979) "Hispanotherium fauna" in Iberian Middle Miocene, its importance and paleogeographical meaning. Ann. Geol. Pays Hellén. (*hors série*) *I:*19–26.

Azzaroli, A. (1982) Insularity and its effects on terrestrial Vertebrates: Evolutionary and biogeographic aspects. In E Montanaro-Gallitelli (ed.): Palaeontology, Essential of Historical Geology. Venice: Modena Press, pp. 193–213.

Berggren, WA and Van Couvering, JA (1974) The late Neogene:Biostratigraphy, geochronology and paleoclimatology of the last 15 million years in marine and continental sequences. Palaeogeogr., Palaeoclimat., Palaeoecol. *16:*1–216.

Bernor, RL (1983) Geochronology and zoogeographic relationships of Miocene Hominoidea. In RL Ciochon and RS Corruccini (eds.): New Interpretations of Ape and Human Ancestry. New York: Plenum, pp. 21–64.

Bizon, G, and Müller C (1977) Remarks on some biostratigraphic problems in the Mediterranean Neogene. In B Biju-Duval and L Montadert (eds.): International Symposium on the Structural History of the Mediterranean Basins, Split 1976. Paris: Technip, pp. 381–390.

Bizon, G, and Müller C (1979) Remarks on the Oligocene/Miocene boundary based on results obtained from the Pacific and the Indian Ocean. Ann. Géol. Pays Hellén. (*hors série*) *I:*101–111.

Blow, WH (1969) Late Middle Eocene to recent planktonic foraminiferal biostratigraphy. In P Bronnimann and HH Renz (eds.): First International Conference on Planktonic Microfossils, Geneva, 1967. Leiden: Brill, pp. 199–421.

Bohlin, B (1946) The fossil mammals from the Tertiary deposit of Taben-buluk, Western Kansu. Part II: Simplicidentata, Carnivora, Artiodactyla, Perissodactyla and Primates. Sino-Swedish Exp. VI: 4. Pal. Sin., n.s. *C(8b):*1–259.

Buchbinder, B, and Gvirtzman, G (1976) The breakup of the Tethys Ocean into the Mediterranean Sea, the Red Sea and the Mesopotamian basin during the Miocene: A sequence of fault movements and desiccation events. Abst. 1st Congr. on Pacific Neogene Stratigraphy pp. 32–35.

Conroy, GC, and Bown, TM (1974) Anthropoid origins and differentiation: The Asian question. Yrbk. Phys. Anthropol. *18:*1–6.

Coryndon, SC, and Savage, RJG (1973) The origin and affinities of African mammal faunas. In Organisms and Continents Through Time. Special papers in palaeontology (12) Syst. Assoc. Publ. *9:*121–135.

Dineur, H (1981) Le genre *Brachyodus*, Anthracotheriidae (Artiodactyla, Mammalia) du Miocène inférieur d'Europe et d'Afrique. Mém. Sci. Terre, Univ. Paris VI, Thèse 3ème cycle, pp. 1–180.

Douvillé, H (1911) Les Foraminifères dans le Tertiaire des Philippines. Philippine Sci., Manila *VI(2):*53–80.

Drooger, CW (1979) Marine connections of the Neogene mediterranean, deduced from the evolution and distribution of larger Foraminifera. Ann. Géol. Pays Hellén., (*hors série*) *I:*361–369.

Gabunia, LK (1979) Biostratigraphic correlations between the Neogene land mammal faunas of the East and Central Paratethys. Ann. Géol. Pays Hellén. (*hors série I:*413–423.

Ginsburg, L (1974) Les faunes de Mammifères burdigaliens et vindoboniens des bassins de la Loire et de la Garonne. Mém. Bur. Rech. Geol. Min. *78:*153–167.

Ginsburg, L, and Mein, P (1980) *Crouzelia rhodanica*, nouvelle espèce de Primate catarrhinien et essai sur la position systématique des Pliopithecidae. Bull. Mus. Natl. Hist. Nat., Paris, 4e sér., *2*, sect. C (2):57–85.

Gvirtzman, G and Buchbinder B (1977) The desiccation events in the eastern Mediterranean during Messinian times as compared with other Miocene desiccation events in basins around the Mediterranean. B Biju-Duval and L Montadert (eds): International Symposium on the Structural History of the Mediterranean Basins, Split 1976. Paris: Technip, pp. 411–420.

Hamilton, WR, Whybrow, PJ, McClure, HA (1978) Fauna of fossil mammals from the Miocene of Saudi Arabia. Nature *274:*248–249.

Hámor, G, and Ravasz-Baranyai, L (1979) K/Ar dating of Miocene pyroclastic rocks in Hungary. Ann. Géol. Pays Hellén. (*hors série*) *II:*491–500.

Haug, E (1908–1911) Traité de Géologie. II. Les Périodes Géologiques. Paris: Librairie Armand Colin, pp. 1397–2024.

Heybroek, F (1965) The Red Sea Miocene Evaporite basin. In: Salt Basins Around Africa, Meeting Inst. Petroleum Geol. Soc., London, pp. 17–40.

Kerdany, MT (1968) Note on the planktonic zonation of the Miocene in the Gulf of Suez region U.A.R. Giorn. di Geol. *35:*157–166.

Madden, CT, and Van Couvering, JA (1976) The Proboscidean datum event:Early Miocene migration from Africa. Géol. Soc. Am. Abstr. Programs, pp. 992–993.

Martin, K (1914) Wann löste sich das Gebiet des Indischen Archipels von der Tethys? Samml. Geol. Reichsmuseum Leiden *9:*337–355.

Martìni, E (1971) Standard Tertiary and Quaternary calcareous nannoplankton zonation. In A Fárinacci (ed): Proc. II Planktonic Conference, Rome, 1970. Rome: Tecnoscienza, pp. 739–785.

McKenna, MC (1973) Sweepstakes, filters, corridors, Noah's arks, and beached Viking funeral ships in paleogeography. In DH Tarling and SK Runcorn (eds): Implications of Continental Drift to the Earth Sciences, Vol. 1. New York: Academic Press, pp. 295–308.

Mein, P (1975) Résultats du groupe de travail des Vertébrés. Report on activity of the R.C.M.N.S. working groups (1971–1975), Bratislava. I.U.G.S. Commission on stratigraphy, subcommission on Neogene Stratigraphy, pp. 78–81.

Mein, P (1979) Rapport d'activité du groupe de travail des Vertébrés. Mise à jour de la biostratigraphie du Néogène basée sur les Mammifères. Ann. Géol. Pays Hellén. (*hors série*) *III:*1367–1372.

Pickford, M (1983) Sequence and environments of the Lower and Middle Miocene Hominoids of Western Kenya. In RL Ciochon and RS Corruccini (eds): New Interpretations of Ape and Human Ancestry. New York: Plenum, pp. 421–440.

Pickford, M, and Tassy, P (1980) A new species of *Zygolophodon* (Mammalia, Proboscidea) from the Miocene hominoid localities of Meswa bridge and Moroto (East Africa). N. Jb. Geol. Paläont. Mh., *4:*235–251.

Raza, SM, Barry, JC, Pilbeam, D, Rose, MD, Shah, SMI, and Ward, S (1983) New Hominoid primates from the middle Miocene Chinji Formation, Potwar Plateau, Pakistan. Nature *306:*52–54.

Rögl, F, Steininger, FF, and Müller, C (1978) Middle Miocene salinity crisis and paleogeography of the Paratethys (Middle and Eastern Europe). In K Hsü, L Montadert et al. (eds): Initial Reports of the Deep Sea Drilling Project. Washington, DC: US Govt Printg Ofc, *42*(1):985–990.

Rögl, F, and Steininger, FF (1983) Vom Zerfall der Tethys zu Mediterran and Paratethys. Die neogene Paläogeographie und Palinspastik des zirkum-mediterranen Raumes. Ann. Naturhist. Mus. Wien *85/A:*135–163.

Rouchy, J-M (1982) La genèse des évaporites messiniennes de Méditerranée. Mem. Mus. Natn. Hist. Nat. sér. C, Sci. Terre. *50:*1–267.

Russell, DE, and Zhai RJ (in press) Mammals and stratigraphy:The Paleogene of Asia. Mém. Extra. Palaeovertebrata, Montpellier.

Sondaar, PY (1977) Insularity and its effect on mammal evolution In MK Hecht, PC Goody, and BM Hecht (eds): Major Patterns in Vertebrate Evolution. New York: Plenum, pp. 671–707.

Stocklin, J (1968) Salt deposits of the Middle East. In RB Mattox et al. (eds): Saline Deposits. Geol. Soc. Amer., Spec. Papers *88:*157–181.

Szalay, FS, and Delson, E (1979) Evolutionary history of the Primates. New York: Academic Press, pp. 1–580.

Tassy, P (1982) Les principales dichotomies dans l'histoire des Proboscidea (Mammalia): Une approche phylogénétique. Géobios, mém. spéc. *6:*225–245.

Tassy, P (in press) Nouveaux Elephantoidea (Mammalia) dans le Miocène du Kenya. Essai de réévaluation systématique.

Teilhard de Chardin, P (1926) Description de Mammifères tertiaires de Chine et de Mongolie Ann. Pal. *15:*1–52.

Thomas, H, Taquet, P, Ligabue, G, and Del'Agnola, C (1978) Découverte d'un gisement de Vertébrés dans les dépôts continentaux du Miocène moyen du Hasa (Arabie Saoudite). C.R. somm. Soc. géol. Fr. *(2):*69–72.

Thomas, H, Sen, S, Khan, M, Battail, B, and Ligabue, G (1982) The Lower Miocene fauna of As Sarrar (Eastern province, Saudi Arabia). Atlal, Saudi Arabian Archaeol. *5. III:*109–136.

Thomas, H (1984) Les Bovidae anté-hipparions des Siwaliks inférieurs (Plateau du Potwar, Pakistan). Mém. Soc. géol. Fr. N.S., 1983 (*145*), pp. 1–68.

Umbgrove, JHF (1929) Tertiary sea-connections between Europe and the Indo-Pacific area. Fourth Pacific Sci. Congr., Batavia-Bandoeng, Java 1929, pp. 1–14.

Van Couvering, JA (1972) Radiometric calibration of the European Neogene. In WW Bishop and JA Miller (eds): Calibration of Hominoid Evolution. Edinburgh: Scot. Acad. Press, pp. 247–271.

Van Couvering, JAH, and Van Couvering, JA (1976) Early Miocene mammal fossils from East Africa:Aspects of geology, faunistics and paleoecology. In GL Isaac and ER McCown (eds): Human Origins, Louis Leakey and the East African Evidence. Menlo Park: WA Benjamin, pp. 155–207.

Williams, JH (1951) Bill l'éléphant. Paris: Hachette, pp. 1–256.

Ancestors: The Hard Evidence, pages 51–59

Patterns of Hominoid Evolution

David Pilbeam
Peabody Museum, Harvard University, Cambridge, Massachusetts 02138

ABSTRACT Although results based on molecular and morphological data are still not entirely congruent, it seems highly probable that the close molecular similarity of humans and African apes reflects genealogy. Orangutans probably shared an ancestor with humans and African apes that was about twice as old as the human–African ape common ancestor. Direct fossil evidence for the earliest hominids and their immediate ancestors is lacking. New fossil data from Asia indicate that orangutan ancestors may have diverged 12–15 million years (m.y.) ago, suggesting that hominids split from African apes 5–10 m.y. ago, probably in Africa. A judicious use of paleontological and neontological data sets limits on what hominid ancestors might have looked like.

INTRODUCTION

Where, when, and how did humans and their ancestors acquire their very large brains, peculiar movement patterns, odd teeth, and odder behaviors that set them apart from other apes? Answers to those questions exist in abundance and diversity, yet there is no general agreement. This essay addresses some of the issues and is concerned broadly with the evolution of the primate superfamily Hominoidea, more specifically with the origin of hominids.

The hominoids have traditionally been subdivided into three groups: humans and their ancestors and relatives—the Hominidae; the large or great ape species of *Pan* (chimpanzee), *Gorilla*, and *Pongo* (orangutan) grouped in Pongidae; and the small or lesser ape species of *Hylobates* (gibbons and siamangs), the Hylobatidae. Apes live today in the forests of Southeast Asia (*Pongo, Hylobates*) and the African forests (*Gorilla, Pan*) and woodlands (*Pan*).

I shall concentrate on the nature and time of origin of the very earliest pre-*Australopithecus* hominids ("proto-hominids") and on the last common ancestor ("LCA") of hominids and the most closely related ape. Direct fossil evidence for these stages is nonexistent or unrecognized, and they therefore have to be "reconstructed" entirely indirectly.

For a tentative hominoid evolutionary story we have a reasonable "end" and "beginning" but a poor "middle"; a pity, since that is where the action was. Apes have probably lived in tropical forests and woodlands of Africa and Asia since at least the early Pliocene (about 5 m.y. ago), although their fossil record is essentially nil. Small brained, bipedal hominids with small canine teeth lived in Africa at least 3.7 m.y. ago, probably as much as 5 m.y. ago, prior to the dispersal of large-brained *Homo*. This is the "end" to be explained. A reasonable "beginning" can be found in the excellent samples of Early Miocene (about 22–17 m.y.) primitive hominoids of East Africa (Andrews, 1978). Hominoids probably evolved in Afro-Arabia and were confined there until at least 17 m.y. ago. Following the initial closing then of the Tethys Sea which separated Africa from Eurasia, hominoids spread to Eurasia, although it is not clear exactly when after 17 m.y. the dispersal happened (Bernor, 1983; Thomas, 1985). Significant evolution occurred during the Middle Miocene (about 17–11 m.y.), Late Miocene (about 11–5 m.y.), and Pliocene (about 5–2 m.y.), although these periods are unfortunately rather poorly sampled (Pilbeam, 1983).

There are four major questions to be answered: What is the branching sequence for living hominoids? What are branching times? How might ancestors be recognized, or what did they look like? Finally, what explains the evolutionary changes? Central both to reconstructing hominoid branching sequences and to recognizing hominid ancestors is our ability to interpret different kinds of similarities between species. The terminology and many of the concepts of phylogenetic systematics or cladism are useful in thinking about evolutionary problems whether or not one is a cladist (Wiley, 1981). Thus, the ability to differentiate synapomorphies (shared-derived similarities) from symplesiomorphies (shared primitive ones) is critical to elucidating branching sequences (unless evolutionary rates do not fluctuate), as well as to recognizing hypothetical ancestors (morphotypes). Clearly homoplasies (convergences, parallelisms) provide no useful information, nor do derived features unique to single lineages (autapomorphies).

For example, the early hominid *Australopithecus africanus* is linked to *Homo sapiens* in Hominidae because they share, among other derived features, adaptations to bipedalism (Howell, 1978). The small brain of *A. africanus*, although similar to those of apes, is a retained primitive feature, and we do not use the similarity to link the two groups together. For more than a decade after discovery, *A. africanus* was not widely accepted as a hominid because, in the terminology of cladism, scholars disagreed about early hominid morphotypes (expected LCAs) and also about which character states were primitive and which were derived (their "polarities"). As the fossil record expanded, different ideas about morphotypes and polarities gradually converged, and *A. africanus* came to be linked by almost everyone with *Homo* (Reader, 1981).

Morphotypes, synapomorphies, and the rest cannot be discovered "in reality" (Cartmill, 1981). Statements about them are hypotheses that may, given enough information, come to be treated as highly plausible. Alternative hypotheses are always possible, both about the nature of similarities and about the nature of phylogenies, and it is important that the alternatives and the reasons for them are clearly spelled out.

BRIEF HISTORY

As a way of putting this account into the context of the many diverse scenarios of hominoid evolution, I will briefly summarize two quite different versions.

Until the mid-1970s I, along with some others, believed that a proto-hominid did indeed exist: *Ramaphithecus*, a hominoid named in 1934 and based originally on a Late Miocene upper jaw fragment from India. The name has since been applied to jaw fragments and isolated teeth collected from Kenya to Hungary to China, ranging in age from about 15 m.y. to 7 m.y. ago. *Ramapithecus* resembles *Australopithecus* in some jaw and tooth traits that were interpreted as hominid synapomorphies (although the concepts and terminology of cladism were not then widely used); hence hominids were assumed to have diverged from apes 15 m.y. or more ago. The LCA of *Ramapithecus* and pongids was reconstructed as being unlike any living hominoid, but as generalized, with a mixture of apelike, monkeylike, and unique features. I, along with most other paleobiologists, have substantially modified my views recently.

A radically different interpretation of hominid origins grew from a study of the primate molecular record. Comparisons of proteins, first indirectly through comparative immunology, then through amino acid sequencing, restriction endonuclease mapping, then later of DNA itself through DNA hybridization and nucleotide sequencing, showed a surprising pattern. Humans and African apes were closely similar, while orangutans were different, with gibbons more different still. It has become generally agreed that such patterns reflect evolutionary patterns. Genealogically, hominids are odd African apes (Goodman, 1963; Sarich and Wilson, 1967a; Sibley and Ahlquist, 1984).

Some workers went further and claimed that certain features of the molecular patterns showed that molecular evolutionary rates were constant between and along lineages and thus "clocklike," in marked contrast to the fluctuating evolutionary rates of morphological characters. Molecular differences (distances) between albumins of living species were used to estimate divergence times; given knowledge from the fossil record of the age of one branching point in a group, others could be calculated without recourse to the fossil record using degrees of molecular difference between living species (Sarich and Wilson, 1967a). Initially using an estimate of about 30 m.y. for the ape-Old World monkey split, the inferred pattern for hominoids had gibbons diverging about 12 m.y. ago, orangutans 10 m.y., while chimpanzees, gorillas, and humans shared an ancestor around 5 m.y. ago (the distances human–chimp and human–gorilla were 1/6 that of human–Old World monkey) (Sarich and Wilson, 1967b). Later, the calibration technique was changed, an average rate of albumin evolution was calculated, and new hominoid dates generated. Divergence ages were: gibbons, 10 m.y.; organutans, 8 m.y.; and chimpanzees, gorillas, and humans, 4 m.y. (Sarich and Cronin, 1975).

Clearly these two view of hominoid evolution are incompatible. They cannot both be correct, although both could be wrong. They differ because molecular and morphological data gave incongruent results about branching sequence and timing. Why?

WEIGHING THE EVIDENCE

Recall three of the four questions posed earlier. What is the hominoid branching sequence? What are branching times? What were ancestors like? Now, what can different kinds of evidence tell us about these questions? I shall divide evidence rather arbitrarily into three categories: neontological, paleontological (both of them "morphological"), and molecular.

Neontological Evidence

Until recently, hominoid branching sequences have been inferred from comparative anatomical studies of living species. In the consensus view the great apes were a monophyletic group, with hominids and hylobatids divergent. Differences of genealogical opinion reflected different hypotheses concerning synapomorphies. Usually pongids were linked together by many supposed synapomorphies: including, for example, simian shelves, limb proportions, wrist joint morphology, lumbar vertebral anatomy, genital morphology, and chromosome number (Kluge, 1983).

Since it became clear from molecular studies that hominids were probably the closest relatives of African apes, a number of morphological and morphometric studies have demonstrated a similar pattern (Oxnard, 1981; Corruccini et al., 1979). Most recently, a number of supposed synapomorphies linking humans and orangutans has been proposed (Schwartz, 1984).

The extent to which branching sequences derived this way are correct depends of course on the plausibility of the hypothesized synapomorphies. Teasing apart homoplasies and homologies and separating symplesiomorphies from synapomorphies is difficult given only the neontological record. Cladists have made various attempts to systematize procedures for identifying likely synapomorphies (Wiley, 1981). Given the low diversity (few surviving species) of living hominoids, the relatively great morphological disparity between surviving species, the poorly sampled hominoid fossil record (Pilbeam, 1983), and the impossibility of specifying deductively "correct" analytic procedures (Cartmill, 1981), it has proved difficult to come up with unambiguous branching sequences so far.

In addition to an improved fossil record, a clearer understanding of both function and ontogeny will make it easier to select relevant characters and to infer character state polarities, and we can expect such inferences to improve greatly as our knowledge of ontogeny becomes more securely synthesized with molecular biology (Alberch et al, 1979). However, we will still need a historical record to check the inferences.

On balance, most modern workers would link humans and African apes as closest relatives. This position clearly owes much to the molecular record.

Neontological data have also been used, in a somewhat different way, in phylogenetic reconstructions. Comparative anatomists and paleontologists have combined characteristics of living hominoids in almost every conceivable way to synthesize chimeric ancestral stages. Gibbonlike, oranglike, chimplike, uniquely different from everything, they have gamboled through the scenarios essentially unconstrained. The fossil record has been elastic enough, the expectations sufficiently robust, to accommodate almost any story.

Paleontological Evidence

The hominoid fossil record is improving, but like most fossil records it still tells a story that is frustratingly difficult to interpret, although some paleontologists have been reluctant to admit this. However, unlike neontological and molecular data, only fossils give direct information about ancestors that differed from descendants (and therefore often from expectations). Who would have deduced from study of only living primates a hominid ancestor with enormous cheek teeth? But, given that ancestors are different, how are we to match actual fossils with predicted ancestors; how and to what extent do we recognize morphotypes? The problem grows as ancestors become more different from living species.

To reconstruct a dated branching sequence from the fossil record we need to be able to bracket each shared ancestor by specifying when a splitting event had and had not occurred, and to order the ancestors relative to each other. Determining when an event had occurred requires recognition of new apomorphies; which is, in principle, feasible. Deciding that it had not yet occurred is much more difficult. Thus *Australopithecus afarensis* and species of *Homo* have enough inferred synapomorphies to satisfy us that hominids had diverged from apes at least 3.7 m.y. ago (Johanson and White, 1979); how much earlier cannot be judged, since morphological rates of evolution cannot be extrapolated into the past because they are known to fluctuate. Virtually no primate branch points are sufficiently densely sampled for us to be able to bracket them tightly.

Unfortunately, the hominoid fossil record is still dominated by jaws and teeth. In living mammals these body parts often differ little between species. They are also often unreliable for sorting species into higher taxa because homoplasies are common and it is often difficult to determine polarities. Probably too few morphological characters have been used (inferred numbers of species and rates of evolution are, in part, functions of morphological complexity) for us to recognize true diversity. As jaws and teeth are perceived in the future to be more complex (for example, as we study not just thin or thick enamel, but ameloblast morphology and ontogeny, enamel histology and disposition, dentine and root morphology, etc.), we may well see inferred species diversity increase. This may happen too as our sampling of body parts improves.

Decisions about character state polarities will always be a function of our understanding of the living record, neontological and molecular. Recently, much progress has been made in understanding function in extant species, and our understanding of the molecular biology of development is starting to grow. The more we know about ontogeny and function in living systems, the more sensible our inferences about homologies and polarities can be. The fossil record is of course inadequate for many of the demands placed upon it. Unfortunately, much that is critical in the past is extremely enigmatic, or lost. However, it is our only source of historical information.

Happily, one paleontological area in which tremendous progress has been made concerns "contextual" information. Dating of rocks and contained fossils steadily improves, and we know much more about plant and animal communities of which hominoids were and were not part (Bernor, 1983; Behrensmeyer, 1982). This provides indirect insight into hominoid adaptations, inasmuch as we can understand the dynamics of living communities.

Molecular Evidence

The molecular record cannot give us past morphological states, but it can tell us something about branching sequences and possibly relative divergence times.

At about the same time it was suggested that molecular evolutionary rates fluctuated much less than morphological ones (Sarich and Wilson, 1967a,b), indeed hardly at all, it was hypothesized that nucleotide substitutions—the ultimate molecular "events"—accumulated stochastically, because most were "unseen" by selection (Kimura, 1968). Hence rates of change, whether measured directly as nucleotide substitutions or indirectly (by protein sequencing, for example) would have an average rate that is constant over given time periods. Early immunological work claimed to document this empirically by the "regularity test"' (Sarich and Wilson, 1967a), the observation that the immunological distances between a more distantly related outlier species and all members of a more closely related set (for example, gibbons versus all other hominoids) were essentially equal. (Departures from equality were considered as stochastic "noise.") It was argued that this pattern could appear only if molecular evolutionary rates were essentially constant.

Since the 1960s, new techniques have been developed permitting ever closer assays of nucleotide sequences. Do these techniques give us enough information to reconstruct and date branching sequences, or if not to give us reasonable estimates? Can we measure the actual amounts of accumulated change along various segments of a lineage, and how are these related to the differences between living species that we actually measure?

To reconstruct branching sequences cladistically, we would ideally like to recognize likely synamorphies, which of course is not possible with measures such as immunological distances (Sarich and Cronin, 1975) or DNA hybridization values (Sibley and Ahlquist, 1984), which do not provide data on character states. Such information is potentially available only from protein and DNA sequences. Clustering techniques employed so far have usually not been designed to identify or use synapomorphies. However, it is in fact hard to recognize more than a few likely synapomorphies in nucleotide or amino acid sequences because mutation reduces their number quite rapidly over time (McKenna, 1979). Nucleotide changes are not irreversible (back mutations are numerous), not always divergent (convergence is frequent), and thus not strictly additive (new events eradicate evidence of previous ones). These factors make phylogeny reconstruction difficult and the extent to which lost information can be recovered is unclear.

Despite these cautionary notes a consistent picture does emerge: humans are very similar, both overall and in hypothesized synapomorphies, to African apes, with orangs about twice as different (Sarich and Wilson, 1967b; Sibley and Ahlquist, 1984). I shall assume that this reflects the phyletic pattern, and there does seem to be a strong consensus on this point (Templeton, 1983; Wilson et al., 1984; Hasegawa et al, 1984; Goodman et al., 1983; although see Schwartz, 1984).

What about molecular rates of evolution? They do seem to fluctuate less than morphological rates—an initial surprise—but how much less? It has been claimed that no external (non-molecular) information is necessary to demonstrate, through the regularity rest, that molecular rates are constant (Sarich and Wilson, 1967a). Objections can be raised to these claims (see Friday, 1981), among the most important of which concerns the type of clustering algorithm used to partition phenetic distances into estimates of phyletic change along segments of an evolutionary tree (Farris, 1972). Some of the more widely used clustering techniques assume rate constancy; methods that do not show that inferred rates can fluctuate threefold or more (Farris, 1972; Friday, 1981; Swofford, 1981; personal calculations). For DNA hybridization data results are better; rates vary only about 25% between fastest and slowest (Sibley and Ahlquist, 1984). Not enough DNA sequence data are yet available, although preliminary analyses of hominoid mitochondrial DNA sequences suggest small fluctuations (Hasegawa et al., in press).

I shall assume here that information from the most complete data set, DNA hybridization, accurately reflects catarrhine branching patterns. Humans and chimpanzees are closest relatives with gorillas a little more distant; orangs are about twice as different from humans and African apes as humans are from African apes. Further, I shall calibrate the "clock" with a "real" paleontological date (actually, a range); namely, the likely divergence of orangutans estimated at 12–15 m.y. ago. This is a moderately well bracketed estimate, and it has the further desirable property of falling among higher primate branch points and therefore close to those in which we are interested.

THE HOMINOID FOSSIL RECORD

Hominoids were confined to Afro-Arabia in the Early Miocene; we have a good record from Kenya and Uganda (Andrews, 1978). The oldest undoubted cercopithecoid monkeys are some 20 m.y. old, although their divergence from hominoids may perhaps be somewhat earlier (Delson, 1979). Several hominoid species grouped into five or six genera document a diverse Early Miocene array of mostly arboreal and mainly forest-adapted primitive hominoids, including species of *Proconsul* and *Rangwapithecus*, and the small *Limnopithecus*, *Dendropithecus*, and *Micropithecus*. They show a blend of ape, monkey, and unique features and may document the split of hylobatids from large hominoids by 20 m.y. ago or more.

Hominoid species spread out of Africa and Arabia beginning some 17 m.y. ago; they are poorly sampled in the Middle Miocene of Africa, Asia, and eastern Europe but probably document the initial split of large hominoids into two groups: a mainly Asian radiation ultimately producing *Pongo* and a mainly African radiation containing the LCA of *Pan*, *Gorilla*, and *Homo* (Ward and Pilbeam, 1983). In the Late Miocene the hominoid record is more densely sampled, particularly between about 10 m.y. and 7 m.y., although even there the best known species, *Sivapithecus indicus* from South Asia, is barely as well known as the most poorly known Plio-Pleistocene hominid species (Howell, 1978; Badgley et al., 1984). Unfortunately, the Late Miocene of Africa is essentially a blank; that is the likely time of origin and location of protohominids, African ape precursors, and their ancestors (LCAs). Since the record is blank, reconstruction of these states requires educated guesswork, by judiciously blending characters from *Australopithecus*, the African apes, *Proconsul*, and *Sivapithecus* and its relatives.

The story is best told from the end.

Pliocene Speculations

The Pliocene ancestors of *Pan*, *Gorilla*, and *Pongo* species probably lived essentially where the living species do, although some earlier orangutans may have been ecologically more like chimpanzees (Smith and Pilbeam, 1980). Pliocene hominids lived in woodland and savannah habitats of the eastern half (at least) of Africa (Howell, 1978). Early australopithecines are anatomically reasonably well known. They had small brains and a unique blend of skull and dental features. In a few cranial and tooth characters *Australopithecus afarensis* resembles *Pan* and *Gorilla*. The total pattern of the early hominid dentition is unique: canines are low-crowned and somewhat incisor-like in both sexes and show marked size dimorphism. The cheek teeth are large with thick enamel caps; they are set in robust mandibles. These hominids were bipeds, though they differed from later *Homo* in several interesting features, perhaps reflecting some arboreal climbing abilities; no evidence of knuckle-walking is present in the forelimb. One can summarize early hominids as being generalized omnivores—mainly frugivores with diets containing hard food items—still committed to trees, if not to forests, for some food and some shelter. They were probably obligate bipeds yet possibly still skilled arborealists (Stern and Sussman, 1983). Speculations about diet and subsistance behavior, ecology, and social and sexual behavior are numerous and I shall neither summarize nor increase them.

Late Miocene Hominoids

The Late Miocene Eurasian hominoid record is improving, although the bulk of the material is still jaws and teeth. Hominoids are found in Northwest Europe, Southeast Europe, Southwest Asia, and East Asia; habitats are generally difficult to reconstruct, but likely were forest and forest-woodland (Bernor, 1983). Apes seem to have been absent from a belt of perhaps more open or seasonal habitats, running from South Central Asia to Northern China, a belt that expanded during the Late Miocene. Much local extinction of Eurasian hominoids occurred about 7 m.y. ago (Bernor, 1983; Badgley et al., 1984).

In Northwest Europe, species that have been classified as *Dryopithecus*, *Hispanopithecus*, *Rudapithecus*, and *Bodvapithecus* (most of which are probably *Dryopithecus*) occupied mainly forest habitats (Bernor, 1983; Pilbeam, 1983). They are known mostly from jaws, teeth, and very few facial parts and isolated postcranials. Some features of the palate and lower face resemble those in Early Miocene *Proconsul* species (Ward and Pilbeam, 1983). Tooth enamel may sometimes be thicker than in *Pan*, and teeth, especially canines, may have been markedly size dimorphic. Forelimb bones indicate ape-like climbing abilities while an interesting femur suggests that bipedalism may have occurred more frequently in positional behaviors than in living pongids (Pilbeam, 1983).

Hominoid species of the Late Miocene living from Greece to China, south and east of the Paratethys Sea (separating them from *Dryopithecus* species to the north) show some dental and facial resemblances that may be synapomorphies. An intriguing sample of jaws and teeth from Greece 9 or 10 m.y. old probably represents one very dimorphic species, *Ouranopithecus macedoniensis* (de Bonis and Melentis, 1978). Enamel is thick, and some features of some postcanine teeth resemble those in early *Australopithecus* (de Bonis et al., 1981). Without postcranial bones to gauge body size, the significance of canine to postcanine tooth size ratios is hard to determine. Some dental and possibly facial features are also like those of other Asian hominoids, but without more material it is difficult to judge the affinities of this species.

From Turkey and from India and Pakistan have come some fine new specimens of *Ramapithecus* and particularly *Sivapithecus* (Badgley et al., 1984). The taxonomy and relationships of these genera are still unclear, even the large sample of jaws and teeth from 8 m.y. levels in the Potwar Plateau of Pakistan being insufficient to determine categorically the number of species being sampled; clearly, there are at least two. Although dentally generally similar in isolated teeth, *Ramapithecus* and *Sivapithecus* are different in dental proportions, mandibular morphology, and anterior tooth anatomy. Being conservative, I retain the two genera for the moment.

A new 8 m.y. old facial specimen of *Sivapithecus indicus* from Pakistan and a somewhat less complete one from Turkey have greatly expanded our knowledge of this genus (Andrews and Tekkaya, 1980; Pilbeam, 1982). Some of the similarities of the premaxilla, palate, orbit, face, and jaw joint between *Sivapithecus* and *Pongo* are probably synapomorphies. They differ from *Proconsul* and *Dryopithecus* (which form a second cluster), and *Pan*, *Gorilla*, and early *Australopithecus* (forming a third cluster) (Ward and Pilbeam, 1983). In the few facial parts in which they are comparable, *Ramapithecus* resembles *Sivapithecus* (Lipson and Pilbeam, 1982; Ward and Pilbeam, 1983). *Sivapithecus* and *Pongo* also differ in many characters, and strengthening or weakening of hypotheses about their relatedness can only come with recovery of more fossil material and better knowledge of ontogeny and functional anatomy.

Sivapithecus and *Ramapithecus* have thick enamelled teeth set in robust jaws (features originally interpreted as synapomorphies with *Australopithecus*), and anterior teeth that show heavy wear; however, they are dentally unique in total morphological pattern. New hominoid postcranial material from Pakistan cannot be unequivocally assigned to species but comes from generalized arboreal apes capable of a broad range of positional behaviors; again they are uniquely different from other hominoids (Rose, 1983). Ecologically *Sivapithecus* and *Ramapithecus* were probably basically arboreal forest-woodland apes, perhaps coming to the ground more than *Pongo*, subsisting on basically frugivorous diets containing perhaps some hard food items.

As noted earlier, the Late Miocene of Africa is poorly known, and then only in East Africa. The few known hominoid fossils are quite incomplete (Howell, 1978; Pilbeam, 1983).

Middle Miocene Hominoids

Middle Miocene Asian and Central European hominoids, mostly known from teeth, come from Czechoslovakia, Turkey, Pakistan, India, and China between about 15 and 11 m.y. ago; they are basically similar to later forms (but note that specimens are mostly teeth; see, for example, Andrews and Tobien, 1977). The Pakistan material, at least, documents the definite appearance of the hypothesized apomorphous *Sivapithecus* facial features by 12 m.y. (Raza et al., 1983); faunal and paleogeographic data suggests that *Sivapithecus*, *Ramapithecus*, and certain other characteristic Middle Miocene mammals migrated from Afro-Arabia during the period between 18 and 12 m.y. ago, shifts made possible by episodic closings of the eastern Tethys Sea (Thomas, 1985).

These Middle and Late Miocene hominoids may be part of an adaptive radiation of forest and forest-woodland species ranging from Europe southeast of the Paratethys Sea to China, analogous to species of *Macaca* or the several genera of African Papionini. They are still poorly known, and as new material is discovered it is quite possible that their apparent homogeneity will evaporate. A preliminary examination of the Lufeng hominoid material (see Wu et al., 1983) suggests to me that these fossils may not represent *Sivapithecus* or *Ramapithecus*. The discovery of one very good specimen of *S. indicus* changed perspective on that species considerably, and there are surely similar surprises to come.

From the Middle Miocene of Africa and Arabia come both *Proconsul*-like and *Limnopithecus*-like species, along with *Ramapithecus*- and *Sivapithecus*-like forms (Andrews and Walker, 1976; Pickford, 1981). Specimens are as yet too fragmentary to determine numbers of species or their exact relationships to European and Asian taxa, although material may document both the initial large hominoid radiation and the split into "African" and "Asian" clades.

SUMMARY

Hominoids diversified in Africa at least 20 m.y. ago as a group of rather odd arboreal omnivores. They spread from Africa between 18 and 12 m.y. ago at a time when significant geographic, climatic, and habitat changes occurred. As the modern geographic shape of the world emerged, climates began to cool and became more seasonal, while in some areas forests were replaced by woodlands and forest-woodland mosaics. In Eurasia, where they are modestly known, hominoids occupied a variety of forest, woodland, and mosaic habitats; they were probably eclectic and mostly herbivorous omnivores, exploiting perhaps harder foods than their competition (and probably their descendants). At least in Eurasia, they preceded or outcompeted cercopithecoid monkeys until the latest Miocene (Andrews, 1981). Around 7 m.y. ago, hominoids became locally extinct in most areas, probably in response to habitat and/or climate changes. Further major climatic, faunal, and floral changes occurred from the end of the Miocene on, a period during which many modern mammalian taxa evolved and spread.

If synapomorphies are correctly identified, the orangutan is the sole living descendant of the once successful *Sivapithecus* group, which was not ancestral to later African hominoids. The enigmatic Chinese Pleistocene hominoid *Gigantopithecus blacki* may also be similarly derived. Both *Pongo* and (perhaps) *Gigantopithecus* evolved significantly since the Late Miocene; at least one of the African hominoid lineages, the Hominidae, changed markedly, and the African apes also probably evolved.

With our meagre knowledge of Late Miocene tropical Africa, and that only for the East, we can but speculate about the nature of prehominids and early hominids. However, if the branching sequence and timing suggested here is approximately correct, tropical Africa is the likeliest place for hominid, *Pan*, and *Gorilla* ancestors to have diverged, between about 10 and 5 m.y. ago. To recognize these ancestors some notion of what they might look like would help. Some of the following are, I think, likely:

a) African Late Miocene hominoids were diverse and were distributed throughout tropical Africa;

b) they occupied forest or forest-woodland habitats;

c) they were as or more arboreal than *Pan*;

d) they were not as frequent ground quadrupeds as *Pan*, nor as well adapted to quadrupedal terrestrial locomotion (knuckle walking);

e) they were not as bipedal as *Australopithecus*;

f) they were adapted to the same kinds of diet as were *Sivapithecus* and *Australopithecus* and shared some dental adaptations, such as large and thick enamelled cheek teeth;

g) they resembled *Sivapithecus*, *Ramapithecus*, and *Proconsul* as much as they did *Pan* or *Australopithecus*;

h) they are likely also to have differed significantly from all four groups.

CONCLUSIONS

Where, when, and how did hominids evolve? Without a decent hominoid record, the answers should remain quite tentative, but a judicious use of molecular and morphological evidence suggests tropical Africa 5–10 m.y. ago for place and time. Direct answers to "how" are impossible at present and will have to wait until we have a tropical African Late Miocene fossil record at least as good as the Asian one, and one that documents African ape lineages—not just Hominidae. Indirect reconstruction of ancestral states by extrapolation from descendants, possible ancestors, and probable cousins is hazardous, but suggests that the earliest hominids and prehominids may have been rather different from traditional expectations. They probably differed from living African apes in many significant features: in ecology; in feeding and positional behavior; and in dental, facial, and postcranial anatomy. They might, for example, have been more bipedal than African apes. Unless we at least entertain these possibilities, we may fail to recognize ancestors when we find them.

LITERATURE CITED

Alberch, P, Gould, S, Oster, G, and Wake, D (1979) Size and shape in ontogeny and phylogeny. Paleobiology *5*:296–317.

Andrews, P (1978) A revision of the Miocene Hominoidea of East Africa. Bull. Brit. Mus. (Nat. Hist.) 30:85–224.

Andrews, P (1981) Hominoid habitats of the Miocene. Nature 289:749.

Andrews, P, and Tekkaya, I (1980) A revision of the Turkish Miocene hominoid *Sivapithecus meteai*. Paleont. *23*(1):85–95.

Andrews, P, and Tobien, H (1977) New Miocene locality in Turkey with evidence on the origin of *Ramapithecus* and *Sivapithecus*. Nature, *268*:699–701.

Andrews, P, and Walker, AC (1976) The primate and other fauna from Fort Ternan, Kenya. In G Isaac and E McCown (eds): Human Origins. Benjamin, pp. 279–306.

Badgley, C, Kelley, J, Pilbeam, D and Ward, S (1984) The paleobiology of South Asian Miocene Hominoidea. In JR Lucaks (ed): The Peoples of South Asia: The Biological Anthropology of India, Pakistan, and Nepal. New York: Plenum.

Behrensmeyer, AK (1982) The geological context of human evolution. Ann. Rev. Earth Planet Sci., *10*: 34–60.

Bernor, R (1983) Geochronology and zoogeographic relationships of Miocene Hominoidea. In RL Ciochon and RS Corruccini (eds): New Interpretations of Ape and Human Ancestry. New York: Plenum, pp. 21–64.

Cartmill, M (1981) Hypothesis testing and phylogenetic reconstruction. Z. zool. Syst. Evolut.-forsch. *19*:73–96.

Corruccini, RS, Cronin, JE, and Ciochon RL (1979) Scaling analysis and congruence among anthropoid primate macromolecules. Hum. Biol. *57*:167–185.

de Bonis, L, Johanson, D, Melentis, J, and White, T (1981) Variations metriques de la denture chez les Hominoides primitifs: Comparaison entre *Australopithecus afarensis* et *Ouranopithecus macedoniensis*. C.R. Acad. Sci. Paris *292*:373–376.

de Bonis, L, and Melentis, J (1978) Les primates hominoides du Miocene Superieur de Macedoine. An. Paléontol. (Vert.) *64*:185–202.

Delson, E (1979) *Prohylobates* (Primates) from the early Miocene of Libya: A new species and its implications for cercopithecid origins. Geobios *12*:725–733.

Farris, JS (1972) Estimating phylogenetic trees from distance matrices. Amer. Nat. *106*:645–668.

Friday, AE (1981) Hominoid evolution: The nature of biochemical evidence. In: C Stringer (ed): Aspects of Human Evolution. London: Taylor and Francis, pp. 1–23.

Goodman, M (1963) Serological analysis of the systematics of recent hominoids. Hum. Biol. *35*:377–424.

Goodman, M, Braunitzer, G, Stangl, A, and Schrank, B (1983) Evidence on human origins from haemoglobins of African apes. Nature, *303*:546–548.

Hasegawa, M, Yano, T-a, and Kishino, H (1984) A new molecular clock of mitochondrial DNA and the evolution of hominoids. Proc. Jap. Acad. *60*:95–98.

Howell, FC (1978) Hominidae. In VJ Maglio and HBS Cooke (eds): Evolution of African Mammals. Cambridge, Massachusetts: Harvard, pp. 154–248.

Johanson, D, and White, T (1979) A systematic assessment of early African hominids. Science, *202*:321–330.

Kimura, M (1968) Evolutionary rates at the molecular level. Nature *217*:624–626.

Kluge, AG (1983) Cladistics and the classification of the great apes. In RL Ciochon and RS Corruccini (eds): New Interpretations of Ape and Human Ancestry. New York: Plenum, pp. 151–177.

Lipson, S, and Pilbeam, D (1982) *Ramapithecus* and hominoid evolution. J.Hum. Evol. *11*:545–548.

McKenna, M (1979) Molecular mammalogy. Systematic Zool. *28*:109–113.

Oxnard, CE (1981) The place of man among the primates: Anatomical, molecular, and morphometric evidence. Homo *32*:149–176.

Pickford, M (1981) Preliminary Miocene biostratigraphy for Western Kenya. J. Hum. Evol. *10*:73–98.

Pilbeam, D (1982) New hominoid skull material from the Miocene of Pakistan. Nature *295*:232–234.

Pilbeam, D (1983) Hominoid evolution and hominid origins. Pont. Acad. Sci. Scr. Var. *50*:43–61.

Raza, SM, Barry, JC, Pilbeam, D, Rose, MD, Shah, SMI, and Ward S, (1983)New hominoid primates from the Middle Miocene Chinji Formation, Potwar Plateau, Pakistan. Nature *305*:52–54.

Reader, J (1981) Missing Links. Boston, Little Brown.

Rose, MD (1983) Miocene hominoid postcranial morphology: Monkey-like, ape-like, neither, or both? In RL Ciochon and RS Corruccini (eds): New Interpretations of Ape and Human Ancestry. New York: Plenum, pp. 405–417.

Sarich, V, and Cronin, J (1975) Molecular systematics of the primates. In M Goodman and RE Tashian (eds): Molecular Anthropology. New York: Plenum, pp. 141–169.

Sarich, V, and Wilson, A (1967a) Rates of albumin evolution in primates. Proc. Natl. Acad. Sci *58*:142–148.

Sarich, V, and Wilson, A (1967b) Immunological time scale for hominid evolution. Science *158*:1200–1203.

Schwartz, JH (1984) The evolutionary relationships of man and orangutans. Nature *308*:501–505.

Sibley, C, and Ahlquist, J (1984) The phylogeny of hominoid primates, as indicated by DNA–DNA hybridization. J. Mol. Evol. *20*:2–11.

Smith, R, and Pilbeam, D (1980) Evolution of the orangutan. Nature *284*:447–448.

Stern, J, and Sussman, R (1983) The locomotor anatomy of *Australopithecus afarensis*. Amer. J. Phys. Anthrop. *60*:279–317.

Swofford, DL (1981) On the utility of the distance Wagner procedure. In VA Funk and DR Brooks (eds): Advances in Cladistics. New York: N.Y. Botanical Garden, pp. 25–43.

Templeton, A (1983) Phylogenetic inference from restriction endonuclease cleavage site maps with particular reference to the evolution of humans and the apes. Evolution *37*:221–244.

Thomas, H (1985) The Early and Middle Miocene land connection of the Afro-Arabian plate and Asia: A major event for hominoid dispersal? In E. Delson (ed): Ancestors: The Hard Evidence. New York: Alan R. Liss, Inc. pp.

Ward, SC, and Pilbeam, D (1983) Maxillofacial morphology of Miocene hominoids from Africa and Indo-Pakistan. In RL Ciochon and RS Corruccini (eds): New Interpretations of Ape and Human Ancestry. New York: Plenum, pp. 211–238.

Wiley, EO (1981) Phylogenetics. New York: Wiley.

Wilson, GN, Knoller, M, Saura, LL, and Schmickel RD (1984) Individual and evolutionary variation of primate ribosomal DNA transcription initiation regions. Mol. Biol. Evol. *1*:221–237.

Wu, R, Xu, Q, and Lu, L (1983) Morphological features of *Ramapithecus* and *Sivapithecus* and their phylogenetic relationships—morphology and comparison of the crania. Acta Anthropol. Sinica *2*:1–10.

Ancestors: The Hard Evidence, pages 60–62

Paleoenvironments, Stratigraphy, and Taphonomy in the African Pliocene and Early Pleistocene

Anna K. Behrensmeyer and H.B.S. Cooke
Paleobiology Department, Smithsonian Institution, Washington, D.C. 20560 (A.K.B.), and 2133 154th Street, Whiterock, British Columbia, Canada V4A 4S5 (H.B.S.C.)

The four papers in the following section provide a brief but nonetheless highly informative glimpse of contextual studies relating to the hominid fossil record. The scope of such research is increasingly broad and multidisciplinary, but ultimately it rests upon the basic stratigraphy, sedimentology, radiometric dating, and paleontology of the localities where hominids have been found. From the factual evidence provided by these fields, researchers are now exploring a wide range of problems that vary in scale from the activities of hominids at particular fossil sites to the context of human evolution in relation to global-scale tectonic and climatic changes over the past five million years.

Of the evidence available from contextual studies, age-dating of hominid fossils or artifacts is probably most important. Impressive numbers of workers, institutions, and funding agencies have devoted large amounts of time and energy to building the time framework for human evolution. This has resulted not only in relatively secure dating of the important hominid-producing localities, but also in substantial advances in methods of dating and correlation that can be applied to the sedimentary situations in which hominids are found.

The Lake Turkana Basin has provided challenging stratigraphic and correlation problems for a large number of geologists since the inception of the Kerio Valley, Omo, and Koobi Fora Research Projects in the late 1960s. Radiometric dating techniques, faunal similarities, paleomagnetic stratigraphy, and volcanic ash correlations have all been used in attempts to tie together various isolated hominid-producing areas around the basin. The paper by Brown et al. presents the latest evidence relating to this goal and demonstrates that after nearly 20 years of work it appears to be close at hand. By using a combination of tuff composition signatures (trace elements and shard shape) and radiometric dating, Brown and his coworkers have established that many tuffs in the Omo and East Turkana areas are the same, and that there are also links to West Turkana and the Hadar area in Ethiopia. A major new element in these correlation studies is the linking of the terrestrial record with deep sea cores taken from the Indian Ocean. Deciphering relationships between stages of hominid evolution, faunal and floral change, rift valley history, and regional- to continental-scale environmental changes depends on a firm time framework, and the horizons for this now appear to be expanding rapidly.

The contribution by Brown et al. also demonstrates the level of detail and documentation required of geochronological information. There have been a number of controversies generated over the ages of hominid-bearing deposits, and workers are well aware of the need for extreme caution in all stages of the dating process. The "quality control" that this has generated in creating the time framework for human evolution can be viewed as a very positive outcome of these controversies.

Aside from the age of the fossil hominids, there is also strong interest in many different aspects of their paleoecology: paleogeography, dietary and habitat preferences, climatic restrictions and responses to climatic change, interactions with other organisms and hominid species, and social and cultural behavior. The evidence bearing on such problems is drawn from sedimentology, faunal and floral remains, stratigraphic and spatial relationships of the fossil and cultural evidence, and processes that can be observed in modern ecosystems. Careful taphonomic study of such evidence has become standard practice in paleoecological reconstructions, and research relating to human evolution has helped to generate a new phase of growth in the field of taphonomy.

The skeletal remains of hominids and other animals in the South African cave deposits have been a continuing source of taphonomic puzzles since the discovery of the Taung skull in 1924. Brain's exemplary long-term investigations of the processes responsible for bone and artifact accumulations in the caves have contributed greatly to our understanding of their significance in terms of early human behavior. His contribution shows how the absence of evidence (i.e., cutmarks) in the bone accumulations of the "orange" and "brown" members at Swartkrans becomes highly significant when contrasted with the overlying "Early Stone Age" member, where cutmarks on bones are common. Understanding of taphonomic processes causing damage to bone surfaces has increased rapidly in the past decade. Combined with Brain's careful analysis of the cave deposits, this now permits informative comparisons where formerly the absence of cutmarks might have been regarded simply as a "sampling problem." In answer to a question during the discussion, Brain noted that since cutmarks on bones are known from other deposits comparable in age to the orange and brown members, their absence in these units (which contain both bone and stone artifacts) emphasizes the potential variability in early hominid behavior.

The influence of environmental change on early human evolution in Africa has been discussed for some time (e.g., Coppens, 1978). Paleoenvironmental and evolutionary questions raised by Vrba provoked considerable interest among participants at the symposium. She proposes an evolutionary pulse affecting bovids, hominids, and probably other mammalian groups between 2.5 and 2.0 m.y. BP, in synchrony with a structural change in African vegetation from more closed to more open (savanna) habitats. This is linked, in turn, to global climatic changes associated with the onset of northern hemisphere glacial conditions. She proposes *Australopithecus africanus* as the more closed-habitat form, and the later *A. robustus, A. boisei,* and *Homo habilis* as more dry- and open-habitat hominids. A radiation of hominids is thus viewed as a direct consequence of vegetation change. Vrba also believes that the emergence of the basal hominid stock may have occurred in concert with a major pulse of bovid evolution and environmental change around 5 m.y. BP.

The actual correspondence of these changes in time was brought out as a matter of concern during the question period. The timing of both the bovid and hominid evolutionary "pulses" is based on first appearances of new forms, and such evidence is notoriously subject to sampling problems. *A. boisei*, for instance, may have evolved earlier than 2.5 m.y. BP but is only represented in the fossil samples after that time. Vrba defended her analysis of bovid radiations on the basis of convergence of the ancestral forms at about 2.5 and 5.0 m.y. The hominids are harder to judge using this line of evidence because there is so much less fossil material. Yet it is an intriguing proposal based on what evidence is available. A question from the audience brought out the possibility that an increase in intelligence also might have accompanied the shift to more open habitats by making early *Homo* a better "ecological generalist."

A further problem with Vrba's hypothesis involves the nature and scale of vegetational change. She points out how apparently synchronous this was in East and South Africa (within a half-million years or so). These areas represent very different latitudinal, tectonic, and local conditions of climate; vegetation is under strong local environmental control today and undoubtedly was in the past as well. Does the apparent change between 2.5 and 2.0 m.y. really reflect an overriding global effect, or is it a fortuitous coincidence of similar (but not unique) trends that can be accounted for by climatic fluctuations of more regional scale?

The answer lies in whether the pollen, geochemical, and mammalian evidence actually reflect local, regional, or continental levels of resolution. The case for continental-scale trends is strongest when there are similar trends demonstrated by many independent types of data in different fossil localities. At this point, the different types of evidence do indicate drier conditions between about 2.5 and 1.7 m.y. BP. But there was undoubtedly a mosaic of different vegetation types available throughout this time, and the specific adaptive pressures on the hominids to become more "open-habitat" animals remain obscure. The evidence for habitats used by the different East African hominids can be interpreted in a variety of ways, one of which is that the robust forms were more closely associated with closed rather than open habitats. (See Boaz, 1977; Behrensmeyer, 1978, 1984 for evidence bearing on this problem.)

The overview of faunas, dating, and paleoecology at Laetoli provided in the paper by Harris illustrates what can be done when an interdisciplinary approach is applied to an extraordinary hominid locality. The unusual fossil preservation at Laetoli, including especially the abundant footprints, provides a unique glimpse of life on the Pliocene savanna of East Africa. The paleoecological reconstruction reveals dry open and bush habitats similar to those of the modern Serengeti. Harris notes that this ecological situation is so different from that of other East African hominid localities that it affects faunal correlations; the absence of certain species of Pliocene pigs and proboscideans could reflect ecological rather than temporal differences.

The new correlations to the Hadar sequence now indicate that there may be an average age difference of about 0.5 m.y. between the Hadar faunas (including the hominids) and the earlier Laetoli fauna. Thus some (but not all) of the problems discussed by Harris in correlating these two areas using key fossil taxa may, in fact, be due to the age discrepancy. This also points out the limits to resolution in faunal correlations between distant, ecologically different areas, in this case on the order of 0.5 m.y. (assuming that the present radiometric ages are correct).

The presence of *Australopithecus afarensis* at Laetoli is at variance with Vrba's hypothesis concerning the adaptations of the earlier hominids to more closed-vegetation habitats. It is, of course, possible that *A. africanus* was different in this respect from its presumed ancestor. If so, then there must have been other periods of habitat shift among the hominids, and this would diminish the overall significance of Vrba's "pulse" of environmentally generated change and speciation between 2.5 and 2.0 m.y. BP. Laetoli serves an important role in showing that most other hominid sites in East Africa may have a strong "humid bias," as Harris terms it, and that our view of hominid paleoecology throughout the Plio-Pleistocene is strongly affected by the limited environments in which fossils have been preserved.

The problems of age-dating and paleoecology are well-represented in these four papers. All are grounded in data derived from rocks and fossils, but they vary in the scope of the problems and levels of interpretation. Brain's paper on new taphonomic evidence from two South African cave sites raises intriguing questions about hominid behavior but remains cautious in interpretation. Vrba's proposed link between climate change and speciation in hominids and bovids, on the other hand, represents an attempt to reach beyond the basic information in search of global-scale cause and effect. These two contributions reflect very different but complementary approaches to contextual studies in human evolution. Information such as that reported by Brain, Harris, and Brown et al. in their contributions demonstrates presently accepted levels of interpretation based on the available primary data from rocks and fossils. Vrba's focus on large-scale theoretical problems underlines the need to seek more evidence that will bear specifically on these problems. In combination, the papers outline a number of important directions for future research.

LITERATURE CITED

Behrensmeyer, AK (1978) The habitat of Plio-Pleistocene hominids in East Africa: Taphonomic and microstratigraphic evidence. In C Jolly (ed): Early Hominids of Africa. London: Duckworth, pp. 165–190.

Behrensmeyer, AK (in press) Taphonomy and the paleoecologic reconstruction of hominid habitats in the Koobi Fora Formation. In Y Coppens (ed): Paris: Foundation Singer-Polignac.

Boaz, NT (1977) Paleoecology of early hominids in Africa. Kroeber Anthro. Soc. Pap. No. *50*:37–62.

Coppens, Y (1978) Evolution of the hominids and of their environment during the plio-pleistocene in the lower Omo Valley, Ethiopia. In WW Bishop (ed.): Geological Background to Fossil Man. Edinburgh: Scottish Academic Press, pp. 499–506.

Ancestors: The Hard Evidence, pages 63–71

Ecological and Adaptive Changes Associated With Early Hominid Evolution

E.S. Vrba
Transvaal Museum, Pretoria 0001, South Africa

ABSTRACT This paper investigates the habitat preferences of early Hominidae. Census data of extant antelopes suggest that today the frequencies of antelope tribes are excellent indicators of gross vegetation cover. On the basis of phylogenetic and temporal distributions of antelope morphologies, I argue that antelope frequencies may also provide reliable paleoecological evidence in early hominid-associated assemblages. Use of this criterion in the Transvaal hominid succession indicates a change from relatively more closed vegetation and mesic environments associated with *Australopithecus africanus* to more open, arid ones associated with early *Homo* and *Australopithecus robustus*. I argue, as I essentially did already in Vrba (1974, 1975), as follows: 1) The particularly marked environmental change, between 2.5 and 2.0 million years (m.y.) ago, was extensive across Africa. It certainly is evident in both eastern and southern Africa. 2) It involved spread of open grassland at the expense of shrinking forests and woodlands, probably caused by a global reduction in temperature and associated alterations in rainfall. 3) The environmental changes *caused* the evolutionary changes in hominids that are observed in eastern and southern Africa over this time period. 4) From the perspective of mammalian evolution in general, I suggest some adaptive changes that may have accompanied this remarkable ecological switch in hominid evolution. 5) Mammalian evolution in Miocene–Recent Africa occurred in concerted pulses. That is, diverse lineages including the Hominidae responded to the same widespread oscillations in temperature, rainfall, and vegetation cover by synchronous waves of speciation and extinction.

INTRODUCTION

The richest and most spectacular modern, large mammal fauna on earth inhabits the African savanna. There is evidence that the origin of the extensive savanna—today occupying some 65% of Africa and ranging from dense, moist woodlands and thickets through open shrublands to vast treeless grasslands (Huntley and Walker, 1982)—dates to the end of the Miocene. Many of the typical monophyletic groups of savanna mammals are first recorded near the Miocene–Pliocene boundary, together with the Hominidae. The human family was among the "founder members" and for most of its history an endemic part of the African savanna biota.

What were the physical and biotic changes that caused hominid evolution? Which environmental variables were of major importance in the ecology of our early relatives, and how might hominids have been adapted to these variables? How did the nature and breadth of habitat preferences differ among hominid species? The few available hominid fossils on their own cannot take us very far towards answering these questions. Instead, the changes in man's family tree need to be studied as an integral part of wider biotic, climatic, and geological evolutionary rhythms.

That is the approach I have taken. In particular, I have used data on antelopes (Bovidae) and on environmental variation in space and time to generate some, and test other, hypotheses of hominid paleobiology. Bovids are especially useful for such analyses: They are by far the most numerous large mammals in most African Miocene–Recent fossil assemblages. Using skull differentiation among extant species as a guide, fossil morphologies are usually readily identifiable to species level. Most clades have undergone a rapid turnover of species since the Miocene and are still in a phase of radiation. They are dominant in the modern large mammal fauna (some 70 species of African Bovidae comprise nearly 40% of hoofed mammals in the world), and their biology has been intensively studied. The vast majority of extant antelope species are narrowly habitat-specific. As a result, bovid data can contribute to the study of hominid chronology and paleoecology (Vrba, 1974, 1975, 1980a, 1982, in press).

Furthermore, early Hominidae were large-bodied, mobile, herbivorous (at least to a large extent), endemic, savanna mammals—all features shared with the abundant antelopes that evolved together with them. It is thus hardly surprising to find that hominid evolutionary pulses—in lineage splitting (speciation), phenotypic change, and extinction—have followed rhythms similar to those in several antelope phylogenies (Vrba, 1984a). This evidence implies not only synchrony, but also in some cases broad ecological analogy, of evolutionary response to the same widespread oscillations in temperature, rainfall, and vegetation cover. I suggest that the evolutionary patterns of Bovidae and other mammal groups can be used to generate hypotheses on the causes of hominid evolution and on the nature of early hominid adaptations.

ANTELOPES AS PALEOECOLOGICAL INDICATORS

To assess which, if any, bovid taxa may be reliable habitat indicators in the Plio-Pleistocene, I first asked: Are extant African Bovidae, at the generic and tribal level, significantly associated with each other and with particular habitats? To answer this question I analyzed modern census data from 16 wildlife areas in subsaharan Africa. The nine tribes of Bovidae analyzed all originated either near the end of the Miocene or previously. There are morphological reasons for hypothesizing that most are monophyletic, *sensu stricto*. A multidimensional graphical technique called correspondence analysis was performed on these data (Vrba, 1980a; Greenacre and Vrba, 1984). It can be called an objective method of statistical analysis in the sense that it does not *a priori* presume any structure (or causative factor) underlying the data. Instead, it reveals any non-random structure *a posteriori*. In this particular case it primarily grouped bovid taxa with each other and with game areas. It secondarily showed associations between bovid taxa and independently-included ecological variables like vegetation cover, rainfall, altitude, soil nutrient status, etc. A remarkable and consistent association of Alcelaphini (the hartebeest-wildebeest-blesbuck group) and Antilopini (the gazelle-springbuck group) with open grassland was demonstrated. In areas with a combination of low altitude and low rainfall (0–400 mm mean annual) as well as those of high altitude with medium rainfall (400–800 mm mean annual) the resultant vegetational physiognomy is a low ratio of wood- to grass-cover. In such areas alcelaphines plus antilopines never account, in this data set, for less than 65% of the total antelope frequency; they never amount to more than 30% in predominantly bush-covered areas.

One remarkable result of the correspondence analysis is that it clearly pinpoints *gross vegetational physiognomy* (i.e., as defined by proportions of wood- versus grass-cover, not by plant species) as of primary importance (accounting for 40% of the variation) in determining tribal distribution of modern antelopes. In fact, the main result is that a particular tribe might be strongly associated with an overall physiognomy of characteristic wood–grass-cover, while in contrast it may range across diverse categories of soil nutrient status and primary productivity, geography, and even to some extent rainfall and altitude (Greenacre and Vrba, 1984).

A second interesting result is that, with few exceptions, the associations with gross vegetational physiognomy are evident at the level of whole monophyletic groups, the tribes. The extant distribution suggests that, in spite of a high species turnover (individual clades may show up to 30 speciation events in 5 m.y.), speciation hardly ever results in entry into a significantly different habitat of grass–wood ratio. It seems that once an ancestral species of a clade is established as (for example) an open grassland grazer, its descendant species are highly unlikely to switch to a browsing, predominantly woodland existence. From a perusal of the literature, I suggest that such constraints on "switching" basic vegetational habitats are general for diverse mammal groups.

Was there such an association between Alcelaphini/Antilopini and open habitat, and between other groups and more woody environs, early on in the histories of these groups, or is it a recent phenomenon? A number of lines of evidence suggest that the habitat associations are ancient: 1) the phylogenetic distributions of extant habitat preferences mostly assort strictly according to monophyletic groups, and the latter had originated before the Early Pliocene; 2) the phylogenetic and temporal distribution of morphological characters strongly supports a notion of fidelity throughout the Plio-Pleistocene of modern tribal habitat associations; and 3) the Miocene–Recent speciation and extinction curves of Alcelaphini-plus-Antilopini are inverse to those of other tribes combined. 4) Furthermore, wherever the hypothesis has been tested so far in the fossil record it has been found supported. For instance, Kappelman (in press) reports that, in the lower Olduvai strata, high alcelaphine plus antilopine frequencies coincide with high frequencies of murids (Jaeger, 1976) and pollen types (Bonnefille, 1979) that indicate open habitat. Thus, I have concluded that the "Alcelaphini-plus-Antilopini criterion" is a reliable indicator of gross vegetation cover in the early hominid record.

ENVIRONMENTS OF HOMINID SUCCESSIONS IN THE TRANSVAAL AND EAST AFRICA

I applied the criterion of the proportions of alcelaphine plus antelopine individuals, out of all bovids present, to the various hominid-associated assemblages in South Africa. The percentages of Alcelaphini plus Antelopini from Makapansgat Member 3 and Sterkfontein Member 4 are very much lower than those of later assemblages, such as those from Swartkrans Member 1 and Sterkfontein Member 5. These results obtained for various Transvaal fossil assemblages, together with the associated hominid species, are represented in Figure 1. (The Kromdraai australopithecine stratum, KBE Member 3, is not included because very few bovid specimens are present in this assemblage; see Vrba, 1981).

After careful and detailed consideration of taphonomic biases, I have concluded (Vrba, 1980a) that this difference in representation of antelope taxa represents a real change in the vegetation and climatic conditions of the Plio-Pleistocene transition in the Transvaal. One important consideration is that the same change, across the same time interval, is evident in East African stratigraphic sequences. For instance, on the environments of the Omo Shungura Formation, Howell (1978, p. 228) notes:

> Vegetation in the earlier time range (2.5 m.y.) on pollen evidence (Bonnefille, 1976), 'included closed and/or open woodland, tree/shrub grassland, grassland, and some shrub thicket and shrub steppe' (Carr, 1976). By contrast, 0.5 m.y. later . . . total diversity of plant taxa is markedly reduced, species of mesic habitat affiliation decrease markedly, as does the percentage of pollen from woody plants, and the proportion of pollen with plains habitat affiliation increases markedly. These changes suggest 'more xerophytic plant community types of grassland, tree/shrub steppe' (Carr, 1976). This rather profound environmental change is wholly corroborated by the microvertebrate fossil record (Jaeger and Wesselman, 1976). A robust australopithecine (*A. boisei*) appears for the first time in the Omo succession at about the time of this environmental change, as does also an early species of genus *Homo*.

By the "2.5 to 2.0 m.y." interval Howell meant Shungura Members C to E, now redated to "earlier than 2.52 to about 2.4" (Brown et al., 1985). This time level falls between ST 4 and later Transvaal strata (Fig. 1). Thus, the East African change, both as regards environment and hominid evolution, parallels the southern evidence across the same time level.

It is important to recognize clearly the limits of what I am claiming here. The analysis of extant antelopes suggests that tribal distribution is *primarily* determined by gross height and spacing of bush and tree cover. The latter is secondarily related to combinations of temperature, rainfall, and other factors (Greenacre and Vrba, 1984). If this is

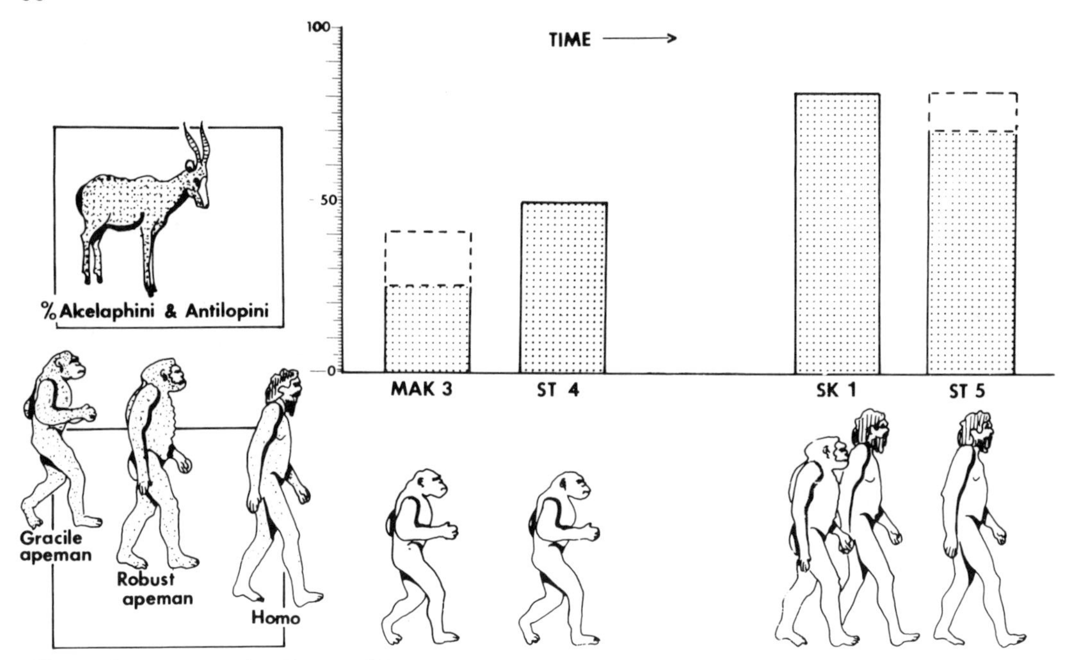

Fig. 1. Associations in four Transvaal limestone cave members from three cave formations. The members are arranged from left to right according to my estimate of chronological sequence. (MAK = Makapansgat Limeworks; ST = Sterkfontein; SK = Swartkrans; numbers refer to members.) Top: histograms of minimum numbers of bovid individuals of Alcelaphini plus Antilopini, calculated as percentages of the total bovid assemblages. Dotted lines denote uncertainty, due, for instance, to possible admixture from other members. Bottom: Hominid species. Note added in proof: Recent work suggests ST 5 predates SK 1.

valid, then the bovid change in the Transvaal succession tells us that a gross vegetational change occurred, from proportionally more bush and tree cover somewhere near 3 m.y. ago to more dominant open grassland close to 2 m.y. ago. However, any conclusion from these data on how this change correlates with temperature and rainfall must be secondary and more speculative. There are lines of evidence that suggest that for any one area a change to more open habitat very probably correlates with reduced temperature, and *vice versa*, and less certainly with reduced rainfall (reviewed in Brain, 1981; Vrba, in press). Secondly, precisely of what nature and magnitude was the alleged vegetation change? I have argued (Vrba, in press) that the change in the Transvaal before 2 m.y. ago was not of large scale, such as from dense woodland/forest to treeless grassland. Rather the change was probably more subtle, occurring somewhere along the vegetational sliding scale *within* the spectrum of moderately open to open savanna environments. The bovid data are not at variance with Butzer's (1971) conclusion that each of the breccias in question, including Sterkfontein Member 4, basically represents colluvial sediments compatible only with an incomplete mat of vegetation.

In sum, the southern African evidence indicates a more vegetationally open and probably more arid environment for the Swartkrans apeman and *Homo*, postdating a wetter and more bush-covered period during which Sterkfontein *Australopithecus africanus* lived in the area. Butzer's (1974) suggestion of a relatively humid environment for the Taung *A. africanus* is in agreement with the Transvaal evidence. As pointed out above, East African data suggest a similar environmental change associated with the same phase in hominid evolution. What we know of the environments associated with *Australopithecus afarensis/africanus* in eastern Africa is not at variance with a notion that the pre-2.5 m.y. hominids lived in areas with a relatively higher wood–grass ratio and more mesic conditions than their descendant species (reviewed in Vrba, in press).

I conclude that we have a strong suggestion, of a broad and unprecise nature but nonetheless reasonably supported, of a difference in habitat specificity between early hominid species as a whole, not just between their local populations. I have argued elsewhere (Vrba, 1980b) that significant evolutionary change is a direct function of environmental change. Thus I see the correlation between different habitats and different hominid species as likely to indicate causation: a widespread change in temperature, rainfall, and vegetation cover *caused* the evolution of the hominid phenotypes currently included in the species *Homo habilis, Australopithecus robustus*, and *A. boisei.*

A separate question is whether the postulated environmental change caused *the particular lineage splitting event* that led to *Homo* on the one hand and "robust" australopithecines on the other. One argument would disagree: White et al. (1981) see the Hadar and Laetoli hominids as belonging to the directly ancestral species of the crucial splitting event, with *A. africanus* already diverged from the ancestry of *Homo.* They argue that both Makapansgat and Sterkfontein *A. africanus* belong in the 2.0–2.5 m.y. time range, and place the branching that gave rise to *Homo* earlier (see Fig. 2). Other anthropologists, notably Tobias (e.g., in press), place the split between 2.5 and 2.0 m.y. ago, with *A. africanus* as the direct ancestor. If the latter view is correct, then the widespread environmental change which I have discussed may have been the cause of the major branching event in hominid evolution: as woodland habitats shrank and fragmented, the preferred habitats of *A. africanus* did as well. Isolated populations and speciation in allopatry may have given rise to earliest *Homo* and to one or more robust australopithecine species.

I await with interest the further developments in the crucial debate between different arguments on hominid phylogeny. As one who has attempted to contribute to early hominid chronology (e.g., Vrba, 1975, 1982) I can observe the following. The confidence that we can now feel regarding the current conflicting reports on the dating of Sterkfontein Member 4, Makapansgat Member 3, and Hadar leaves something to be desired. But in regard to the debate on hominid phylogeny I do not regret that: one can cite the cogent argument that what is crucial to phylogenetic analysis is character analysis and not the temporal placement of taxa.

I suggest that, whatever the outcome of the debate on hominid phylogeny, we have good evidence of a causal environmental influence on hominid evolution between the times represented by *A. africanus* finds on the one hand, and the Swartkrans Member 1 and Sterkfontein Member 5 hominids on the other.

MIOCENE-RECENT EVOLUTIONARY PULSES

Figure 2 compares a curve of first records for bovid species, from Late Miocene-Recent African strata south of the Sahara, with a phylogenetic tree of Hominidae following widespread consensus on the timing of speciation events. A number of taphonomic and chronologic biases are inherent in my procedures for scoring speciation frequencies of Bovidae (Vrba, 1984). In Figure 3 a part of the hominid tree is compared with oxygen isotope curves from the deep sea record. I am assuming that these climatic data, especially those of Thunell and Williams (1983), represent global changes in temperature. Although most sources agree with the estimates in Figure 3 of earliest records in the *Homo* and robust lineages, some would propose different hominid phylogenies (e.g., the alternative supported by White et al. [1981], represented by dashed lines in Fig. 2).

In spite of the uncertainties, a comparison of Figures 2 and 3, also with additional data on terminal Miocene cooling (review in Brain, 1981), suggests certain hypotheses for further testing. During the end of the Miocene, when temperatures plunged to a low point unprecedented during the entire preceding Tertiary, a number of bovid tribes appeared for the first time. They include the essentially African tribes Alcelaphini, Aepycerotini, Reduncini, Hippotragini and Tragelaphini. Thus, global climatic change may have coincided with hominid origin (Brain, 1981) and perhaps also with a spectacular bovid radiation into new "adaptive zones" (Fig. 2). Near 2.5 m.y. ago global temperatures again plunged to a low point. A peak in bovid origination seems to be synchronous with this cooling event and perhaps also with the hominid speciation event that gave rise to *Homo* and robust australopithecines (Figs. 2, 3). Intriguingly, a wave of antelope evolutionary activity may also coincide with the origin of *Homo sapiens* (Fig. 2), and the global mean "cooling step" at 0.9 m.y. could be synchronous with extinction of the robust australopithecines (Fig. 3). The fossil records of other mammals (e.g., Maglio and Cooke, 1978) tentatively suggest that the evolutionary peaks in Figure 2 may be paralleled in other lineages. In fact the patterns point to the possibility of widespread climatic changes as major causes of evolutionary turnover.

I predict that we may find, as data that are more numerous and more chronologically secure become available, that evolution occurs in concerted pulses that involve many groups.

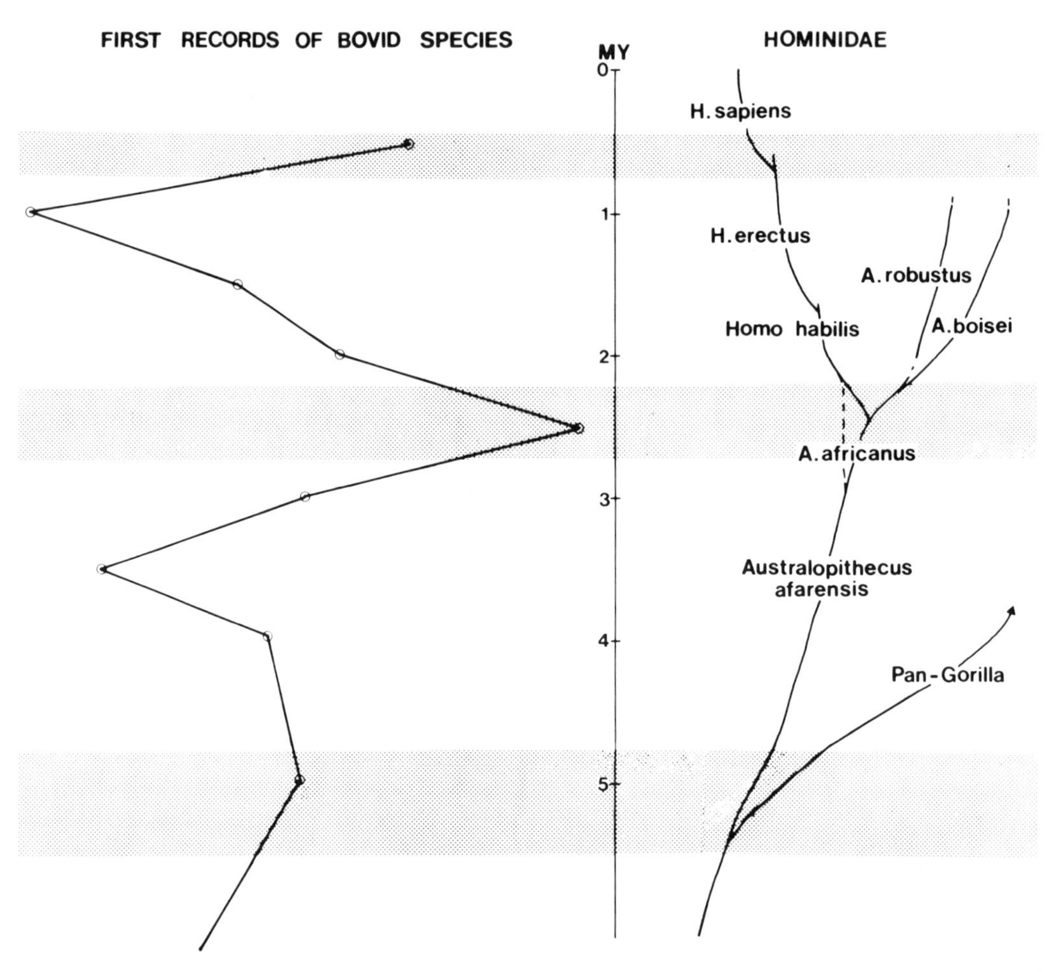

Fig. 2. Rates of origination of all species of Bovidae known from subsaharan Africa, Late Miocene–Recent, and a phylogenetic tree of Hominidae (dashed alternative for origin of *Homo* after White et al., 1981; see text).Stippled bands represent approximate timing of major evolutionary events in bovid and hominid radiation. Chronology updated after Brown et al. (1985). my, Millions of years.

Diverse lineages in the biota should respond by synchronous waves of speciation and extinction to global temperature extremes and attendant environmental changes.

EVOLUTION IN ECOLOGICAL CHARACTERS OF HOMINIDAE

It is likely that hominids at some point underwent a significant ecological "switch," in terms of vegetation cover and yearround availability of moisture; and it is possible that this switch coincided with a speciation event that led to *Homo* on the one hand and *A. robustus/boisei* on the other. If so, it was a remarkable and rare event in the context of Miocene–Recent mammal evolution (e.g., Vrba, 1980a; Greenacre and Vrba, 1984). An examination of similar events in other phylogenies suggests hypotheses of changes in adaptive characters that may have occurred in the hominids.

Resource breadth

Extant survivors of lineages that entered more open, arid habitats typically are either more generalist in breadth of resource use than their more numerous, bush-loving relatives (i.e., they can subsist in the ancestral and in the new environments), or they are specialists on open habitats (see Vrba, 1984 for some examples). The hominine lineage may have taken the first evolutionary route, the robust one the second.

There are of course numerous publications on diet and dietary breadth in early hominid

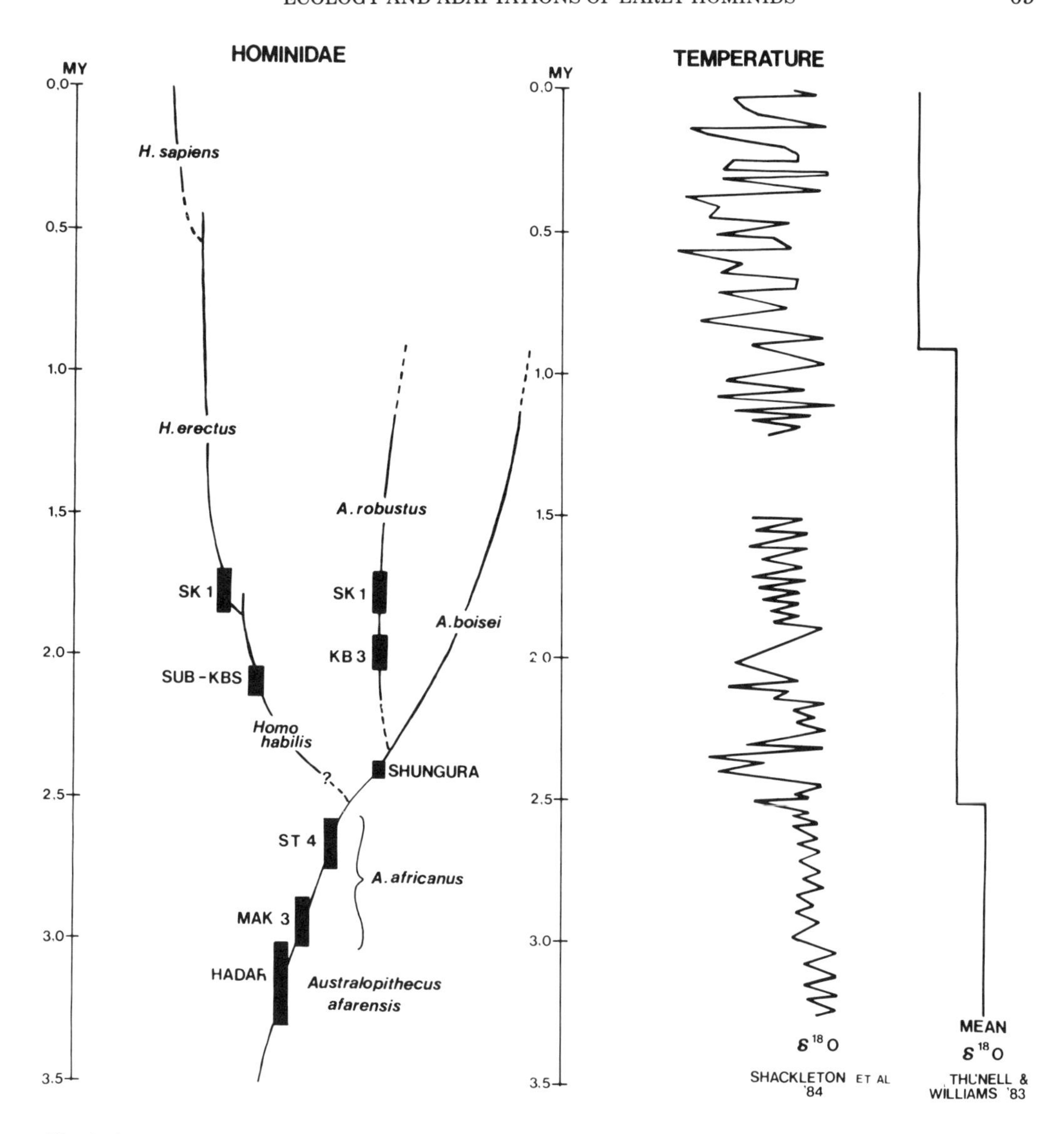

Fig. 3. A comparison, over the time range 3.5 m.y. to Recent, of the hominid tree with oxygen isotope data from the deep sea record. Solid rectangles represent chronological estimates (after Brown et al., 1985, for East Africa, and my biochronology for South Africa) for hominid occurrences from particular strata, including earliest records of the *Homo* and robust lineages. The climatic curve on the left is adapted form Shackleton et al. (1984, Figs. 3, 4). It refers to deep sea oxygen isotope data from a North Atlantic site. The curve on the right, adapted from Thunell and Williams (1983, Fig. 5), represents a schematic summary of mean oxygen isotope values from deep sea sites in the Pacific, Caribbean, Mediterranean, and off Northwest Africa. I am assuming that these oxygen isotope curves represent changes in temperature.

species (reviewed in Grine, 1981). One early "dietary hypothesis" is that of Robinson (1963): the cranial, mandibular, and dental differences between the Transvaal apemen reflect the fact that *A. africanus* was an omnivore living in drier, more open conditions, while the robust apemen were specialist, extreme herbivores living in a wetter and vegetationally more luxurious environment. Up to the early 1970s Robinson's conclusions were the dominant views on environment and trophic adaptation of Transvaal hominids. On the basis of alcelaphine and antelopine proportions, I wrote in 1975 (*contra* Robinson, 1963) that we "should consider whether [the musculature of robust apemen] was so massive and the molars proportionally so large, because their 'vegetables' were of the tough grassland type." Since then there has been agreement with this suggestion from many different kinds of analyses. For instance, recently Grine (1981) reviewed many hypotheses of early hominid diet and ecology and added new data on the scanning electron microscopic details of occlusal wear of Transvaal hominid teeth. His findings suggest "that the 'robust' forms employed more crushing and puncture-crushing activity and relied more heavily upon Phase II activity than did the 'gracile' australopithecines . . . It is concluded that the 'robust' hominids habitually, or at least seasonally, triturated harder, more resistant and perhaps smaller food objects than were masticated by the 'gracile' australopithecines" (p. 203). He considers that robusts were adapted to xeric environments, where many of the available food staples are relatively tough. In principle, hypotheses of dietary breadth are testable via several kinds of analyses, such as Grine's (1981) on toothwear, and trace element and isotopic analyses of fossils.

An attendant prediction is that the geographic distributions of habitat specialists should be more restricted and more susceptible to environmental fluctuations than those of generalists. Thus we may (or may not) find in one particular stratigraphic column that *A. robustus* (or *A. boisei*) was at different times absent and present in the area (as the geographic distribution shrank and expanded) in predictable synchrony with environmental changes as independently deduced, while *Homo* persisted through the oscillations.

Mobility

As Baker (1978) observed, most animals are "migratory" to some extent. But it is undeniable that those living in open, arid habitats, where resources tend to be patchy in space and/or time, invariably have a greater tendency to seasonal and more extensive movement (see also Wiens, 1976). Did one or both of the hominid lineages that diverged during the Pliocene migrate seasonally across ecotonal margins? That hypothesis has been tested and found supported in species of hypsodont herbivores. At present I cannot think of any way to test it in early hominid species.

A related hypothesis is that apemen in the *A. afarensis/africanus* lineage (if really in habitats where trees were more prevalent and larger) were relatively more arboreal and less bipedal than their descendants (although each species both climbed and walked). There are of course several anatomical studies that have tried to test this. Among them, in a 1979 paper I noted on the basis of the STS 7 scapula of *A africanus* that this Sterkfontein hominid, although apparently a biped, may have been more adapted to tree-climbing than later hominids.(Incidentally, Stern and Susman, 1983, p. 280, group me among those who "view that the evidence is equivocal because the functional significance of morphologic differences between australopithecines and humans is moot." This is inaccurate. In fact, I concluded that "gracile australopithecines had a greater potential for climbing, hanging, reaching and arm-swinging than has modern man" (Vrba, 1979, p. 128). I await with interest the further developments in the debate on this issue, recently rekindled in regard to the Hadar hominids by workers such as Stern and Susman (1983, on the "arboreal side") and Lovejoy et al. (1982, on the "bipedal side").

Social behavior

It is well-known that evolution towards life in more vegetationally open and arid environments is invariably accompanied by fundamental changes in social behaviour, whatever group of mammals one may be considering. Furthermore, the evolutionary responses tend to be similar from group to group (e.g., Krebs and Davies, 1978; Estes, submitted). Thus I welcome a hypothesis like Lovejoy's (1981) (whether it may be found true in detail or not) that focusses on the role of reproductive and social behaviour in hominid evolution. Can we perceive traces in the hominid record after 2.5 m.y. of features such as increased gregariousness, more emphasis on visual communication and reduced sexual dimorphism, etc.? Such features occur convergently in diverse mammal lineages—including primates—that evolved into open habitats.

CONCLUSION

Near 2.5 million years ago a particularly marked and widespread environmental change occurred in Africa. It involved an increase in open grasslands at the expense of wood and tree cover, probably resulting from a global reduction in temperature and associated changes in rainfall. The environmental changes caused the evolutionary changes in hominids that are observed in both eastern and southern Africa over the same time period. This suggestion was essentially put forward by Vrba (1974, 1975) and has since received empirical support from various new data.

Some prevalent models of evolution (such as allopatric speciation, and particularly punctuated equilibria) predict the following: 1) Evolutionary events are a direct function of environmental change (Vrba, 1980b). 2) Thus, speciations and extinctions across diverse lineages should occur as concerted pulses in predictable synchrony with changes in the physical environment, chiefly in global temperature. The Miocene–Recent African record of the abundant Bovidae supports these predictions, and the literature suggests that data on diverse other faunal and floral groups may do so as well. Thus Hominidae were probably "founder members" of the biota of the extensive African savanna, together with many other phylogenetic groups. Similarly, the origin of *Homo* was not an isolated event. Instead it was part of a wave of evolutionary activity that was forced upon the biota by a common environmental cause. The perspective of mammalian evolution in general suggests some adaptive changes that may have accompanied the Late Pliocene ecological switch in hominid evolution.

LITERATURE CITED

Baker, RR (1978) The Evolutionary Ecology of Animal Migration. London: Hadder and Stoughton.

Bonnefille, R (1979) Méthode palynologique et reconstitutions paléoclimatiques au Cénozoique dans le Rift Est Africain. Bull. Soc. géol. France sér. (7) *2*:331–342.

Brain, CK (1981) The evolution of man in Africa: Was it a consequencce of Cainozoic cooling? Annex. Transv. Geol. Soc. S. Afr. *84*:1–19.

Brown, FH, McDougall, I, Davies, T, and Maier, R (1985) An integrated Plio-Pleistocene chronology for the Turkana Basin. In E Delson (ed): Ancestors: The Hard Evidence. New York: Alan R. Liss, Inc., pp. 82–90.

Butzer, KW (1971) Another look at the australopithecine cave breccias of the Transvaal. Am. Anthrop. *73*:1197–1201.

Butzer, KW (1974) Paleoecology of South African australopithecines—Taung revisited. Current Anthrop. *15*:367–388.

Estes, RD (Submitted) To be or not to be sexually dimorphic.

Greenacre, MJ, and Vrba, ES (1984) A correspondence analysis of biological census data. Ecology *65*:984–997.

Grine, FE (1981) Trophic differences between "gracile" and "robust" australopithecines: A scanning electron microscope analysis of occlusal events. S. Afr. J. Sci. *77*:203–230.

Howell, FC (1978) Hominidae. In VJ Maglio and HBS Cooke (eds): Evolution of African Mammals. Cambridge, Massachusetts: Harvard University Press, pp. 154–248.

Huntley, BJ, and Walker, BH (eds) (1982) Ecology of Tropical Savannas. New York: Springer–Verlag.

Jaeger, JJ (1976) Les rongeurs (Mammalia, Rodentia) du Pléistocéne Inférieur d'Olduvai Bed I (Tanzania). 1ére Partie: Les Muridés. In RJG Savage and SC Coryndon (eds): Fossil Vertebrates of Africa, Vol. Four. London: Academic Press, pp. 57–120.

Kappelman, J (in press) Climatic change during the Plio-Pleistocene at Olduvai Gorge, Tanzania. S. AF. J. Sci.

Krebs, JR, and Davies, NB (eds) (1978) Behavioural Ecology: An Evolutionary Approach. London: Blackwell.

Lovejoy, CO (1981) The origin of man. Science *211*:341–350.

Lovejoy, CO, Johanson, DC, and Coppens, Y (1982) Hominid lower limb bones recovered from the Hadar formation: 1974–1977 collections. Am. J. Phys. Anthropol. *57*:679–700.

Maglio, VJ, and Cooke, HBS (eds) (1978) Evolution of African Mammals. Cambridge, Massachusetts: Harvard University Press.

Robinson, JT (1963) Adaptive radiation in the australopithecines and the origin of man. In FC Howell and F Bourliere (eds): African Ecology and Human Evolution. Chicago: Aldine, pp. 385–416.

Shackleton, NJ, Backman, J, Zimmerman, H, Kent, DV, Hall, MA, Roberts, DG, Schnitker, D, Baldauf, JG, Desprairies, A, Homrighausen, R, Huddlestun, P, Keene, JB, Kaltenback, AJ, Krumsieck, KAO, Morton, AC, Murray, JW, and Westberg-Smith, J (1984) Oxygen isotope calibration of the onset of ice-rafting and history of glaciation in the North Atlantic region. Nature *307*:620–623.

Stern, JT, and Susman, RL (1983) The locomotor anatomy of *Australopithecus afarensis*. Am. J. Phys. Anthrop. *60*:279–317.

Thunell, RC, and Williams, DF (1983) The stepwise development of Pliocene-Pleistocene paleoclimate and paleoceanographic conditions in the Mediterranean: Oxygen isotope studies of DSDP Sites 125 and 132. In JE Meulenkamp (ed): Reconstruction of Marine Paleoenvironments. Utrecht Micropal. Bull. *30*:111–127.

Tobias, PV (in press) Punctuational and phyletic evolution in the hominids. In ES Vrba (ed): Species and Speciation, Transvaal Museum Monographs. Pretoria.

Vrba, ES (1974) Chronological and ecological implications of the fossil Bovidae at the Sterkfontein Australopithecine site. Nature *250*:19–23.

Vrba, ES (1975) Some evidence of chronology and palaeoecology of Sterkfontein, Swartkrans and Kromdraai from the fossil Bovidae. Nature *254*:301–304.

Vrba, ES (1979) A new study of the scapula of *Australopithecus africanus* from Sterkfontein. Amer. J. Phys. Anthrop. *51*:117–129.

Vrba, ES (1980a) The significance of bovid remains as indicators of environment and predation patterns. In AK Behrensmeyer and AP Hill (eds): Fossils in the Making: Vertebrate Taphonomy and Paleoecology. Chicago: The University of Chicago Press, pp. 247–271.

Vrba, ES (1980b) Evolution, species and fossils: How does life evolve? S. Afr. J. Sci. *76*(2):61–84.

Vrba, ES (1981) The Kromdraai Australopithecine site revisited in 1980: Recent investigations and results. Ann. Transv. Mus. *33*:18–60.

Vrba, ES (1982) Biostratigraphy and chronology, based particularly on Bovidae of southern African hominid-associated assemblages: Makapansgat, Sterkfontein, Taung, Kromdraai, Swartkrans; also Elandsfontein (Saldanha), Broken Hill (now Kabwe) and Cave of Hearths. Prétirage, 1er Cong. Internat. Paléontol. Hum. Nice: CNRS, pp. 707–752.

Vrba, ES (1984) Evolutionary pattern and process in the sister-group Alcelaphini—Aepycerotini (Mammalia: Bovidae). In N Eldredge and SM Stanley (eds): Living Fossils. New York: Springer–Verlag, pp. 62–79.

Vrba, ES (in press) Palaeoecology of early Hominidae, with special reference to Sterkfontein, Swartkrans and Kromdraai. In Y. Coppens (ed): l'Environnement des Hominidés. Paris: Singer–Polignac.

White, TD, Johanson, DC, and Kimbel, WH (1981) *Australopithecus africanus*: Its phyletic position reconsidered. S. Afr. J. Sci. *77*:445–470.

Wiens, JA (1976): Population responses to patchy environments. Ann. Rev. Ecol. Syst. 7:81–120.

Ancestors: The Hard Evidence, pages 72–75

Cultural and Taphonomic Comparisons of Hominids From Swartkrans and Sterkfontein

C.K. Brain
Transvaal Museum, Pretoria, South Africa 0001

ABSTRACT Sterkfontein Member 5 has yielded several *Homo* fossils, especially the partial cranium of STW 53, as well as a number of stone artefacts attributed to the early Acheulean and some three bone implements. Animal bones show chop- and cut-marks, suggesting hominid meat-eating, and it is likely that the hominids occupied the entrance area of the cave. At Swartkrans, new excavations have revealed a more complex stratigraphy than previously described, with the deposit that yielded all earlier-reported *Homo* and *A. robustus* isolated from the rest as a "hanging remnant." Equivalent to that is the lower bank of member 1, or "orange member" of informal current usage, which has yielded large samples of bone for taphonomic analysis, including bone points (digging tools, as confirmed by experiment and microscopy) as well as stone artefacts and hominids. Above an erosional break lies the "brown member" (or Member 2, older phase), which has produced fossils of both hominid genera along with stone and bone artefacts. Still higher in the sequence is the "Early Stone Age" member, which yields the first evidence of hominid-produced cut-marks and burnt bone. A robust australopithecine tooth may be intrusive from an older layer. A third erosion surface separates this level from the little-studied "Middle Stone Age" member.

INTRODUCTION

The Pliocene and earliest Pleistocene sites of the Blaauwbank (Sterkfontein) Valley in the Transvaal have long been known as the source of one of the greatest collections of early hominids (and associated fauna) in the world. Paleontological study of this material has centered on systematic and paleobiological questions, but my own efforts have concentrated on understanding the origin of the bone accumulations and the inferences we can draw from them about the lifeways of the early hominids (Brain, 1981). An interesting taphonomic comparison can be drawn between the Member 5 fossiliferous horizon at Sterkfontein, where I have not worked personally for many years, and several levels in the Swartkrans site nearby.

STERKFONTEIN

Recent work at this oldest of the several sites in the valley has been carried out by Professor Phillip Tobias, Alun Hughes, and their team. In terms of the topic of this brief survey, our attention may be focussed particularly on Member 5, a deposit that rests unconformably on Member 4, source of the large *Australopithecus africanus* sample. Member 5 contains a large number of stone artefacts, which, according to Stiles and Partridge (1979), are of an early Acheulean industry, similar to those found in Bed II at Olduvai Gorge, above the Lemuta Member, approximately 1.5 million years old. One especially interesting point about these artefacts is that 90% of the collection studied consists of cores or choppers, with many flakes detached, but only 10% of the artefacts are flakes, indicating that the tool-making was performed away from the cave or its catchment area. At present this is distinctive and unexplained.

It seems likely that the entrance area of the Sterkfontein cave was occupied by groups of *Homo,* most probably of the species represented by the Member 5 cranium STW 53. At least some of the bones found in this member are thought to represent human food remains, and, in connection with the bovid bones, Vrba (1975) has suggested that the low proportion of juvenile animals represented suggests that the meat was scavenged rather than actively hunted. Chop- and cut-marks on some of the bones suggest that hominid feeding on meat occurred in the catchment area of the cave. The Member 5 environment is reconstructed as an open grassland with limited woody cover, during a period warmer and drier than now. Three bone tools are known from the Member 5 deposit and these have been interpreted (Brain, 1981) as digging instruments used to unearth edible bulbs on the rocky, open hillsides.

SWARTKRANS

Cultural and taphonomic aspects of the Swartkrans deposits are currently being investigated. The original hominid sample of Broom and Robinson came exclusively from the "*hanging remnant*" of Member 1—a large block of fossiliferous breccia adhering to the cave's north wall and undercut across its entire width by an erosional space. Below this remnant, and to the south, a complex arrangement of sediments rests on the cave floor, and the key to the understanding of these deposits is the realisation that the cave has been repeatedly subjected to cycles of deposition and erosion, the effects of which have been superimposed (Brain, 1982a). The first filling of the cave with Member 1 sediment was followed by an interlude of erosion within the cave, during which the middle levels of the deposit were washed down into lower caverns, leaving the *hanging remnant* along the north wall and the "lower bank" on the cave floor further to the south. The "lower bank" is currently being referred to informally as the *orange member,* which merges into a cone of travertine-cemented stones close to where the original entrance is thought to have been. The space created by the first erosion cycle was subsequently filled by calcified sediment referred to as the *brown member.* Space created by the next erosion cycle was filled with sediment informally referred to as the *Early Stone Age member.* Finally, a large erosional space on the west and north side of the cave was filled by sediment known informally as the *Middle Stone Age member.* The stratigraphic situation may therefore be summarised as in Figure 1. A summary of some conclusions on the cultural and taphonomic aspects of the Swartkrans members follows.

Hanging remnant of Member 1

Although this has produced a large hominid and faunal sample that has been analysed from various aspects (Brain, 1981), the sample is strongly biased in favour of identifiable, often cranial specimens. The hardness of the matrix has meant that a really representative bone assemblage for taphonomic interpretation has never been prepared. Bone or stone artefacts have not been found *in situ,* but I strongly suspect that this is because insufficient breccia has been treated in acid for the recovery of *all* stone and bone fragments. Nonetheless, the sample of some 80 or so robust australopithecine individuals and several of *Homo* (see Clarke, 1985) from this unit is of major importance.

Lower bank of Member 1: The "orange member"

Large volumes of this sediment have proved sufficiently soft to allow excavation with recovery of all bone and interesting stone pieces. A very large sample of great value for taphonomic interpretation has been assembled and is now being studied. Carnivore tooth marks are reasonably common on the bone pieces, but hominid-induced cut marks have not been seen on any. Hominid remains occur throughout the excavated levels; these have not been studied in detail yet, but they are listed (except for very recent finds) in Brain (1982b). Stone artefacts occur sporadically through all excavated levels, as do occasional bone tools, similar to those from Sterkfontein Member 5. A total of 25–30 bone tools have been recovered from the several Swartkrans *in situ* members. Many of these have been broken, but the worn points are quite diagnostic.

They can be duplicated experimentally by taking a piece of long bone and digging out edible bulbs on a rocky hillside. It takes some 20 minutes of digging to get out a bulb, and

Hanging remnant
of Member 1

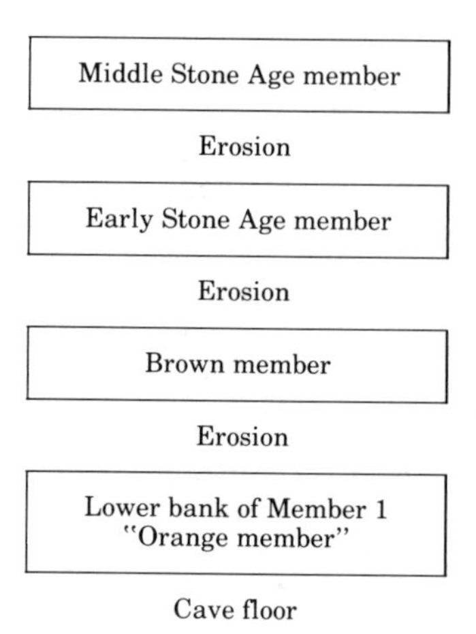

Fig. 1. Summary of Swartkrans stratigraphy.

after several hours or days of use the bones start to acquire this remarkable wear on the ends. When one examines this wear under the scanning electron microscope, there are two patterns visible: a crisscross of wear marks leading down to the tip and a series of crossing marks. In the modern analogues, which show the same patterns as the fossils, the latter marks are caused by pulling the bone against sharp stones in the soil while digging out the bulb. Extremely similar objects continue at Swartkrans from the "orange member" through the two succeeding units, suggesting a long continuation of this food-gathering practice.

The "brown member"

This deposit coincides very largely with Member 2 of Butzer (1976) and rests unconformably on the "orange member." Stone and bone artefacts occur sporadically throughout the excavated depth (about 4 metres) and again, a very large bone sample has been assembled, though this has not been studied in detail yet. Hominid remains (both *A. robustus* and *Homo* sp.) occur throughout the entire depth (Brain, 1982b), and these await detailed study. Carnivore tooth marks are fairly common on the bone pieces, but hominid-induced cut-marks have not been observed.

The Early Stone Age member

Excavation in this deposit has only been going on since the beginning of 1984, so the sample is still very small. Stone artefacts are not common, but those that are present appear to be Acheulean. Several bone tools similar to those in the two older members have been found.

A point of particular interest is that several bone pieces have already been found showing unmistakable hominid-induced cut-marks. Furthermore, a number of pieces of bone appear to have been burnt. Neither of these features has been observed on any of the very numerous bone pieces (ca. 100,000) examined from the two older Swartkrans members.

The only hominid remains found so far in this member are a beautiful upper third molar of a robust australopithecine (which may conceivably have been recycled from one of the earlier deposits) and two phalanges, one of which shows carnivore tooth marks on it.

An intriguing question

The presence of stone and bone artefacts in the "orange" and "brown" members strongly suggests that the hominids who made these tools came to the Swartkrans cave with this simple equipment. They presumably sheltered there at night, and some of them fell prey to the carnivores that frequented the cave. *Although these hominids had stone artefacts, why is there no evidence of meat-eating, in the form of cut-marks on bone, from either of the earlier Swartkrans members?* And yet, as soon as one passes into the third level, the so-called Early Stone Age member, there is evidence of meat-eating and also of deliberate use of fire (though this remains to be proved conclusively). The sudden appearance of meat-eating evidence at Swartkrans may be a result of the acquisition of fire, permitting *Homo* groups not only to overnight in the cave, but also to sit by the fire and process (?hunted) meat. Detailed study of the remains will allow us to evaluate this hypothesis more carefully in the coming year.

ACKNOWLEDGMENTS

I would like to thank the authorities at the American Museum of Natural History, in particular Ian Tattersall and Eric Delson, for inviting me to participate in the "Ancestors" symposium.

LITERATURE CITED

Brain, CK (1981) The Hunters or the Hunted? An Introduction to African Cave Taphonomy. Chicago: University of Chicago Press.

Brain, CK (1982a) Cycles of deposition and erosion in the Swartkrans cave deposit. Paleoecology of Africa *15:*27–29.

Brain, CK (1982b) The Swartkrans site: Stratigraphy of the fossil hominids and a reconstruction of the environment of early *Homo*. Prétirage, 1er Cong. Internat. Paléontol. Hum. Nice: CNRS, pp 676–706.

Butzer, KW (1976) Lithostratigraphy of the Swartkrans Formation. S. Afr. J. Sci. *72:*136–141.

Clarke (1985) *Australopithecus* and early *Homo* in Southern Africa. In E Delson (ed): Ancestors: The Hard Evidence. New York: Alan R. Liss, Inc., pp. 171–177.

Stiles, DN and Partridge, TC 1979 Results of recent archaeological and palaeoenvironmental studies at the Sterkfontein extension site. S. Afr. J. Sci. *75:*346–352.

Vrba, ES (1975) Some evidence of chronology and palaeoecology of Sterkfontein, Swartkrans and Kromdraai from the fossil Bovidae. Nature *254:*301–304.

Ancestors: The Hard Evidence, pages 76–81

Age and Paleoecology of the Upper Laetolil Beds, Laetoli, Tanzania

John M. Harris
Division of Earth Sciences, Los Angeles County Museum of Natural History, Los Angeles, CA 90007

ABSTRACT The fortuitous preservation at Laetoli of a terrestrial biota in wind-reworked tuffaceous sediments provides a rare glimpse of the diversity of life in the Pliocene savannas of eastern Africa. The fauna, flora, and footprints from the upper unit of the Laetolil Beds have been dated radiometrically at between 3.5 to 3.8 million years (m.y.), an estimate consistent with the correlative evidence provided by the fossil mammals. The biota apparently represents open grassland with scattered trees, with an annual rainfall of between 60–100 cm occurring in a single wet season. This pattern is comparable to that found in the area today, contrasting strongly with most other Pliocene rift valley riverine faunas, the predominance of which may well reflect sampling bias.

The Pliocene site of Laetoli in Tanzania (3°12′ South, 35°10′ East) is perhaps best known for its hominid fossils and footprints, although it has yielded a wealth of other animal and plant fossils and a profusion of prints and trails (Leakey et al., 1976; Leakey and Hay, 1979; White, 1977, 1980, 1981). The upper unit of the Laetolil Beds, dated at between 3.5 and 3.8 m.y., documents a paleoenvironmental setting rather different from the fluviatile and lacustrine deposits encountered elsewhere in the East African rift system. The following discussion of the age and environmental setting of the upper Laetolil Beds draws heavily on the unpublished researches of more than 30 colleagues participating in the International Laetoli Research Project.

The Laetolil Beds are exposed over an area of about 1,600 km^2 in the southern part of the Serengeti Plain of Tanzania, where they overlie the Precambrian basement in the southern part of the Eyasi Plateau. The Laetolil Beds are disconformably overlain by the Ndolanya Beds. At some localities the Ndolanya Beds are capped by the Ogol Lavas that have been dated at about 2.4 m.y. (Drake and Curtis, in press); at others they are overlain by the Naibadad Beds, which correlate with the lower part of Bed I at Olduvai (Hay, in press). These older units are in turn discomformably overlain by the Ngaloba Beds, which have yielded Acheulian artifacts and the cranium of an early *Homo sapiens* (L.H. 18).

The Laetolil Beds have been divided into two units (Hay, in press). The lower unit is widely exposed to the southeast of Laetoli; its lower portion comprises a mudflow deposit and both aeolian and water-worked tuffs; its upper portion is composed of water-worked tuffs, lapilli tuffs, and conglomerates. The lower unit of the Laetolil Beds has yielded a modest vertebrate fauna and a number of fossil termitaries. The upper unit of the Laetolil Beds, in which the hominid fossils and footprints were preserved, represents an accumulation of airfall and wind-reworked tuffs originating from Sadiman, a volcano located some 35 km east of the Laetoli site. Achieving a maximum thickness of about 59 m, the upper unit decreases in thickness from northeast to southwest at a regular, exponential rate. These tuffs were deposited on the uplifted and dissected peneplain existing prior to earth movements that resulted in the formation of the Eyasi Fault. During the accumulation of this unit, the land surface was of low relief and mantled by vegetation that was, at least seasonally, insufficient to prevent extensive aeolian transportation of sand-sized ash particles. The rarity of water-laid sediments in the sequence of the upper Laetolil Beds is significant. Caliche paleosols encountered in the sequence provide an additional line of evidence for an arid or semiarid climate. Phillipsite requires an alkaline environment and the soil may have had a pH of 9.5, at least seasonally, as a result of a dry climate and a supply of carbonatite ash as in the Serengeti today.

The age of the upper unit of the Laetolil Beds has been estimated using standard K–Ar dating techniques (Drake and Curtis, in press). Six biotite dates from Tuff 8 at Locality 1 average 3.46 ± 0.12 m.y. A biotite-bearing airfall tuff—exposed north of Locality 1, occuring about 50 meters below Tuff 7, and marking the boundary of the upper and lower units of the Laetolil Beds—has been dated at about 3.76 ± 0.03 m.y. Dates of intermediate age have been obtained from the intervening horizons within this succession.

Although the holotype specimen of *Australopithecus afarensis* was collected from Laetoli, most of the specimens attributed to this taxon are from Hadar. The relative ages of the hominid-bearing faunas from these two localities are thus of some interest. The upper portion of the Laetolil Beds appears radiometrically older than both the lower portion of the Omo Shungura Formation and the fossiliferous portion of the Pliocene sequence at Koobi Fora, but younger than Lothagam and Kanapoi, correlations that are supported by suid evidence. As Brown et al. (1985) have shown, the Sidi Hakoma Tuff, which underlies the hominid-bearing portion of the Hadar sequence, is correlative with the Shungura B and Tulu Bor Tuffs of the Lake Turkana Basin and is thus radiometrically younger than the upper part of the Laetolil Beds. Unfortunately, because of apparent differences in the habitats represented at Laetoli and Hadar, the faunal correlative evidence is less precise than one might have wished.

At Hadar, three suids are present in the Sidi Hakoma Member (the lowest fossiliferous portion of the sequence)—*Nyanzachoerus kanamensis*, *Notochoerus euilus*, and a species that has been variously interpreted as *Potamochoerus* (Harris and White, 1979) or an ancestral form of *Kolpochoerus* (Cooke, 1978). Although *Notochoerus euilus* is a dentally conservative species, some third molars from both Laetoli and the Sidi Hakoma Member have a massive and blocky appearance that Harris and White (1979) interpret as primitive for the species. This would in turn suggest temporal equivalence of these portions of the succession at the two localities. *Kolpochoerus/Potamochoerus* teeth from the two localities are dentally identical. Although persisting at younger horizons in the Lake Turkana Basin, *Nyanzachoerus kanamensis* has not been recorded so far from Laetoli, where its absence may be primarily of paleoenvironmental significance (Harris, in press, b). The presence of the archaic white rhino *Ceratotherium praecox* in the upper Laetolil Beds at Laetoli, and at the lower (but not upper) portions of the Hadar sequence (Guerin, in press), supports the correlation afforded by *Notochoerus euilus*.

The only elephant now recognized from the upper unit of the Laetolil Beds is *Loxodonta exoptata*, which Beden (1983) interprets as a derived form of the stem loxodont *L. adaurora*. Although an advanced subspecies (*L. adaurora kararae*) persists at younger horizons at Omo and Koobi Fora, at both Hadar and Koobi Fora the nominate (primitive) subspecies of *L. adaurora* is found at horizons below those yielding *Loxodonta exoptata*. It is thus possible to interpret the upper Laetolil Beds as younger than the lowest fossiliferous portions of the Hadar and Koobi Fora successions on the basis of distribution of *L. exoptata*. The correlative significance of the presence of *L. exoptata* is, however, open to question. In the parallel *Elephas* lineage, *Elephas ekorensis* is interpreted as morphologically more primitive than, and ancestral to, *Elephas recki brumpti*. Yet, in the Omo succession, *Elephas ekorensis* is the only elephantid from the Usno Formation, while its presumed daughter species *E. recki brumpti* occurs at the temporally equivalent portion of the nearby Shungura Formation (Beden, 1979). Given the great linear distance and apparent environmental dissimilarities between Hadar and Laetoli, the presence or absence of *Loxodonta exoptata* might lack precise temporal significance.

Other large mammals in the fauna are less useful for correlative purposes. The Laetoli hipparions require detailed systematic revision before any correlation might be attempted. Deinothere (Harris, in press, a) and giraffid species appear ubiquitous in eastern Africa at this point in time, though it is interesting that only large and small giraffines are present at Laetoli and Hadar, whereas a third form of intermediate size is present in younger horizons at Omo and Koobi Fora. The bovids (Gentry, in press) and cerecopithecoid species (Leakey and Delson, in press) reflect rather different habitats at the two localities to the detriment of their correlative potential. The only rodent taxa shared by Laetoli and Hadar belong to the ubiquitous genera *Tatera* and *Hystrix* (Denys, in press). Laetoli shares a number of savanna-adapted rodents with Omo and with Pliocene sites in South Africa, but the potential chronological significance of the distribution of rodent taxa in sub-Saharan Africa has yet to be fully explored. Given the differences of habitats represented at the two localities, available faunal correlative evidence suggests that the Laetoli and Hadar faunas are of broadly similar age.

The eastern semiarid to arid part of the Serengeti Plain provides an excellent model for interpreting the paleoenvironments represented at the Laetoli site. Today this area is a grassland savanna with scattered brush and *Acacia* trees. Rainfall is highly seasonal and averages 50 cm per year. The plain is blanketed with a deposit of ash, most of which erupted from Oldonya Lengai about 3000 years ago. This ash deposit has been extensively transported by wind and is presently reworked during the dry season (Hay, personal communication).

Some 80 taxa of plants are represented by pollen from the upper Laetolil Beds (Bonnefille and Riollet, in press). That 60 of these are still found today within a 30 km radius of Laetoli emphasises the essentially modern aspect of the Late Pliocene paleoflora—at least at the generic level. Grass rootlets, fossil leaves, and calcite woodcasts—the latter ranging in size from branches to twigs, some of which bear thorns—have been recovered from the part of the sequence containing the Footprint Tuff. Unfortunately, the internal structure of these fossils has been totally replaced by calcite, precluding their further identification. The Laetoli paleoflora is compatible with an altitude of 1500–1800 m and with an arid climate with a substantial dry season. The pollen spectra suggest an open savanna vegetation with a very diverse herbaceous component in which grasses were abundent.

The dozen species of terrestrial mollusc recorded from Laetoli include at least three new species (Verdcourt, in press). By comparison with their living relatives, the fossil molluscs suggest a scenario of grassland with scattered trees and a rainfall somewhere between 60 and 100 cm annually. An alternative scenario—denser woodland with a narrow belt of evergreen forest—would fit the fossil molluscan evidence but is less concordant with the geological and other paleontological evidence.

Invertebrates are also represented in the Laetolil Beds by trace fossils—termitaries, ova of solitary Hymenoptera, and a possible insect trail. The termitaries, which occur also in the lower Laetolil Beds, provide remarkable documentation of the activities of higher termites (Sands, in press), a group whose previous fossil history has been restricted to younger strata in West Africa. The hymenopteran cocoons and brood cells constitute the first fossil remains of solitary bees from Africa (Ritchie, in press). Unfortunately, the paleoecological significance of these trace fossils is limited by the lack of current knowledge of the ecology and behavior of their extant representatives.

Eighty species of fossil vertebrates have been recovered from the Laetolil Beds. Because of the nature of accumulation of the sediments, and in contrast to most other Tertiary vertebrate fossil localities in eastern Africa, there are no aquatic vertebrates (fish, freshwater chelonians, crocodiles, hippos) documented from the upper portion of the Laetolil Beds, while water-frequenting animals (such as reduncine bovids) are rare or absent. A number of the represented taxa provide information about formerly prevailing paleoecologic conditions.

The physiological requirements of vertebrate ectotherms make them more sensitive to climatic conditions than endotherms. Two new chelonian species from Laetoli confirm a climate that was not radically different from the area today. The growth rings observed in the scutes suggest an annual cycle of single wet and dry seasons while the presence of a large new species of *Geochelone* suggests a semiarid climate and temperatures that remained above freezing (Meylan and Auffenberg, in press). The snake species represented at Laetoli (Meylan, in press) similarly suggest an open dry savanna, while all the birds are present, at least at generic level, in the extant avifauna of East Africa (Watson, in press).

Insectivores are very rare elements of the fauna (Butler, in press); whether they were actually living at Laetoli or brought in as prey species is debatable. The distribution of extant *Rhynchocyon* species (represented at Laetoli by two specimens only) is restricted to habitats containing shaded leaf litter.

The presence of a new *Galago* species at Laetoli (Walker, in press) confirms the nearby presence of trees, but the predominant cercopithecid is *Parapapio* (rather than *Theropithecus* as at Hadar, Omo, and Koobi Fora) (Leakey and Delson, in press). Papionins are also the dominant cercopithecids at southern African Pliocene sites, and their presence at Laetoli could be consistent with a comparatively open habitat with low bush cover.

The fossil rodents also show notable affinities with southern Africa (Denys, in press) and could be indicative of a dry *Acacia* savanna. The abundance of murids at Hadar and the absence there of dry savanna genera suggest more humid paleoenvironmental conditions than at Laetoli. It is possible to infer relative continuity in savanna habitats between Laetoli and southern Africa, less so between Laetoli and localities in Ethiopia. It is conceivable that the Ethiopian highlands provided a barrier to interchange of small mammals, for the Hadar fauna is characterized by a number of taxa with strong Asian affinities, only one of which extends as far south as Omo and none of which occur at Laetoli. The completeness of preservation of some of the pedetid rodents and lagomorphs suggests entrapment and burial in their burrows during accumulation of the ash (Davies, in press a,b). It would appear from this context and from its anatomy that *Serengetilagus* was, unlike the extant African hares, a burrowing lagomorph.

The composition of the population of small carnivores from Laetoli is comparable to that found in the region today (Petter, in press). The larger carnivores further confirm the establishment, by Laetolil Beds time, of a savanna ecosystem broadly comparable to that of today (Barry, in press). However, the role of the sabretoothed felids in such an ecosystem remains to be determined.

The paleoecological significance of the sole elephant species from the Laetolil Beds—*Loxodonta exoptata*—is uncertain. The apparent absence of *Elephas* from Laetoli and the subsequent comparative rarity of *Loxodonta* in East Africa prior to the Late Pleistocene presumably reflect differences in the preferred habitats of these two genera. It is tempting to infer that *Loxodonta* species were better adapted to drier and less wooded conditions than *Elephas* species.

At Laetoli, as at most East African Pliocene and Pleistocene localities, the white rhino (*Ceratotherium*) is more frequent than the black (*Diceros*). Whether this reflects their real abundance or their social structure (extant white rhinos, by being more gregarious and less territorial than black rhinos, might perhaps offer greater sampling potential) remains to be determined. All three perissodactyl groups—equids, chalicotheres, and rhinos—present at Laetoli could be interpreted as consistent with dry wooded or bush savanna.

Of the suids, the similarity of the smaller Laetoli species to extant bushpigs implies a similar preferred habitat, while the larger hypsodont *Notochoerus euilus* would appear to be better adapted to open grassland conditions. The apparent absence of *Nyanzachoerus kanamensis*, a brachyodont large species present at all other East African sites of broadly comparable age, would be consistent with comparably drier conditions at Laetoli.

The presence of large numbers of giraffes would suggest wooded savanna or nearby woodland. It is possible that the preponderance of the small species (*Giraffa stillei*) might reflect the prevailing size of trees and bushes (Harris, in press, c).

As Gentry (in press) has shown, alcelaphines, neotragines, and antilopines together constitute 60% of the identifiable bovids from the upper Laetolil Beds and are indicative of more open (non-wooded) environments. The less common tragelaphines, bovines, and cephalophines are indicative of closed habitats. At localities at which alcelaphines are abundant, *Praedamalis deturi* and a new species of hippotragine(?) are also common, though these last two species are not usually found in association. It is not known if these two species, like the alcelaphines, are representative of open country or if they indicate an admixture of thicker vegetation from more closed habitats. The proportion of *Madoqua* remains to those of other bovids in the Laetolil Beds varies considerably between the different localities. Fossil dik diks are particularly abundant at Localities 1, 10W, and 16, where they amount to 29%, 35%, and 31% respectively of all the bovoid remains (M.D. Leakey, personal communication). Extant dik diks are mainly browsers, live in *Acacia* bush and thorn scrub habitats, and avoid open grassland.

The variation in distribution of the fossil *Madoqua* material at Laetoli may conceivably represent non-uniform distribution of the thicker vegetation.

The indication of single annual wet and dry seasons, occurring today and as suggested by the fossil chelonian evidence, receives some support from the sedimentary sequence of Tuff 7—the Footprint Tuff (Hay, in press). On the land surface immediately below the Footprint Tuff, the presence of grass is indicated by small fossilised rootcasts, but this land surface must have been essentially devoid of grass when the tuff accumulated, because the upright blades would have disrupted the delicate layering preserved in the ash. Grassland grazed down to the stubble but with local concentrations of arboreal debris (twigs and branches) would appear to have been the scenario at the onset of deposition of the Footprint Tuff and is today characteristic of much of the Serengeti towards the end of the dry season.

Hay interprets the Footprint Tuff to have been deposited during a short span of time, probably a few weeks, at the time of year marking the end of the dry season and the onset of the annual rains. Horizons of rain prints in the lower unit of the Footprint Tuff could have been produced by short-lived showers at the onset of the rainy season. The upper unit accumulated over a period with heavy rains, presumably the early part of the wet season, and widespread erosion at the base of the upper unit and between superjacent layers 2 and 3 is attributable to heavy showers. Once the ash became fully saturated, it would no longer preserve rain prints; this would account for the rarity of rainprinted horizons in the upper unit. Moreover, wet—as opposed to damp—ash might account for the lesser clarity of footprints at the higher levels of the Footprint Tuff.

The stratigraphic distribution of the different types of footprints is in accordance with such a dry to wet season hypothesis. Prints from the lower unit of the Footprint Tuff are primarily of lagomorphs, guinea fowl, rhinoceros, and other animals that might be expected to remain in grassland savanna during the dry season. Footprints of larger bovids are more common in the upper unit, and the prints of proboscideans, equids, cercopithecids, and hominids are restricted to the upper unit. It is conceivable that the greater diversity of identifiable footprints in the upper unit records migration into or through the Laeoli area at the onset of the rainy season. If so, it would confirm that the seasonal migrations in evidence today had become established prior to the Pliocene. Many of the tracks exhibit normal walking gait, and preexisting game trails were utilised; thus it is possible, at the juncture represented by the footprint levels, that the ash showers were not sufficiently heavy to disrupt established patterns of game movement.

The fortuitous preservation at Laetoli of a terrestrial biota in non-waterlaid sediments provides a rare glimpse of the diversity of life on the Pliocene African savanna. Ecological conditions at Laetoli three and a half million years ago were not too dissimilar from those of the area today. How widespread such conditions were in sub-Saharan Africa at that point in time is difficult to judge, but it is possible that the Neogene fluviatile and lacustrine deposits of the rift valley systems have provided samples of fossil populations that are unduly biased towards the humid edge of the spectrum.

LITERATURE CITED

Barry, JC (in press) Large carnivores (Canidae, Hyaenide, Felidae) from Laetoli. In MD Leakey and JM Harris (eds): The Pliocene Site of Laetoli, Northern Tanzania. Oxford: Oxford University Press.

Beden, M (1979) Phylogénie des Eléphantidés: l'Apport des grandes sites du Plio-Pleistocène d'Afrique orientale. Bull. Soc. geol. Fr. Paris, ser. 7 *21:*271–276.

Beden, M (1983) Family Elephantidae. In JM Harris (ed): The Fossil Ungulates: Proboscidea, Perissodactyla and Suidae. Koobi Fora Research Project, Vol. 2. Oxford: Clarendon Press, pp. 40–129.

Beden, M (in press) Fossil Elephantidae from Laetoli. In MD Leakey and JM Harris (eds): The Pliocene Site of Laetoli, Northern Tanzania. Oxford: Oxford University Press.

Bonnefille, R, and Riollet G (in press) Palynological spectra from the upper Laetolil Beds. In MD Leakey and JM Harris (eds): The Pliocene Site of Laetoli, Northern Tanzania. Oxford: Oxford University Press.

Brown, FH, McDougall, I, Davies, T, and Maier, R (1985) An integrated Plio-Pleistocene chronology for the Turkana Basin. In E Delson (ed): Ancestors: The Hard Evidence. New York: Alan R. Liss, Inc., pp. 82–90.

Butler, PM (in press) Fossil insectivores from Laetoli. In MD Leakey and JM Harris (eds): The Pliocene Site of Laetoli, Northern Tanzania. Oxford: Oxford University Press.

Cooke, HBS (1978) Plio-Pleistocene Suidae from Hadar, Ethiopa. Kirtlandia *29:*1–63.

Davies, C (in press, a) Fossil Pedetidae from Laetoli. In MD Leakey and JM Harris (eds): The Pliocene Site of Laetoli, Northern Tanzania. Oxford: Oxford University Press.

Davies, C (in press, b) Note on the fossil Lagomorpha from Laetoli. In MD Leakey and JM Harris (eds): The

Pliocene Site of Laetoli, Northern Tanzania. Oxford: Oxford University Press.

Denys, C (in press) Fossil rodents (other than Pedetidae) from Laetoli. In MD Leakey and JM Harris (eds): The Pliocene Site of Laetoli, Northern Tanzania. Oxford: Oxford University Press.

Drake, R, and Curtis GH (in press) K–Ar geochronology of the Laetoli fossil localities. In MD Leakey and JM Harris (eds): The Pliocene Site of Laetoli, Northern Tanzania. Oxford: Oxford University Press.

Gentry, AW (in press) Pliocene Bovidae from Laetoli. In MD Leakey and JM Harris (eds): The Pliocene Site of Laetoli, Northern Tanzania. Oxford: Oxford University Press.

Guerin, C (in press) Fossil Rhinocerotidae from Laetoli. In MD Leakey and JM Harris (eds): The Pliocene Site of Laetoli, Northern Tanzania. Oxford: Oxford University Press.

Harris, JM (in press a) Fossil Deinotheriidae from Laetoli. In MD Leakey and JM Harris (eds): The Pliocene Site of Laetoli, Northern Tanzania. Oxford: Oxford University Press.

Harris, JM (in press b) Fossil Suidae from Laetoli. In MD Leakey and JM Harris (eds): The Pliocene Site of Laetoli, Northern Tanzania. Oxford: Oxford University Press.

Harris, JM (in press c) Fossil Giraffidae and Camelidae from Laetoli. In MD Leakey and JM Harris (eds): The Pliocene Site of Laetoli, Northern Tanzania. Oxford: Oxford University Press.

Harris, JM and White TD (1979) Evolution of the Plio-Pleistocene African Suidae. Trans. Am. Phil. Soc. *69:*1–128.

Hay, RL (in press) Geology of the Laetoli area. In MD Leakey and JM Harris (eds): The Pliocene Site of Laetoli, Northern Tanzania. Oxford: Oxford University Press.

Leakey, MD and Hay, RL (1979) Pliocene footprints in the Laetolil Beds at Laetoli, Northern Tanzania. Nature *278:*317–323.

Leakey, MD, Hay, RL, Curtis, GH, Drake, RE, Jackes, MK and White, TD (1976) Fossil hominids from the Laetolil Beds. Nature *262:*460–466.

Leakey, MG, and Delson, E (in press) Fossil Cercopithecidae from the Laetolil beds. In MD Leakey and JM Harris (eds): The Pliocene Site of Laetoli, Northern Tanzania. Oxford: Oxford University Press.

Meylan, PA (in press) Fossil snakes from Laetoli. In MD Leakey and JM Harris (eds): The Pliocene Site of Laetoli, Northern Tanzania. Oxford: Oxford University Press.

Meylan, PA, and Auffenberg, W (in press) The Chelonians of the Laetolil beds. In MD Leakey and JM Harris (eds): The Pliocene Site of Laetoli, Northern Tanzania. Oxford: Oxford University Press.

Petter, G (in press) Small carnivores (Viverridae, Mustelidae, Canidae) from Laetoli. In MD Leakey and JM Harris (eds): The Pliocene Site of Laetoli, Northern Tanzania. Oxford: Oxford University Press.

Ritchie, JM (in press) Trace fossils of burrowing Hymenoptera. In MD Leakey and JM Harris (eds): The Pliocene Site of Laetoli, Northern Tanzania. Oxford: Oxford University Press.

Sands, WA (in press) Ichnocoenoses of probable termite origin. In MD Leakey and JM Harris (eds): The Pliocene Site of Laetoli, Nothern Tanzania. Oxford: Oxford University Press.

Verdcourt, B (in press) Mollusca from the Laetolil and Upper Ndolanya beds. In MD Leakey and JM Harris (eds): The Pliocene Site of Laetoli, Northern Tanzania. Oxford: Oxford University Press.

Walker, AC (in press) Fossil Galaginae from Laetoli. In MD Leakey and JM Harris (eds): The Pliocene Site of Laetoli, Northern Tanzania. Oxford: Oxford University Press.

Watson, GE (in press) Pliocene bird fossils from Laetoli. In MD Leakey and JM Harris (eds): The Pliocene Site of Laetoli, Northern Tanzania. Oxford: Oxford University Press.

White, TD (1977) New fossil hominids from Laetoli, Tanzania. Am J. Phys. Anthropol. *46:*197–130.

White, TD (1980) Additional fossil hominids from Laetoli, Tanzania: 1976–1979 specimens. Am. J. Phys. Anthropol. *53:*487–504.

White, TD (1981) Primitive hominid canine from Tanzania. Science *213:*348–349.

Ancestors: The Hard Evidence, pages 82–90

An Integrated Plio-Pleistocene Chronology for the Turkana Basin

F.H. Brown, I. McDougall, T. Davies, and R. Maier
Department of Geology, University of Utah, Salt Lake City, Utah 84112 (F.H.B.) and Research School of Earth Sciences, Australian National University, Canberra, Australia (I.M., T.D., R.M.)

ABSTRACT Here we report new K/Ar age measurements on materials from the northern Turkana Basin and on recently established correlations between tuffs in the northern part of the basin. We then frame a general chronology for the Plio-Pleistocene deposits of the basin as a whole. A significant change in the chronology of the central part of the Shungura Formation is suggested with Tuff G dated at 2.3 Ma (million years ago), Tuff F at 2.36 Ma, Tuff D at 2.52 Ma, and Tuff B-10 at 2.95 Ma. The Hasuma Tuff at Koobi Fora is typified and correlated to Shungura Tuff C. A previous date on Tuff K actually dated material of Tuff L, leaving no dates for Tuff K. The Tulu Bor Tuff at Koobi Fora (= Shungura Tuff B) is estimated to be 3.35 Ma old by interpolation, and it has been located in deep sea cores younger than 3.5 Ma. The underlying Lokochot Tuff (= Shungura Tuff A) is less clearly placed in time (probably 3.4–3.55 Ma); a correlative tuff overlies most of the Lothagam Faunal sequence in the Kerio Valley, southwest of Lake Turkana. An unamed sequence west of Turkana includes numerous tuffs correlated with those of the Shungura and Koobi Fora Formations.

INTRODUCTION

Sediments of Plio-Pleistocene age are exposed over much of the Turkana Basin of Northern Kenya and southern Ethiopia (Fig. 1). Hominid fossils have been collected from the Usno and Shungura Formations of the Lower Omo Valley, the Koobi Fora Formation east of Lake Turkana, exposures of sediment that correlate with the Shungura Formation discovered by Richard Leakey west of the lake, and the Lothagam Group and Kanapoi Formation in the lower Kerio Valley.

Here we report on new tephra correlations between various areas, on new K/Ar measurements of materials from the Shungura Formation, and on initial K/Ar measurements of materials from the west side of Lake Turkana. We also utilize information that is presently in the process of publication (McDougall, 1985). An extensive review of earlier work is not given; rather, we concentrate on framing a chronology for the Plio-Pleistocene deposits in the basin as a whole.

STRATIGRAPHY

The stratigraphy of the Shungura, Usno, and Mursi Formations has been published by de Heinzelin et al. (1976) and Butzer (1976), and that of the Koobi Fora region has been set forth by Vondra and Bowen (1978) and Findlater (1976), with subsequent modifications by Brown and Cerling (1982). The stratigraphy of Lothagam and Kanapoi has been treated by Powers (1980); that of the sediments correlated with the Shungura Formation west of Lake Turkana is as yet unpublished, but results of a preliminary survey are incorporated in Figure 2, which shows schematic stratigraphic sections of the formations.

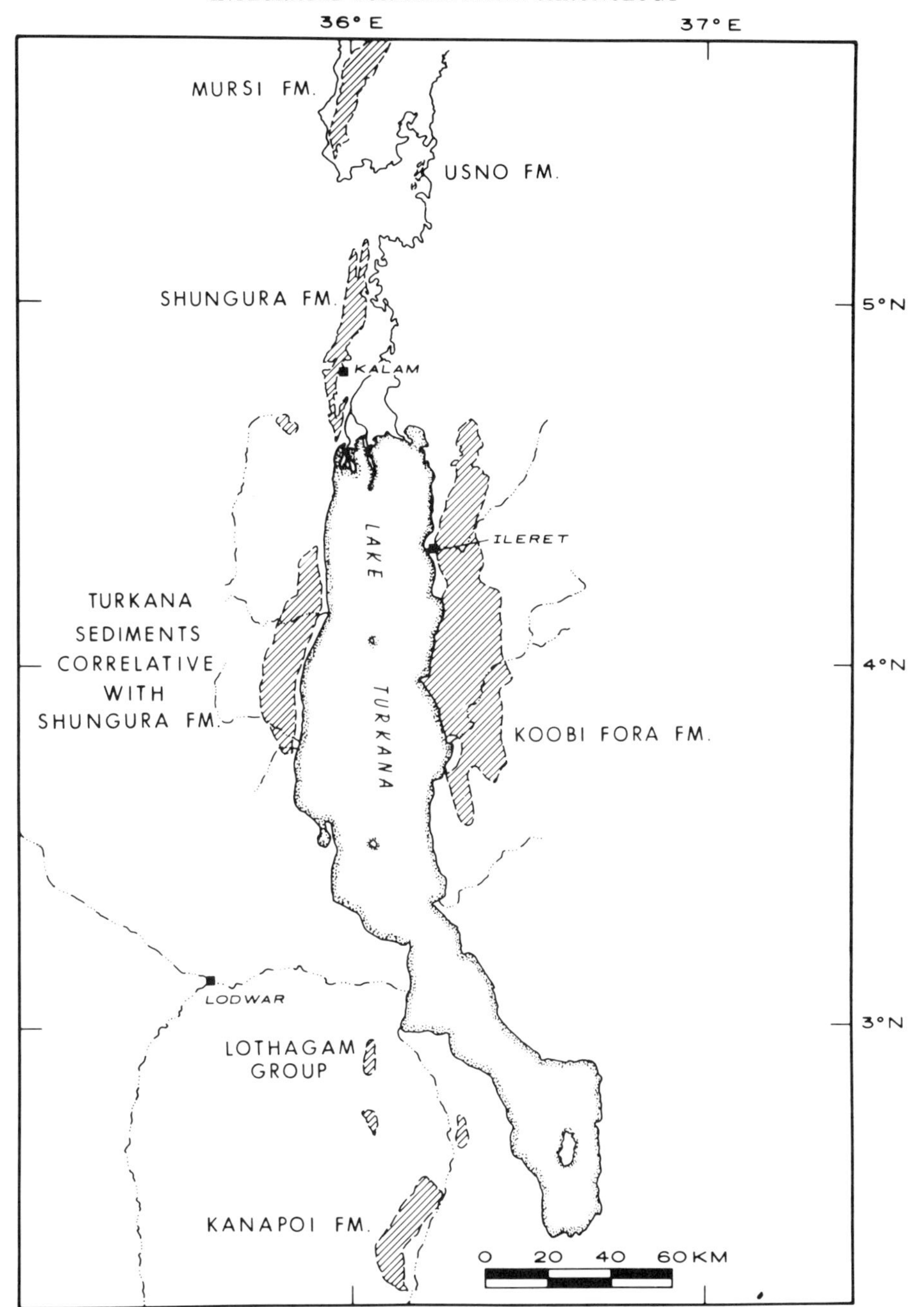

Fig. 1. Sketch map of Turkana Basin showing principal areas of Plio-Pleistocene outcrop (hatched areas).

TEPHRA CORRELATIONS

Correlations established between the Koobi Fora and Shungura Formations by Cerling and Brown (1982) and by subsequent investigations are shown in Figure 2, along with newer correlations established to other areas. Supporting data for previously unreported correlations are given in Table 1.

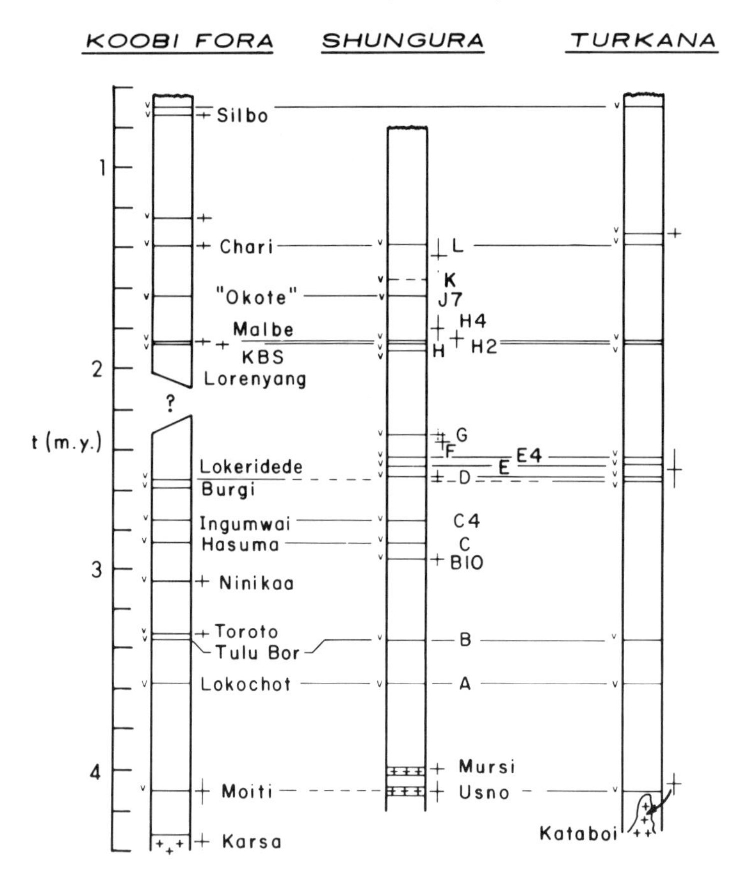

Fig. 2. Diagram showing ages and errors of dated horizons in the Turkana Basin (crosses on right of columns), and correlations between discrete areas. t, Time, in millions of years (m.y.); v, tuff,

Findlater (1978) defined the Hasuma Tuff as a tuff that crops out extensively to the south of Bura Hasuma Hill in the Allia Bay region of Koobi Fora. Here we designate the tuff exposed 1 km southwest of the junction of the Asuma Forest Road with the main parks road at Koobi Fora as the type Hasuma Tuff, noting that this locality is one specified as containing the Hasuma Tuff by Findlater. Here the tuff is about 1.5 m thick, is deposited in a channel, and is composed of medium gray, fine sand-sized glass shards. It lies within a sequence of fluvial deposits about 10 m above the Ninikaa Tuff and 14 m below the Burgi Tuff. It may be correlated with Tuff Cα of the Shungura Formation (Table 1).

As the deposits west of Lake Turkana have not been described in detail, we refer to tephra in that area by their designations in the Shungura or Koobi Fora Formations. Tuffs A, B, D, E, E-4, H-2, H-4, and L of the Shungura Formation have been identified west of the lake, in addition to the Moiti Tuff and a tuff of the Guomde Formation that were otherwise known only from the Koobi Fora region (Table 1). Tuff A is also known from the lower Kerio Valley.

CHRONOLOGICAL CONTROLS

The fossiliferous deposits of the Kerio Valley are believed to be amongst the oldest in the Turkana Basin, but only a few K/Ar age measurements exist on rocks of this region. The Lothagam faunas lie above a basalt dated at 8.5 ± 0.2 Ma (recalculated to new constants), and those of the lower part of Lothagam are older than a basalt sill dated at 3.8 ± 0.4 Ma (recalculated) (Behrensmeyer, 1976). Faunas from the upper part of Lothagam do not appear to be clearly constrained by the age of the sill. The only data

TABLE 1. X-ray fluorescence analyses of glass separates from tuffs in the Turkana Basin

Area	Tuff	Sample	Fe_2O_3[1]	CaO[2]	Ba	Mn	Nb	Ti	Y	Zn	Zr
Turkana	Moiti	82–723	3.09	0.25	400	817	113	1280	113	248	1039
Koobi Fora	Moiti	81–487	2.98	0.22	410	712	115	1182	113	227	1066
Kerio Valley	Lokochot	Loth-11M	3.70	0.18	116	881	147	1117	135	250	1396
Turkana	Lokochot	81–532	3.59	0.18	103	806	146	1209	133	244	1369
Koobi Fora	Lokochot	80–295	3.58	0.19	123	837	152	1263	141	239	1403
Turkana	α-Tulu Bor	81–544	1.35	0.49	399	280	75	1156	54	55	281
Koobi Fora	α-Tulu Bor	80–228	1.31	0.58	419	274	77	1482	54	57	298
Turkana	β-Tulu Bor	82–717	1.63	0.32	118	389	85	879	61	90	366
Koobi Fora	β-Tulu Bor	80–179	1.56	0.32	135	387	83	930	63	79	366
Shungura	Tuff C-α	ET-25	6.17	0.26	76	2647	197	3584	125	279	1317
Koobi Fora	Hasuma	GHC820	6.35	0.14	21	2914	196	3654	128	319	1338
Turkana	Tuff above Burgi	81–539	3.34	0.17	5	1378	165	1431	90	166	1197
Koobi Fora	Tuff above Burgi	80–277A	3.51	0.15	5	1341	163	1377	92	181	1209
Turkana	Tuff D	81–527A	3.55	0.17	5	1348	183	1158	102	184	1407
Shungura	Tuff D	F–146	3.36	0.13	5	1270	183	1189	101	179	1398
Turkana	Tuff E	82–776	6.10	0.20	35	2382	165	2277	127	276	1336
Shungura	Tuff E	ET–45	6.11	0.20	7	2402	168	2276	128	279	1338
Turkana	Tuff E-4	82–779	4.30	0.19	5	1447	152	1504	111	223	1213
Shungura	Tuff E-4	780–1A	4.17	0.13	5	1523	149	1556	103	205	1248
Turkana	Tuff H2	82–775	3.05	0.18	27	869	196	1178	129	227	1233
Koobi Fora	KBS	77–17	3.09	0.23	30	843	194	1297	129	253	1239
Turkana	Tuff H4	82–790	5.10	0.33	36	1983	160	3142	94	236	1137
Koobi Fora	Malbe	80–192	4.71	0.36	42	1942	153	3229	91	242	1091
Turkana	Tuff L	82–786	2.90	0.19	225	696	99	1105	100	143	1074
Koobi Fora	Chari	77–23	2.81	0.18	223	564	103	1093	102	190	1085
Turkana	Tuff above Silbo	81–508	3.00	0.18	5	950	99	1028	79	153	866
Koobi Fora	Tuff above Silbo	50–143	3.01	0.22	5	973	104	1080	79	153	874

[1]Fe_2O_3 = Total iron expressed as Fe_2O_3 (wt%).
[2]CaO in wt%; all other elements in ppm.

on the Kanapoi formation are ages on the Kalokwanya Basalt, which overlies the sediments. This basalt has yielded ages ranging from 2.6 to 3.1 Ma (Behrensmeyer, 1976; Fitch and Miller, 1976), and, as it is of reversed polarity (Powers, 1980), it is likely that it lies within either the Mammoth or Kaena event. The faunal collections from Kanapoi and Ekora may be considerably older than the overlying basalt. Dr. Alan Walker collected a sample of tuff from a point roughly midway between Lothagam and Kanapoi (approximate coordinates 2°43′N lat., 36°5′E long.) that has proven to correlate with the Lokochot Tuff of Koobi Fora. From interpretation of aerial photos, this tuff is believed to overlie most of the deposits at Lothagam, but it may belong to the youngest part of the sequence there (known informally as Lothagam-3). It is utterly distinct from the tuff described at Kanapoi by Patterson et al. (1970) and Powers (1980).

Brown and Nash (1976) reviewed potassium–argon data available for the Shungura Formation up until that time. New data that are incorporated here were obtained at the Australian National University in 1983. McDougall (1985) has reviewed earlier K/Ar and $^{40}Ar/^{39}Ar$ determinations on material from the Koobi Fora Formation (here considered as most of the Plio-Pleistocene sediments of the Koobi Fora region).

New K/Ar data, obtained by methods described by McDougall (1985) but distinct from those reported in his paper, are given in Table 2. These pertain largely to the Shungura Formation with a few determinations on materials from west of the lake. Paleomagnetic data are also available for many horizons, and these are used as constraints in placing some horizons. All of these data are summarized in Figure 2.

The Karsa Basalt, dated at 4.35 ± 0.05 Ma (McDougall, 1985), has the oldest relevant age in the Koobi Fora region and establishes a maximum age for the Koobi Fora Formation. On the west side of Lake Turkana near Kataboi, several basalt dikes intrude sediments that lie stratigraphically below the Moiti Tuff.An age of 4.05 ± 0.06 Ma on one of these dikes (81–793) is taken as evidence that the sediments below the Moiti Tuff at this locality are greater than 4 Ma old.

The Mursi Basalt overlies the fossiliferous sediments of the Mursi Formation and has been previously dated at 4.15 ± 0.2 Ma (Brown and Nash 1976). No new data are available for this flow, but a basalt from the sequence mapped as the Mursi Formation by Davidson (1983) has yielded an age of 3.99 ± 0.04 Ma (NK-46, Table 2), confirming that the Mursi Formation is greater than 4 Ma old.

The Usno Basalt is defined as Member U-1 of the Usno Formation. Two internally inconsistent ages on this flow were reported by Brown and Lajoie (1971). Two new determinations have been made on the same sample (WS-1, Table 2), which are internally consistent and markedly older than previous determinations. This basalt lies stratigraphically below all fossiliferous exposures of the Usno Formation, and provides a maximum age of 4.10 ± 0.06 Ma for the fossil faunas there.

The oldest tuff within the Koobi Fora Formation is the Moiti Tuff, dated at 4.10 ± 0.07 Ma by McDougall (1985); the age is regarded as a maximum age because of the small size of samples available and the consequent possibility of contamination. Nonetheless, this age is consistent with age measurements on other materials from the region.

The Toroto Tuff, which lies 4–10 m above the Tulu Bor Tuff in the southern part of the Koobi Fora region, has yielded an age of 3.32 ± 0.02 Ma (McDougall, 1985), which confines the Lokochot Tuff and the Tulu Bor Tuff to the interval 3.32–4.10 Ma. By interpolation on the basis of stratigraphic thickness the Tulu Bor Tuff (= Tuff B) is placed at 3.35 Ma.

For the interval between the Tulu Bor Tuff (= Tuff B) and the KBS Tuff (= Tuff H2), there are seven dated horizons: the Toroto and Ninikaa Tuffs at Koobi Fora, and Tuffs B-10, D, D-4, F and G in the Shungura Formation. These data are shown in Figure 2. Linkages between these two areas are provided by the Hasuma Tuff (defined here, Table 1) with Tuff C and the Ingumwai Tuff with Tuff C-4 of the Shungura Formation. We have estimated the ages of Tuffs C and C-4 by interpolation between the ages established for Tuff B-10 (Table 2) and Tuff D (Table 2) using stratigraphic thicknesses in the Shungura Formation.

The age of Tuff B-10 of the Shungura Formation appears to be very near 2.95 Ma, and a reversed magnetozone corresponding to the Kaena subchron should accordingly be found near this level. As no such reversed magnetozone was observed in this part of the for-

TABLE 2. Potassium–argon analytical data on basalt and anorthoclase separates from tuffs in the northern Turkana Basin[1]

Lab. no	Field no.	wt (gm)	$\%K^+$	$^{40}Ar_{rad}$ ($\times 10^{-11}$ moles/gm)	$^{40}Ar_{rad}$ (%)	Age and Error (Ma)
Mursi Formation Basalt						
82–24	NK-46	3.008	0.926, 0.916	0.639	54.0	3.99 ± 0.04
Usno Formation Basalt						
83–23–1	WS–1	7.000	0.691, 0.698	0.492	24.8	4.11 ± 0.06
82–23–2	WS–1	7.001	0.691, 0.698	0.489	23.6	4.08 ± 0.06
Kataboi Basalt						
83–19–1	81–793	7.001	0.679, 0.684	0.483	26.8	4.08 ± 0.05
83–19–2	81–793	7.001	0.679, 0.684	0.478	23.2	4.04 ± 0.06
Tuff B–10 (Shungura)						
83–5–1	B–10	1.500	5.052, 5.046	2.570	83.1	2.93 ± 0.03
83–5–2	B–10	1.001	5.052, 5.046	2.616	82.0	2.98 ± 0.03
Tuff D (Shungura)						
83–14–1	780–14D	1.001	5.129, 5.274	2.336	40.5	2.57 ± 0.03
83–14–2	780–14D	1.001	5.219, 5.274	2.279	83.2	2.50 ± 0.03
83–15	208–12	1.002	5.144, 5.177	2.257	90.5	2.52 ± 0.03
83–16	ET-40	1.000	5.157, 5.158	2.284	84.8	2.55 ± 0.03
83–7		1.500	5.526, 5.555	2.301	78.5	2.42 ± 0.03
83–8		1.000	5.151, 5.116	2.235	83.4	2.58 ± 0.03
83–6–1		1.502	5.304, 5.338	2.315	84.3	2.51 ± 0.03
83–6–2		1.001	5.304, 5.338	2.338	80.8	2.53 ± 0.03
Tuff D–4 (Shungura)						
83–12	780–Ei	1.502	5.237, 5.235	2.547	75.8	2.80 ± 0.03
Tuff F (Shungura)						
83–364–1		0.601	6.305, 6.220	2.527	87.2	2.32 ± 0.04
83–364–2		0.641	6.305, 6.220	2.596	86.2	2.39 ± 0.04
Tuff G (Shungura)						
83–10–1		1.502	5.846, 5.772	2.341	86.5	2.32 ± 0.03
83–10–2		1.203	5.846, 5.772	2.349	87.8	2.33 ± 0.03
83–31–1		0.509	5.855, 5.848	2.373	81.3	2.34 ± 0.03
Tuff H2 (Shungura)						
83–13–1		1.501	5.139, 5.115	1.629	74.9	1.83 ± 0.02
83–17–2		1.502	5.139, 5.115	1.633	75.5	1.84 ± 0.02
83–13–3		1.500	5.139, 5.115	1.646	78.7	1.85 ± 0.02
Unnamed tuff above Chari Tuff near Nariokotome, Turkana						
83–11	82–771A	1.601	5.096, 5.114	1.179	74.7	1.33 ± 0.02
83–12	82–771B,D	1.001	4.823, 4.802	1.111	49.0	1.33 ± 0.02
Unnamed tuff above Chari Tuff, Ileret, Koobi Fora Region						
83–3–1	82–835	1.500	4.653, 4.658	1.023	71.1	1.27 ± 0.01
83–3–2	82–835	1.501	4.653, 4.654	0.994	73.6	1.23 ± 0.01

[1] $\lambda_{\epsilon} + \lambda_{\epsilon'} = 0.581 \times 10^{-10}\ a^{-1}$; $\lambda_{\beta} = 4.962 \times 10^{-10}\ a^{-1}$; $^{40}K/K = 1.167 \times 10^{-4}$ mol mol^{-1}.

mation by Brown et al. (1978), it is likely that the Kaena event is not recorded, or was not sampled. The reversed magnetozone in submember B-2 of the Shungura Formation, which was formerly thought to represent both the Mammoth and Kaena subchrons, probably represents only the Mammoth subchron.

New measurements of the age of Tuffs D, F, and G are reported in Table 2; they are older than ages reported by Brown and Nash (1976). Comparison of data from the three laboratories involved (U.C. Berkeley, Lamont, and A.N.U.) shows that the argon determinations at Lamont were low by about 15% on average. Excluding the Lamont data, the mean age of Tuff D (N = 13) is 2.52 ± 0.11 Ma; the mean of the new determinations by themselves is 2.52 ± 0.05 Ma, which we regard as the best estimate of the age of Tuff D.

A new measurement on a finer fraction of the anorthoclase separated from a sand in submember D4 of Member D of the Shungura Formation yielded 2.80 ± 0.03 Ma, probably because of detrital contamination. We do not use this age in construction of the chronology shown in Figure 2.

New feldspar separates were prepared from an analyzed pumice from Tuff F of the Shungura Formation. Duplicate measurements on the newly separated material yielded a mean age of 2.36 ± 0.05 Ma, which is considerably older than found earlier.

Three new K/Ar measurements on two different samples have been made on anorthoclase from Tuff G, which yield a mean age of 2.33 ± 0.03 Ma. As mentioned, this is markedly older than the previous estimate of the age of Tuff G.

If, as it now appears, the interval between Tuff D and Tuff G corresponds to 2.3–2.5 Ma, the identification of the normal magnetozone in upper Member F as a Reunion event by Brown et al. (1978) must be erroneous, as all estimates of the age of the Reunion events are less than 2.2 Ma. Hence the normal magnetozones shown by Brown et al. (1978) within Member G (G-4–G-8; G-13) are more likely to represent the Reunion events. Using recent estimates of the age of the Reunion events (Mankinen and Dalrymple, 1979; McDougall, 1979), the base of the lacustrine sequence (G-14) in the Shungura Formation is about 2 Ma old. The normal magnetozone in upper Member F may possibly represent the X-anomaly of Heirtzler et al. (1968), near 2.3 Ma in age.

The hiatus shown in the section at Koobi Fora (Figure 2) is of uncertain duration, but it must be < 0.6 Ma in length. The principal evidence for its existence is the lack of any tephra layers at Koobi Fora that correlate with Tuffs E–G of the Shungura Formation. This is so, despite the fact that some tephra layers in this interval are known from the west side of Lake Turkana.

From the KBS Tuff to the top of the section in the Koobi Fora region there are five dated horizons—the KBS and Malbe Tuffs, both very near 1.88 ± 0.02 Ma; the Chari Tuff at 1.39 ± 0.01 Ma; a tuff between the Chari Tuff and Silbo Tuff in the Ileret Area dated at 1.25 ± 0.02 Ma; and the Silbo Tuff dated at 0.74 ± 0.01 Ma. All of the above dates are from McDougall (1985), except that on 82–835, for which the data are presented in Table 2.

Correlations have been made from the Koobi Fora region to the Shungura Formation at the levels of Tuff H2, H4, J7, and L. Relationships between the "Okote," Koobi Fora, and Middle Ileret Tuffs are still not completely clear. The correlation shown on Figure 2 is between *an* "Okote" Tuff and Tuff J7 of the Shungura Formation (Cerling and Brown, 1982). Dates on materials from the Shungura Formation are consistent with, but considerably less precise than those obtained on materials from the Koobi Fora region by McDougall (1985), hence the latter ages are applied to Tuffs H2, H4, and L. Previously, an age had been reported for Tuff K (Brown and Nash, 1976), but analysis of a glass separate from this material (Kar Olo) shows that the pumice actually derives from Tuff L of the Shungura Formation. There is no date on Tuff K.

K/Ar ages were determined on two anorthoclase separates from pumice clasts from a tuff near Nariokotome on the west side of Lake Turkana. This tuff lies about 12 m above the Chari Tuff. The determinations are in good agreement at 1.33 ± 0.02 Ma and consistent with its stratigraphic position.

DISCUSSION

The chronology developed above is applicable principally to faunas collected from the Turkana Basin, although even in that region control is poor for sites in the lower Kerio Valley. The recent discovery of several of these same ash horizons in the Gulf of Aden by Sarna et al. (1985) and in the Tulu Bor Tuff at Hadar (Brown, 1982; but see

Aronson et al. 1983) makes it likely that the chronology may be applied over much broader regions. The discovery of the Tulu Bor Tuff within the *Discoaster tamalis* zone in DSDP core 231 (Sarna et al., 1985) clearly places it <3.5 Ma, and it is hence very close in age to that of the Toroto Tuff.

There is still some difficulty in interpreting the chronology below the level of the Tulu Bor Tuff. The best estimate of the age of the correlative of the Lokochot Tuff in the Gulf of Aden cores is 3.55 Ma. Tuff A of the Shungura Formation (also correlative with the Lokochot Tuff) appears to be very near a polarity transition, in which case it should be somewhat younger (near 3.42 Ma). Still, it may be that the upper (normally magnetized) part of the Lokochot Tuff has been remagnetized, and that the reversed polarity of the lower part indicates that it was deposited during the Gilbert Chron ($\geqslant$3.42 Ma). As most hominid remains collected in the Turkana Basin are derived from levels above the Tulu Bor Tuff, these chronological problems apply to very few of the specimens.

Although our knowledge of the Kanapoi-Ekora region is limited, we have analyzed a tuff from that area that is so different in composition from the other tuffs of the Turkana Basin that it almost certainly reflects derivation from a distinct source. The trachytic volcanoes northwest of Lake Baringo described by Webb and Weaver (1976) were active from 3–5 Ma ago and are compositionally similar to the Kanapoi ash. This ash horizon may also exist in the Baringo Basin, and, if identified there, will provide an important link between that area and the Turkana Basin.

Discrepancies between the chronology of the Shungura Formation and that of the Koobi Fora Formation that were the subject of much debate have now largely disappeared. Even so, considerable refinement of the dating of the stratigraphic sequences in the Lake Turkana region is still possible, and work toward this end is presently in progress.

ACKNOWLEDGMENTS

The work reported herein was materially supported by the Research School of Earth Sciences of the Australian National University and by NSF grant BNS-8210735. R. Rudowski is thanked for his help with mineral separation. Suzanne Zink efficiently and cheerfully typed the manuscript and made numerous changes. Paul Onstott drafted the illustrations. The continued good will of the government of Kenya and logistic support of the National Museum of Kenya is gratefully acknowledged.

LITERATURE CITED

Aronson, JL, Walter RC, and Taieb, M (1983) Correlation of Tulu Bor Tuff at Koobi Fora with the Sidi Hakoma Tuff of Hadar. Nature *306:*209–210.

Behrensmeyer, AR (1976) Lothagam Hill, Kanapoi, and Ekora: A general summary of stratigraphy and faunas. In Y Coppens, FC Howell, GL Isaac, and REF Leakey (eds): Earliest Man and Environments in the Lake Rudolf Basin. Chicago: University of Chicago Press, pp. 163–170.

Brown, FH (1982) Tulu Bor Tuff at Koobi Fora correlated with the Sidi Hakoma Tuff at Hadar. Nature *300:*631–632.

Brown, FH, and Cerling, TE (1982) Stratigraphical significance of the Tulu Bor Tuff of the Koobi Fora Formation. Nature *299:*212–215.

Brown, FH, and Lajoie, KR (1971) Radiometric age determinations on Pliocene/Pleistocene formations in the lower Omo basin, southern Ethiopia. Nature *299:*483–485.

Brown, FH, and Nash, WP (1976) Radiometric dating and tuff mineralogy of Omo group deposits. In Y Coppens, FC Howell, GL Isaac, and REF Leakey (eds): Earliest Man and Environments in the Lake Rudolf Basin. Chicago: University of Chicago Press, pp. 50–63.

Brown, FH, Shuey RT, and Croes, MK (1978) Magnetostratigraphy of the Shungura and Usno Formations, southwestern Ethiopia: New data and comprehensive reanalysis. Geophys. Jour. Roy. Astr. Soc. *54:*519–538.

Butzer, KW (1976) The Mursi, Nkalabong, and Kibish Formations, Lower Omo Basin, Ethiopia. In Y Coppens, FC Howell, GL Isaac, and REF Leakey (eds): Earliest Man and Environments in the Lake Rudolf Basin. Chicago: University of Chicago Press, pp. 12–23.

Cerling, TE, and Brown, FH (1982) Tuffaceous Marker Horizons in the Koobi Fora Region and the Lower Omo Valley. Nature *299:*216–221.

Davidson, A (1983) The Omo River Project: Reconnaissance Geology and Geochemistry of Parts of Ilubabor, Kefa, Gemu Gofa and Sidamo, Ethiopia. Bulletin no. 2., Ministry of Mines and Energy, Ethiopian Institute of Geological Surveys.

de Heinzelin, J, Haesaerts, P, and Howell, FC (1976) Plio-Pleistocene formations of the lower Omo basin with particular reference to the Shungura Formation. In Y Coppens, FC Howell, GL Isaac, and REF Leakey (eds): Earliest Man and Environments in the Lake Rudolf Basin. Chicago: University of Chicago Press, pp. 24–49.

Findlater, I (1976) Tuffs and the recognition of isochronous mapping units in the East Rudolf succession. In Y Coppens, FC Howell, GL Isaac, and REF Leakey (eds): Earliest Man and Environments in the Lake Rudolf Basin. Chicago: University of Chicago Press, pp. 94–104.

Findlater, I (1978) Isochronous surfaces within the Plio-Pleistocene sediments east of Lake Turkana. In WW Bishop (ed): Geological Background to Fossil Man. Edinburgh: Scottish Academic Press, pp. 415–420.

Fitch, FJ, and Miller, JA (1976) Conventional potassium-argon and argon-40/argon-39 dating of volcanic rocks from East Rudolf. In Y Coppens, FC Howell, GL Isaac, and REF Leakey (eds): Earliest Man and Environments in the Lake Rudolf Basin. Chicago. University of Chicago Press, pp. 123–147.

Heirtzler, R, Dickson, GO, Herron, EM, Pitman III, WC, and Le Pichon, X (1968) Marine magnetic anomalies, geomagnetic reversals and motions of the ocean floor and continents. J. Geophys. Res. *73:*2119.

Mankinen, EA, and Dalrymple, GB (1979) Revised geomagnetic polarity time scale for the interval 0–5 m.y. BP. Jour. Geophys. Res. *84:*615–626.

McDougall, I (1979) The present status of the geomagnetic polarity time scale. In MW McElhinny (ed): The Earth: Its Origin, Structure and Evolution. London: Academic Press, pp. 543–566.

McDougall, I (1985) K-Ar and $^{40}Ar/^{39}Ar$ dating of the hominid bearing Plio-Pleistocene sequence at Koobi Fora, Lake Turkana, northern Kenya. Geol. Soc. Amer. Bull. *96:*159–175.

Patterson, B, Behrensmeyer, AK, and Sill, WD (1970) Geology and fauna of a new Pliocene locality in northwestern Kenya. Nature *226:*918–921.

Powers, D (1980) Geology of Mio-Pliocene sediments of the lower Kerio Valley, Kenya. Unpublished Ph.D. Dissertation, Princeton University.

Sarna, AM, Meyer, C, Roth, PH, and Brown, FH (1985) Correlation of tephra horizons from East African hominid sites to deep sea cores. Nature *313:*306–308.

Vondra, C, and Bowen, BE (1978) Stratigraphy, sedimentary facies and paleoenvironments, East Lake Turkana, Kenya. In WW Bishop (ed): Geological Background to Fossil Man. Edinburgh: Scottish Academic Press, pp. 395–414.

Webb, PK, and Weaver, SD (1976) Trachyte shield volcanoes: A new volcanic form from South Turkana, Kenya. Bull. Volcanol. *40:*294–312.

Ancestors: The Hard Evidence, pages 91–93

Pliocene Hominids

Michael H. Day
Division of Anatomy, United Medical and Dental Schools of Guy's and St Thomas's Hospitals, University of London, London, England

This session was opened by Phillip Tobias, whose paper begins with an unequivocal statement: "Palaeo-anatomical studies are the most important single means of sorting assemblages of fossils." This clarion call to consider more precisely and meticulously the anatomy of our fossils echoes the statement of Le Gros Clark twenty years before, that the total morphological pattern, and not its individual units, provides the reliable evidence on which zoological relationships can be determined. Implicitly and explicitly morphometric methods (Oxnard, 1975) and Hennig's (1966) taxonomic system were relegated to being simply regarded as adjunct methods for interpreting morphological observations that make some contribution to the formulation of taxonomic and phylogenetic conclusions. It was suggested that all too often far-reaching taxonomic and phylogenetic claims are made that are based upon insufficient morphological comparison of one trait, one region, or one complex of the body.

This firm statement set the tone for the session, and as the day proceeded it was clear that this call for restraint was not in the least unjustified. It is possible to erect a cladogram on small anatomical differences that may well be within the range of normal variation of the species, yet declare this as a basis for phylogenetic speculation; on the other hand, it is possible by *gestalt* perception, almost alone, to declare a taxonomic evaluation and erect a phylogeny for consideration. Examples of both of these approaches can be found in the papers within this section if one looks carefully.

Tobias's plea was succeeded by a consideration of seven out of 148 features from the brain, the cranium, the mandible, and the teeth as discriminators between hominid taxa. He showed that each trait can separate in some instances and not in others. This argument was used to warn Olson (1978, 1981) and Ward and Kimbel (1983) against incautious claims from restricted areas of study. (It is not often that opponents get their heads knocked together before the fight begins!)

Following the opening paper there was a sharp debate between Todd Olson and the Berkeley group, represented on this occasion by Bill Kimbel. Olson's view of the Hadar and Laetoli material is expressed by his recognition of *Paranthropus* and *Homo* as two genera present in the Pliocene in Africa that are derived from a previous common ancestor. This position leads him to identify two species at Hadar and Laetoli, *contra* the single species position (*A. afarensis*) held by the Berkeley group. Olson's views are based on occipital anatomy, dental anatomy, and details of the nasal sutural patterns. There is no doubt that once the Hadar material was fully published the range of size variation in the sample was seen to be very large and may be greater than that usually acceptable as being due to sexual dimorphism. In addition, the relevant Hadar finds have skull details that appear to recall the anatomy of robust australopithecines. This is denied by Kimbel et al., who suggest that it is based on a misinterpretation by Olson of cranial base anatomy, in particular his view of the occipitomastoid crest and the digastric muscle's origin. In effect, this paper was a set-piece rebuttal of Todd Olson's known views and a restatement of the new "single species hypothesis" of hominid origins (that species being *A. afarensis*). This line of argument was continued by White in terms of the dental similarities between the Hadar and Laetoli samples.

The cranial and dental evidence for australopithecine evolution was discussed further by Michel Sakka (the occipital region),

Fred Grine (the deciduous dentition), and Yoel Rak (the facial region). Rak's view of the australopithecine face was encapsulated by his selection of five features of the masticatory system that he believes show that *A. africanus* is on the robust line rather than that which leads on to early *Homo*. These features include dishing of the face, infraorbital inclination, zygomatico-alveolar crests, palatal retraction, and, most dramatically, the anterior pillars of the face that border the pyriform aperture of the nose. In the discussion Yoel Rak confirmed his view that the anterior pillars of the face correlate with proportionate dental size but are not features of KNM ER 1470. Fred Grine was also asked if his deciduous dental studies supported the view that *A. africanus* is a sister group of the robust australopithecine lineage or of the *Homo* lineage. He suggested that it is inappropriate to form a sister group relationship between *A. africanus* and early *Homo* since the retention of primitive characters may not indicate any special phylogenetic relationship. Michel Sakka, having described the complexity of the occipital musculature of the great apes, was reluctant to hazard any functional reasons other than the complex nature of head movement. Unfortunately, his manuscript was not submitted for publication in this volume.

Ron Clarke reported on his earlier reconstruction of SK 847 (the composite cranium) and suggested it may merit identification as a new subspecies of *Homo erectus*, being regarded as the root of this species. Both SK 847 and ER 3733, Clarke said, have convex cranial bases, as compared to the "unusual concave or depressed" base he observed in Asian and later African *erectus* specimens (see also discussions by Maier and Laitman, in succeeding sections). Clarke regards his reconstruction of Stw. 53 as "almost identical" with Olduvai Hominid 24, and he thus assigns Stw. 53 to *Homo habilis*. He noted that in some *A. africanus* faces one can see the premaxillary suture running alongside the nasal aperture, while in Stw. 53 it parallels the nasal bones and then curves into the nasal cavity. This is another feature in which *H. habilis* would be intermediate between *A. africanus* and later *Homo*, especially *H. sapiens* (with no premaxilla visible on the face). In sum, Clarke now regards Sterkfontein as having both *A. africanus* (in Member 4) and *H. habilis* (in Member 5), while Swartkrans (later in time) has yielded both *Paranthropus* and *H. erectus* from Member 1 (the "hanging remnant") and perhaps from younger horizons (see also Brain, in the preceding section).

Phylogenetic questions were approached by Henry McHenry, who discussed the position of *A. africanus* and the succeeding phylogenetic lines leading to *Homo* and *A. robustus/boisei*. He favoured the position of *A. africanus* as the point of divergence of the two lines, thus contributing to both, rather than a placement on the robust line alone. This is despite postcanine megadontia that seems to make *A. africanus* specialised in the direction of *A. robustus/boisei*. *Homo habilis* is thus seen to evolve from a species with a complex of traits associated with well-developed postcanine megadontia; subsequently, the whole *Homo* lineage can be characterised by dental reduction.

The postcranial contributions came from the Stony Brook (SUNY) group and from Paris. Randy Susman stated clearly and without any equivocation that the Laetoli footprints are acceptable as clear evidence of bipedalism. It is a great relief to us all that the battles of the past do not have to be refought all over again. There seems to be no disagreement that the earliest evidence of bipedalism is in the form of footprints the shape of which reveals functions that are close to those of modern man. One conflict of evidence still arises, however. The footprints show short toes, but the Hadar specimens provide longer fingers and toes and curved phalanges, yet straight metacarpals. Susman et al. do not consider these features to be relicts of previous arborealism and feel that they must indicate a level of arboreal adaptation and presumably activity in the Hadar hominids.

In discussion, the two postcranial morphologies described from Hadar and referred to previously by Susman and his colleagues were questioned in terms of their range of variation in relation to sex. Jack Stern stated that the range of variation, if it is sexual dimorphism, is greater in some aspects of the Hadar sample than in any other extant primate. Despite this, he inclined to the belief that the anatomical dimorphism *is* great in this sample and that it corresponds to both behavioural and sexual dimorphism. This is a simple and clear statement that the range of variability in this group is larger than that of any other known modern primate group—and if this is special pleading, so be it. Brigitte Senut, whose paper also covered the work of Christine Tardieu, clearly did not agree, and the question remains unsolved.

The questions that concern Pliocene hominids raised at this meeting therefore relate to the credibility of *A. afarensis* as a single taxon from two sites widely differing in both time and space. Evidence was adduced either way, and some of it was rebutted. The occipito-mastoid region, the nasal sutures, and the venous sinuses of the posterior cranial fossa have all attracted attention in this regard in recent years. The postcranial evidence was not rebutted, however, and it was left for us all to decide whether the special pleading for a wider range of Hadar sexual dimorphism than previously known in primates is a convincing position to sustain.

The general taxonomic position of the australopithecines seems as confused as ever. While some incline to *A. africanus* as being ancestral to the robust line (Rak, Kimbel et al.), others (McHenry, Clarke) favour a central position for *A. africanus* as perhaps ancestral to both the *Homo* lineage and the robust line of australopithecines. Olson, and less definitively Clarke, have resuscitated *Paranthropus* to receive the robust species, in the former case allocating *A. africanus* (including *H. habilis*) to *Homo*. The specific identity of the Kromdraai and Swartkrans robust forms was questioned by Grine, following past suggestions by Clark Howell (e.g., 1978). In terms of functional analysis, the unequivocal bipedalism of the Pliocene early hominids combined with their possible arboreal adaptations also remains an area for further speculation and research.

We can do worse than take Tobias's homily to heart: let us look at the anatomy more closely, let us consider features in groups, as complexes and as a pattern, and then—and only then—come to balanced judgments on functional and taxonomic issues.

LITERATURE CITED

Hennig, W (1966) Phylogenetic Systematics. Urbana: University of Illinois Press.

Howell, FC (1978) Hominidae. In VJ Maglio and HBS Cooke (eds): Evolution of African Mammals. Cambridge: Harvard University Press, pp. 154–248.

Olson, TR (1978) Hominid phylogenetics and the existence of *Homo* in Member 1 of the Swartkrans Formation, South Africa. J. Hum. Evol. *7*:159–178.

Olson, TR (1981) Basicranial morphology of the extant hominoids and Pliocene hominids: The new material from the Hadar Formation, Ethiopia, and its significance in early human evolution and taxonomy. In CB Stringer (ed): Aspects of Human Evolution. London: Taylor and Francis, pp. 99–128.

Oxnard, CE (1975) Uniqueness and Diversity in Human Evolution. Morphometric Studies of Australopithecines. Chicago: University of Chicago Press.

Ward, SC, and Kimbel, WH (1983) Subnasal alveolar morphology and the systematic position of *Sivapithecus*. Am. J. Phys. Anthropol. *61*:157–171.

Ancestors: The Hard Evidence, pages 94–101

Single Characters and the Total Morphological Pattern Redefined: The Sorting Effected by a Selection of Morphological Features of the Early Hominids

Phillip V. Tobias

Department of Anatomy, School of Medicine, University of the Witwatersrand, Parktown 2193 South Africa

ABSTRACT Of 148 traits at the head end of a hominid, listed elsewhere by the author, seven are here selected. They are the inion/opisthocranion relationship, recesses of the mandibular fossa, curvature of the anterior wall of the mandibular fossa, mandibular robusticity, maxilloalveolar index, Frisch's M^3/M^1 mesiodistal diameter index, and cranial capacity (updated). Data are presented on the efficacy of each character in sorting hominids. It is shown that the characters express themselves variously among hominid taxa. For example, mandibular robusticity fails to separate *Homo habilis* from australopithecines, but distinguishes these groups from *H. erectus* and *H. sapiens*. Five other traits distinguish *H. habilis* from australopithecines, while one separates *H. habilis* from *A. africanus* but not from *A. robustus* and *A. boisei*. Hence a classification and phylogeny based on a single character, complex, or region would differ according to the trait or complex used. In many recent studies on hominid fossils, broad systematic and phylogenetic inferences have been drawn from single characters, complexes, or regions, in virtual disregard of evidence from other regions. Two pleas are made: 1) more careful palaeoanatomical and comparative studies are needed on hominid fossils, and 2) systematic and evolutionary inferences should be based on as much of the total morphological pattern as is preserved (Le Gros Clark's original approach being modified, where applicable, by cladistic, genetic, ontogenetic, and functional considerations).

INTRODUCTION

Palaeo-anatomical studies are the most important single means of sorting assemblages of fossils. Another strategy of immense value is the deployment of molecular data from living creatures, but the applicability of this approach is limited to only two phases. The first is the origin of the Hominidae from the latest common ancestor of hominids and pongids (Tobias, 1975a; Pilbeam, 1983; Greenfield, 1983; Doolittle, 1983), and the second is the most recent phase of hominid evolution, from the late Middle and Late Pleistocene through the present (cf. Nei and Roychoudhury, 1982). Between these extremes—that is, between about 5 million years (m.y.) and 0.125 m.y. BP—the molecular approach has little to contribute, save for the study of remnants of proteins in the fossils themselves (Lowenstein, 1983).

Since most African fossil hominids are dated between these time brackets, we are obliged, in assessing their affinities, to depend on palaeo-anatomical studies. Many analytical refinements for the handling of morphological observations, metrical and non-metrical, are available. These include: distance statistics, discriminant and cluster analyses, and other morphometric strategies (cf. Oxnard, 1975); attempts to determine which traits are primitive (plesiomorphic) and which derived (apomorphic), which uniquely so and which as shared between related groups; the evaluation of functional, genetic, and ontogenetic aspects; and the use of Le Gros Clark's (1964) concept of the *total morphological pattern.*

Whatever analytical refinements we use, critical and informed observation is the essential starting point of any study of fossils. Since claims have often been based on inadequate or superficial morphological comparisons and phylogenies erected on questionable anatomical grounds, my paper makes a twofold call: first, for more precise and meticulous observations and measurements, and more careful and incisive comparisons, of the fossils; and second, for restraint in the drawing of systematic and phylogenetic conclusions from studies on a single trait, complex, or region of the body.

At the 1978 Nobel Symposium, a list of 148 morphological variables of the head-parts, which may be gleaned from the fossil record and which I had found useful in helping to distinguish among the early hominids, was given for the first time (Tobias, 1980). The list was described as "comprehensive though not exhaustive" and included both single traits and clusters of features. They were distributed as follows: brain (endocast), 12; cranium, 94; mandible, 18; and teeth, 24. Not all of the 148 features were necessarily independent variables, from a genetic or epigenetic developmental viewpoint; nevertheless, it was hoped that the list might prove helpful in the study, description, and comparison of the many new hominid skulls that have been emerging from the soil of the Old World.

In this paper, I have selected seven features from that list, one on the brain (endocast), three on the cranium, one on the mandible, and two on the dentition. This is an ad hoc selection, spanning the major cranial regions. For each feature, some data on its variable expression in different fossil hominids are presented, in order to show how the trait helps to discriminate among various hominid taxa.

THE RELATIONSHIP BETWEEN INION AND OPISTHOCRANION

The position of inion represents the highest extent in the median plane of the nuchal muscular attachment area. Opisthocranion is a point on the calvaria in the mid-sagittal plane, which lies furthest from glabella.

When crania are held in the Frankfurt Horizontal, the relative positions of these two points are seen to vary among different hominoids. In pongid crania inion is higher than opisthocranion. On a selection of early hominid crania, the positions are as follows:

Australopithecus africanus Sts 5: inion coincident with opisthocranion.
MLD 1: inion lower than opisthocranion (Dart); inion coincident with opisthocranion (Robinson).
MLD 37/38: inion lower than opisthocranion (Dart); inion approximately coincident with opisthocranion (Tobias).

A. boisei OH 5, KNM-ER 406, KNM-ER 1805: the points coincide.

Homo habilis OH 13, OH 16, OH 24: inion lower than opisthocranion.
KNM-ER 1470: inion appears to lie below opisthocranion.
KNM-ER 1813: inion lower than opisthocranion.

Homo erectus (including KNM-ER 3733): inion and opisthocranion coincide.

H. sapiens soloensis (Ngandong): inion and opisthocranion coincide.

H. sapiens rhodesiensis (Kabwe): inion and opisthocranion coincide.

H. sapiens neanderthalensis: in most, inion lies lower than opisthocranion.

Modern *H. sapiens*: inion almost always lower than opisthocranion.

Dart (1948, 1962) claimed a separation between the points in two Makapansgat specimens, MLD 1 and MLD 37/38, but Robinson (1954) suggested that the isolated parieto-occipital fragment, MLD 1 (on which the opisthocranion can be only estimated and not precisely determined), could better be oriented with a steeply-sloping planum nuchale, as in Sts 5, in which position inion and opisthocranion coincide. I agree with Robinson's interpretation and have determined that the same probably applied to MLD 37/38 as well (Tobias, 1985). The two points certainly coincide in *A. robustus* and *A. boisei* crania, as in *H. erectus*, including KNM-ER 3733 (see figures in Leakey and Walker, 1976), and in the Ngandong and Kabwe (Broken Hill) crania (Weidenreich, 1943, 1945).

If we accept Robinson's and my views on MLD 1 and MLD 37/38, it seems that in *Australopithecus* inion generally coincides with opisthocranion. In the bigger-brained group, assigned to *H. habilis*, opisthocranion, with cerebral expansion, has seemingly risen above inion, as in crania of *H. sapiens*. The development of a heavy occipital torus, as in *H. erectus*, accompanies expansion of the nuchal muscles, and this morphogenetic process appears to carry inion upward, apparently secondarily, to coincide with opisthocranion. A polarity analysis here may go badly awry if the underlying morphogenetic events are disregarded.

RECESSES OF THE MANDIBULAR FOSSA

In the mandibular fossa of *H. erectus pekinensis*, Weidenreich (1943, p. 47) spoke of "a very strange feature obviously peculiar to *Sinanthropus*"; namely, the extension of the posteromedial part of the mandibular fossa as a narrow recess, which looks like a cleft between the squama and tympanic plate. He called it the *medial recess of the mandibular fossa*. He construed it to be the consequence of the medial wall's convexity, rather than concavity, and related this, in turn, to the anteroposterior compression of the fossa and the degree of perpendicularity of the tympanic plate.

In australopithecines, too, there is a convex medial wall, only this is formed by the bulge of the entoglenoid process. It is especially clear in Sts 5, Sts 19, MLD 37/38, OH 5, and KNM-ER 406. Because the anteroposterior diameter of the fossa is very large, and since the medial part of the tympanic plate is anteriorly directed, a recess is present, though of a very different character from that in *H. erectus*. It extends forward medially between the entoglenoid process and the anteriorly directed part of the tympanic plate, as a definite recess for which I have proposed the name the *anteromedial recess of the mandibular fossa*. This feature is not homologous with the medial recess of *H. erectus pekinensis*, nor has it been described previously, though a groove in the medial half of the mandibular fossa of Sts 5 was noted by Broom and Robinson (1950, p. 21).

In the position of the anteromedial recess of *Australopithecus*, pongids have only a fissure, as is true of *H. erectus* of Zhoukoudian and Olduvai, as well as *H. sapiens*. Of all the australopithecine crania examined, TM 1517 of Kromdraai shows the anteromedial recess to the least degree.

In *H. habilis* (as represented by the putatively female cranium of OH 24), there is a clear medial recess, and no anteromedial recess—that is, the structure resembles that of *H. erectus* rather than that of *Australopithecus*. In the position that would be occupied by an anteromedial recess there is only a simple fissure. The relevant parts are not preserved in OH 13 and OH 16.

Partly, at least, the structure of this part of the mandibular fossa of OH 24 seems to be a morphogenetic consequence of anteroposterior compression of the fossa. This has crowded the tympanic forward towards the entoglenoid, obliterated the anteromedial recess and converted it into a fissure, and created between the tympanic and the entoglenoid convexity a medial recess exactly like that in the Zhoukoudian crania.

THE MEDIAL END OF THE ANTERIOR WALL OF THE MANDIBULAR FOSSA

A complex of features characterises the anterior wall of the mandibular fossa of all hominids, ancient and modern (Tobias, 1967, Ch. 4). Variations affect chiefly the slope of the wall, the degree to which its lateral end forms a posteriorly directed convexity, and the amount of curvature in the wall, especially the degree to which its medial end turns posteriorly.

In three *H. habilis* crania from Olduvai—OH 13, OH 16, and OH 24—the medial part of the anterior wall turns sharply posteriorly, so that this part of the wall faces more laterally than posteriorly. This medial recurvation is strong, too, in *A. robustus* and *A. boisei* (e.g., TM 1517 from Kromdraai, OH 5 from Olduvai, and KNM-ER 406 from Ileret), but it is weakly developed in *A. africanus* (e.g., Sts 5 and MLD 37/38). In *H. erectus* and in modern man this feature is variable, though most recent human crania examined have only a weak medial curvature of the anterior wall, so that the medial part of the surface faces posterolaterally, rather than laterally. Nevertheless, in some modern human crania a strong medial recurvation of the anterior wall is encountered, of as marked a degree as in *H. habilis* crania.

The morphogenetic aspects of this variate remain to be determined. The relationship between the head of the mandible and the size and form of the mandibular fossa does not offer a ready explanation here, since hominids with large mandibles and expanded condylar processes (e.g., *A. boisei*) and some with modest jaws and reduced condylar processes (e.g., *H. habilis*) share the trait, whereas *A. africanus*, with mandibles of intermediate sturdiness, barely partakes

TABLE 1. Mean robusticity index of mandible at M_1

Australopithecus		
A. boisei of East Africa		ca. 72.3 (n = 16)
A. robustus of Swartkrans		ca. 66.9 (n = 5–7)
A. robustus of Kromdraai		66.3 (n = 1)
A. africanus of Transvaal		ca. 63.1 (n = 7)
Hominids of Hadar		61.5 (n = 10)
Laetoli		61.1 (n = 1)
Homo habilis		
Of Olduvai		67.3 (n = 2)
Mixed "early *Homo*" of East Turkana		69.2 (n = 9)
Homo erectus and miscellaneous		
H.e. pekinensis		59.6 (n = 6)
H.e. of Africa		58.8 (n = 7)
"*P. dubius*" and *Meganthropus*		58.1 (n = 2)
H.e. lantianensis		58.1 (n = 1)
H.e. (?) of Europe		57.9 (n = 2)
H.e. erectus of Java		54.2 (n = 3)
Homo sapiens neanderthalensis		49.3 (n = 8)
Homo sapiens sapiens	♂	♀
Fushun Chinese	38.4	41.7
Koreans	39.4	42.4
South African Negro	46.2	47.0

of it. In this feature, from the scanty available evidence, *H. habilis* appears to have a structure somewhat closer to that of *A. robustus* and *A. boisei* than to that of *A. africanus*.

MANDIBULAR ROBUSTICITY

The robusticity index of the mandible is usually taken at the mesiodistal mid-tooth position of M_1, relating the width to the height of the corpus of the mandible at that point. Such an index may be determined at a variety of positions from the symphysis menti to the M_3, and patterns of variation along the tooth-row have been studied.

Table 1 gives mean values of the robusticity index at M_1 for site samples and populations of early and recent hominids. The data are grouped in Table 2. Read from above downwards, the ranges of sample means show a definite tendency towards a reduction in the means, *Australopithecus* having the highest values (i.e., the most robust mandibles) and modern man the lowest values. However, the *H. habilis* mandibles of Olduvai and the mixed "early *Homo*" mandibles of East Turkana have sample means that fall within the range of sample means for *Australopithecus*. In other words, despite their generally smaller construction, *H. habilis* jaws are as robust as australopithecine ones, as assessed by the robusticity index at M_1.

One morphogenetic factor that may be relevant is the relationship of the corpus mandibulae to the size of the roots in it. Although the tooth crowns of *H. habilis*, especially

TABLE 2. Robusticity index of mandible at M_1

Taxon	Grouped ranges of sample means
Australopithecus	61.5–72.3
H. habilis	67.3–69.2
H. erectus	54.2–59.6
H. sapiens (Late Pleistocene: Africa and Europe)	46.8–49.3
H. sapiens sapiens	38.4–47.0

from P_3 to M_1, are reduced as compared with those of *A. africanus*, it appears that the roots have remained virtually as sturdily developed in *H. habilis*. This may account for the retention of a rugged mandibular body in *H. habilis*.

THE MAXILLO-ALVEOLAR INDEX

The definitions of the maxillo-alveolar breadth and length employed here are those of Flower (1881). A number of South and East African hominid maxillae have been directly measured by me and their indicial values have been determined (Tobias, 1985).

The values of the index in *A. africanus* crania are 91.3 in Sts 5, ca. 97.0 in Sts 17, and 98.5 in Sts 53, the mean being 95.6% (n = 3). The values in *A. robustus crassidens* are 92.5 in SK 46, 95.9 in SK 48, and 99.0 in SK 79, with a mean of 95.8 (n = 3), virtually identical with the *A. africanus* mean. For *A. boisei* we have values of 95.0 in OH 5, ca. 92.8 in Chesowanja 1, and 93.0 in KNM-ER 406 (mean 93.6, n = 3). The lower mean for *A. boisei* reflects the excessive lengthening of its maxillary alveolar arch. On several specimens only one of the two measurements required for this index can be taken. Hence, the percentage ratio of the mean dimensions has been calculated for each taxon. The values are 97.2 for *A. africanus* (n = 3–6), 97.1 for *A. robustus* (n = 5–6), and 90.2 for *A. boisei* (n = 3–4). These ratios confirm the distinction in maxillo-alveolar proportions between *A. africanus* and *A. robustus* on the one hand and *A. boisei* on the other.

The readings for nine crania of *Australopithecus* thus range from 91.3 to 99.0. OH 24 with ca. 113.7 is the only early pre-*erectus* hominid with an index value of over 100. It approximates the readings in *H. erectus*, for which Weidenreich (1945) cited readings for Sangiran 4 of 104.0 (unrestored) and 116.0 (restored) and for *H. erectus pekinensis* (reconstructed) of 107.6. Similarly, values for three fossil *H. sapiens* (including Neandertal crania) are 101.2 to 116.2 (Weidenreich, 1943). The means for modern human populations range from 108.2 to 126.0 (Martin, 1928). Thus, the high value in OH 24—well over 100.0%—clearly separates it from the australopithecine crania and aligns it with the central tendencies of samples of crania of *Homo*.

FRISCH'S M^3/M^1 LENGTH INDEX

Frisch's (1965) Index expresses the mesiodistal diameter of M^3 as a percentage of that of M^1, giving a gauge of the degree of reduction of the third molar. Three Olduvai *H. habilis* specimens (OH 13, OH 16, and OH 24) have values of 96.9, 90.1, and 91.9, respectively. The Omo specimen L 894-1 of the Upper Shungura Formation, which Boaz and Howell (1977) assigned to *H. habilis* (or *H. modjokertensis*), has a value of 96.8, while KNM-ER 1813 from East Turkana (Day et al., 1976)—which I believe also belongs to *H. habilis*—has a value of 95.1. Thus, all five of these *H. habilis* specimens show M^3 reduction.

On the other hand, the australopithecines show values greater than 100.0. Four Sterkfontein specimens that possess both relevant teeth, on one or on both sides, have values ranging from 102.0 to 115.3, with a sample mean of 110.6% (n = 4). If one uses the ratio of means, the value is 106.3 (n = 13–17). For *A. robustus* of Swartkrans, the mean value is 107.1 (n = 5), while the ratio of the means for *A. robustus* (from both Kromdraai and Swartkrans) is 110.3 (n = 18). The mean for *A. boisei* is 109.4% (n = 3).

The *H. habilis* group, which is so distinct in this regard from the australopithecines, is more closely related to *H. erectus* with values of 87.8 for Sangiran 4 and 88.1 for Zhoukoudian (ratio of means), and to recent *H. sapiens* for which a sample of populations yields ratios of means that range from 84.3 to 89.7%.

UPDATE ON CRANIAL CAPACITY

Revised estimates have been made of the cranial capacities of the four Olduvai *H. habilis* crania, OH 7, OH 13, OH 16, and OH 24. The revisions include estimates of the "adult values" for the first three, which were of juvenile status at the time of death. Moreover, other evidence—detailed elsewhere (Tobias, 1985)—has led to the conclusion that OH 7 and OH 16 were male and OH 13 and OH 24 female. These revisions are summarized in Table 3.

Two determinations for East Turkana specimens that I consider to belong to the same taxon, *H. habilis*, are available. They are 750 cm^3 for KNM-ER 1470 and 510 cm^3 for KNM-ER 1813 (Holloway, 1983). If those two specimens do belong to the same taxon as the four from Olduvai, they widen the sample range and the variability of *H. habilis* cranial capacities. The sample range for the six specimens from Olduvai and East Turkana extends from 510 to 750 cm^3, and the combined sample mean is 640 cm^3.

If ER 1470 is adjudged male and ER 1813 female, as seems most probable on evidence other than that of cranial capacity, the mean for three putative males from Olduvai and East Turkana is 687 cm^3 and for three presumptive females 592 cm^3. The index of sexual dimorphism based on these small samples is 13.8% (of the putative male mean). This index compares well with corresponding values of 13.8 for orang-utan, 14.8 for gorilla, and 10.6 for 67 modern human populations and is twice as great as the value of 6.9 for chimpanzee (Tobias, 1971a, 1971b, 1975b).

The *H. habilis* mean of 645 cm^3 for four Olduvai *H. habilis* crania, or of 640 cm^3 for six Olduvai and East Turkana presumptive *H. habilis* crania, is appreciably greater than the mean for *A. africanus* crania. The mean for six "adult" cranial capacities attributed to *A. africanus* is 442 cm^3 (Holloway, 1970) or 441 cm^3 (Tobias, 1975b). The standard deviation for *A. africanus* has been computed at 21.6 cm^3 (Tobias, 1985). The *H. habilis* mean of 640 cm^3 (n = 6) is 199 cm^3 or 9.2 S.D.s greater than that for *A. africanus* (n = 6). For individual specimens of *H. habilis*, the standardized deviations from the *A. africanus* sample mean are: KNM-ER 1813, + 3.2 S.D.s; OH 24, + 7.1 S.D.s; OH 16 +

TABLE 3. Revised estimates of cranial capacity for Olduvai Homo habilis

Specimen	Putative sex	Capacity as estimated (cm^3)	"Adult" capacity estimated (cm^3)
OH 7	♂	647	674
OH 13	♀	659[1]	673[1]
OH 16	♂	622	638
OH 24	♀	594	594
Mean for two presumptive males	♂	—	656
Mean for two presumptive females	♀	—	633.5
Olduvai mean (n = 4)	♂ + ♀	—	645

[1]These are mid-values between two new estimates of present capacity and two of "adult value" (Tobias, 1985).

9.1 S.D.s; OH 13, + 10.7 S.D.s; OH 7, + 10.8 S.D.s; KNM-ER 1470, + 14.3 S.D.s.

Whether we consider just the four Olduvai crania or all six of these specimens as representing *H. habilis*, the newest data confirm that the mean cranial capacity of *H. habilis* was a quantum jump ahead of that of *A. africanus*. On present samples, the mean was 45–46% greater than that of *A. africanus*.

The data for some other hominid taxa are 530 cm^3 for *A. robustus crassidens* of Swartkrans (n = 1), 514 cm^3 for East African *A. boisei* (n = 4), 865 cm^3 for *H. erectus* of Africa (n = 5), 883 cm^3 for *H. erectus erectus* of Indonesia (n = 7), 1,043 cm^3 for *H. erectus pekinensis* (n = 5), and 1,151 cm^3 for *H. sapiens soloensis* of Ngandong (n = 5) (based on data in Weidenreich, 1943; Tobias 1975b, 1985; Holloway, 1976, 1983).

DISCUSSION

It has been shown that seven variables at the head-end of hominid individuals differ *inter se* in their capacity to discriminate among hominid taxa: i) The inion-opisthocranion relationship appears to distinguish *H. habilis* from *A. africanus* and *A. boisei*, and *H. habilis* from *H. erectus*, but fails to sort *A. africanus* and *A. boisei* from *H. erectus*, or from the *rhodesiensis* or *soloensis* subspecies of *H. sapiens*. ii) Recesses in the medial part of the mandibular fossa appear to sort *Australopithecus* spp. from *H. habilis* and *H. erectus*, but do not differentiate *H. habilis* from *H. erectus*. iii) The degree of recurvation of the medial end of the anterior wall of the mandibular fossa distinguishes *H. habilis* from *A. africanus* but not from *A. robustus* or *A. boisei*. iv) The robusticity of the corpus mandibulae at M_1 fails to sort *H. habilis* from *Australopithecus* spp., but sorts *H. erectus* from *H. habilis* and from Late Pleistocene *H. sapiens*. v) The maxillo-alveolar index effectively sorts *H. habilis* from *Australopithecus* spp. and helps distinguish *A. boisei* from *A. robustus* and *A. africanus*. vi) Frisch's M^3/M^1 mesiodistal diameter index separates *H. habilis* from *Australopithecus* spp. and relates *H. habilis* more closely to *H. erectus* and *H. sapiens*. vii) Cranial capacity data distinguish *H. habilis* from both *A. africanus* and *H. erectus* and moderately separate *A. africanus* from *A. robustus* and *A. boisei*.

If one were to rely on any one of these seven single traits to construct the pattern of systematic interrelationships and a phylogenetic scheme, one would obtain appreciably different results according to which trait one chose. Clearly, it is imperative to base inferences as to relationships and phylogeny upon all available traits that can contribute in a biologically meaningful manner—and the seven explored here have been selected from an *incomplete* list of 148 morpho-markers!

This point is worth stressing at a period when there has been a dramatic multiplication of hominid fossils, the number of African early hominid individuals having nearly quadrupled from a middle estimate of 143 in 1957 to well over 500 in 1984. This largesse has proved a somewhat mixed blessing. On the positive side, it has made possible detailed studies on specific anatomical regions, for each of which a goodly sample is available. In this volume alone are reported the results of two of the many recent doctoral studies carried out in the Anatomy Department of the Witwatersrand University (Clarke, 1985; Grine, 1985), and similar work is ongoing elsewhere. These studies, by revealing the total morphological pattern of each region or structural complex, in the sense of Le Gros Clark (1964), are invaluable for the advancement of palaeoanthropology.

A danger lies, however, in the manner in which the results of such very specialized studies may be interpreted. To an increasing degree investigators are tending (mistakenly, as I think) to *draw conclusions of a broad systematic and phylogenetic nature* from results obtained in their anatomically or functionally restricted areas of study, such results being considered in virtual isolation from the mass of available data on other parts of the body. Striking recent examples are provided by the use of evidence from the cranial base (Olson, 1978, 1981, 1985) and from the subnasal region and nasal cavity—the morphology of which (considered on its own), it is claimed (I believe mistakenly), "effectively removes [*A. africanus*]. . . from consideration as an ancestor of *Homo*" (Ward and Kimbel, 1983, p. 166)—to assign specimens to taxa and to erect phylogenetic schemata.

Both in method and logic, such an approach is questionable. At the least, if the investigator has reason to believe that one region or complex provides a more valid basis for classificatory and phylogenetic inferences than other regions or the totality of regions, it behooves him to validate such belief. Otherwise the very thorough and meticulous nature of an intensive regional study may impart a semblance of validity and rigour to the broader systematic and evolutionary inferences drawn. There is, too, a further danger—already manifest in published works—that different broad inferences will be drawn from different regions of the body.

CONCLUSIONS: TOTAL MORPHOLOGICAL PATTERN REITERATED AND REDEFINED

To avert this unfortunate and confusing state of affairs, it is essential that the broad systematic and evolutionary inferences should be based, not on single characters, complexes, and regions considered in isolation, but on the totality of available characters, complexes, and regions. In other words, such broad inferences and phylogenetic constructs should be founded on an updated version of what Le Gros Clark (1964) called the total morphological patttern, or on as much of it as is available.

As Le Gros Clark wrote (1964, p. 16)," . . . it is doubtful whether any single structural detail or measurement by itself can be accepted as providing a clear-cut distinction which would permit a positive identification of a single specimen of a fossil hominoid." On the contrary (p. 49), ". . . it can hardly be emphasized too strongly that, in assessing the taxonomic position of a fossil specimen, account must be taken of the total morphological pattern (and not its individual units) which provides the reliable morphological evidence on which zoölogical relationships can be determined."

Le Gros Clark had been using this term and concept as early as 1950, the year in which the original German edition of Hennig's *Grundzüge einer Theorie der Phylogenetischen Systematik* appeared. The concept was developed in the first edition (1955) of Le Gros Clark's *The Fossil Evidence for Human Evolution*, well before the first English version of Hennig's work (1966) was published. Hence, although Le Gros Clark was at pains to distinguish between "primitive or generalized" and "divergent or specialized" morphological characters, his concept of the "total morphological pattern" essentially antedates the emergence of "cladistics." For "cladistics," as a perception of the methods and goals of systematic biology, gained ground fairly slowly and only *after* the English version of Hennig's work appeared. It is appropriate now to refine and redefine Le Gros Clark's original concept, as follows:

> The assessment of the phylogenetic and systematic status of a fossil specimen, or of a series of specimens, should be based, not on individual characters considered in isolation, but on the basis of the total pattern which they present in combination (or on as much of it as is preserved and available). Such a total pattern should be determined for each available region or structural complex, as well as for the organism as a whole, where the preserved remains permit, or on as much of the whole organism as is available. Moreover, the appraisal of the characters making up the regional combinations and patterns should be tempered, where possible or where applicable, by the determination of the polarity states of traits, as well as by genetic, ontogenetic, and functional considerations.

ACKNOWLEDGMENTS

I express my appreciation to the American Museum of Natural History, especially its Director, Dr. Thomas D. Nicholson, and Drs. I. Tattersall and E. Delson, for inviting me to participate in the "Ancestors" symposium and for generous hospitality. I am grateful to the late Dr. L.S.B. Leakey, Dr. Mary Leakey, and Mr. Richard E.F. Leakey for allowing me to study East African fossils and to Dr. C.K. Brain for granting me access to the Transvaal Museum collection of fossil hominids. My gratitude is expressed to my research assistants, Valerie Strong and Marie Luise Betterton.

LITERATURE CITED

Boaz, NT, and Howell, FC (1977) A gracile hominid cranium from Upper Member G of the Shungura Formation, Ethiopia. Am. J. Phys. Anthrop. *46:*93–108.

Broom, R, and Robinson, JT (1950) Man contemporaneous with the Swartkrans ape-man. Am. J. Phys. Anthrop. *8:*151–156.

Clarke, RJ (1985) *Australopithecus* and early *Homo* in southern Africa. In E Delson (ed): Ancestors: The Hard Evidence. New York: Alan R. Liss, Inc., pp. 171–177.

Dart, RA (1948) The Makapansgat proto-human Australopithecus prometheus. Am. J. Phys. Anthrop. *6:*259–284.

Dart, RA (1962) The Makapansgat pink breccia Australopithecus skull. Am. J. Phys. Anthrop. *20:*119–126.

Day, MH, Leakey, REF, Walker AC, and Wood, BA (1976) New hominids from East Turkana, Kenya. Am. J. Phys. Anthrop. *45:*369–436.

Doolittle, RF (1983) Molecular biology and the study of primate evolution. Pontif. Acad. Sci. Scripta Varia *50:*141–150.

Flower, WH (1881) On the cranial characters of the natives of the Fiji Islands. J. Anthrop. Inst. *10:*161.

Frisch, JE (1965) Trends in the Evolution of the Hominoid Dentition. Bibliotheca Primatol. *9*. Basel: Karger.

Greenfield, LO (1983) Recent advances and suggestions for expansion of the field of human origins. Pontif. Acad. Sci. Scripta Varia *50:*29–41.

Grine, FE (1985) Australopithecine evolution: The deciduous dental evidence. In E Delson (ed): Ancestors: The Hard Evidence. New York: Alan R. Liss, Inc., pp. 153–167.

Hennig, W (1966) Phylogenetic Systematics. Urbana: University of Illinois Press.

Holloway, RL (1970) New endocranial volumes for the australopithecines. Nature *227:*199–200.

Holloway, RL (1976) Some problems of hominid brain endocast reconstruction, allometry, and neural reorganization. In Colloq. VI, IX Congr. Union Internat. Sci. Prehist. Protohist. pp. 69–119.

Holloway, RL (1983) Human brain evolution: A search for units, models and synthesis. Can. J. Anthrop. *3:*215–230.

Leakey, REF, and Walker, AC (1976) *Australopithecus, Homo erectus* and the single species hypothesis. Nature *261:*572–574.

Le Gros Clark, WE (1950) South African fossil hominoids. Nature *165:*893.

Le Gros Clark, WE (1955) The Fossil Evidence for Human Evolution, 1st ed. Chicago University Press.

Le Gros Clark, WE (1964) The Fossil Evidence for Human Evolution, 2nd ed. Chicago University Press.

Lowenstein, JM (1983) Fossil proteins and evolutionary time. Pontif. Acad. Sci. Scripta Varia *50:*151–162.

Martin, R (1928) Lehrbuch der Anthropologie, 2nd ed. Jena:Gustav Fischer.

Nei, M, and Roychoudhury, AK (1982) Genetic relationship and evolution of human races. In MK Hecht, B Wallace, and CT Prance (eds): Evolutionary Biology 14:1–59.

Olson, TR (1978) Hominid phylogenetics and the existence of *Homo* in Member 1 of the Swartkrans Formation, South Africa. J. Hum. Evol. *7:*159–178.

Olson, TR (1981) Basicranial morphology of the extant hominoids and Pliocene hominids: The new material from the Hadar Formation, Ethiopia, and its significance in early human evolution and taxonomy. In CB Stringer (ed): Aspects of Human Evolution. London: Taylor and Francis, pp. 99–128.

Olson, TR (1985) Cranial morphology and systematics of the Hadar Formation hominids and "*Australopithecus*" *africanus*. In E Delson (ed): Ancestors: The Hard Evidence. New York: Alan R. Liss, Inc., pp. 102–119.

Oxnard, C (1975) Uniqueness and Diversity in Human Evolution: Morphometric Studies of Australopithecines. Chicago: University of Chicago Press.

Pilbeam, D (1983) Hominoid evolution and hominid origins. Pontif. Acad. Sci. Scripta Varia *50:*43–61.

Robinson, JT (1954) The australopithecine occiput. Nature *174:*262.

Tobias, PV (1967) Olduvai Gorge. Volume II. The Cranium and Maxillary Dentition of Australopithecus (Zinjanthropus) boisei. London: Cambridge University Press.

Tobias, PV (1971a) The Brain in Hominid Evolution. New York: Columbia University Press.

Tobias, PV (1971b) The distribution of cranial capacity values among living hominoids. Proc. III Internat. Congr. Primatol., Zurich, 1970. Vol. 1:18–35.

Tobias, PV (1975a) Long or short hominid phylogenies—Palaeontological and molecular evidences. In F Salzano (ed): The Role of Natural Selection in Human Evolution. Amsterdam: North Holland Publishing Co., pp. 89–118.

Tobias, PV (1975b) Brain evolution in the Hominoidea. In RH Tuttle (ed): Primate Functional Morphology and Evolution. The Hague: Mouton Publishers, pp. 353–392.

Tobias, PV (1980) A survey and synthesis of the African hominids of the Late Tertiary and Early Quaternary periods. In L-K Königsson (ed): Current Argument on Early Man. Oxford: Pergamon Press, pp. 86–113.

Tobias, PV (1985) Olduvai Gorge, Volume IV. Homo habilis and other Hominid Remains from Beds I and II. (in preparation).

Ward, SC, and Kimbel, WH (1983) Subnasal alveolar morphology and the systematic position of *Sivapithecus*. Am. J. Phys. Anthrop. *61:*157–171.

Weidenreich, F (1943) The skull of *Sinanthropus pekinensis*. Palaeont. Sinica *127:*1–486.

Weidenreich, F (1945) Giant early man from Java and South China. Anthrop. Papers Amer. Mus. Nat. Hist. *40:*1–134.

Ancestors: The Hard Evidence, pages 102–119

Cranial Morphology and Systematics of the Hadar Formation Hominids and "*Australopithecus*" *africanus*

Todd R. Olson
Sophie Davis School of Biomedical Education, City College of New York, New York, New York 10031

ABSTRACT Details of cranial morphology in the extant hominoids and Pliocene hominids are described that indicate the presence of independently derived specializations of the occipitomastoid and nasal regions in the *Paranthropus* and *Homo* lineages. Also identified are a variety of taxonomically relevant features that unite the gracile Pliocene hominids from South Africa, including the Taung specimen, with the *Homo* lineage. Functional studies of these regions suggest that the cranial base has been primarily influenced by selective pressures for upright posture and that both the basicranial and nasal regions in *Paranthropus* have been secondarily affected by the evolution of dental/masticatory specializations.

A phylogenetic analysis of the A.L. 333-45 and -105 crania from the Hadar Formation demonstrates the presence in these specimens of basicranial and/or nasal specializations that identify them as members of the *Paranthropus* clade. The total pattern in the dentitions of specimens collected with these crania make it possible to identify them as *Paranthropus africanus* (Weinert, 1950). Following Schmid (1983), the skull of A.L. 288-1 is recognized as a form of "*Australopithecus*," and the complete specimen is designated as the lectotype of *Australopithecus africanus aethiopicus* (Tobias, 1980b). The hypothesis that the morphological dichotomy within the Hadar hominid sample is best explained by sexual dimorphism is questioned, and the alternative hypothesis that it represents the initial radiation of the *Paranthropus* and *Homo* lineages is adopted.

INTRODUCTION

The collection and detailed study of the Pliocene hominids from Laetoli and Hadar (Johanson et al., 1978; Johanson and White, 1979; Johanson and Edey, 1981) has resulted both in the reappraisal of the importance of the taxon "*Meganthropus*" *africanus* and in a major reevaluation of the affinities of "*Australopithecus*" *africanus*. Initial studies of the Hadar and Laetoli hominids by other than the principal workers have been critical of these interpretations and suggested alternative hypotheses for this material. The homogeneity of the Hadar hominids has been questioned by Coppens (1977, 1983), Olson (1981), Stern and Susman (1983), Senut (1983), Schmid (1983), Tardieu (1983), and Senut and Tardieu (1985). That the Hadar and Laetoli fossils should be included in the same taxon has been questioned on morphological grounds by Tobias (1980a,b). The new species status of "*A. afarensis*" has been questioned based on the morphology of the specimens by Boaz (1979, 1983), Tobias (1980b), and Kennedy (1980) and on the basis of nomenclatural procedures by Day et al. (1978), Tobias (1980b), and Logan et al. (1983). That the Hadar and Laetoli hominids are more primitive than all other known hominid taxa has been questioned in almost all of the above studies, in particular by Boaz (1979, 1983), Tobias (1980b), and Olson (1981). In addition to being contrary to the views of most paleoanthropologists when it was suggested (Johanson and White, 1979), the proposal that "*Australopithecus*" *africanus* was not involved in either the ancestry or early evolution of the *Homo* lineage has been rejected subsequently by Boaz (1979, 1983), Kennedy (1980), Tobias (1980b), Bonde (1981), Grine (1981), Schmid (1983), Wolpoff (1983), and McHenry (1984).

CRANIAL MORPHOLOGY AND THE EVALUATION OF EARLY HOMINID PHYLOGENETICS

In reviewing the major morphological and phylogenetic problems involved in the study of human evolution, Le Gros Clark (1964, 1971) has pointed out the fallacy involved in assigning equal taxonomic value to all morphological data. Le Gros Clark (1964, p. 27) further suggested that "it is of the utmost importance that . . . particular attention should be given to those characters whose taxonomic relevance has been duly established by comparative anatomical and paleontological studies." The principle of taxonomic relevance, which Le Gros Clark (1964) described as identifying the morphological features of a taxon that are sufficiently unique and consistent to distinguish the members of that taxon from those of related groups, is very similar to Hennig's (1966) concept of apomorphy.

Recently, Tobias (1985) has suggested that the emphasis in human paleontological analysis should be shifted away from this principle and towards the concept of "total morphological pattern." Tobias's (1985) redefinition of this concept is inconsistent with the methods of systematic analysis described by Le Gros Clark (1964) and Hennig (1966) in two basic ways. First, Tobias (1980a, 1985), in constructing a list of 148 craniodental variables, does not consider that only a small number of these features may be of taxonomic relevance or apomorphic in the Pliocene hominids. Second, the objective of Tobias's (1985) methodology is to determine what is better termed the total phenetic similarity of two or more taxa, rather than to identify the total morphological pattern of one, as described by Le Gros Clark (1964, 1971). Farris (1977, 1980) and Wiley (1981) have discussed the problems inherent in trying to generate long lists of morphological characters that are then summed into a total measure of morphological pattern. Tobias's (1985) criticism of studies such as those of Olson (1978, 1981) on Pliocene hominids and Ward and Kimbel (1983) on *Sivapithecus* avoids the more central issue of taxonomic relevance, which Le Gros Clark (1964) considered of such importance that he recognized it as one of his principles of systematic analysis.

The high taxonomic relevance of the basicranium has been long recognized in the study of vertebrate phylogeny, primarily because this part of the skull is only minimally influenced by extrinsic/non-genomic factors (Laitman, 1977; Laitman and Heimbuch, 1982; Olson, 1981). As a result, the basicranium tends to be highly conservative in its morphology relative to other parts of the body that interact more directly with the external environment. It is a central tenet of most systematic methodologies (e.g., Le Gros Clark, 1964, 1971; Simpson, 1961; Hennig, 1966) that evolutionarily conservative characters are important and taxonomically relevant indicators of phylogenetic affinity.

The basicranium in the great apes, fossil hominids, and modern *Homo sapiens* has been described by Olson (1978, 1981) in studies of the significance of cranial base morphology in early human evolution. The results of these studies indicate that the configuration, orientation, and size of the mastoid process, origin of the digastric muscle, and occipitomastoid area in the four great ape species are remarkably consistent and probably little removed from the hypothesized ancestral hominoid condition (Fig. 1). In contrast, the morphology of these features in *Paranthropus* (Tobias, 1967) and *H. sapiens* (Williams and Warwick, 1980) differs both when compared with each other and when each taxon is compared with the primitive pattern of the great apes (Figs. 1, 3). With regard to this dichotomy in hominid basicrania, the re-analysis of this region in the A.L.333-45 cranium presented here supports its identification as the most primitive member of the *Paranthropus* lineage (Olson, 1981). New evidence from the facial region of the A.L.333-105 juvenile cranium further supports the conclusion that the Hadar Formation contains fossils that possess a series of derived characters which ally them with *Paranthropus* and distinguish them from specimens of *Homo (Australopithecus) afri-*

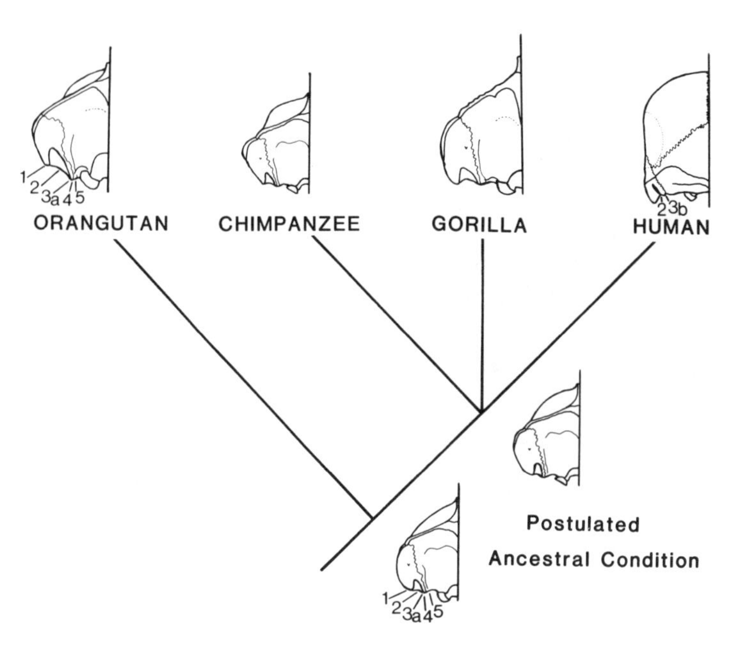

Fig. 1. Cladogram of extant hominoid taxa illustrating basicranial morphology in *norma occipitalis* with postulated ancestral conditions. Figures drawn to scale. Structures indicated: (1) mastoid process, (2) site of origin of the digastric muscle, (3a) occipitomastoid crest in great apes, (3b) human juxtamastoid eminence, (4) location of occipital artery, (5) inferior nuchal line.

canus and from later members of the *Homo* clade. Based on the similarity between some of the Hadar specimens and the Laetoli hominids (including the original Garusi specimen), as well as the distinctiveness of this combined sample from *P. robustus*, these specimens are attributed to the species *Paranthropus africanus*, which was first described by Weinert (1950) after his study of the Garusi material.

MATERIALS

The material examined in this study includes the fossil material and the mature and immature extant hominoid crania detailed in Olson (1981). In addition, 71 *H. sapiens* specimens in the collections of the American Museum of Natural History were examined. This extraordinary A.M.N.H. series of 496 human skulls from Berg in Austria has been described extensively by Shapiro (1929) and used by Howells (1973) as a reference population in his study of cranial variation.

In addition, dissections were conducted on the mastoid and suboccipital regions in two specimens each of *Gorilla* and *Pan troglodytes* and one specimen of *Pongo*. Through August of 1983, 47 dissections have been done on human cadaver material in the Gross Anatomy Laboratory at the Sophie Davis School of Biomedical Education.

BASICRANIAL MORPHOLOGY OF THE EXTANT HOMINOIDS

Mastoid Process

The size, orientation, and morphology of the mastoid processes in the gorilla, chimpanzees, and orang-utan are remarkably similar and considerably different from the condition seen in *H. sapiens.* The mastoid process in the great apes is typically small, located well medial to the lateral margin of the supramastoid crest (Fig. 1). In contrast to the condition seen in the great apes, the mastoid in *H. sapiens* is typically large, inflated laterally, and projecting in a prominent finger-like projection.

Temporal Origin of the Digastric Muscle

In view of the confusing anatomical nomenclature of the occipitomastoid region

(Hublin, 1978), it is necessary to revise the description presented in Olson (1981) of this region in the hominoids. In *Gorilla*, *Pan*, and *Pongo*, the digastric muscle has a broad, fleshy non-tendinous origin from a shallow fossa, ranging from oval to V-shaped in outline, on the medial surface of the mastoid process of the temporal bone. This depression lies between the tip of the mastoid laterally and the true occipitomastoid crest medially. The occipitomastoid suture marks the ridge of an occipitomastoid crest that runs along the medial border of the attachment for the digastric muscle. In most specimens, particularly the great ape males, the medial margin of the fossa is raised and contributes to this crest.

Dissections on human cadaver material have revealed that the derived nature of the origin of the human digastric muscle differs significantly from the great apes. In humans the origin is usually identified as arising from the mastoid notch of the temporal bone. Forty-seven dissections have confirmed the observations of Taxman (1963) that the digastric arises by a strong tendon from the lateral surface of the juxtamastoid eminence (Rouviere, 1954), while the mastoid notch serves as the site of insertion of the longissimus capitis muscle. The root of the digastric tendon is associated with the posteromedial and medial (= juxtamastoid eminence) margins of the mastoid notch.

Occipitomastoid Crest, Suture, and Occipital Groove in the Great Apes

Among the great apes, the anatomical relationships of the occipitomastoid crest, the occipitomastoid suture, and the occipital groove are remarkably similar (Fig. 1). The occipitomastoid crest is a pneumatized ridge the apex of which follows the occipitomastoid suture that joins the temporal and occipital bones. In a number of adult crania, both the medial border of the digastric attachment (homologous to the juxtamastoid crest in *Homo*) and the lateral margin of the superior oblique insertion are raised as small, closely associated ridges paralleling the occipitomastoid suture atop the occipitomastoid crest. The groove formed between these two ridges is termed the occipital groove and has been described as the site of the occipital artery (Sakka, 1972, 1973). However, dissections of this area failed to document the presence of the occipital artery in this groove. The occipital artery came the closest to occupying the groove in *Pongo*. In the specimen dissected, the left occipital artery came within 4 mm of the apex of the occipitomastoid crest. The path of the artery paralleled exactly the course of the left occipital groove because it passed between the heads of the two muscles that created the margins of the groove.

Occipitomastoid and Juxtamastoid Crests in *Homo sapiens*

Taxman (1963), Sakka (1972, 1973), and Hublin (1978, 1983) have concluded that the term occipitomastoid crest is a misnomer when it is applied in human anatomy (e.g., Weidenreich, 1943; Olson, 1978, 1981; Williams and Warwick, 1980) to the bony crest separating the mastoid notch from the occipitomastoid suture. The results of this study fully confirm the conclusion that it is not homologous with the true occipitomastoid crest of the great apes and that Rouviere's (1954) "juxtamastoid eminence" is a more appropriate name for this structure. The series of human dissections done in this study support Taxman's (1963) conclusion that the juxtamastoid eminence provides a mechanical advantage to the digastric muscle by enlarging the surface area available for the attachment of its tendon.

Without an appreciation of the above relationships, it is difficult to interpret the variation seen in modern human crania. Variation observed in this region reflects not only differences in the juxtamastoid eminence associated with the size of the digastric muscle but also differences in the degree of cresting that occurs along the lateral margin of the superior oblique muscle. Taxman's (1963) six plates provide an excellent illustration of the common variations in this region. Both the juxtamastoid eminence and the lateral margin of the superior oblique insertion can be prominent structures separated by a well-defined groove (Taxman, 1963, Plates 1–2). The juxtamastoid eminence may be well defined, while the lateral border of the attachment for the superior oblique muscle is poorly defined (Plates 3–4) or undeveloped (Plates 5–6). Taxman (1963) also documented variation in the shape of the juxtamastoid eminence.

The variability seen in the occipital groove on the human basicranium also warrants comment, since the same relationship exists between the groove and the occipital artery as was described previously in the great apes. Taxman's (1963) study indicates considerable variation for the feature identified as the groove of the occipital artery. In all cases, the occipital groove is formed by the juxtamastoid eminence laterally and typically, but not always, by a more medially located crest formed by the superior oblique muscle. The occipital artery was observed on the cranial base in 24 of the 47 human cadavers dissected. In no individual was the artery found in the floor of the occipital groove, nor was the size of the groove related to the size of the artery. The prominence of the occipital groove was invariably determined by the size of the digastric and superior oblique muscles. Interestingly, the individual in which the occipital artery came the closest to the base of the skull was an elderly female with muscular atrophy and almost no palpable juxtamastoid eminence or muscle scar for the superior oblique.

While variation and/or nomenclature have created problems for workers interested in this area of the basicranium, the occipitomastoid region remains an extremely relevant source of data about early human evolution and the phylogenetic relationships of fossil hominid taxa.

Significance of Hominoid Basicranial Morphology

Both Le Gros Clark (1964) and Robinson (1958) have suggested that the differences in the nuchal and suboccipital regions that distinguish the great apes and humans are correlated with postural adaptations. Le Gros Clark (1964) concluded that the inferiorly directed foramen magnum and relatively small horizontal nuchal planum in hominids are associated with erect posture, while the posteriorly directed foramen magnum and large vertical nuchal planum of the great apes are reflections of their basically pronograde posture. The results of this study indicate that the evolution of upright posture was the single most important factor in shaping the derived morphology of the hominid occipitomastoid region.

The laterally positioned mastoid processes in *Homo* are a specialization that allows for the efficient support and rotation of the cranium about an inferiorly directed foramen magnum. The large size and pronounced projection of the human mastoid process and juxtamastoid eminence are functionally correlated with the reorientation and concomitant reduction in the size of the horizontal nuchal surface that facilitates upright posture. The relocation of the nuchal planum to the inferoposterior surface of the cranium is associated with a relative decrease in the total surface area of this region. The typical response of the musculoskeletal system to such changes is the development of prominent processes and depressions that maintain bony surface area while decreasing the circumference of the muscle. A large mastoid process, a deep and narrow mastoid notch, a prominent juxtamastoid eminence, a frequently large crest along the lateral border of the superior oblique muscle, and a well developed inferior nuchal line are all part of this adaptive resculpturing of the basicranium associated with the evolution of upright posture in the *Homo* lineage.

While considerable uncertainty remains about the precise locomotor characteristics of the pre-bipedal hominid ancestor, it seems likely that 1) it had a pronograde posture and 2) the posture of the common ancestor of humans, gorillas, and chimpanzees was also pronograde. If all the hominoids are assumed to have arisen from a pronograde ancestor (Fig. 10, Node 1), one may conclude from the close similarity of the basicranial morphology of all the extant pronograde great apes that the following cranial characteristics are symplesiomorphic for the hominoids: a relatively large, vertically inclined nuchal planum; small uninflated mastoid processes inflected well under the cranium; ovoid and shallow mastoid notches; low rounded occipitomastoid crests contiguous with the occipitomastoid suture; and a poorly developed inferior nuchal line (Fig. 1).

NASAL MORPHOLOGY OF THE EXTANT HOMINOIDS

Characteristically, in great apes of both sexes, nasion and glabella are closely approximated on the supraorbital torus (Fig. 2). The nasal bones are widest immediately superior to the piriform aperture and taper to an apex at nasion. The maxillary processes of the frontal bone project inferiorly beside the nasal bones, leaving the frontomaxillary sutures in a transverse plane below nasion. The phylogenetic significance of the similarities between the great ape taxa is further enhanced by several factors: 1) the presence of this pattern in the orang-utan with its highly derived interorbital region;

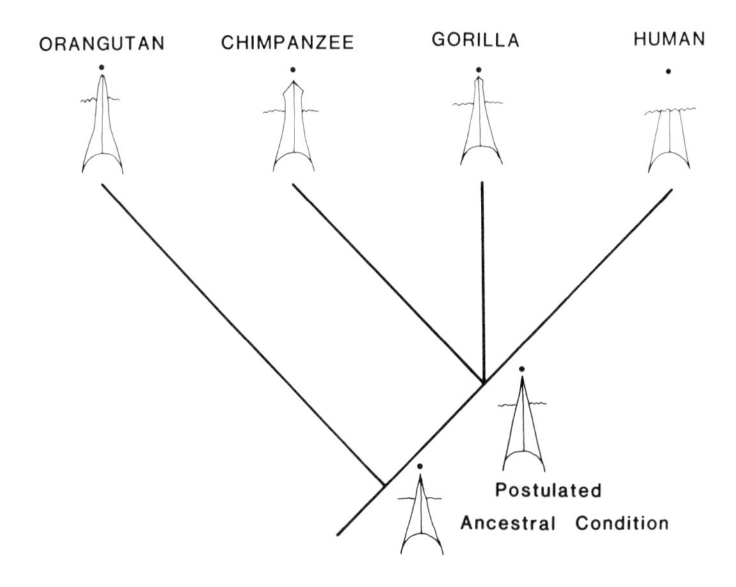

Fig. 2. Cladogram of the extant hominoid taxa illustrating *norma frontalis* of nasal bones and superior margin of piriform aperture. Figures are drawn to scale and indicate the differences between the nasal outlines and the relative positions of nasion and glabella (●).

2) the existence of this morphological pattern in the Miocene *Sivapithecus* skull described by Pilbeam (1982); and 3) the limited ontogenetic variability observed in these features. The infant great ape skulls compared in this study varied little from each other with regard to the above features. This last point suggests that the morphological similarity in the nasal region of the great apes is symplesiomorphic.

The nasal bones in *H. sapiens* retain the primitive hominoid condition of a roughly triangular profile, with their greatest breadth being immediately superior to the piriform aperture and their lateral margins converging towards nasion. They are, however, derived in the reduction/absence of the maxillary processes of the frontal bone as seen in a frontal view, and in the location of nasion relative to glabella (Fig. 2). In *H. sapiens*, the frontonasal and frontomaxillary sutures are linear and continuous below an anteriorly projecting supraorbital torus, and nasion is removed from its primitive association with the supraorbital torus and glabella (Fig. 2). The ontogenetic and comparative evidence for the primitiveness and consistency of these features within the *H. sapiens* clade is compelling. Fetal human crania exhibit the adult condition even in the absence of a supraorbital torus (Williams and Warwick, 1980). An ontogenetic series of 23 crania from the Berg village collection indicates the presence of a highly conservative pattern in these derived features, in which the range of variation in the total pattern between fetal and old adult individuals is very small. The ontogenetic evidence from this region in *H. sapiens* is consistent with the developmental data from other hominoid taxa, and it indicates that the morphology of the nasal region in the hominoids is of considerable taxonomic relevance in the determination of the phylogenetic affinities of both adult and immature fossil hominid skulls.

CRANIAL ANATOMY OF THE PLIOCENE HOMINIDS

Basicranial Morphology

Prior to the discovery of the Hadar fossils, the Pliocene hominids had been described as consisting of two cranially distinct forms (Robinson, 1958; Tobias, 1967; Clarke, 1977; Olson, 1978). With regard to the basicranium, "*A.*" *africanus* (in Fig. 3d–f) is derived in the direction of *H. sapiens* (Clarke, 1977; Olson, 1978, 1981), and *Paranthropus* (Fig.

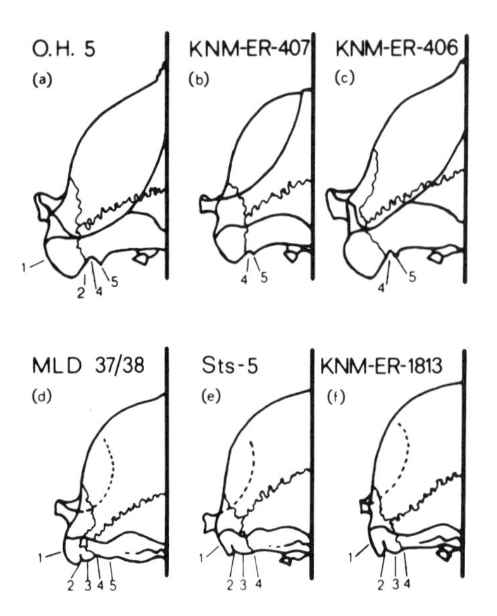

Fig. 3. *Norma occipitalis* of specimens of *Paranthropus* from Olduvai Gorge (a) and Koobi Fora (b, c); and of *Homo (Australopithecus)* from Makapansgat (d), Sterkfontein (e), and Koobi Fora (f). Figures represent reconstructions based upon the occipital morphology of the entire cranium. Structures indicated: (1) mastoid process, (2) mastoid notch or site of origin for the digastric muscle, (3) juxtamastoid eminence, (4) location of occipital artery, (5) inferior nuchal line. Scale is 4 cm.

3a–c) has been described (Tobias, 1967; Olson, 1978, 1981) as autapomorphic for hominids.

While the mastoid processes are enlarged in both groups, the paranthropine mastoids differ morphologically from those in the *Homo* lineage, and they are more extensively expanded. On the medial surface, the paranthropine mastoid process is inflated to the point where the occipitomastoid suture, the primitive occipitomastoid crest, and the area of origin for the digastric muscle are incorporated into the base of the process. Inferiorly, the tips of the mastoids project well below the level of the occipital condyles, while laterally the specialized mastoids project beyond the margins of the supramastoid crests. Not only do the basicranial features of *Paranthropus* and *Homo* distinguish these taxa from each other, they also independently distinguish each taxon from the common hominid ancestor and from the more ancient ancestor shared by humans and the African apes (Olson, 1981).

The functional basis for the occipitomastoid characters that typify the *Homo* lineage seems most directly associated with the condition created by the inferior reorientation of the nuchal planum and its consequent reduction in overall surface area. However, in contrast to the members of the *Paranthropus* clade, mastoid enlargement in the early forms of *Homo* was insufficient to compensate for the reduction in overall nuchal surface area. Consequently, the digastric muscle became increasingly restricted to the medial aspect of the mastoid notch and evolved a strong tendinous origin that produced the juxtamastoid eminence at its site of attachment to the cranial base. This complex morphological pattern is present in several crania from the Pliocene and Early Pleistocene (Fig. 3d–f). Both Sts 5 and MLD 37/38 have narrow and deep mastoid notches bounded medially by a juxtamastoid eminence ridge that forms the medial surface of the notch. A juxtamastoid eminence is also present in KNM-ER 1813 (Fig. 3f) from the Koobi Fora Formation, which specimen is variously identified as either representing "*Australopithecus*" *cf. africanus* (Olson, 1978; Walker, 1981) or "*Homo habilis*" (Howell, 1978; White et al., 1981; Cronin et al., 1981).

Both the juxtamastoid eminence and a distinct occipitomastoid crest are absent in *Paranthropus*. The juxtamastoid eminence is absent because it has not evolved in this lineage, and the primitive hominoid occipitomastoid crest is absent, as a distinctive structure, because it became incorporated into the medial wall of the greatly inflated mastoid process.

In their interpretation of the A.L.333-45 cranium, Johanson and White (1979, p. 323) identify "a host of primitive features" that characterize this specimen and distinguish it from both *Paranthropus* and early forms of *Homo*. Among the characters identified are: 1) compound nuchal crests on either side of the occiput, and 2) heavily pneumatized and projecting mastoid processes that give the cranial base a concave transverse profile. While the first character is definitely a primitive hominoid feature, it is not a primitive hominid characteristic (Olson, 1981). The second character is primitive neither for hominoids nor for hominids. The basicranial features of the A.L.333-45 cranium described by Kimbel et al. (1982) differ considerably from the primitive hominoid pattern

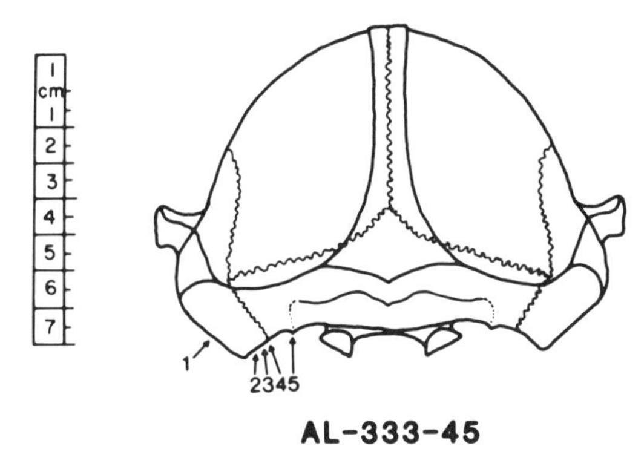

Fig. 4. *Norma occipitalis* of Hadar cranium A.L.333-45. Reconstruction revised from Olson (1981) to decrease amount of concavity in cranial base. Structures indicated: (1) mastoid process, (2) area of insertion of digastric muscle, (3) occipitomastoid crest, which is incorporated into mastoid process, (4) location of occipital artery, (5) inferior nuchal line.

(Fig. 1). The mastoid processes are heavily pneumatized and greatly enlarged in all directions. While the lateral inflation of the mastoids is more extensive than typically observed in any of the extant great apes, the tip of each mastoid retains the primitive appearance of being inflected under the basicranium (Figs. 1, 4). The extent of both the medial and inferior inflation of the mastoid processes in A.L.333-45 also exceeds the characteristic primitive condition of the great apes. In addition, the extensive inferior inflation of the mastoids is a feature that distinguishes this specimen from the primitive hominoid condition of small and only slightly projecting processes (Fig. 1).

The morphology of the occipitomastoid region in A.L.333-45 includes highly specialized features that distinguish it from all hominines and ally it with the paranthropines. The enlarged mastoid processes and concave cranial base are more highly developed than in any normal hominine crania, and they represent a radical departure from the presumed primitive hominoid condition. Since Olson (1981) first suggested that A.L.333-45 was a paranthropine, two independent studies of this specimen (Holloway, 1983; Falk and Conroy, 1983) have arrived at the same conclusion, based on the presence of an accessory venous sinus system that is found in all *Paranthropus* specimens that preserve the relevant anatomy. In fact, all four crania from Afar Locality 333—A.L.333-45, -105, -114, and -116—are reported by Kimbel et al. (1982) as possessing this distinctive *Paranthropus* pattern. Kimbel (1984) has attempted to deny the phylogenetic significance of this character. However, the fact that it is diagnostic of all *Paranthropus* crania with this region preserved, rarely found in any extant ape taxa, and absent in all relevant specimens of "*Australopithecus*" *africanus* and "*Homo habilis*" argues strongly in support of Falk and Conroy's (1983) conclusion about the diagnostic importance of this character.

Nasal Morphology

The structures around and within the anterior nasal aperture have been recognized (Robinson, 1953; Tobias, 1967; Clarke, 1977; Olson, 1978; Rak, 1983) as useful in establishing phylogenetic relationships among the early hominids. While the hominid basicranial specializations discussed previously have been associated primarily with the evolution of orthograde posture, the configuration of the nasal bones in the early hominids is associated with a different set of selective pressures (Olson, 1978). In addition, the nasal region of the skull has a totally different ontogenetic history from the skull base, which is part of the chondrocranium (Williams and Warwick, 1980). The fact that both the basic adaptive and ontogenetic factors influencing the morphology of the basicranium and the configuration of the nasal bones are distinct increases the taxonomic relevance of the nasal region when it is used in conjunction with cranial base characters to establish phylogenetic relationships.

A comparative study of the nasal region in the extant hominoids indicates additional reasons why specializations of the nasal region would be extremely useful in phylogenetic studies of fossil hominids. First, the basic configuration of the nasal region and the relationship of nasion to glabella are established during late fetal development, and second, postnatal growth produces only slight changes in the relative position of the nasal, maxillary, and frontal bones. Furthermore, the condition observed in *Homo sapiens* is derived with respect to all extant ape taxa making it possible 1) to identify a member of the *Homo* lineage and 2) to reconstruct the primitive condition of the last common ancestor of the human and African ape clades (Fig. 2).

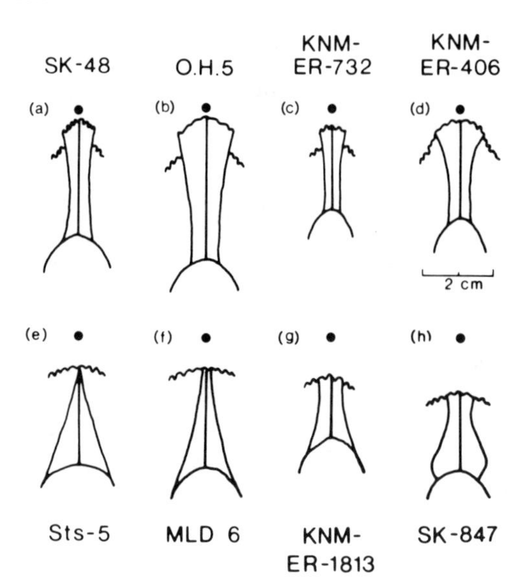

Fig. 5. *Norma frontalis* of nasal bones and superior margin of piriform aperture in specimens of *Paranthropus* from Swartkrans (a), Olduvai Gorge (b), and Koobi Fora (c, d), of *Homo (Australopithecus)* from Sterkfontein (e), Makapansgat (f), and Koobi Fora (g), and of *Homo (Homo)* from Swartkrans (h). Illustrations are reconstructions that indicate the differences in nasal profiles and in the relative positions of nasion and glabella (●) and the plane between the points where the two frontomaxillary sutures meet the nasal bones.

The taxonomic relevance of nasal morphology in early hominid systematics is further enhanced by the existence in *Paranthropus* of a third hominoid pattern, which is derived when compared to either the symplesiomorphic condition seen in extinct apes (Pilbeam, 1982) and extant great apes, or the autapomorphic condition in *Homo*. In *Paranthropus*, the primitive hominoid relationship between nasion and glabella and the frontonasal and frontomaxillary sutures is retained, while the lateral outline of the nasal bones is derived. Nasion and glabella in *Paranthropus* are closely approximated superior to a plane between the points where the frontomaxillary sutures meet the nasal bones, and the nasomaxillary sutures are either parallel or divergent superiorly (Fig. 5). The frontal profile of paranthropine nasal bones has taken on what can be described as a keystone outline.

While the frontal profile of the nasal bones in *Paranthropus* is derived relative to the postulated common hominid ancestor, the morphology observed in such specimens as Sts 5 and MLD 6 (Fig. 5e–f) retains the primitive hominoid profile but is derived in the location of nasion relative to glabella. In Sts 5, MLD 6, and later more advanced specimens, such as KNM-ER 1813 and SK 847 (Fig. 5g–h), as well as all undoubted members of the *Homo* lineage, the nasal bones appear truncated below the supraorbital torus with the frontomaxillary and frontonasal sutures oriented in basically a horizontal plane below the variably inflated glabellar region of the supraorbital torus. The extent of this superior truncation of the nasal region by the frontal bone in Sts 5 and MLD 6 appears to be somewhat less derived than the condition observed in later *Homo* taxa. However, in both specimens, nasion is at the base of the supraorbital torus in a position well inferior to glabella.

The entire nasal region has been described in one Hadar specimen—A.L.333-105—collected at the same locality that produced the A.L.333-45 previously described as sharing derived basicranial characters with later specimens of *Paranthropus*. The morphological interpretation and measurements used to reconstruct the A.L.333-105 specimen (Fig. 6a) are those given by Kimbel et al. (1982). However, where Kimbel et al. (1982, p. 478) have concluded that the size and form of the nasal bones in A.L. 333-105 are "unusual," this study suggests that the nasal morphology of this specimen is consistent with the evidence from its venous sinus drainage pattern (Kimbel, 1984) and that A.L. 333-105 is best interpreted as an early member of the *Paranthropus* clade. The form of the nasal bones in A.L.333-105 is "unusual" in that it is inconsistent with the claim of Johanson and White (1979, p. 321) that the Hadar hominids "are demonstrably more primitive than those of hominid specimens from other sites." Comparison of the A.L.333-105 nasal morphology with the postulated primitive hominid condition (Fig. 2) indicates that it is derived. Kimbel et al. (1982, p. 478) state that nasion is situated a short distance below glabella, that a maximum combined nasal breadth of 5 mm occurs about 15 mm above rhinion, and that "below maximum breadth the nasals gradually constrict to a combined breadth of 2.5 [mm] and then expand again to a combined breadth of 4.7 at rhinion." Fortunately, the immature age of this specimen does not affect the taxonomic relevance of its nasal morphology. Facial

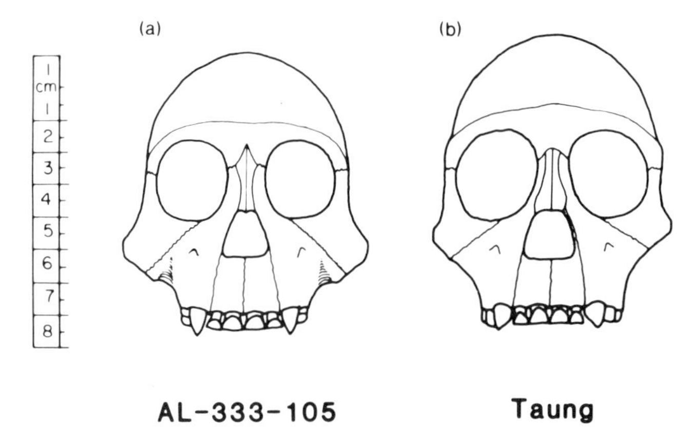

Fig. 6. *Norma frontalis* reconstructions of A.L.333-105 from Hadar (a) and Taung child (b) crania.

growth would only further reduce the combined nasal breadth at the inferior half of the nasal bones, where they are overlapped laterally by the frontal processes of the maxillary bones (Hollinshead, 1982). Thus, it is reasonable to conclude that the nasal profile of an adult A.L.333-105 would bear an even closer resemblance to known adult specimens of *Paranthropus* (Fig. 4a–d) than does the actual condition observed in the juvenile skull.

The presence in A.L.333-105 of basicranial features characteristically found in specimens of *Paranthropus* (Holloway, 1983; Kimbel, 1984) as well as a derived paranthropine nasal profile sheds light on Tobias's (e.g., 1973, 1978, 1980b) suggestion that the Taung skull might be a member of the *Paranthropus* lineage. Grine (1981) has indicated that the dentition of the Taung infant does not possess the derived features that characterize immature specimens of *Paranthropus*. Grine's (1981) observations on the Taung dentition are even more significant in view of Tobias's (1973) opinion that this specimen may represent the terminal phase of paranthropine evolution. One could reasonably argue that the earliest paranthropines may not exhibit the full suite of dental specializations seen in later forms, but it is unlikely that the terminal individuals of this clade would exhibit none of the dental features seen in its ancestors. This same argument is relevant to the observations of Tobias (1967), Holloway (1983), and Kimbel (1984) that the characteristic pattern of cranial venous sinus drainage seen in *Paranthropus* is absent in the Taung specimen. Furthermore, the presence of this pattern in the A.L.333-105 cranium of approximately the same ontogenetic age would seem to vitiate any arguments for the paranthropine affinities of the Taung specimen based on the appearance of characteristic robust features at a later stage in its development. Finally, while the absence of derived dental or basicranial features does not preclude the inclusion of the Taung specimen in *Paranthropus*, the evidence from the nasal region is more conclusive. The Taung nasal morphology (Fig. 6b) exhibits the apomorphic condition seen in the gracile Pliocene forms and later members of the *Homo* lineage.

CRANIAL CHARACTERISTICS OF "*AUSTRALOPITHECUS*" *AFRICANUS*

Johanson and White (1979) concluded, primarily on the basis of their interpretation of the chronological placement of the South African "australopithecine" sites, that the gracile specimens are intermediate between the unspecialized hominids from Hadar and Laetoli and the highly specialized paranthropines (Fig. 7). Their morphological analysis, further elaborated by White et al. (1981), supposedly identified features that: 1) established an ancestor-descendant relationship between the gracile and robust early hominids from South Africa, and 2) removed the gracile South African hominids from

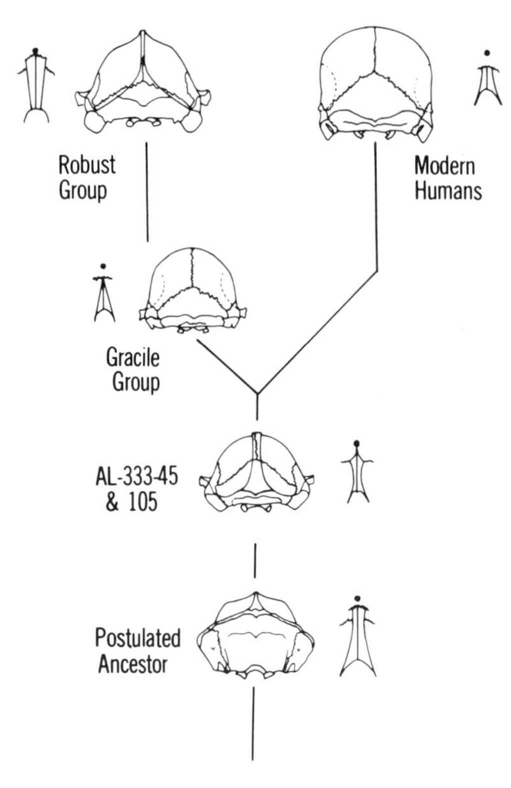

Fig. 7. Ancestor–descendant relationships of hominid taxa according to the phylogenetic hypothesis of Johanson and White (1979). Postulated ancestor included from Figures 1 and 2. Crania and nasal regions drawn to independent scales. Symbols (e.g., ●) as in Figures 1 and 2.

consideration as either the common ancestor of all later hominids or as an early member of the *Homo* lineage.

The morphological evidence presented by White et al. (1981) in support of the identification of "*Australopithecus*" *africanus* as the sister-group of the paranthropines was based largely on the following studies: 1) the work of White (1977a) on dental and mandibular morphology, 2) the subsequently published study of facial anatomy by Rak (1983), and 3) the study of the cranial base by Kimbel (in White et al., 1981). White et al. (1981, p. 466) recognized in their analysis that the dental, mandibular, and facial characteristics that they identify as shared derived features uniting the gracile and robust South African hominids all are part of a "morphological complex represent[ing] an adaptation to generating and withstanding increased amounts of vertical occlusal force . . . ." Since all of these characters have been interpreted as a single adaptive complex, it becomes increasingly likely that the character similarities in these two groups are homoplastic or nonhomologous. In fact, the morphological traits listed by White et al. (1981) are strikingly similar to the dental/facial complex identified by Jolly (1970) as part of the seed-eater model for the differentiation of the early hominids. Jolly (1970) postulated just such a parallel evolution in this adaptive complex of features during the initial phases of evolution in the *Homo* and *Paranthropus* lineages.

With regard to the features of the cranial base presented by White et al. (1981) to support their hypothesis about the phylogenetic affinities of "*A.*" *africanus*, Kimbel (in Lewin, 1984, p. 478) now believes "that the cranial base of *africanus* is in part derived in the direction of all later hominids." This is consistent with the evidence from the cranial base presented in other studies (Clarke, 1977; Olson, 1978) that indicates "*A.*" *africanus* shares derived features with members of the *Homo* lineage and is not uniquely derived in the direction of *Paranthropus*.

Probably the most widely accepted hypothesis concerning the phylogenetic position of "*A.*" *africanus* (e.g., Boaz, 1983; Grine, 1981; Tobias, 1980a; Wolpoff, 1983; McHenry, 1984) is that this taxon is the common ancestor of *Paranthropus* and *Homo* (Fig. 8). Boaz (1983), Tobias (1980a), and Wolpoff (1983) do not consider the Hadar and Laetoli specimens to be specifically different from "*A.*" *africanus*, while Grine (1981) and McHenry (1984) would place the Hadar and Laetoli specimens in a different species from "*A.*" *africanus*. The inclusion of the Hadar and Laetoli material with the gracile specimens from South Africa on the basis of apparent dental similarities and the placement of this group as the common ancestor of *Paranthropus* and *Homo* create a number of problems when the evidence from the basicranial and nasal regions is considered. The initial problem is that the inclusion of the Hadar hominids in "*A.*" *africanus* would extend both the range and nature of the variation in these regions beyond that observed in any living primate analogy. Second, the cranial morphology of the Hadar specimens (such as A.L.333-45 and -105) and "*A.*" *africanus*

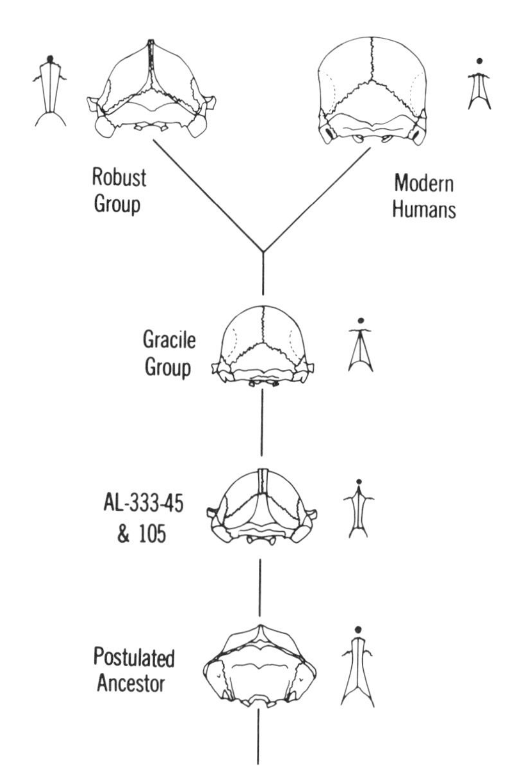

Fig. 8. Ancestor–descendant relationships of hominid taxa according to the phylogenetic hypotheses of Tobias (1980b), Boaz (1983), and Wolpoff (1983). Postulated ancestor included from Figures 1 and 2. Crania and nasal regions drawn to independent scales. Symbols as in Figure 7.

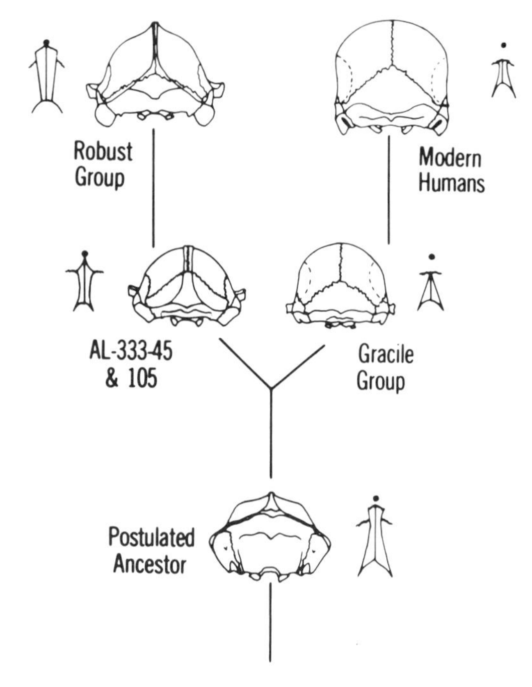

Fig. 9. Ancestor–descendant relationships of hominid taxa according to the phylogenetic hypotheses of Robinson (1972), Clarke (1977), and Olson (1978). Postulated ancestor included from Figures 1 and 2. Crania and nasal regions drawn to independent scales. Symbols as in Figures 7 and 8.

(from Taung, Sterkfontein, and Makapansgat) are derived in different ways from their postulated ancestor. To arrange these two groups as chronospecies (Grine, 1981; McHenry, 1984) is consistent with the differences in cranial morphology (Fig. 8), but it creates some difficulties in reconciling the initial derivation of the *Paranthropus*-like cranial morphology in A.L.333-45 and -105 and the subsequent evolution of the *Homo*-like basicranium and nasal regions in "*A.*" *africanus*, which they place as the primitive ancestor of the *Paranthropus* and *Homo* lineages.

The third and most parsimonious phylogenetic arrangement of "*A.*" *africanus* (Robinson, 1963, 1972; Jolly, 1970; Clarke, 1977; Olson, 1978) is based on the interpretation that its morphological similarities with members of the *Homo* lineage are shared-derived characters, which preclude its being the common ancestor of all subsequent hominids (Fig. 9).

HADAR FORMATION CRANIA

The first cranial fragment collected in the Hadar Formation was a left temporal bone, A.L.166-9, recovered in 1973. Following the initial study of A.L.166-9, Taieb et al. (1974) concluded that it most closely resembled the highly specialized temporal region of the robust "australopithecines." Subsequently, Johanson and White (1979) concluded from their study of the enlarged collection from the Hadar Formation that the morphology of the cranial specimens was more primitive than that previously described for any fossil hominid taxon. Two of the features of the Hadar crania were purported by Johanson and White (1979) to be primitive hominid characteristics: heavy pneumatization of the lateral part of the cranial base, and a concave nuchal planum that is longer than the occipital breadth. These features do not appear to represent the primitive hominid condition (Fig. 1) and are features identified by Tobias (1967) and Olson (1981) as characterizing *Paranthropus*. The evidence from the basicranium of A.L.333-45 and the nasal region of A.L.333-105 presented in this study supports the recent observations on the venous sinus drainage pattern by Holloway (1983) and Falk and Conroy (1983) that indicate the existence in several Hadar crania of paranthropine features.

The phylogenetic analysis (Hennig, 1966) of the morphology of the A.L.333-45 and -105 crania identifies derived features in the venous sinus system, nasal region, and basicranium that identify these specimens as members of the *Paranthropus* clade (Fig. 9). To argue that the mastoid and nasal morphology of A.L.333-45 and A.L.333-105, respectively, represent the primitive hominid pattern (Figs. 7, 8) implies an unlikely evolutionary history for these regions. Presumably, Johanson and White (1979) would derive early hominid basicranial and nasal morphology from a pattern similar to that found in all of the African apes (Figs. 1, 2). Their characterization of the Hadar crania as primitive is at odds with the postulated primitive conditions based on out-group comparisons with the great apes. The independent derivation of the comparatively smaller and less inflated mastoids and *Homo*-like nasal regions seen in "*Homo habilis*" (e.g., KNM-ER 1470) and in "*A.*" *africanus* (e.g., Sts 5) seems even less likely in view of Johanson and White's (1979) proposal that would then derive the paranthropine features from the *Homo*-like morphology of "*A.*" *africanus* and not from the Hadar specimens which they more closely resemble (Fig. 7). In the absence of a hypothetical basis for this sequence of evolutionary reversal and parallelism, it appears to be a most implausible sequence of events. A more parsimonious hypothesis is to assume that the specialized similarities seen in some of the Hadar crania and the paranthropines are synapomorphic rather than nonhomologous.

The A.L.288-1 or "Lucy" specimen has been reconstructed by Schmid (1983), who concluded that the morphology of its cranium was inconsistent with Johanson and White's (1979) conclusion that it is more primitive than all known fossil hominids. In view of the cranial evidence for the presence of *Paranthropus* within the Hadar sample, Schmid's (1983) statement that the A.L.288-1 cranium does not differ from the skulls known of *Australopithecus africanus* takes on added significance.

MORPHOLOGICAL DIVERSITY WITHIN THE HADAR FORMATION HOMINIDS

Johanson and White (1979, p. 325) have argued that the extremely variable collection of fossil hominids from the Hadar Formation constituted a single species that was characterized by a substantial level of sexual dimorphism. They also suggested that an alternative interpretation of this material "would be that some of smaller individuals, particularly the partial 'Lucy' skeleton, represent a distinct lineage contemporary with the majority of the Hadar hominids." While this alternative was rejected by Johanson and White (1979), a growing number of subsequent studies of the Hadar hominids cited at the beginning of this paper are concluding that the variability observed in this material is interspecific rather than intraspecific in nature.

The existence of a robust and a gracile hominid group in the Hadar Formation is also suggested by the degree and pattern of dental variability present in this material (Coppens, 1977; Olson, 1981). In terms of the dental features that specifically distinguish the gracile and robust Hadar groups, the gracile taxon is typically primitive in having retained far more of the characteristics believed to have been present at Node 1 of Figure 10.

NOMENCLATURE AND SYSTEMATIC AFFINITIES OF THE HADAR HOMINIDS

The analysis of the cranial morphology of A.L.333-45 and -105 demonstrates the existence of apomorphic features that identify these individuals as members of the *Paranthropus* clade. An extensively pneumatized and projecting mastoid process, the incorporation of the occipitomastoid crest into the expanded medial wall of the process, an enlarged occipital-marginal venous sinus system, and keystone shaped nasal bones are all derived features that characterize the descendants of Node 2 in Figure 10. These characters evolved in the paranthropine lineage following their cladogenesis at Node 1. These features are characteristically associated with the paranthropine clade, and they differ considerably from the primitive con-

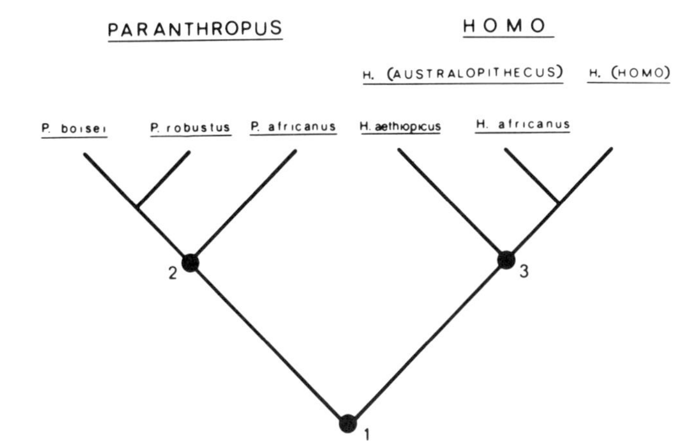

Fig. 10. Cladogram illustrating the phylogenetic relations of Pliocene hominid taxa. The genus *Homo* is divided into the subgenera *Australopithecus* and *Homo*. Each dichotomy is defined by those unique characters that ally the descendants of that node and/or differentiate each clade from its sister group. The significance of the numbered nodes is discussed in the text.

dition postulated for the common ancestor of humans and the African apes prior to Node 1.

The unprecedented natural concentration of fossil hominid remains at the Afar Locality 333, including the A.L.333-45 and -105 crania, and the extraordinary nature of the event that led to their deposition (Johanson, 1976), allow a number of observations to be made about this assemblage that might otherwise be considered unwarranted. While it is difficult either to verify or to refute the assertion that they constitute a "family" group, the geological data strongly suggest that this is a biocenotic assemblage, and the morphological evidence, particularly from the adult dentitions, support the interpretation that it is a single species. The adult dental material of this assemblage is of further importance because it permits the specific identification of this group.

Morphologically, the dental material from the Afar Locality 333 differs only slightly from the pattern seen in the AL.400-1a mandible, which Johanson and White (1979, p. 323) have demonstrated to be strikingly similar" to the Laetoli fossil hominids described by White (1977b). As there is no convincing morphological evidence to suggest that the Laetoli hominid remains, including the Garusi I maxillary fragment, represent more than one species, all of these specimens as well as those from the Afar Locality 333 are interpreted as a single species. A comparative study of the other Hadar Formation hominid dental remains indicates that many of these specimens (A.L.145-35, 166-9, 188-1, 200-1a, 207-13, 266-1, 277-1, and 400-1a) should also be included in this robust taxon.

Of the specimens included in this species-group, the A.L.333-45 and -105 crania are particularly important because their morphology establishes the phylogenetic affinity of this species with *Paranthropus*, and it reduces the possibility that this species was ancestral to either "*Australopithecus*" *africanus* (*sensu stricto*) or to the *Homo* clade. The classification of this species in a genus-group that does not include *Australopithecus* (Dart, 1925) necessitates the restoration of Weinert's (1950) original nomen, *Meganthropus africanus*, as the valid species-group name for this taxon. As a junior subjective synonym, *Australopithecus afarensis* is rejected. Following Day et al. (1978), *Paranthropus africanus* (Weinert, 1950) is thus the valid name of the robust Hadar taxon, based on the inclusion of Weinert's (1950) Garusi holotype in its hypodigm. *Paranthropus africanus* is distinguished from *P. robustus* by its more primitive dentition (Remane, 1954; Şenyürek, 1955). The dentition of *P. africanus* retains the primitive hominoid rectangular dental arcade in which the postcanine teeth are nearly parallel, and its premolars are distinctly bicuspid, rather than molariform. The P_3 of *P. africanus* is unique in having a large buccal cusp and only a small lingual cuspule (Wolpoff, 1979). This is considerably more primitive than the pattern seen in *P. robustus*, where the buccal and lingual cusps of the P_3 are subequal in size. The canines and incisors of *P. africanus* are also relatively larger than those found in *P. robustus*. Another primitive feature of the anterior dentition of *P. africanus* is that a diastema is typically present on both the upper and lower jaws. In terms of known cranial characteristics, the basicranium of A.L.333-45 differs from those of *P. robustus* and *P. boisei* primarily in having mastoid processes that are not as laterally inflated (Figs. 3a–c, 4).

Unfortunately, the specimens attributable to the gracile Hadar taxon are much less numerous than those of *Paranthropus africanus*. Although cranial material is associated with the A.L.288-1 specimen, there is, as yet, no good material of the basicranial or nasal regions known from this group. Schmid (1983) concluded from his reconstruction of A.L.288-1 that the skull of "Lucy" possessed specialized features also observed in "*A.*" *africanus*. At the present time, the majority of specimens assigned to the gracile group consists of either upper or lower jaws. The dental characteristics of specimens A.L.128-23, 198-1, 288-1, 311-1, and 411-1 are almost universally primitive in their appearance and, as Coppens (1977) and Andrews (personal communication) have pointed out, are very similar to specimens of *Sivapithecus/Ramapithecus*.

The taxonomic identity of the gracile group from the Hadar Formation has been previously in doubt (Olson, 1981). The first available species-group name for this taxon is *Australopithecus africanus aethiopicus* (Tobias, 1980b). Tobias (1980b, p. 16) concluded from a study of the Pliocene hominids "that the Laetoli and Hadar hominids belong to the same lineage as that represented by the hominids of Makapansgat Members 3 and 4 and of Sterkfontein Member 4." In addition, Tobias (1980b) proposed that "the Laetoli and Hadar hominids cannot be separated from *A. africanus* and that they represent two new subspecies of that species. Since '*A. afarensis*' is tied to a Laetoli specimen as holotype, only the Laetoli specimens should be designated *A. africanus afarensis* . . . and the Hadar fossils *A. africanus aethiopicus.*" It is clear from this statement that Tobias (1980b) proposes "*A. a. aethiopicus*" as a new species group name for the Hadar hominids. Given the above statements and the inclusion by Tobias (1980b, p. 14) of diagnostic characters, *A. a. aethiopicus* must be considered available from its date of publication as a species-group name for the taxon containing its type specimen, and it *cannot* be interpreted as conditionally proposed.

In his description of *A. africanus aethiopicus*, Tobias (1980b) unfortunately did not identify a holotype from among the Hadar Formation type-series. The A.L.288-1 associated skeleton is here designated as the lectotype of *A. africanus aethiopicus*. A full description of this specimen is given by Johanson et al. (1982a,b). The A.L.128-23, 198-1, 311-1, and 411-1 specimens are designated as paralectotypes. Johanson et al. (1982b,c), Kimbel et al. (1982), and White and Johanson (1982) provide a description of these specimens. The suggestion by Tobias (1980b), that the A.L.199-1 gracile palate differs from "*A. africanus aethiopicus*" because the bidental width of the M^2s is greater than the width of the M^3s, is inconsistent with reconstruction of the AL.199-1 specimen (Johanson and White, 1979).

The designation of the A.L.288-1 individual as the lectotype of *A. africanus aethiopicus* and the association of its cranial material with "*Australopithecus*" *africanus* and the *Homo* lineage by Schmid (1983) make possible the specific identification of the gracile Hadar hominids as *Homo (Australopithecus) aethiopicus*. At this time, *Homo (Australopithecus) aethiopicus* cannot be distinguished from *Homo (Australopithecus) africanus* on the basis of cranial morphology (Schmid, 1983). The most obvious differences that distinguish these two taxa from their common ancestor at Node 3 (Fig. 10) are in their postcanine dentitions, and these differences indicate that *H. aethiopicus* is the less derived taxon. Tooth rows of *H. aethiopicus* are typically straight, while those of *H. africanus* are convex laterally. The unicuspid P_3 in *H. aethiopicus* is more primitive than the bicuspid tooth of *H. africanus*. The lower first and second molars in *H. aethiopicus* are characterized by enlarged hypoconulids, which are not typical of *H. africanus*. It is also possible that *H. aethiopicus* may prove to be more primitive postcranially than *H. africanus*.

CONCLUSIONS

The phylogenetic interpretation of the basicranial and nasal morphology of the great apes and the Pliocene hominids presented in this paper confirms Robinson's (1953, 1963, 1972) conclusions about the distinctiveness and affinities of *H. (Australopithecus) africanus* (Dart, 1925) and *Paranthropus* (Broom, 1938). In addition, the analysis of the Hadar hominids indicates that the A.L.333-45 and -105 crania possess a series of unique features of high taxonomic relevance that identify them as primitive members of the *Paranthropus* clade. In view of the cranial evidence for *Paranthropus* in the Hadar Formation, it is suggested that the extreme morphological and metrical differences in the dentitions from Hadar can be more parsimoniously interpreted as evidence for the cladogenesis of the *Paranthropus* and *Homo* lineages. Based on a comparison of the dental assemblage associated with the A.L.333 crania and the recently enlarged Laetoli sample, Weinert's (1950) "*Meganthropus*" *africanus* is recognized as the most primitive species of *Paranthropus*. While all of the Laetoli and many of the more complete Hadar fossils are specifically attributed to *Paranthropus africanus* (Weinert, 1950), the cranial and dental morphology of other Hadar specimens, including A.L.288-1—designated here as the lectotype of *Australopithecus africanus aethiopicus* Tobias, 1980b—identify them generically as primitive members of the *Homo* clade and specifically as *Homo (Australopithecus) aethiopicus*. This interpretation of the Hadar hominids is similar to the systematic scheme initially proposed by Taieb et al. (1974). The basicranial and nasal characteristics of the paranthropines and hominines contain a number of derived features that distinguish these two clades not only from each other but also from their postulated common ancestor at Node 3 in Figure 10.

The existence of a second hominid species at Hadar is suggested by both the degree of morphological variation and, more importantly, by the total morphological pattern exhibited in the more complete individual specimens from this site. In terms of the distinct dental and cranial features that distinguish *Paranthropus africanus* from *Homo (Australopithecus) aethiopicus, H. aethiopicus* is typically more primitive in having retained more of the characteristics from its common ancestor with the paranthropines. It should be emphasized that neither Hadar hominid taxon has departed appreciably in dental morphology from their common ancestor, hence the difficulties in separating the dental remains of these two groups. The many symplesiomorphic features in these two taxa suggest that the Hadar Formation samples a period very close in time to the origin of the *Paranthropus* and *Homo* lineages (Fig. 11). They also suggest that the *Paranthropus* clade was initially the most highly specialized, while the early forms of *Homo* retained a more generalized morphology and may have been more reliant upon behavioral specializations that provided the basis for the subsequent emergence of culture in this group.

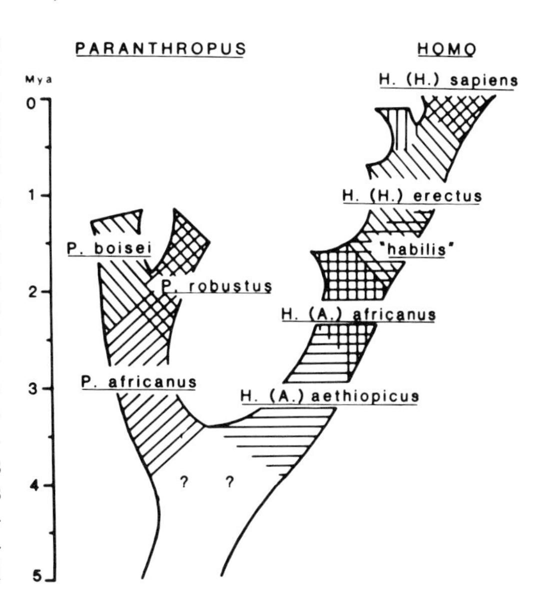

Fig. 11. Phylogeny of the Pliocene and Pleistocene hominids. M.y.a., millions of years ago.

ACKNOWLEDGMENTS

I wish to acknowledge the assistance and generosity of P. Andrews, C.K. Brain, D.C. Johanson, M.D. Leakey, R.E.F. Leakey, C.B. Stringer, and P.V. Tobias in permitting me to study the fossil hominid material in their care. Thanks are also due to Laurie Volk for preparing the figures in his paper, Frank Pace for photographing them for publication, and to Anne Sweazey, Warren Kinzey, John Barthelme and especially Eric Delson for their editorial comments on this manuscript.

Finally, I want to express my appreciation to the American Museum of Natural History and particularly to its Director, T.D. Nicholson, for the hospitality extended to all participants in the "Ancestors" Symposium. I wish to extend my deepest appreciation to E. Delson, I. Tattersall, and J. van Couvering for organizing this unique gathering of humans and for inviting me to participate.

LITERATURE CITED

Boaz, NT (1979) Hominid evolution in eastern Africa during the Pliocene and early Pleistocene. Ann. Rev. Anthropol. *8*:71–85.

Boaz, NT (1983) Morphological trends and phylogenetic relationships from Middle Miocene hominoids to Late Pliocene hominids. In RL Ciochon and RS Corruccini (eds): New Interpretations of Ape and Human Ancestry. New York: Plenum, pp. 705–720.

Bonde, N (1981) Problems of species concepts in palaeontology. In J Martinelli (ed): Concept and Method in Paleontology. Barcelona: University of Barcelona, pp. 19-34.

Broom, R (1938) The Pleistocene anthropoid apes of South Africa. Nature *142:*377–379.

Clarke, RJ (1977) The cranium of the Swartkrans hominid SK-847, and its relevance to human origins. Unpublished Ph.D. thesis, University of the Witwatersrand, Johannesburg.

Coppens, Y (1977) Evolution morphologique de la première prémolaire inférieure chez certains Primates supérieurs. C.R. Acad. Sci. [D] (Paris) *285:*1299–1302.

Coppens, Y (1983) Systematique, phylogenie, environnement et culture des australopitheques, hypotheses et synthese. Bull. Mém. Soc. Anthropol. (Paris), Sér. XIII, *10:*273–284.

Cronin, JE, Boaz, NT, Stringer, CB, and Rak, Y (1981) Tempo and mode in hominid evolution. Nature *292:*113–122.

Dart, RA (1925) *Australopithecus africanus*, the man-ape of South Africa. Nature *115:*195–199.

Day, MH, Leakey, MD, and Olson, TR (1978) On the status of *Australopithecus afarensis*. Science *207:*1102–1103.

Falk, D, and Conroy, GC (1983) The cranial venous sinus system in *Australopithecus afarensis*. Nature *306:*779–781.

Farris, JS (1977) On the phenetic approach to vertebrate classification. In MK Hecht, PC Goody, and BM Hecht (eds): Major Patterns in Vertebrate Evolution. New York: Plenum, pp. 823–850.

Farris, JS (1980) The information content of the phylogenetic system. Syst. Zool. *27:*483–519.

Grine, FE (1981) Trophic differences between 'gracile' and 'robust' australopithecines: A scanning electron microscope analysis of occlusal events. S. Afr. J. Sci. *77:*203–230.

Hennig, W (1966) Phylogenetic Systematics. Urbana: University of Illinois Press.

Hollinshead, WH (1982) Anatomy for Surgeons, 3rd ed., Vol. I. Philadelphia: Harper & Row.

Holloway, RL (1983) Cerebral brain endocast pattern of *Australopithecus afarensis* hominid. Nature *303:*420–422.

Howell, FC (1978) Hominidae. In VJ Maglio and HBS Cooke (eds): Evolution of African Mammals. Cambridge, Massachusetts: Harvard University Press, pp. 154–248.

Howells, WW (1973) Cranial variation in man: A study by multivariate analysis of patterns of difference among recent human populations. Pap. Peabody Mus. *67:*1–259.

Hublin, JJ (1978) Le torus occipital transverse et les structures associées: Évolution dans le genre *Homo*. Thèse 3eme cycle, University of Paris VI.

Hublin, JJ (1983) Les superstructures occipitales chez les prédécesseurs d'*Homo erectus* en Afrique: Quelques remarques sur l'origine du torus occipital transverse. Bull. Mém. Soc. Anthropol. (Paris) Sér. XIII, *10:*303–312.

Johanson, DC (1976) Ethiopia yields first "Family" of early man. Natn. Geogr. Mag. *150:*790–811.

Johanson, DC, and Edey, MA (1981) Lucy—the beginnings of humankind. New York: Simon and Schuster.

Johanson, DC, Lovejoy, CO, Kimbel, WH, White, TD, Ward, SC, Bush, ME, Latimer, BM, and Coppens, Y (1982a) Morphology of the Pliocene partial hominid skeleton (A.L.288-1) from the Hadar Formation, Ethiopia. Am. J. Phys. Anthropol. *57:*403–451.

Johanson, DC, Taieb, M, and Coppens, Y (1982b) Pliocene hominids from the Hadar Formation, Ethiopia (1973–1977): Stratigraphic, chronologic, and paleoenvironmental contexts, with notes on hominid morphology and systematics. Am. J. Phys. Anthropol. *57:*373–402.

Johanson, DC, and White, TD (1979) A systematic assessment of early African hominids. Science *202:*321–330.

Johanson, DC, White, TD, and Coppens, Y (1978) A new species of the genus *Australopithecus* (Primates: Hominidae) from the Pliocene of Eastern Africa. Kirtlandia *28:*1–14.

Johanson, DC, White, TD, and Coppens, Y (1982c) Dental remains from the Hadar Formation, Ethiopia: 1974–1977 collections. Am. J. Phys. Anthropol. *57:*545–603.

Jolly, CJ (1970) The seed-eaters: A new model of hominid differentiation based on a baboon analogy. Man *5:*5–26.

Kennedy, GE (1980) Paleoanthropology. New York: McGraw Hill.

Kimbel, WH (1984) Variation in the pattern of cranial venous sinuses and hominid phylogeny. Am. J. Phys. Anthropol. *63:*243–263.

Kimbel, WH, Johanson, DC, and Coppens Y (1982) Pliocene cranial remains from the Hadar Formation, Ethiopia. Am. J. Phys. Anthropol. *57:*453–499.

Kimbel, WH, White, TD, and Johanson, DC (1984) Cranial morphology of *Australopithecus afarensis*: A comparative study based on a composite reconstruction of the adult skull. Am. J. Phys. Anthropol. *64:*337–388.

Laitman, JT (1977) The ontogenetic and phylogenetic development of the upper respiratory system and basicranium in man. Ph.D. dissertation, Yale University.

Laitman, JT, and Heimbuch, RC (1982) The basicranium of Plio-Pleistocene hominids as an indicator of their upper respiratory system. Am. J. Phys. Anthropol. *59:*323–343.

Le Gros Clark, WE (1964) The Fossil Evidence of Human Evolution, 2nd ed. Chicago: University of Chicago Press.

Le Gros Clark, WE (1971) The Antecedents of Man, 3rd ed. Chicago: Quadrangle Books.

Lewin, R (1984) Ancestors worshipped. Science *244:*477–479.

Logan, TR, Lucas, SG, and Sobus, JC (1983) The taxonomic status of *Australopithecus afarensis* Johanson in Hinrichsen 1978 (Mammalia, Primates). Haliksa'i: Univ. New Mexico Contrib. Anthropol. *2:*16–27.

McHenry, HM (1984) Relative cheek tooth size in *Australopithecus*. Am. J. Phys. Anthropol. *64:*297–306.

Olson, TR (1978) Hominid phylogenetics and the existence of *Homo* in Member I of the Swartkrans Formation, South Africa. J. Hum. Evol. *7:*159–178.

Olson, TR (1981) Basicranial morphology of the extant hominoids and Pliocene hominids: The new material from the Hadar Formation, Ethiopia, and its significance in early human evolution and taxonomy. In CB Stringer (ed): Aspects of Human Evolution. London: Taylor and Francis, pp. 99–128.

Pilbeam, D (1982) New hominoid skull material from the Miocene of Pakistan. Nature *295:*232–234.

Rak, Y (1983) The Australopithecine Face. New York: Academic Press.

Remane, A (1954) Structure and relationships of *Megan-*

thropus africanus. Am. J. Phys. Anthropol. *12:*123–126.

Robinson, JT (1958) *Telanthropus* and its phylogenetic significance. Am. J. Phys. Anthropol. *11:*445–501.

Robinson, JT (1958) Cranial cresting patterns and their significance in the Hominoidea. Am. J. Phys. Anthropol. *16:*397–428.

Robinson, JT (1963) Adaptive radiation in the Australopithecines and the origin of man. In FC Howell and JF Bourliere (eds): African Ecology and Human Evolution. Chicago: Aldine, pp. 385–416.

Robinson, JT (1972) Early Hominid Posture and Locomotion. Chicago: University of Chicago Press.

Rouviere, H (1954) Anatomie Humaine, 7th ed., tome I. Paris: Masson et Cie.

Sakka, M (1972) Anatomie comparée de l'écaille de l'occipital (squama occipitalis P.N.A.) et des muscles de la nuque chez l'homme et les Pongidés, Pt. 1. Mammalia *36:*696–750.

Sakka, M (1973) Anatomie comparée de l'écaille de l'occipital (squama occipitalis P.N.A.) et des muscles de la nuque chez l'homme et les Pongidés, Pt. 2. Mammalia *37:*126–191.

Schmid, P (1983) Eine Rekonstruktion des Skelettes von A.L.288-1 (Hadar) und deren Konsequenzen. Folia Primatol. *40:*283–306.

Senut, B (1983) Les hominides Plio-Pleistocenes: Essai taxinomique et phylogenetique a partir de certains os longs. Bull. Mém. Soc. Anthropol. (Paris) Sér. XIII, *10:*325–334.

Senut, B, and Tardieu, C (1985) Functional aspects of Plio-Pleistocene hominid limb bones: Implications for taxonomy and phylogeny. In E Delson (ed): Ancestors: The Hard Evidence. New York: Alan R. Liss, Inc., pp. 193–201.

Şenyürek, M (1955) A note on the teeth of *Meganthropus africanus* Weinert from Tanganyika Territory. Türk Tarih Kur. Bell. *73:*1–55.

Shapiro, HL (1929) Contributions to the craniology of Central Europe. I. Crania from Greifenberg in Carinthia. Anthropol. Pap. Am. Mus. Nat. Hist. *31:*1–120.

Simpson, GG (1961) Principles of Animal Taxonomy. New York: Columbia University Press.

Stern, JT, and Susman, RL (1983) The locomotor anatomy of *Australopithecus afarensis*. Am. J. Phys. Anthropol. *60:*279–317.

Taieb, M, Johanson, DC, Coppens, Y, Bonnefille, R, and Kalb, J (1974) Découverte d'Hominidés dans les séries plio-pléistocènes d'Hadar (Bassin de l'Awash; Afar, Éthiopie). C.R. Acad. Sci. [D] (Paris) *279:*735–738.

Tardieu, C (1983) L'articulation du genou des primates catarhiniens et hominides fossiles: Implications phylogenetique et taxinomique. Bull. Mém. Soc. Anthropol. (Paris) Sér. XIII, *10:*355–372.

Taxman, RM (1963) Incidence and size of the juxtamastoid eminence in modern crania. Am. J. Phys. Anthropol. *21:*153–157.

Tobias, PV (1967) The cranium and maxillary dentition of *Australopithecus (Zinjanthropus) boisei*. In LSB Leakey (ed): Olduvai Gorge, Vol. 2. Cambridge: Cambridge University Press.

Tobias, PV (1973) Implications of the new age estimates of the early South African hominids. Nature *246:*79–83.

Tobias, PV (1978) The South African australopithecines in time and hominid phylogeny, with special reference to the dating and affinities of the Taung skull. In CJ Jolly (ed): Early Hominids of Africa. London: Duckworth, pp. 45–84.

Tobias, PV (1980a) A survey and synthesis of the African hominids of the late Tertiary and early Quaternary periods. In LK Konigsson (ed): Current Argument on Early Man. Oxford: Pergamon, pp. 86–113.

Tobias, PV (1980b) "*Australopithecus afarensis*" and *A. africanus*: Critique and an alternative hypothesis. Palaeont. Afr. *23:*1–17.

Tobias, PV (1985) Single characters and total morphological patterns redefined: The sorting effected by a selection of morphological features of the early hominids. In E Delson (ed): Ancestors: The Hard Evidence. New York: Alan R Liss, Inc., pp. 94–101.

Walker, A (1981) The Koobi Fora hominids and their bearing on the origins of the genus *Homo*. In BA Sigmon and JS Cybulski (eds): *Homo erectus* Papers in Honor of Davidson Black. Toronto: University of Toronto Press, pp. 191–215.

Ward, SC, and Kimbel, WH (1983) Subnasal alveolar morphology and the systematic position of *Sivapithecus*. Am. J. Phys. Anthropol. *61:*157–171.

Weidenreich, F (1943) The skull of *Sinanthropus pekinensis:* A comparative study on a primitive hominid skull. Palaeont. Sinica, new ser. D, 10 (whole series no. 127):1–485.

Weinert, H (1950) Über die Neuen Vor- und Frühmenschenfunde aus Afrika, Java, China and Frankreich. Z. Morphol. Anthropol. *42:*113–148.

White, TD (1977a) Anterior mandibular corpus of early African Hominidae. Ph.D. dissertation, University of Michigan, Ann Arbor.

White, TD (1977b) New hominids from Laetoli, Tanzania. Am. J. Phys. Anthropol. *46:*197–230.

White, TD, and Johanson, DC (1982) Pliocene hominid mandibles from the Hadar Formation, Ethiopia: 1974–1977 collections. Am. J. Phys. Anthropol. *57:*501–544.

White, TD, Johanson, DC, and Kimbel, WH (1981) *Australopithecus africanus*: Its phyletic position reconsidered. S. Afr. J. Sci. *77:*445–470.

Wiley, EO (1981) Phylogenetics. New York: John Wiley.

Williams, PL, and Warwick, R (1980) Gray's Anatomy, 36th ed. Philadelphia: Saunders.

Wolpoff, MH (1979) Anterior dental cutting in the Laetolil hominids and the evolution of the bicuspid P_3. Am. J. Phys. Anthropol. *51:*233–234.

Wolpoff, MH (1983) Australopithecines: The unwanted ancestors. In KJ Reichs (ed): Human Origins. Washington, D.C.: University Press of America, pp. 109–126.

NOTE ADDED IN PROOF

Kimbel et al. (1984) appeared subsequent to the submission of this manuscript. The issues raised with regard to the methodology, terminology, data, and conclusions of Olson (1978, 1981) will be discussed in a forthcoming publication.

Ancestors: The Hard Evidence, pages 120–137

Craniodental Morphology of the Hominids From Hadar and Laetoli: Evidence of "*Paranthropus*" and *Homo* in the Mid-Pliocene of Eastern Africa?

William H. Kimbel, Tim D. White, and Donald C. Johanson
Institute of Human Origins, Berkeley, California 94709 (W.H.K., D.C.J.); Departments of Anthropology and Biology, Kent State University, Kent, Ohio 44242 (W.H.K.); Department of Anthropology, University of California, Berkeley, California 94720 (T.D.W.)

ABSTRACT Since the naming of the Pliocene hominid species *Australopithecus afarensis* in 1978, diverse alternative systematic revisions of the Hadar and Laetoli craniodental remains attributed to this taxon have been formulated. In this paper we critically examine one of these interpretations, that offered by Olson (1981). On the basis of the A.L.333-45 mastoid region, Olson perceives the presence of "*Paranthropus*" at the Hadar site. His examination of the Hadar/Laetoli dentognathic remains suggests the presence of "*Paranthropus*" and "*Homo*." Olson's two-taxon interpretation cannot be sustained either metrically or morphologically. Furthermore, we contend that Olson has misinterpreted the morphological variation that exists in the mastoid region of both extant hominoids and Plio-Pleistocene hominids, resulting in his erecting a typological set of criteria to distinguish early hominoid taxa. In light of all the evidence, the *A. afarensis* cranial base exhibits a primitive, rather than derived, morphological pattern.

INTRODUCTION

Since the naming of the Pliocene hominid species *Australopithecus afarensis* (Johanson et al., 1978), paleoanthropologists have devised a plethora of alternative systematic interpretations of the 3.0–3.6 million-year-old craniodental remains from Hadar and Laetoli attributed to this taxon. In this presentation we critically review the detailed systematic revision proposed by Olson (1981).[1]

In contrast to the phylogenetic scheme of Johanson and White (1979; see also White et al., 1981; Rak, 1983; Kimbel et al., 1984), wherein *A. afarensis* is seen as a primitive stem hominid ancestral to the *Homo* clade on the one hand and the *A. africanus*-"robust" *Australopithecus* clade on the other, Olson (1981) concludes that *A. afarensis* actually represents two species that he distinguishes at the generic level: *Paranthropus africanus* (Weinert)[2] and *Homo* (*Australopithecus*) sp. indet.[3] Two issues are involved here. Olson's division of the combined Hadar/Laetoli sample into two taxa implies that the morphological and metrical variation observed in the collection exceeds that charac-

[1]In this paper, we do not specifically address Olson's (1985) restatement of his views, which are presented in the preceding article; however, see p. 136 for one comment.

[2]Olson's *Paranthropus* = *A. robustus* (Broom) + *A. boisei* (Leakey). The origin of the binomen *Paranthropus africanus* is traced to Weinert's (1950) assignment of the Garusi I maxillary fragment to *Meganthropus africanus*.

[3]Olson's *Homo* (*Australopithecus*) = *A. africanus* (Dart) + *H. habilis* (Leakey, Tobias, and Napier), distinct from *Homo* (*Homo*), which includes *H. erectus* (Dubois), *H. neanderthalensis* (King), and *H. sapiens* Linnaeus. See Olson (1978).

terizing samples of any other hominoid species. The other issue concerns the identification of valid shared-derived traits (synapomorphies) linking discrete subsets of the Hadar/Laetoli collection to either "*Homo*" or "*Paranthropus*."

The primary evidence marshalled by Olson (1981) for the presence of "*Paranthropus*" in the East African mid-Pliocene is the cranial base morphology of Hadar specimen A.L.333-45. Olson identified several features in the mastoid region of this partial calvaria that he interpreted as unique specializations shared with "*P.*" *robustus* (Swartkrans and Kromdraai) and "*P.*" *boisei* (Olduvai Gorge and Koobi Fora): "Extensively pneumatized and projecting mastoid processes, the incorporation of the occipitomastoid crest into the expanded medial wall of the process and the origination of the digastric muscle from a shallow oval fossa located entirely on the anteromedial surface of the mastoid . . ." (1981, p. 116).

The second line of evidence adduced by Olson pertains to dental, palatal, and mandibular morphology, upon which basis two taxa are diagnosed (no cranial base specimens are assigned by Olson to *Homo*). All specimens from Laetoli (including Garusi I) and Afar Locality 333/333w, plus A.L.145-35, 188-1, 200-1a, 207-13, 266-1, and 400-1a are attributed to *Paranthropus africanus* because they allegedly share with "*P.*" *robustus* the following derived characters (relative to Olson's "*Homo*" specimens): reduced relative canine size, incipient premolar molarization, enlarged molars, and a relatively reduced "inferior mandibular torus" (sic). *Paranthropus africanus* is distinguished from the younger "*P.*" *robustus* sample by four primitive retentions: relatively larger incisors and canines, bicuspid rather than molariform premolars, high frequency of $I^2/\underline{C}$ and $\overline{C}/P_3$ diastemata, and a rectangular dental arcade with nearly parallel postcanine tooth rows. (These traits are among those used by Johanson et al. (1978) to diagnose *A. afarensis.*)

In contrast, *Homo* sp. indet. at Hadar (A.L.128-23, 198-1, 199-1, 288-1, 311-1, and 411-1) is defined by Olson almost exclusively on the basis of primitive dental and mandibular morphology. Only one character rescues this hominid taxon from "waste basket" (paraphyletic *sensu* Hennig) status: the inferior angulation of the anterior palatal surface in A.L.199-1, which alledgedly "shows the beginnings of the specialized condition more fully derived in the later hominines" (1981, p. 121).

The major implication of Olson's study is that the common ancestor of the "*Paranthropus*" and *Homo* lineages predates 4.0 million years. "It also suggests that the *Paranthropus* lineage was initially the most highly specialized morphologically, while the early forms of *Homo* retained an even more generalized morphology . . ." (1981, p. 123).

THE EVIDENCE

Cranial Base

Pneumatization and projection of the mastoid process. Of the A.L.333-45 cranium Olson (1981, p. 115) writes: "Its mastoid processes are . . . pneumatized and inflated in all directions with the medial and inferior enlargement being the most notable. While the lateral inflation of the mastoids is . . . more extensive than that seen in any of the extant great apes . . . the tip of each mastoid retains the primitive appearance of being inflected under the basicranium. However, the extent of medial and inferior inflation of the mastoids . . . exceeds the condition seen in any of the great apes."

According to Olson's description, A.L.333-45 exhibits derived features that deviate from the extant great ape condition and ally it with "*Paranthropus*" (e.g., O.H. 5, KNM-ER 406 and 407, TM 1517): the degree of lateral, medial, and inferior inflation of the mastoid processes. This determination is based on a polarity scale along which the female gorilla (whose morphology is accepted by Olson as approximating that of the hypothetical ancestor of African apes and hominids) occupies the primitive pole and "*Paranthropus*" the derived pole. *Homo sapiens* is seen by Olson to exhibit a set of morphological specializations in the mastoid region that depart from both the hypothetical ancestral morphotype and "*Paranthropus*": "[the mastoid process] is typically large, inflated laterally and projecting in a prominent finger-like process. Its overall orientation below the skull is in a parasagittal plane and its lateral margin is frequently inflated beyond the supramastoid crest" (1981, p. 105). Specimens of *A. africanus* (e.g., Sts 5, MLD 37/38) are attributed to *Homo* in part because of their "smaller and less inflated mastoid morphologies" (1981, p. 116) compared to A.L.333-45 and "*Paranthropus*."

As we argued recently (Kimbel et al., 1984, p. 379), Olson's phylogenetic interpretation of the A.L.333-45 mastoid region is mainly influenced by his misinterpretation of ape cranial base morphology: "the small appearance of the mastoid in these forms [apes] is primarily a reflection of the minimal amount of pneumatization in this region" (1981, p. 104), and "the mastoid pneumatization and enlargement . . . [in A.L.333-45] . . . represent a radical departure from the presumed primitive hominoid condition where the mastoids are neither highly pneumatized nor enlarged" (1981, p. 115). That this characterization is simply inaccurate can be substantiated by reference to any ape cranium in which the mastoid region is damaged. In *all* apes the entire temporal squama, including the mastoid process and the supramastoid crest, is invaded by an extensive network of air cells, a fact thoroughly documented over 50 years ago by Hofmann (1926; see also Groth, 1937).

The extent to which temporal bone pneumatization influences the inferior projection of the mastoid process is notoriously variable. The large degree of variation in mastoid projection in the African apes has been recorded by Ashton and Zuckerman (1952) and by Schultz (1950, 1952). While it is true that the mastoids in many hominid skulls project below the cranial base more than in most apes, this is far from universal. The distinctive pyramidal form of the mastoid process in *Homo* does not appear to be the result of an *increase* in pneumatization, as implied by Olson, but rather can be viewed as a consequence of the restriction of temporal pneumatization to the mastoid process *per se*, its increased verticality (versus the primitive condition of inflected mastoids), and a deepening of the digastric fossa (see below).

The crania pictured in Figure 1 illustrate the normal variation encountered in large samples of extant hominoid crania. When compared to the variation observed in the African great apes, the inferior and lateral projection of the mastoids in A.L.333-45 is not remarkable (Fig. 2). Indeed, our study of African ape mastoid morphology reveals that A.L.333-45 is most similar to female gorillas, the basicranial anatomy of which—according to Olson—most closely approximates the presumed hypothetical ancestor of the African apes and hominids!

As we showed elsewhere (Kimbel et al., 1984, p. 379), crania of *A. afarensis* also exhibit variation in the inferior and lateral projection of the mastoid process. While the mastoids in A.L.333-45 project below the level of the occipitomastoid crests (Fig. 2A), in A.L.333-84 and A.L.333-12 the mastoid tip and occipitomastoid crest are of subequal prominence inferiorly (Fig. 2C). In fact, A.L.333-84 and 333-112 agree with Olson's description of ape mastoid morphology: "In many specimens the mastoid process . . . is hardly more than a very short, robust crest . . . the tip of the mastoid rarely projects below the level of the occipitomastoid crest" (1981, p. 104).

As far as lateral inflation of the mastoid is concerned, the A.L.333-84 and 333-45 mastoids project to the edge of the prominent supramastoid crests, while in A.L.333-112 the supramastoid crest is slightly more projecting laterally. African apes (especially gorillas) exhibit similar variation in the lateral projection of the mastoid process, as is evident in Figure 1. Female gorilla crania in which the mastoid protrudes laterally beyond the supramastoid crest are not rare, thus casting doubt on the validity of Olson's categorical claim that the great ape supramastoid crest "is the most prominent structure of the mastoid region in occipital profile" (1981, p. 103).

In view of these facts, we find it difficult to understand exactly what features of the A.L.333-45 mastoid process indicate special affinity with "*Paranthropus*." In fact, the mastoids of "*P.*" *robustus* and "*P.*" *boisei* are highly distinctive and can be distinguished easily from those of other hominoid taxa.

As Olson correctly observes, the enlargement of the mastoid in "*Paranthropus*" is extreme: ". . . the tip of the mastoid . . . projects well below the level of the occipital condyles, while laterally the specialized mastoids project beyond the margins of the supramastoid crest" (1981, p. 110). This morphology is observed in O.H. 5, KNM-ER 406, 407, and 732, Omo 323-76-898, and SK 46, 52, and 83. Contrary to Olson's depiction of A.L.333-45, the mastoid process does *not* project below the occipital condyles. Olson apparently neglected to correct for the upward crushing of the left anterior part of the nuchal plane that was not restored when this specimen was reconstructed (see Kimbel et al., 1984). The right occipital condyle is in the anatomically correct position, and even though the right mastoid tip is abraded, it is clear that it did not project below the condyle.

Mastoid process morphology in "*Paranthropus*" imparts a unique occipital profile to crania of this taxon, as Olson (1978) has previously noted (Fig. 2B). The supramastoid crest is usually a robust, inflated structure, but its lateral projection is exceeded by that of the mastoid, the lateral surface of which bears a distinct mastoid crest denoting the insertions of *mm. splenius capitis, longissimus capitis* and *sternocleidomastoideus.* Consequently, the supramastoid and mastoid crests are always well separated on the *pars mastoidea* of "*Paranthropus*" crania, even in those specimens that have compound temporal/nuchal (T/N) crests (e.g., O.H. 5, Omo 323-76-898; contrary to Olson's claim, compound T/N crests do *not* occur "consistently" in "*Paranthropus*"). In contrast, the compound T/N crest in *A. afarensis* specimens A.L.333-45 and 333-84 extends onto the *pars mastoidea*, and the supramastoid and mastoid crests are not individually discernible until approximately halfway between asterion and porion (Kimbel et al., 1982). Thus, the occipital profile characteristic of "*Paranthropus*" crania is not observed in *A. afarensis.* This difference most likely relates to the differential development of posterior *m. temporalis* fibers in *A. afarensis* (White et al., 1981; Kimbel et al., 1984). In "*Paranthropus*" the anterior fibers of this muscle are emphasized, and the supramastoid crest (which marks the postero-inferior limit of the *m. temporalis* origin) is separated from the mastoid crest by a well-defined supramastoid sulcus (Fig. 2B).

As shown by Clarke (1977), crania of *A. africanus* (*sensu stricto*) also have laterally inflated mastoids. As in "*Paranthropus*," the mastoids of specimens Sts 5 and MLD 37/38 project laterally beyond the supramastoid crests (Fig. 2D). Although Olson (1978) attributes these crania to "*H.*" *africanus*, their mastoids are notably more laterally inflated than those of KNM-ER 1813 (Fig. 2C) or probably SK 847, specimens that are included in "*H.*" *africanus* by Olson but whose mastoids are demonstrably more similar to the relatively uninflated mastoids of some *H. erectus* crania (e.g., KNM-ER 3733, Sangiran 2, *Sinanthropus* Skull III). This is not meant to imply that mastoid morphology like that seen in KNM-ER 1813 would not be expected in some *A. africanus* (*sensu stricto*) specimens. However, we are disturbed by Olson's willingness to accept a large amount of variation in his "*H.*" *africanus*, while at the same time refusing to acknowledge such variation in *A. afarensis* and the extant African apes.

The occipitomastoid crest and suture, occipital groove, and digastric fossa. Olson (1978, 1981) accords much importance to the topographic relationship of the occipitomastoid crest to the occipitomastoid suture and groove for the occipital artery in determining the phylogeny of Plio-Pleistocene hominids. In "*Paranthropus*" the primitive, apelike arrangement of these structures is retained: the occipitomastoid suture and the occipital groove are superimposed on the occipitomastoid crest, which separates the digastric fossa (marking the origin of posterior *m. digastricus*) of the temporal bone from the nuchal plane of the occipital bone.

However, "*Paranthropus*" anatomy differs from the ape pattern in that "On their medial surfaces, the mastoids are inflated to the point that the occipitomastoid crest and the area of origin for the digastric muscle have become incorporated into the processes" (Olson, 1981, p. 110; in his 1978 paper Olson applies this description only to "*P.*" *boisei*, but in 1981 it is viewed as a generic character). This contrasts with the ape condition where "the digastric muscle originates in a very shallow oval to v-shaped fossa [which] lies between the tip of the mastoid process laterally and the occipitomastoid crest medially" (1981, pp. 105–106).

According to Olson (1978, 1981) *Homo* (including *A. africanus* of other authors) is distinguished from "*Paranthropus*" and the extant apes in that the "digastric fossa [is] separated from the occipitomastoid suture and occipital groove by the occipitomastoid crest . . ." (1978, p. 166). Olson (1981) contends that the occipitomastoid crest in *Homo* was displaced laterally so that it lies entirely on the temporal bone, rather than straddling the occipitomastoid suture.

Hadar cranium A.L.333-45 is attributed to the "*Paranthropus*" clade by Olson owing to its primitive occipitomastoid configuration (1981: Figure 5) and its derived pattern of medial mastoid inflation (the incorporation of the digastric origin and occipitomastoid

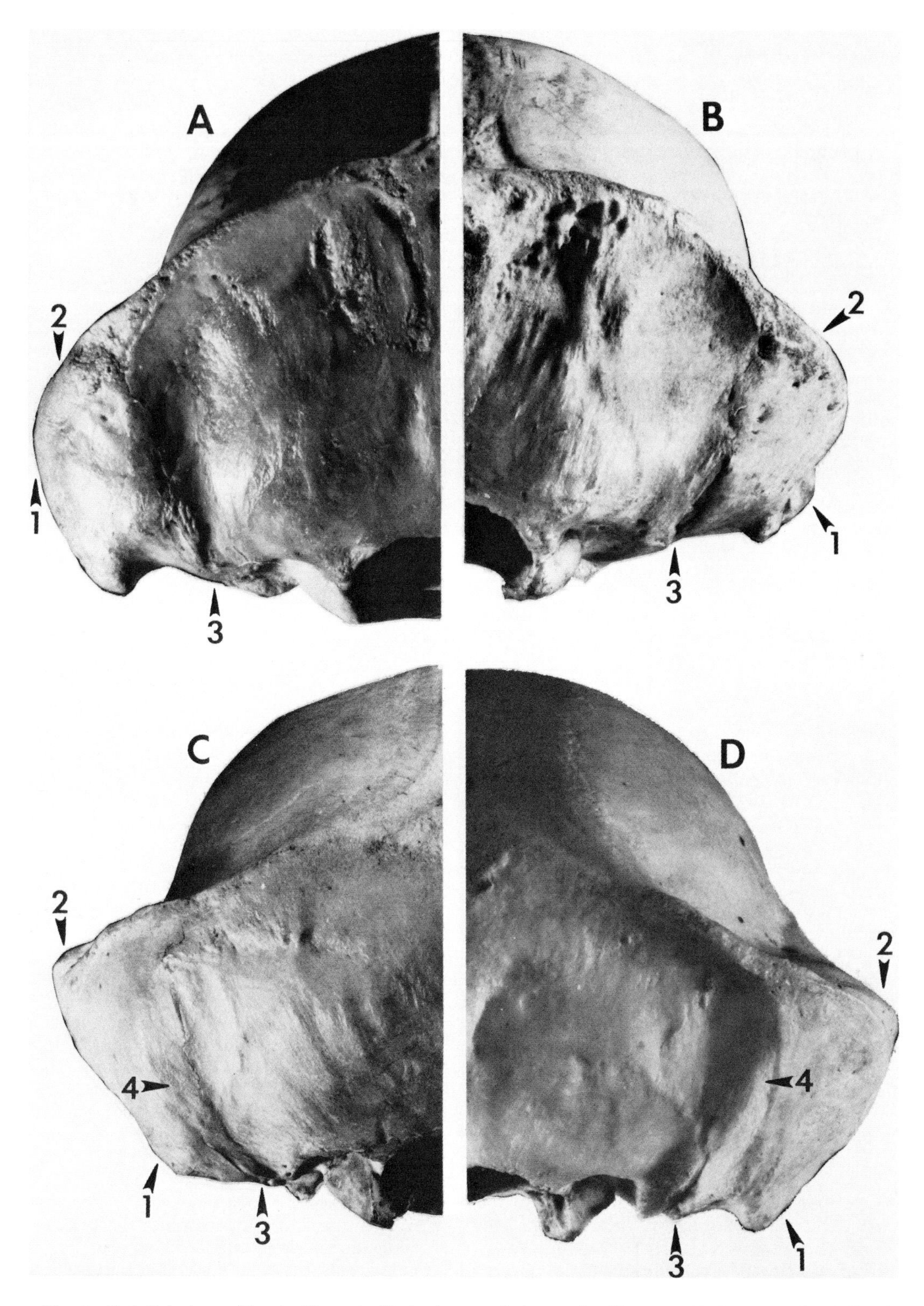

Fig. 1. Occipital views of hominoid crania illustrating variation in mastoid inflation. (A, B) *Gorilla* females. Specimen A has a mastoid process (1) that is inflated lateral to the supramastoid crest (2). Note also that the entire mastoid is inflated inferiorly below the occipitomastoid crest (3), in contrast to the small, nipple-like mastoid process seen on cranium B. (C, D) *Pan troglodytes* males. The mastoid of specimen D is inflated inferiorly below the occipitomastoid crest. Note also that in both crania the superior oblique muscle insertion (4) spans the occipitomastoid suture and becomes confluent with the occipitomastoid crest anteriorly (see also Fig. 3C). All natural size.

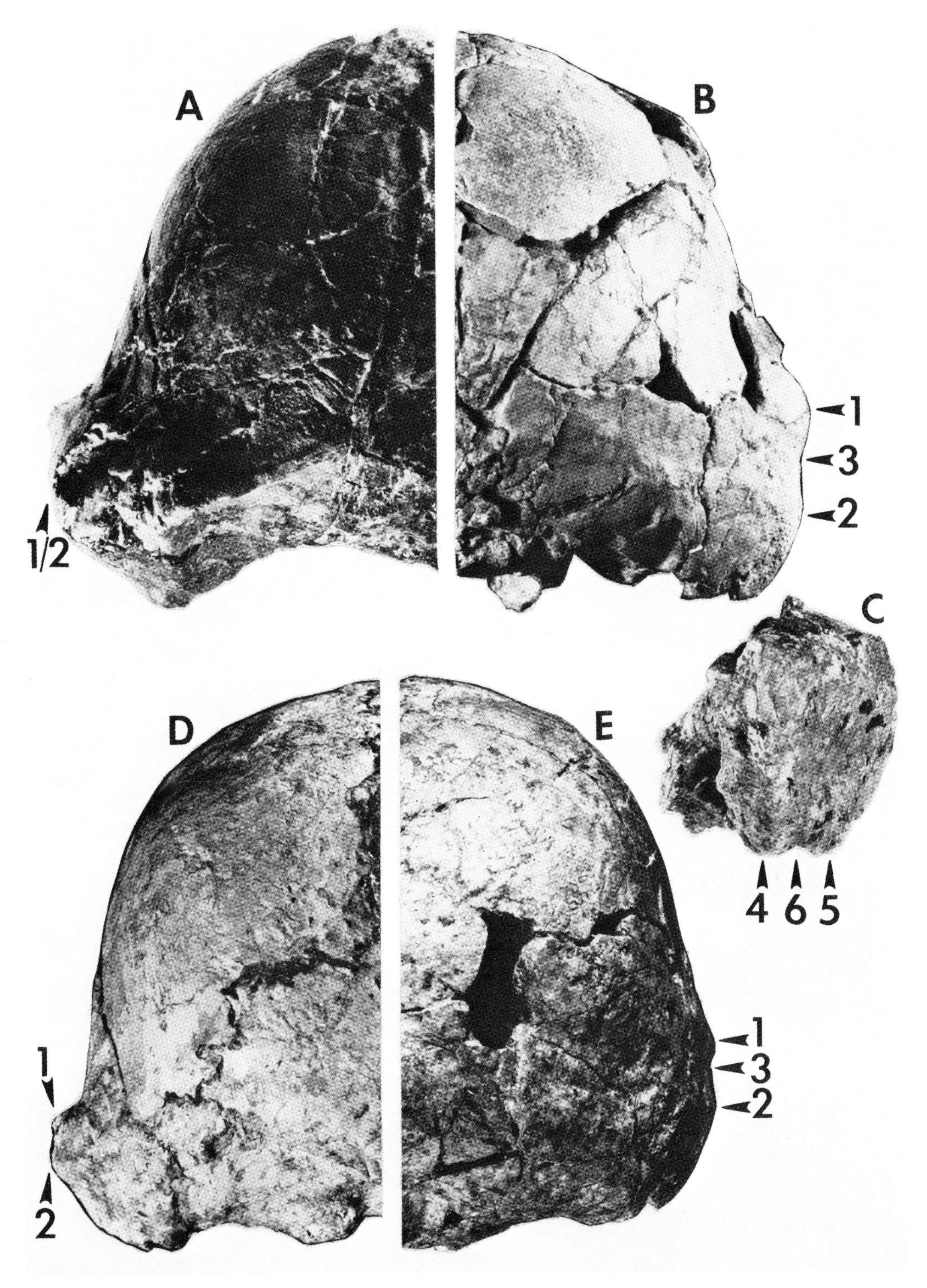

Fig. 2. Occipital views of hominid crania. (A) *A. afarensis*, A.L. 333-45; (B) *A. boisei*, KNM-ER 407; (C) *A. afarensis*, A.L. 333-84. In A and C note that the mastoid (1) and supramastoid (2) crests are merged on the *pars mastoidea*, in contrast to B, where these crests are divided by the supramastoid sulcus (3). In C note the minimal projection of the mastoid tip (5) below the occipitomastoid crest (4) and digastric fossa (6); the fossa is a clearly defined depression between the mastoid tip and the prominent occipitomastoid crest. (D) *A. africanus*, MLD 37/38; (E) *H. habilis*, KNM-ER 1813. Note the much greater lateral inflation of the mastoid region in D; this inflation carries the mastoid crest lateral to the supramastoid crest. In (E) the pneumatization of this entire region is reduced, and the mastoid process is oriented close to the sagittal plane. All natural size.

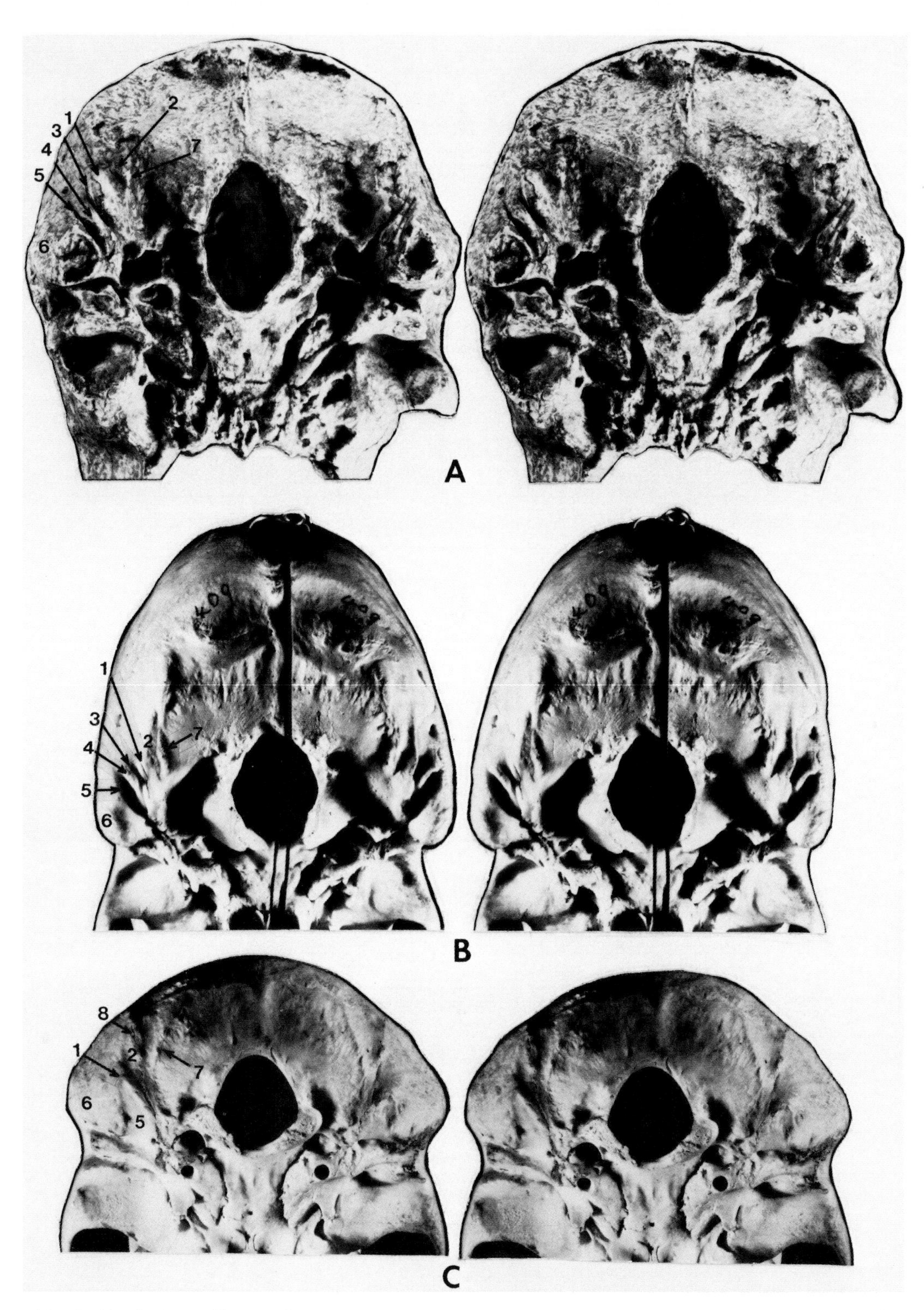

Fig. 3. Basal views of hominoid crania (stereo). (A) Solo XI; (B) modern human; (C) *P. troglodytes*. In A and B note that the occipitomastoid crest and suture (1) are superimposed; the crest forms the lateral margin of the superior oblique insertion (2). Lateral to the crest are the occipital groove (3), the juxtamastoid eminence (4), the digastric fossa (5), and the mastoid process (6). In C note the superior oblique insertion extending well onto the temporal bone; its lateral margin is confluent with the occipitomastoid crest, which is located almost entirely on the temporal bone (compare to MLD 37/38, Fig. 4A). (7) indicates the inferior nuchal line; (8) indicates the occipitomastoid suture on specimens in which it and the occipitomastoid crest are not superimposed. One-half natural size.

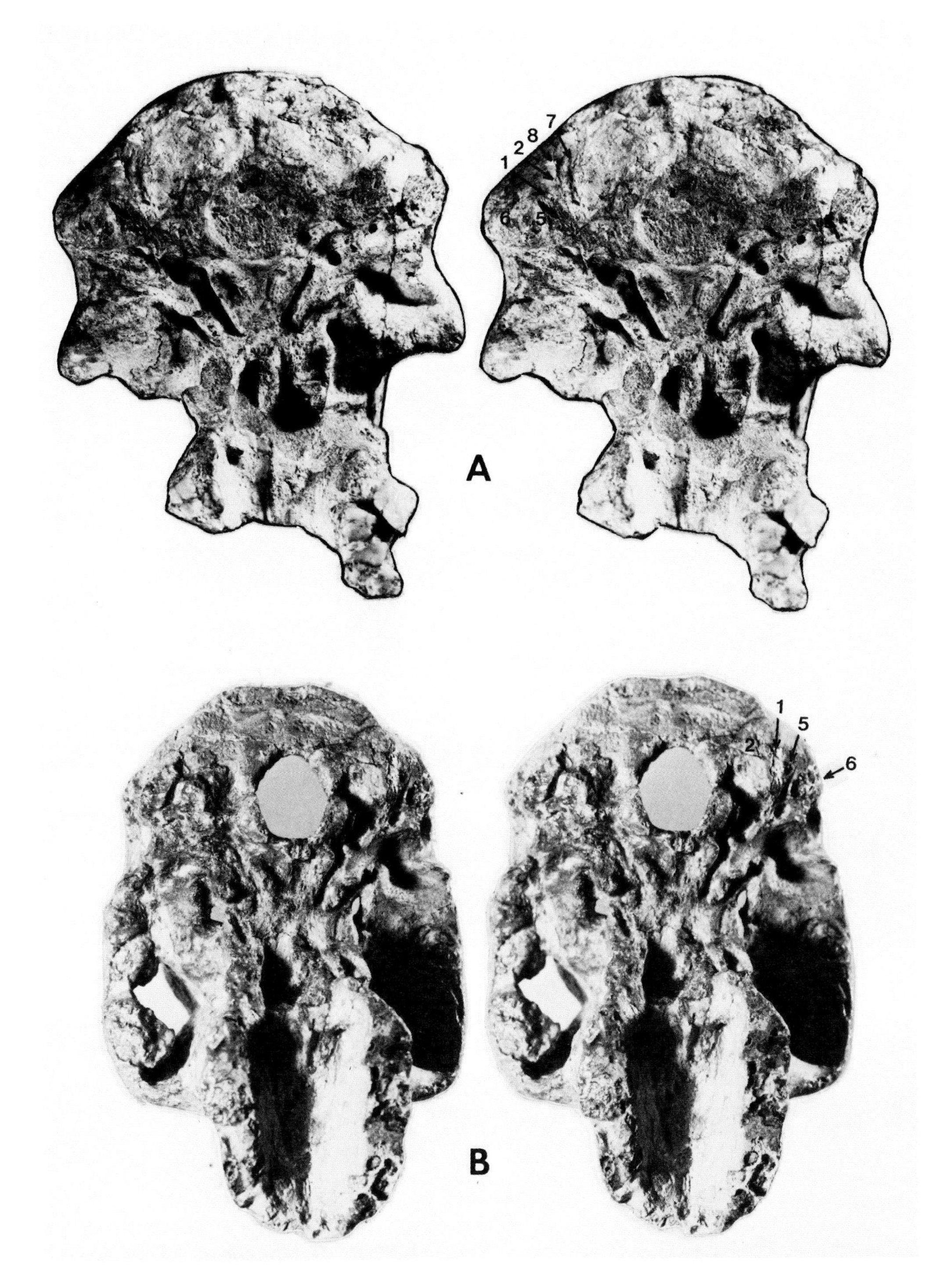

Fig. 4. Basal views of *A. africanus* crania (stereo). (A) MLD 37/38; (B) Sts 5. The superior oblique muscle insertion spans the occipitomastoid suture in MLD 37/38 as in many apes (see Fig. 3C); the occipitomastoid crest is continuous with the lateral margin of this muscle's area of insertion. On Sts 5 the occipitomastoid crest and suture are superimposed. See text for further discussion. Key as in Figure 3. One-half natural size.

crest into the expanded medial face of the mastoid process).

Before we assess the occipitomastoid morphology of *A. afarensis* in the context of these conclusions, it is important to correct an anatomical error apparent in Olson's descriptions of this region in hominoids, namely the homology of the occipitomastoid crest. The term *crista occipitomastoidea* was coined by Weidenreich (1943, p. 64), who applied it to the well-developed bony ridge running along the occipitomastoid suture and bordering the digastric fossa medially in the "*Sinanthropus*" crania. Weidenreich felt that the fissure of the occipitomastoid suture in skull III housed the occipital artery; thus, in this specimen at least, the occipitomastoid crest was divided into mastoid and occipital portions by the suture and the occipital groove (the primitive pattern in Olson's scheme). Later, Weidenreich contrasted the occipitomastoid morphology of the Solo crania with that of "*Sinanthropus*": "In Solo man the mastoid portion of [the occipitomastoid] ridge has shifted so close to the mastoid process that it appears as a proper ridge, which has been called crista paramastoidea, while the occipital portion appears as an independent ridge which seems to be double-edged (crista occipito-mastoidea)" (1951, p. 280). What Weidenreich does not point out is that the narrow, deep sulcus *lateral* to the occipitomastoid crest and suture and *medial* to the "crista paramastoidea" on the Solo crania corresponds to the groove for the occipital artery of modern human anatomy. Nor does Weidenreich mention that the occipitomastoid crest forms the lateral margin of the well-defined spatulate depression for the insertion of *m. obliquus capitis superior*. Figure 3A shows this region in a cast of Solo skull XI.

The structure referred to as the "crista paramastoidea" by Weidenreich is properly referred to as the "juxtamastoid eminence" or "juxtamastoid process" (Rouvière, 1927; Taxman, 1963; Walensky, 1964; Hublin, 1978). Nowhere does Olson mention the presence of the juxtamastoid eminence on human crania. The juxtamastoid eminence is known to be a variable but frequent feature in modern humans, but, more importantly, it often coexists with the ridge buttressing the occipitomastoid suture, the structure to which Weidenreich gave the name *crista occipitomastoidea* (Corner, 1896; Taxman, 1963; Walensky, 1964; de Villiers, 1968). Figure 3B shows a modern human cranium with both an occipitomastoid crest and a juxtamastoid eminence.

The proper identification of the juxtamastoid eminence is highly relevant to the composition of Olson's *Homo* sample, because to warrant inclusion in this genus a specimen must exhibit an *occipitomastoid crest* that separates the digastric fossa from the occipitomastoid suture and the occipital groove. In neither *Homo* cranium pictured in Figure 3 (A,B) is this the case. Instead, it is the juxtamastoid eminence that separates the digastric fossa from the occipital groove, with the occipitomastoid crest and suture bordering the occipital groove medially. Swartkrans Member 1 *Homo* specimens SK 27 and SK 847 are among the earliest crania to exhibit the juxtamastoid eminence; this structure coexists with a true occipitomastoid crest running along the occipitomastoid suture and forming the lateral margin of the *m. obliquus capitis superior* insertion (see also Clarke, 1977).

Although Olson combines MLD 37/38 and Sts 5 with SK 27 and SK 847 in "*H.*" *africanus*, we do not believe these Makapansgat and Sterkfontein crania exhibit morphology homologous with that characterizing the Swartkrans *Homo* specimens. Our examination of more than 200 extant ape crania reveals that the occipitomastoid crest is associated with the lateral margin of the *m. obliquus capitis superior* insertion. Among these crania we found numerous instances where the superior oblique insertion crossed the occipitomastoid suture onto the temporal bone. In these cases a crest was developed on the temporal bone, separating the digastric fossa from the more *medial* superior oblique insertion, occipitomastoid suture, and occipital groove (Figs. 1C,D, and 3C). Where the superior oblique insertion was confined to the occipital bone, the occipitomastoid crest and suture were superimposed or very closely approximated. Only in *Homo*, however, does the occipital groove ever course *lateral* to the occipitomastoid crest and suture. Where this occurred, a juxtamastoid eminence was usually present, separating the digastric fossa from the occipital groove (Fig. 3A,B).

Careful consideration of the nuchal muscle markings on the left side of the MLD 37/38 cranial base (Fig. 4A) indicates that the short ridge bordering the digastric fossa medially is also the anterolateral margin of the superior oblique insertion, which crosses the occipitomastoid suture onto the temporal bone. There is no trace of juxtamastoid eminence (as there is in SK 27 and SK 847). Thus this ridge is a true occipitomastoid crest, but its location on the temporal bone, lateral to the occipitomastoid suture and occipital groove, does not distinguish MLD 37/38 from ape crania where the superior oblique insertion spans the occipitomastoid suture.

Finally, we believe Olson has erroneously depicted the occipitomastoid region of Sts 5 (1981: Figure 4). The occipitomastoid crest is shown as being located entirely lateral to the occipitomastoid suture, a condition ostensibly supporting this specimen's assignment to *Homo*. On the contrary, the occipitomastoid crest on Sts 5 is clearly built up on *both* sides of the suture (Fig. 4B), as in many crania of humans and apes.

In the three adult *A. afarensis* specimens (A.L.333-45, 333-84, 333-112) the occipitomastoid crest is a robust, highly pneumatized ridge located on the temporal bone. The medial margin of the crest is the occipitomastoid sutural face, with which the relatively thin, downturned lateral edge of the nuchal plane articulates (see left side of A.L.333-45, Fig. 5A). Where this region is intact (right side of A.L.333-45, Fig. 5A), the

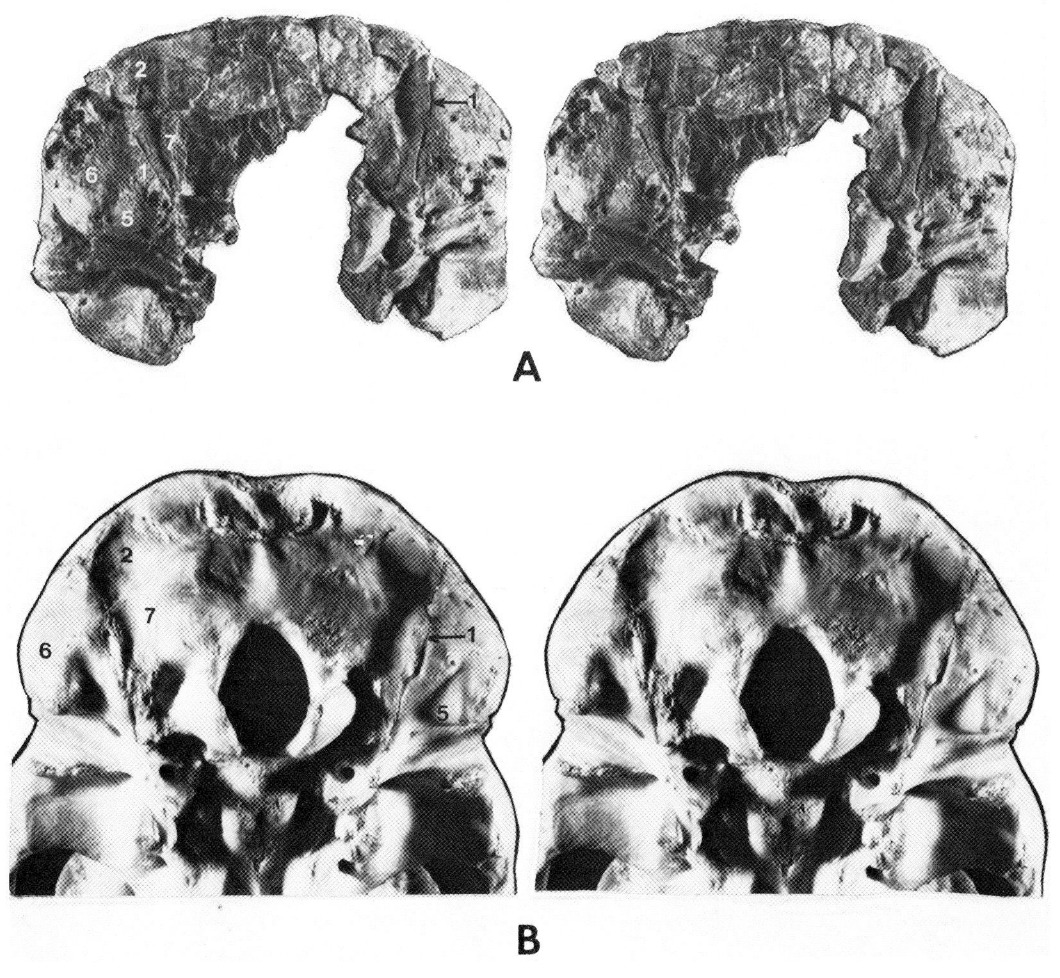

Fig. 5. Basal views of hominoid crania (stereo). (A) *A. afarensis*, A.L.333-45; (B) *G. gorilla*. In both specimens the entire mastoid region is inflated inferiorly below the floor of the digastric fossa and the occipitomastoid crest; the crest is a highly pneumatized ridge, clearly separable from the fossa floor and the mastoid tip. See text for further discussion. Key as in Figure 3. One-half natural size.

occipitomastoid suture appears to divide the occipitomastoid crest.

As already noted, Olson includes A.L.333-45 in "*Paranthropus*" in part because he believes that the occipitomastoid crest is incorporated into the medial wall of the mastoid process and that the digastric fossa is confined to the anteromedial face of the mastoid (see Figure 5 of Olson, 1981). Both statements are misleading.

In basal view (Fig. 5A), the digastric fossa on the left side of A.L.333-45 is seen to be a shallow, triangular, and mediolaterally concave area located between the mastoid process tip and the occipitomastoid crest (on the right, the mastoid tip and digastric fossa are eroded away). In Olson's schematic rendering (1981: Figure 5, an occipital view), the contour of the digastric fossa is portrayed as a nearly flat, superomedially inclined surface from which the occipitomastoid crest is indistinguishable. In fact, however, the occipitomastoid crest is inflated below the floor of the digastric fossa, and thus is a clearly defined entity forming the medial margin of the posterior digastric origin. The occipitomastoid crest in A.L.333-45 is no more "incorporated" into the medial face of the mastoid than it is in the female gorilla cranium pictured in Figure 5B.

Specimens A.L.333-84 (Fig. 2C) and 333-112 exhibit occipitomastoid configurations very similar to that of A.L.333-45. In both individuals the occipitomastoid crest is independent of the mastoid process, and the digastric fossa is a well-defined triangular depression between the crest and the mastoid process tip. Furthermore, the apex of the digastric fossa impinges on the extensive posterolateral face of the mastoid process in all three specimens. There is no factual basis for Olson's claim that the digastric muscle origin is restricted to the anteromedial surface of the mastoid process in the Hadar specimens.

Dentognathic Anatomy[4]

Tooth size variation in the* A. afarensis *sample. Olson (1981, p. 115) states that the "size range present in the Hadar hominid dental sample appears to be greater than that reported for the most dimorphic of contemporary hominid analogies, the gorilla . . ." This unsubstantiated claim is used to bolster Olson's identification of two hominid taxa ("*Paranthropus*" and "*Homo*") in the East African mid-Pliocene.

To test Olson's claim, we computed the coefficients of variation (CV) for the buccolingual breadth measurements of early hominid postcanine teeth reported by White et al. (1981). The results are given in Table 1, where comparative data for extant catarrhines taken from Kay (1982) are also presented. The CV of only one *A. afarensis* postcanine dental metric (buccolingual breadth of M^1) slightly exceeds those of the extant or fossil comparative samples. No *A. afarensis* postcanine tooth CV approaches those of the combined Siwalik Miocene hominoid dental sample that led Kay (1982: Table 3) to distinguish "two medium-sized species" (*Sivapithecus indicus* and *S. sivalensis*). The CV for every *A. afarensis* postcanine metric except M_2 breadth falls below the average postcanine CVs calculated by Gingerich and Schoeninger (1979) for more than 37 primate species (the *A. afarensis* M_2 breadth CV is only 0.38% above the mean primate CV for M_2 breadth.) We therefore reject Olson's assertion that the Hadar hominid dental sample exhibits more size variation than can be accommodated in a single species range.

Relative Canine Size. According to Olson, two hominid taxa can be identified in the Hadar/Laetoli sample on the basis of relative canine size. Although Olson does not answer the important question "Relative to what?," we assume he means relative to postcanine tooth size, since this is one of the major criteria that led Robinson (e.g., 1954) to distinguish two species of South African early hominids.

Data presented in Table 2 test the proposition that two clusters of specimens can be discerned on the basis of canine:postcanine tooth size; i.e., "*Paranthropus*" (with relatively small canines) and "*Homo*" (with relatively large canines). Not only is there no discrimination between Olson's "*Paranthropus*" and "*Homo*" categories, but the two specimens with the largest $\bar{C}/P_3$ ratios (A.L.333w-58 and L.H. 3) are included by Olson in "*Paranthropus*!"

Olson was apparently impressed by the relatively small canine in the A.L.400-1a mandible, a specimen that is indeed at the low end of the Hadar relative canine size

[4]White (1985) provides a detailed comparison of the Hadar and Laetoli dental samples and finds no basis for a species-level distinction between the hominids from these sites.

TABLE 1. Coefficients of variation—Buccolingual breadth of postcanine teeth

Type	P^3	P_3	P^4	P_4	M^1	M_1	M^2	M_2	M^3	M_3
A. afarensis[1]	6 5.97	18 8.35	6 4.78	13 8.10	8 6.57	15 6.75	6 4.35	18 7.48	7 7.15	11 6.96
A. africanus[1]	12 4.28	8 6.60	16 6.72	10 6.14	19 5.09	8 7.21	24 7.85	16 6.38	14 7.42	13 6.43
A. robustus[1]	19 6.44	18 9.43	23 6.62	17 7.56	29 5.83	19 6.29	21 6.01	20 5.44	22 4.22	21 8.04
A. boisei[1]	5 8.89	5 7.16	4 9.61	9 8.96	5 6.37	6 7.56	— —	6 6.56	— —	9 9.49
P. troglodytes[2]	7.0	10.9[3]	6.2	7.2	6.4	6.4	9.1	7.3	10.7	9.5
G. gorilla[2]	7.4	10.3	6.7	8.4	6.1	6.4	6.8	7.1	7.8	8.1
P. pygmaeus[2]	9.0	9.1	7.3	8.3	6.1	4.8	6.7	5.9	7.7	6.6
P. anubis[2]	7.8	17.1	6.6	6.9	6.4	6.9	8.2	8.2	9.9	8.2
C. badius[2]	8.5	17.6	7.4	10.2	5.8	6.6	6.3	7.8	8.9	6.7

[1]Calculated from data in White et al. (1981). Figures above CVs refer to sample size (individuals).
[2]Data from Kay (1982).
[3]The generally higher CVs for nonhominid P_3 breadth reflect the incorporation of this tooth into the dimorphic $\underline{C}/P_3$ complex.

TABLE 2. Canine:postcanine crown area (mm^2) ratios[1]

Specimen	C/P3	C/P4	C/M1	C/M2	C/M3	C/P3-M1
Mandibular						
"*Paranthropus*"						
L.H.3	0.91	0.94	0.67	—	—	0.27
A.L.400–1a	0.70	0.68	0.46	0.36	0.37	0.20
A.L.333w–58	1.20	—	—	—	—	—
"*Homo*"						
A.L.128–23	0.84	0.87	0.54	0.44	—	0.24
A.L.198–1	0.88	0.90	—	0.51	0.44	—
Maxillary						
"*Paranthropus*"						
L.H.3	1.21	—	0.77	—	—	—
L.H.6	0.83	—	0.60	—	—	—
A.L.200–1a	0.91	0.99	0.62	0.50	0.47	0.27
A.L.333–2	0.92	0.93	—	—	—	—
"*Homo*"						
A.L.199–1	0.97	0.98	0.64	0.51	0.55	0.28

[1]Ratios calculated from measurements in White (1977, 1980) and Johanson et al. (1982b). Mesiodistal lengths of postcanine teeth were corrected for interproximal wear. Where a specimen is represented by left and right teeth, the average value is given. Taxonomic definitions of Olson (1981).

range. However, the relative canine size distribution in the entire Hadar/Laetoli sample fails to support any suggestion of taxonomic diversity.

Premolar molarization. Olson (1981, pp. 120–121) lists several features of the maxillary and mandibular premolars that supposedly distinguish between "robust" ("*Paranthropus*") and "gracile" ("*Homo*") categories at Hadar (Table 3).

The maxillary premolar sample of Olson's "*Homo*" category consists of a heavily worn P^3 and P^4 from the same jaw, A.L.199-1. The mesial and distal foveae on these teeth are nearly entirely worn away, and so we are not surprised that Olson reports them as being "small." However, similarly worn teeth in Olson's "robust" category do not differ from A.L.199-1 in the size of the mesial and distal foveae (A.L.333-1, 333w-42, L.H. 5, L.H. 22). Conversely, less heavily worn teeth exhibit larger foveae (A.L.200-la, Garusi I, L.H. 1, L.H. 3, L.H. 6). Maxillary premolars of other early hominid taxa exhibit similar variation, confirming that the size of the mesial and distal foveae varies as a function of occlusal wear (e.g., *A. africanus*: heavily worn P^3 in Stw 13 vs. relatively unworn P^3 in Sts 12; heavily worn P^4 in Stw 45 vs. relatively unworn P^4 in Sts

TABLE 3. Olson's taxonomic differentia of Hadar premolars[1]

	"*Paranthropus*"	"*Homo*"
$P^3 + P^4$	"mesial and distal foveae are well-developed and bounded externally by crests"	"foveae are small and the crests absent"
	crowns "appear much more mesiodistally expanded"	crowns "retain a primitive narrow oval outline"
P_3	"variable-sized second lingual cusp"	lingual cusp absent
P_4	"talonid and mesial margin . . . have become enlarged to the extent that the outline of the tooth is round rather than bucco-lingually oval"	"primitive in retaining . . . oval condition and the small talonid, which accounts for less than half the surface area of the crown"

[1]From Olson (1981, pp. 120–121).

52a). The mechanics of occlusal attrition and the topography of the occlusal foveae ensure that fovea size will decrease and eventually disappear as wear proceeds.

Likewise, Olson's failure to take into account different degrees of occlusal wear renders distinctions based on mesial and distal marginal ridge development invalid. In this case, however, even teeth with *similar* amounts of occlusal wear exhibit great variation in the development of these ridges (*A. africanus*: P^3, Sts 52a vs. Sts 24a; P^4, Sts 52a vs. MLD 6. *A. robustus*: P^3, SK 867 vs. SK 33; P^4, TM 1517a, *left* side vs. TM 1517a, *right* side. *A. afarensis*: P^4, L.H. 1 vs. Garusi I).

In Table 4 we present crown shape indices (mesiodistal length/buccolingual breadth) of Hadar/Laetoli maxillary premolars as a test of Olson's statement that the premolars of the "robust" category are mesiodistally expanded compared to those of the "gracile" category (again, n = 1, A.L.199-1, for the "gracile" sample). When mesiodistal lengths are corrected for interproximal wear, the distinction between A.L.199-1 and the "robust" sample vanishes. The range of shape indices for the *A. afarensis* P^3 is 0.66–0.73 (n = 8) and for P^4 is 0.68–0.79 (n = 6). This represents *less* variation in premolar shape than accepted by Olson for "*P.*" *robustus* (P^3: 0.66–0.84, n = 17; P^4: 0.65–0.77, n = 20) (White et al., 1981).

In ignoring the marked effects of occlusal and interproximal wear on crown topography and shape of the A.L.199-1 teeth, Olson has been led to erect an artificial set of criteria to distinguish two maxillary premolar morphs in the Hadar sample. In reality, no such distinctions exist that cannot be attributed to individual and/or age-related variation.

The $\underline{C}/P_3$ complex is unquestionably one of the most primitive aspects of *A. afarensis* craniodental anatomy (Johanson and White, 1979; White et al., 1981; White, 1981; Johanson et al., 1982b). Thus, *A. afarensis* is the only Plio-Pleistocene hominid species in which the P_3 metaconid is often expressed only as an inflated lingual ridge. Olson contends that presence/absence of the P_3 metaconid is a valid taxonomic discriminator within the Hadar collection, with the "*Homo*" sample consisting of individuals without the P_3 metaconid (A.L.128-23, 198-1, 288-1, 311-1). Unfortunately, the teeth on two of Olson's "*Homo*" mandibles are so heavily worn that no reliable indication of P_3 cusp number remains (A.L.198-1, 311-1). Olson (1981, p. 120) states that the P_3 metaconid is "not present in the smaller specimens," yet A.L.207-13, a small (presumably female) jaw the height and breadth dimensions of which approximate those of A.L.128-23 and 288-1i (White and Johanson, 1982), is placed in "*Paranthropus*" by Olson owing to its relatively strong P_3 metaconid. On the other hand, A.L.277-1, a large male mandible, is assigned to "*Paranthropus*" even though its P_3 is clearly unicuspid (Johanson et al., 1982b).

The size of the P_3 metaconid varies in other early hominid species (*A. africanus*: Stw 14 [small] vs. Sts 52b [large]; *A. robustus*: SK 100 [small] vs. TM 1517b [large]), but only *A. afarensis* has individuals with unicuspid P_3s. We do not believe that the variation in Hadar P_3 metaconids indicates taxonomic diversity, because it should be expected that

TABLE 4. Crown shape indices (m-d/b-l) of Hadar dentitions[1]

	"*Paranthropus*"		"*Homo*"	
P^3	L.H.3	0.66	A.L.199-1	0.66
	L.H.6	0.71		
	L.H.25	0.73		
	Garusi I	0.73		
	A.L.200-1a	0.72		
	A.L.333-1	0.71		
	A.L.333-2	0.73		
P^4	Garusi I	0.73	A.L.199-1	0.68
	A.L.200-1a	0.68		
	A.L.333-1	0.73		
	A.L.333-2	0.79		
	A.L.333w-42	0.75		
P_4	L.H.3	0.92	A.L.198-1	0.91
	L.H.4	0.96	A.L.128-23	0.77
	L.H.14	0.95	A.L.288-1	0.78
	A.L.207–13	0.92		
	A.L.266-1	0.93		
	A.L.277-1	0.88		
	A.L.333-44	0.99		
	A.L.333w-1	0.90		
	A.L.333w-60	0.74		
	A.L.400-1a	0.86		
M_1	L.H. 2	1.01	A.L.128-23	1.01
	L.H. 3	1.02	A.L.288-1	1.13
	L.H.4	1.05		
	L.H. 16	1.01		
	A.L.145-35	0.97		
	A.L.200-1b	1.04		
	A.L.266-1	1.07		
	A.L.333-74	1.00		
	A.L.333w-1a	1.07		
	A.L.333w-12	1.03		
	A.L.333w-60	1.00		
	A.L.400-1a	1.05		
M_2	L.H.4	1.09	A.L.128-23	0.97
	L.H.23	1.11	A.L.198-1	1.00
	A.L.145-35	1.08	A.L.288-1	1.08
	A.L.188-1	1.01		
	A.L.207-13	1.06		
	A.L.266-1	0.93		
	A.L.277-1	1.04		
	A.L.333w-1a	1.07		
	A.L.333w-27	1.09		
	A.L.333w-48	1.04		
	A.L.333w-57	1.18		
	A.L.333w-59	0.97		
	A.L.333w-60	0.99		

[1]Indices calculated from data in White (1977, 1980) and Johanson et al. (1982b). Mesiodistal lengths were corrected for interproximal wear. Where a specimen is represented by left and right teeth, the average value is given.

earlier and more primitive hominids would exhibit variation in this character, with some individuals expressing the metaconid as only a swollen lingual ridge.

Olson suggests that the enlargement of the mesial marginal ridge and expansion of the talonid (to comprise more than half the occlusal surface area) on the Hadar "*Paranthropus*" P_4s results in a round rather than buccolingually oval crown outline. Only two specimens of Olson's "*Homo*" sample, A.L.128-23 and 288-1i, retain enough P_4 occlusal relief to enable assessment of mesial marginal ridge and talonid size. By expressing relative talonid size as "the distance from the deepest point in talonid basin to maximum distolingual crown corner divided by mesiodistal crown length (corrected for interproximal wear) $\times$ 100" we find that *no* Hadar or Laetoli P_4 has a talonid that occupies more than half the crown surface area. In fact, the talonid of the A.L.128-23 P_4 (at 44%) is not only relatively larger than those of "robust" specimens such as A.L.277-1 (40%), 333w-60 (33%), and 400-1a (35%) but is relatively the *largest* P_4 talonid of any *A. afarensis* mandible. The variation in the combined Hadar/Laetoli sample is no greater than that observed in the large samples of *A. africanus* (Sts 52b, narrow talonid vs. Stw 5, wide) or *A. robustus* (SK 88, narrow vs. SK 826, wide).

The width of the mesial marginal ridge also fails to serve as a valid taxonomic character when applied to the Hadar/Laetoli collection. The L.H. 3 P_4 (a supposed "*Paranthropus*" specimen) has an exceptionally narrow mesial marginal ridge, while the A.L.288-1i ("*Homo*") P_4 has one of the widest ridges. Again, the distribution of the trait violates the definitions of the proposed taxa.

Data in Table 4 test Olson's proposition that the Hadar/Laetoli "*Paranthropus*" P_4s have rounder occlusal outlines than those which he assigns to "*Homo*." Although the A.L.128-23 and 288-1i: P_4s are narrower mesiodistally than most of the other *A. afarensis* P_4s, the A.L.333w-60 P_4 is the least rounded tooth in the entire sample. Yet this tooth occurs in one of the largest mandibles in the series and is attributed by Olson to "*Paranthropus*." In contrast, the shape index of the A.L.198-1 P_4 is similar to those of many of Olson's "*Paranthropus*" specimens, but this relatively small mandible is assigned to "*Homo*." Supposed distinctions based on P_4 shape are entirely unjustified, since *A. africanus* and *A. robustus*, species that Olson places in different genera, do not differ in mean P_4 shape index values (White et al., 1981).

Mandibular molar shape and cusp pattern. Olson contends that the M_1 and M_2 crowns of Hadar/Laetoli "*Paranthropus*" are buccolingually enlarged (i.e., square), while these teeth in "*Homo*" specimens are relatively narrow buccolingually (i.e., rectangular). Data in Table 4 show that this alleged distinction is simply false. Apparently, Olson noted the relatively narrow M_1 of A.L.288-1i and then generalized this condition to other specimens and other teeth. In fact, however, the molars of A.L.128-23, a supposed *Homo* mandible, are squarer than those of most of Olson's "*Paranthropus*" specimens, and A.L.333w-57, a "*Paranthropus*," has the narrowest M_2 of any Hadar or Laetoli specimen. The ranges of molar shape indices for *A. africanus* (M_1: 1.01–1.15, n = 7; M_2: 1.03–1.18, n = 14) and *A. robustus* (M_1: 1.01–1.17, n = 19; M_2: 1.02–1.17, n = 20) confirm that this character is highly variable within early hominid species, although Hadar/Laetoli M_1 and M_2 crowns are on average significantly broader than those of *A. africanus* and *A. robustus* (White et al., 1981).

Olson states that "the hypoconulid has been lost in most examples" of Hadar "*Paranthropus*" M_1 and M_2 crowns (1981, p. 121), contributing to a square occlusal profile. We are at a loss to explain how Olson reached this conclusion, since every M_1 and M_2 in the Hadar/Laetoli collection retains the hypoconulid except when this cusp is broken away or worn too heavily to discern (Johanson et al., 1982b). The M_1 and M_2 crowns of *A. afarensis* have a "squared-off" distal occlusal profile because the hypoconulid tends to be mesially appressed (White et al., 1981); this is a feature of specimens assigned by Olson to "*Paranthropus*" (e.g., A.L.333w-60) as well as "*Homo*" (e.g., A.L.128-23).

Relative size of the "inferior mandibular torus". This structure is more commonly called the inferior *transverse* torus of the symphyseal region. According to Olson's taxonomy, Hadar "*Homo*" mandibles have inferior and superior transverse tori that are "subequal" in size (the primitive pattern), while "*Paranthropus*" mandibles exhibit relatively reduced inferior transverse tori (the derived condition).

Unfortunately, among Hadar mandibles attributed by Olson to "*Homo*" only A.L. 288-1i permits accurate assessment of relative torus size. In Figure 6 the symphyseal cross-sections of A.L.288-1i (below) and A.L.400-1a (above; "*Paranthropus*") are compared. We disagree with Olson's observation that the inferior torus of the supposed "*Paranthropus*" jaw is reduced, either absolutely or relative to the superior torus. Indeed, the inferior torus of A.L.288-1i is the relatively smaller in this comparison. Other Hadar "*Paranthropus*" mandibles, such as A.L. 277-1 and 333w-60, also have an inferior transverse torus that is well-developed relative to the superior transverse torus (White and Johanson, 1982). We are not sure how Olson arrived at the conclusion that the Hadar "*Paranthropus*" mandibles have relatively weak inferior tori. We can only speculate that he based this observation on the A.L.266-1 specimen, where *both* tori are very weak, partly owing to extensive abrasion of cortical bone (White and Johanson, 1982).

Sagittal contour of the anterior palatal surface. The contour of the *A. afarensis* palatal surface has been described as "flat and shallow, with little or no shelving of premaxillary portions" (Kimbel, in Johanson et al., 1982a, p. 382). Olson, however, observed the minor inferior flexion of the anterior palatal surface in A.L.199-1 and claimed that "the palatal morphology of [this] specimen . . . would appear to indicate that its phylogenetic affinities are with the *Homo* lineage . . ." (1981, p. 122). Hadar "*Paranthropus*" maxillae (i.e., A.L.200-1a, 333-1, 333-2) are said to retain the primitive condition of flat, horizontal anterior palatal surfaces.

Although the anterior palatal surface of A.L.199-1 is inclined inferiorly slightly more than in the other Hadar maxillae, we disagree that this difference is taxonomically significant. In a small random sample of gorilla (n = 5) and orangutan (n = 2) crania (initially part of the series examined in the study of mastoid morphology), we observed at least as much intraspecific difference in anterior palatal inclination as seen between A.L.199-1 and 200-1a. Olson's "*H.*" *africanus* exhibits even greater variation in this feature (compare SK 47 [an *A. robustus* cranium in our opinion], with a flat, shallow anterior palate to Sts 5 or Sts 52a, with much steeper and deeper anterior palates). Here again, Olson is willing to accept substantial variation as a normal characteristic of other hominoid species, whereas *A. afarensis* is expected to be monomorphic.

SUMMARY AND CONCLUSIONS

In summarizing his analysis Olson writes: "In view of the demonstrable occurrence of *Paranthropus* in the Hadar Formation the

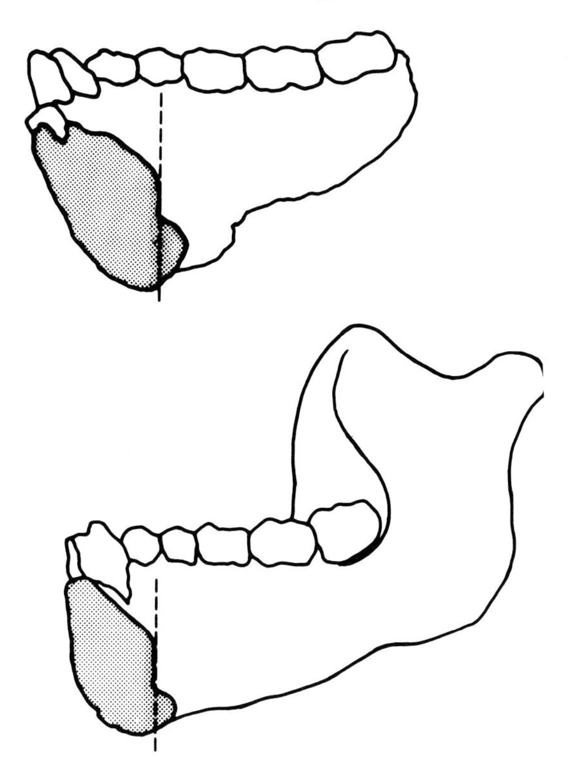

Fig. 6. Outline drawings of *A. afarensis* mandibles A.L.400-1a (top) and A.L.288-1i (bottom), oriented on the occlusal plane. Dotted perpendicular line is tangent to the posteriormost point on the superior transverse torus and demonstrates that the inferior transverse torus in A.L.400-1a is absolutely and relatively as large as that in A.L.288-1i. See text for discussion.

dental remains from these deposits were examined and it is suggested that the extreme morphological and metrical differences which Johanson and White (1979) attributed to sexual dimorphism are much more reasonably interpreted as the earliest evidence for the dichotomy of the *Homo* and *Paranthropus* lineages" (1981, pp. 122–123). We do not believe that this conclusion is warranted when the evidence adduced in its support is subjected to close scrutiny.

We have shown above that the multiple taxon argument is an inadequate explanation of the morphological diversity observed in the Hadar sample. Moreover, Olson is incorrect in his belief that the distribution of traits in Hadar "*Paranthropus*" and "*Homo*" parallel our division of the sample into male and female morphs. For instance, the large male mandible A.L.333w-60 ("*Paranthropus*" = our male morph) has the narrowest P_4 of any Hadar specimen, yet this trait according to Olson should characterize the Hadar "*Homo*" group (= our female morph). Similarly, the male mandible A.L.277-1 ("*Paranthropus*") has a unicuspid P_3, a feature supposedly diagnostic of "*Homo.*" Confusion also arises in Olson's treatment of lower molar shape: the female A.L.128-23 has nearly square molars, although Olson assigns this specimen to "*Homo,*" a taxon diagnosed as having buccolingually narrow molars; the female A.L.333w-57 has the narrowest M_2 of any Hadar/Laetoli mandible, but this specimen is attributed to "*Paranthropus,*" a taxon supposedly characterized by square molars. It is clear, then, that nearly all of the traits that Olson uses to differentiate two species at Hadar cut across his taxonomic and our sexual boundaries.

Olson's analysis of the Hadar/Laetoli sample is plagued by a persistent tendency to generalize from the morphology of a single individual to other specimens without providing any justification for doing so (e.g., relative canine size, mandibular molar shape, inferior transverse torus size). An even more fundamental problem is Olson's failure to acknowledge the effects of individual age on several of his comparisons (e.g., morphology and shape of A.L.199-1 maxillary premolars).

But the major stumbling block faced by Olson is the widely divergent amounts of morphological variation he is willing to recognize in different early hominid species. Thus, his "*H.*" *africanus* includes individuals with such dissimilar facial and occipitomastoid morphologies as Sts 5 and SK 847 (Olson, 1978), but the slight difference in anterior palatal shelving between A.L.199-1 and 200-1a is viewed as diagnostic of two genera.

Finally, our study of fossil hominid and extant hominoid mastoid morphology has revealed several errors in Olson's anatomical analysis of the hominoid cranial base. Specifically, we have shown that to characterize extant hominoid mastoid (including occipitomastoid) morphology as an invariant "type" is to grossly oversimplify the anatomical complexity and variation in this cranial region. We have demonstrated that, in its lateral and inferior inflation, the mastoid process of *A. afarensis* specimen A.L.333-45 does not differ significantly from the condition found to characterize many extant apes, particularly female gorillas—the morph thought by Olson to best approximate the ancestral hominoid pattern. The mastoid morphology of A.L.333-84 and 333-112 fully confirms our conclusions based on A.L.333-45. The mastoid and supramastoid morphology of *A. robustus* and *A. boisei* has been found to diverge sharply from that observed in *A. afarensis* and the extant African apes. This suggests to us that the evolutionary remodelling of the hominoid mastoid region may have more to do with trends in masticatory function than has previously been recognized (Kimbel and Rak, 1985).

We have also shown that Olson has confused nonhomologous structures (occipitomastoid crest and juxtamastoid eminence) in his treatment of hominoid occipitomastoid morphology. This confusion, apparently combined with an insufficient appreciation of normal variation in *Homo* and inaccurate descriptions of some *A. africanus* (*sensu stricto*) crania, results in a taxon "*H.*" *africanus* that includes specimens with occipitomastoid morphologies ranging from the modern human condition (SK 847) to the modern ape condition (MLD 37/38).[5]

We have been unable to find any features of the *A. afarensis* mastoid region that convincingly demonstrate unique specializations shared with *A. robustus* and *A. boisei*. When the range of variation observed in *A. afarensis* and extant hominoids is taken into account, the Pliocene hominid emerges as extremely primitive. This assessment of *A. afarensis* is supported by its asterionic sutural configuration (Kimbel and Rak, 1985) and the proliferation of mastoid pneumatization anteriorly to inflate the base of the temporal squama (Kimbel et al., 1984). Neither of these features is observed in any other hominid species, but both are characteristic of the extant apes. They combine with the morphology of the mastoid region to make the posterior part of the *A. afarensis* cranial vault one of the most strikingly primitive aspects of this species' cranial anatomy.[6]

At the "Ancestors" symposium, Olson (see 1985) argued that the overlapping of nasion and glabella in the juvenile cranium A.L.333-105 supports his identification of "*Paranthropus*" at Hadar. In an earlier paper, Olson (1978) concluded that "*Paranthropus*" retains the primitive, African ape condition of nasion–glabella coincidence, whereas *Homo* is characterized by a derived state in which nasion lies well below glabella. However, as Clarke (1977) has shown, glabella is inflated inferiorly to meet nasion in "*Paranthropus*," but when these points overlap in African apes the nasal bones always extend upward to meet a superiorly situated glabella. Not only does this difference invalidate the alleged symplesiomorphy identified by Olson in "*Paranthropus*" and African apes, but A.L.333-105 exhibits the African ape rather than the "*Paranthropus*" condition.

In conclusion, we have found no evidence of multiple hominid taxa in the mid-Pliocene of eastern Africa. Although the results of this investigation do not contribute directly to the resolution of the debate over the precise position of *A. africanus* in the human evolutionary tree (Johanson and White, 1979; Tobias, 1980; Olson, 1981, 1985; White et al., 1981; Boaz, 1983; Wolpoff, 1983; Rak, 1983; McHenry, 1984; Kimbel et al., 1984), we do not believe that the "hard evidence" provides a basis for excluding this species from the ancestry of *A. robustus/A. boisei*.

ACKNOWLEDGMENTS

We wish to thank Bruce Latimer and Steve Ward for valuable comments and discussion during the preparation of the manuscript. We are also indebted to John Aicher (photography), Anson Laufer (photography), and Luba Dmytryk Gudz (illustration), whose expert skills have proven invaluable to our research during the past five years or more. We thank Jaré Cardinal for typing the manuscript and for providing incisive editorial commentary.

Research reported here was funded in part by grants from the Foundation for Research into the Origins of Man (F.R.O.M.), the National Science Foundation, the GAR Foundation, and the Cleveland Foundation. Finally, we thank Drs. Eric Delson and Ian Tattersall of the American Museum of Natural History for inviting us to participate in the symposium "Paleoanthropology: The Hard Evidence," at which an abbreviated version of this paper was read.

---

[5]Olson (1985) now recognizes the juxtamastoid eminence as a distinct structure, but he incorrectly applies this term to the true occipitomastoid crest on Sts. 5, MLD 37/38, and KNM-ER 1813.

[6]The prevalence of the occipital-marginal venous drainage pattern in *A. afarensis* ostensibly supports Olson's identification of "*Paranthropus*" at Hadar. Detailed discussions of venous drainage patterns and their value for hominid phylogeny reconstruction are available elsewhere (Falk and Conroy, 1983; Kimbel, 1984) and are therefore not reiterated here.

LITERATURE CITED

Ashton, EH, and Zuckerman, S (1952) The mastoid process in the chimpanzee and gorilla. Am. J. Phys. Anthropol. *10*(n.s.):145–151.

Boaz, NT (1983) Morphological trends and phylogenetic relationships from Middle Miocene hominoids to Late Pliocene hominids. In RL Ciochon and RS Corruccini (eds): New Interpretations of Ape and Human Ancestry. New York: Plenum, pp. 705–720.

Clarke, RJ (1977) The Cranium of the Swartkrans Hominid SK 847 and its Relevance to Human Origins. Unpublished Ph.D. Thesis, University of the Witwatersrand, Johannesburg.

Corner, EM (1896) The processes of the occipital and mastoid regions of the skull. J. Anat. Physiol. *30:*386–389.

de Villiers, H (1968) The Skull of the South African Negro. Johannesburg: University of the Witwatersrand Press.

Falk, D, and Conroy, GC (1983) The cranial venous sinus system in *Australopithecus afarensis*. Nature *306:*779–781.

Gingerich, PD, and Schoeninger, M (1979) Patterns of tooth size variability in the dentition of primates. Am. J. Phys. Anthropol. *51:*457–466.

Groth, W (1937) Vergleichend-anatomische Untersuchung zur Frage der Entstehung des Warzenfortsatzes beim Menschen und den Menschenaffen. Gegen. Morph. Jrb. *79:*547–599.

Hofmann, L (1926) Zur Anatomie des Primatenschläfenbeines und seiner pneumatischen Räume unter Berücksichtigung des menschlichen Schläfenbeines. Monat. Ohren. Laryng.-Rhinol. *60:*921–949.

Hublin, JJ (1978) Le Torus Occipitale Transverse et les Structures Associees: Evolution dans le Genre *Homo*. These 3 cycle, Universite Paris VI.

Johanson, DC, Taieb, M, and Coppens, Y (1982a) Pliocene hominids from the Hadar Formation, Ethiopia (1973–1977): Stratigraphic, chronologic and paleoenvironmental contexts, with notes on hominid morphology and systematics. Am. J. Phys. Anthropol. *57:*373–402.

Johanson, DC and White, TD (1979) A systematic assessment of early African hominids. Science *203:*321–330.

Johanson, DC, White, TD, and Coppens, Y (1978) A new species of the genus *Australopithecus* (Primates: Hominidae) from the Pliocene of eastern Africa. Kirtlandia *28:*1–14.

Johanson, DC, White, TD, and Coppens, Y (1982b) Dental remains from the Hadar Formation, Ethiopia: 1974–1977 collections. Am. J. Phys. Anthropol. *57:*545–604.

Kay, RF (1982) *Sivapithecus simonsi*, a new species of Miocene hominoid, with comments on the phylogenetic status of the Ramapithecinae. Int. J. Primatol. *3:*113–173.

Kimbel, WH (1984) Variation in the pattern of cranial venous sinuses and hominid phylogeny. Am. J. Phys. Anthropol. *63:*243–263.

Kimbel, WH, Johanson, DC, and Coppens, Y (1982) Pliocene hominid cranial remains from the Hadar Formation, Ethiopia. Am. J. Phys. Anthropol. *57:*453–500.

Kimbel, WH, and Rak, Y (1985) Functional morphology of the asterionic region in extant hominoids and fossil hominids. Am. J. Phys. Anthropol. *66:*31–54.

Kimbel, WH, White, TD, and Johanson, DC (1984) Cranial morphology of *Australopithecus afarensis*: A comparative study based on a composite reconstruction of the adult skull. Am. J. Phys. Anthropol. *64:*337–388.

McHenry, HM (1984) Relative cheek-tooth size in *Australopithecus*. Am. J. Phys. Anthropol. *64:*297–306.

Olson, TR (1978) Hominid phylogenetics and the existence of *Homo* in Member 1 of the Swartkrans Formation, South Africa. J. Hum. Evol. *7:*159–178.

Olson, TR (1981) Basicranial morphology of the extant hominoids and Pliocene hominids: The new material from the Hadar Formation, Ethiopia and its significance in early human evolution and taxonomy. In CB Stringer (ed): Aspects of Human Evolution. London: Taylor and Francis, pp. 99–128.

Olson, TR (1985) Cranial morphology and systematics of the Hadar Formation hominids and "*Australopithecus*" *africanus*. In E. Delson (ed): Ancestors: The Hard Evidence. New York: Alan R. Liss, Inc., pp. 102–119.

Rak, Y (1983) The Australopithecine Face. New York: Academic Press.

Robinson, JT (1954) Prehominid dentition and hominid evolution. Evolution *8:*324–334.

Rouvière, H (1927) Anatomie Humaine, 2nd ed. Paris: Masson.

Schultz, AH (1950) Morphological observations on gorillas. In WK Gregory (ed): The Anatomy of the Gorilla. New York: Columbia University Press, pp. 227–251.

Schultz, AH (1952) Über das Wachstum der Warzenfortsätze beim Menschen und den Menschenaffen. Homo *3:*105–109.

Taxman, RM (1963) Incidence and size of the juxtamastoid eminence in modern crania. Am. J. Phys. Anthropol. *21*(n.s.):153–157.

Tobias, PV (1980) "*Australopithecus afarensis*" and *A. africanus*: Critique and an alternative hypothesis. Palaeont. Afr. *23:*1–17.

Walensky, NA (1964) A re-evaluation of the mastoid region of contemporary and fossil man. Anat. Rec. *149:*67–72.

Weidenreich, F (1943) The skull of *Sinanthropus pekinensis*. Palaeontol. Sin. New Ser. *D* (*10*):1–484.

Weidenreich, F (1951) Morphology of Solo Man. Anthr. Pap. Amer. Mus. Nat. Hist. *43:*205–290.

Weinert, H (1950) Über die neuen Vor- und Frühmenschenfunde aus Afrika, Java, China und Frankreich. Z. Morph. Anthropol. *42:*113–145.

White, TD (1977) New fossil hominids from Laetolil, Tanzania. Am. J. Phys. Anthropol. *46:*197–230.

White, TD (1980) Additional fossil hominids from Laetolil, Tanzania: 1976–1979 specimens. Am. J. Phys. Anthropol. *53:*487–504.

White, TD (1981) Primitive hominid canine from Tanzania. Science *213:*348–349.

White, TD (1985) The hominids of Hadar and Laetoli: An element by element comparison of the dental samples. In E Delson (ed): Ancestors: The Hard Evidence. New York: Alan R. Liss, Inc., pp. 138–152.

White, TD, and Johanson, DC (1982) Pliocene hominid mandibles from the Hadar Formation, Ethiopia: 1974–1977 collections. Am. J. Phys. Anthropol. *57:*501–544.

White, TD, Johanson, DC, and Kimbel, WH (1981) *Australopithecus africanus*: Its phyletic position reconsidered. S. Afr. J. Sci. *77:*445–470.

Wolpoff, MH (1983) Australopithecines: The unwanted ancestors. In K Reichs (ed): Hominid Origins. Washington, D.C.: University Press of America, pp. 109–126.

Ancestors: The Hard Evidence, pages 138–152

The Hominids of Hadar and Laetoli: An Element-by-Element Comparison of the Dental Samples

Tim D. White
Department of Anthropology, University of California, Berkeley, California 94720

ABSTRACT A detailed comparative analysis of the available dental samples from the Pliocene hominid sites of Hadar in Ethiopia and Laetoli in Tanzania shows that these two site samples are united in their possession of a large number of primitive characters. The minor metric and morphological differences found to distinguish the dental samples are attributable to intraspecific variation and/or sampling error. The specific unity of fossils from Hadar and Laetoli is emphasized by the results of this analysis.

"The question is never whether or not the compared populations are completely identical. Population geneticists have demonstrated conclusively that no two natural populations in sexually reproducing animals are ever exactly alike. To find a statistically significant difference between several populations is, therefore, of only minor interest to the taxonomist; he takes it for granted."
E. Mayr (1969, p. 187).

"When attempting to classify fossil hominids and to determine the relationship between the individual types, there seems to be no other choice but to define first the morphological characters of the latter and to compare their resemblances and their differences. Neither archaeology nor geology can be of aid, still less act decisively on this question. Comparative anatomy alone is capable of furnishing us with the information required for the recognition of primitive and advanced phases of evolution and for ranging them within phylogenetic lines."
F. Weidenreich (1939, p. 64).

INTRODUCTION

When the first hominid was recovered from the Laetolil Beds in 1935, there existed no other adult *Australopithecus* fossils to compare it to. This posed no interpretive problem, however, because the fossil, a worn lower canine, went unrecognized as belonging to a hominoid primate until 1979 (White, 1981a). On the other hand, the maxillary fragment with P^3-P^4 (Garusi I) and the isolated third molar (Garusi II) recovered in 1939 by the Kohl–Larsen expedition were compared to Javan and South African discoveries, and their significance was debated throughout the 1940s and 1950s. Remane's (1951, 1954) and Senyürek's (1955) work on the relatively primitive Garusi I specimen stands today as a monument to the value of careful anatomical analysis in the study of the early hominid dentition. Robinson's (1953, 1955, 1956) conclusions about the same specimen were more cautious: " . . . it seems to me unwise to go any farther at present than identifying it as an australopithecine" (1956, p. 4). Clarke (1977) and Protsch (1981) review the history of systematic interpretations of the 1939 specimens.

The most recent field activity, laboratory analysis, and systematic interpretation of hominid fossils from the Laetolil Beds came as a result of Dr. Mary Leakey's Laetoli expeditions between 1974 and 1979. I provided anatomical descriptions of this material, representing a total of 24 individuals, in 1977 and 1980. The recovery of these Laetoli hominids between 1974 and 1979 coincided with the accumulation of a substantially larger collection of Pliocene hominids from the Ethiopian site of Hadar. Comparative study of the Laetoli and Hadar collections began in earnest during 1975, when that year's Hadar fossils (including the first fossils from Afar Locality 333) were brought through Nairobi. The author joined Don Johanson, Tom Gray, Phillip Tobias and Bernard Wood for informal comparisons. Important similarities between the fossil samples were noted. In 1976 I wrote about the Laetoli hominids: "Much of the recently discovered comparable fossil hominid material from the Hadar region of Ethiopia shows strong similarity to the Laetolil specimens, and further collection combined with detailed comparative analysis of material from both localities is essential for the further understanding of human origins" (Leakey et al., 1976, p. 466).

By 1978 this further discovery and analysis led to our description of *A. afarensis* as a hominid species comprising the then-available samples from the Laetolil Beds and Hadar Formation (Johanson et al., 1978). In this publication and a subsequent one concerned with phylogenetic considerations (Johanson and White, 1979), we pooled the site samples, designating the species holotype as L.H. 4 "because of its distinctive, diagnostic morphology and because it has previously been fully described and illustrated (White, 1977)." We chose to name the species after the place from which most of the fossils derived (the Afar). The choice of a Laetoli specimen as holotype served to stress the morphological similarities we had discovered between the two site samples.

Simpson (1961, p. 5) describes a holotype specimen as: " . . . simply something to which a name is attached by purely legalistic convention." He goes on to state: "It should have nothing to do with the nomenclatural processes of defining the species and should have no special role in identifying other specimens . . . it is nominalistic absurdity to confuse a set of objects with the name or symbol for that set." We subscribe to this philosophy and clearly stated the reasoning behind selecting L.H. 4 as the type specimen of *A. afarensis* in 1978. Several subsequent authors, however, have objected to designation of L.H. 4 as a holotype, stating that this choice had been made "in a most curious manner" (Tobias, 1981a), was "questionable" (Day et al., 1980), "typological" and "dangerous" (Mayr, 1982), and even "folly" (Day, 1982). None of these authors provided substantive evidence to show that the site samples differed.

In 1979 Mary Leakey stated about the Laetoli hominids: "They bear considerable resemblances to the material collected from the Afar in Ethiopia" (Leakey and Hay, 1979, p. 320). Mary Leakey has subsequently commented on *A. afarensis* as follows: " . . . the arbitrary application of the same specific name to the hominids from the two localities, which are separated by over 1000 miles, appears to be based on insufficient proof of identity. It would have been desirable for a detailed comparison to be made of such material as is common to both sites" (Leakey, 1981, p. 102). As shown above, application of the name was hardly "arbitrary." As made clear in our publications, our comparative study was certainly "detailed." Given our results and published conclusions, we find it surprising that so many authors have been so quick to question our attribution of Laetoli and Hadar remains to *A. afarensis* without first making their own comparisons of the published data and readily available original specimens. It is apparently an easy activity to postulate taxonomic difference between the sites but a more difficult endeavor to demonstrate it.

Space restrictions in earlier publications (Johanson and White, 1979; White et al., 1981) have prevented us from providing the full results of our 1978 element-by-element comparison between Hadar and Laetoli site samples. The "Ancestors" forum provides a perfect opportunity to present some of my comparative and interpretive results.

Previous work on *A. afarensis* has concentrated on this taxon's phylogenetic status within the early hominids. We have attempted to follow the lead taken by Delson et al. (1977), who expressed the hope that: " . . . a consideration especially of our methodology, and secondarily of our evidence, will persuade future researchers to concentrate on the identification of shared derived characters among taxa and search for sister-groups, rather than on the worship of "ancestors" (p. 276). Now that our work on *A. afarensis* has appeared, and its explicit hypotheses are available for field testing, it seems an appropriate time to present a more detailed version of the comparative work done on the two primary *A. afarensis* site samples, Hadar and Laetoli.

Tobias (1980a, 1981a) has emphatically and repeatedly pointed out the lack of published details on the within- and between-site comparisons made before creation of the nomen *A. afarensis*. The site samples have been assessed independently in an effort to discern morphological change within the Hadar Formation and within the Laetolil Beds. No evolutionary change in the hominids is evident on the basis of the available samples. In fact, similarities in size and morphology between the stratigraphically youngest (A.L.288-1, "Lucy") and the oldest (A.L. 128-23, A.L.129) Hadar hominids are striking. To my knowledge, no author has yet seriously interpreted the Hadar hominids as demonstrative of *in-situ* evolution in the Afar.

Detailed comparison of the two site samples led to a pooling of the Hadar and Laetoli hominid collections in creating *A. afarensis*. Tobias (1978a,b, 1980a,b, 1981a,b) has steadfastly maintained that the Hadar and Laetoli collections belong to *A. africanus*, while Johanson and White (1979), White et al. (1981), and many others have continued to recognize a species distinction. Several authors have suggested that at least two hominid taxa are represented in the mid-Pliocene collections, but no consistent stand regarding the sorting or affinities of these taxa is discernible. Coppens (1983) holds "Pré-*Australopithecus*" to be present at Laetoli but possibly not the maker of the footprints, while Tuttle (1981) finds it difficult to imagine the Hadar foot bones being accomodated in the Laetoli prints (but see White and Suwa, 1985) and Olson (1981, 1985) places all of the Laetoli hominids into "*Paranthropus africanus*" (but see Kimbel et al., 1985).

Several authors appear to have been influenced by the relative dates of the Hadar and Laetoli samples, stating that a temporal gap might result in morphological differences (Day, 1982) or maintaining that the close temporal proximity of the samples makes a case for morphological stasis "unconvincing" (Cronin et al., 1981). These views appear to reflect a persistent and entrenched pan-gradualistic view of the evolutionary process (Eldredge and Tattersall, 1982) rather than a detailed analysis of the available data.

Only three authors have advanced any real evidence for a difference between the Hadar and Laetoli hominid collections. Blumenberg and Lloyd (1983), from a purely dental metric consideration of the site samples, conclude that only four of sixteen measurements show any significant difference and that it " . . . seems justifiable to pool the two samples" (p. 158). Wolpoff (1980) concludes that " . . . the few differences between the Afar and Laetoli dental samples may be the result of a late date for Afar" (p. 137). He goes on to describe differences in the anterior mandibular corpus that he holds to distinguish the site samples—the dental differences he alleges to be distinctive are dismissed below.

Tobias (1980a, 1981a) suggests that subspecific distinctions in paleontology have been insufficiently applied. He considers incomplete dental metric data available to him at the time of writing and concludes that subspecific differences between the Hadar and Laetoli samples might exist. It is clear that this is, however, largely a *post-hoc* judgment heavily influenced by Tobias' emphasis on chronological and geographical data. In this context it seems particularly appropriate to stress the strengths of Weidenreich's previously cited (1939) observations regarding morphological characters. Day states that dating must "colour" the views of workers asking systematic questions concerning Pliocene hominids. Perhaps his claim merely reflects, in turn, a view of evolution already colored by Darwinian gradualism. It is evident that the data most important in an accurate assessment of evolutionary questions concerning geographic subspeciation and morphological stasis or lack thereof are the morphological data themselves.

The relatively upland, poorly watered Pliocene ecology of Laetoli was fundamentally different from that described for Hadar. Certain ubiquitous faunal elements such as *Nyanzachoerus kanamensis* and *Hexaprotodon* are absent from the Laetolil Beds, while common Laetoli forms such as *Madoqua* are absent or rare at Hadar (see also Harris, 1985). These biogeographically significant facts almost certainly reflect environmental rather than chronological differences. Yet hominids existed in both environments and persisted through much of the stratigraphic columns at both localities. This is surely of considerable biological importance.

The specific unity of the Hadar and Laetoli site samples of fossil hominids is therefore an important question that transcends mere nomenclatural considerations. The significance of any differences between the Hadar and Laetoli hominid samples must be judged against both extant and extinct related taxa. Interested readers can gain an appreciation of the nature of the site samples assessed in the present comparison by referring to published descriptions of the Hadar and Laetoli fossils (White, 1977, 1980; Johanson et al., 1982). My purpose here is to provide a detailed, comparative, element-by-element analysis of the hominid dentition from Hadar and Laetoli. A comparison of bony morphology of the cranial and postcranial skeleton is beyond the scope of this presentation but will be published elsewhere and discussed briefly in the concluding remarks.

MATERIALS AND METHODS

All hominid dental evidence from the Laetolil Beds, including the original Garusi specimens (White, 1980), was compared to all available hominid dental remains from the Hadar Formation (Johanson et al., 1982). Abbreviations for mesiodistal and buccolingual are MD and BL, respectively. Measurements are taken as described in White (1977), with estimated corrections for interproximal attrition in the case of canines through the third molars. Corrections published as $\pm$ 1.0 mm are not included. Worn incisors are not accurately correctable, and the raw values given here for worn incisor crowns must be treated accordingly. Numbers of individuals listed in text and tables are accurately reported—where left and right tooth crowns are both available, only the right side crown is measured, unless it is broken. Measured values are taken preferentially over estimated values. It has been our standard procedure in reporting sample sizes *not* to count right and left crowns of one hominid as two individuals, despite insinuations to the contrary made by Kennedy (1979). The MD and BL values for an individual are taken from the same crown where possible.

For P_3s, the plesiomorphic, asymmetric crown profile is difficult to accomodate in standard measuring (as it is also in apes), as reported by White (1977). Tobias (1978a) has apparently misunderstood the situation, suggesting that my P_3 measurements were possibly in error. This is emphatically not the case, as was made obvious in my 1977 paper. In the present work, both standard and minimum–maximum values are provided for P_3s. Figures 1–6 plot all metric data.

In the present paper, comparisons of wear are extended only to the macroscopic level. In comparisons of the Hadar and Laetoli dental elements with other hominid taxa, *A. africanus* is defined as in White et al. (1981), and only the *A. robustus* sample from Swartkrans is used, because of dental differences between this sample and the Kromdraai sample (see Grine, 1981).

THE DECIDUOUS DENTITION

A. Upper Deciduous Dentition:

1) Upper first deciduous incisor: None yet known.

2) Upper second deciduous incisor: One Hadar individual, A.L.333-67.

3) Upper deciduous canine: Three from Hadar and two from Laetoli.

a) Size: More variation is seen in the Laetoli sample.

b) Crown morphology: Similar where preserved.

c) Wear: The specimens are at different wear stages, making wear difficult to compare, but no significant differences are evident.

4) Upper deciduous first molar: Two individuals from each site.

a) Size: More variation is seen in the Laetoli sample.

b) Crown morphology: The mesiobuccal extension of the enamel line and buccal relief in general is more marked in the Laetoli specimens.

c) Wear: No significant differences.

Comment: The two Hadar specimens from 333 are remarkably similar, while the differences in size and occlusal relief between L.H. 3/6 and L.H. 21 are dramatic. Variation in the mesiobuccal enamel line extension seen in the combined site sample is equivalent to that seen in limited samples of modern *Homo sapiens*.

5) Upper second deciduous molar: Three Laetoli and two Hadar individuals.

a) Size: The variation between the three Laetoli individuals (especially L.H. 3 and 21) encompasses the trivial differences between the two individuals from A.L.333.

b) Crown morphology: No differences.

c) Wear: There is a tendency for faceting to develop on Hadar specimens, whereas L.H. 21 lacks this but shows protocone dentin exposure.

Comment: As for dm^1, above.

B. Lower Deciduous Dentition:

1) Lower first deciduous incisor: Only Hadar A.L.333x-25 is known.

2) Lower second deciduous incisor: One Laetoli and two Hadar specimens.

a) Size: The Laetoli specimen represents a slightly larger individual, but there is a greater size difference between A.L.333-76 and -68 than between A.L.333-76 and L.H. 3.

b) Crown morphology: No differences are evident.

c) Wear: The L.H. 3 and A.L.333-68 occlusal wear is virtually identical in dentin exposure and incisal edge bevel.

3) Lower deciduous canine: Two individuals are known from each site.

a) Size: The L.H. 2 specimen is smaller than the Hadar counterparts, but the unmeasurable L.H. 3 specimen was probably larger before postmortem damage. Thus, Hadar would be encompassed within the Laetoli range.

b) Crown morphology: No differences.

c) Wear: The Laetoli specimens are too damaged to compare.

Comment: More variation in size, morphology, and wear is seen in *A. robustus* at Swartkrans (SK 61; SK 62).

4) Lower deciduous first molar: Two individuals are known from each site.

a) Size: The site ranges show nearly complete overlap. Variation at Laetoli (L.H. 2; L.H. 3) is greater than that seen at Hadar.

b) Crown morphology: No sample-specific differences are evident.

c) Wear: Protoconid wear on L.H. 2 does not show the thin planar facet observed on A.L.333-43 but similar variation is seen in modern human populations.

Comment: More variation in size, morphology, and wear is seen in *A. robustus* at Swartkrans (SK 61; SK 63).

5) Lower second deciduous molar: One individual is represented at each site.

a) Size: The L.H. 2 individual is somewhat larger than the Hadar counterpart.

b) Crown morphology: Similar.

c) Wear: Heavier on the Laetoli individual.

Comment: Differences in size, wear, and morphology are less than observed in either *A. africanus* (MLD 2; Sts 24) or in *A. robustus* (SK 61; SK 841).

THE PERMANENT DENTITION

A. Upper Permanent Dentition:

1) Upper central incisor: Two Laetoli and four Hadar specimens are known.

a) Size: In BL dimensions the site samples are equivalent. The unworn L.H. 3 MD dimension is larger than the moderate to heavily worn Hadar counterparts. However, the midcrown MD dimension of the very worn A.L.333x-4 specimen indicates that this tooth was almost certainly as long as L.H. 3.

b) Crown morphology: L.H. 3 has more pronounced lingual relief because of the split basal tubercle, but otherwise the tubercle mass and marginal ridges are equivalent to those seen in A.L.200-1a.

c) Wear: Laetoli specimens unworn.

Comment: A similar degree of size and morphological variation is seen in *A. africanus* (MLD 43; Sts 24) as well as in the Swartkrans sample, even when *Homo* individuals are removed from the latter.

2) Upper lateral incisor: Four Laetoli and six Hadar individuals.

a) Size: There is nearly complete overlap in size between the two site samples, with both small and large individuals seen at each site.

b) Crown morphology: No significant differences are evident.

c) Wear: The Hadar sample varies widely, and three of four Laetoli specimens are unworn.

Comment: The unworn A.L.333x-2 I^2 would have been the most massive tooth in the combined sample, and A.L.198-17b is clearly the smallest. This degree of variation is also seen in *A. africanus* (Sts 52a; TM 1512) but is greater than that seen in the small Swartkrans sample of *A. robustus*.

3) Upper canine: Three Laetoli and five Hadar individuals.

a) Size: The L.H. 3 specimen's MD dimension exceeds all others by about a millimeter, while its BL dimension virtually equals the top of the Hadar range. This results in a shape index showing L.H. 3 at the narrow end of the range, although full crown development might have lowered the index slightly. Crown shape index shows two of the three Laetoli individuals to have slightly less MD crown compression than those from Hadar, but the difference seems insignificant (see below). Crown modules show almost complete overlap between the site samples.

b) Crown morphology: No significant differences are seen between the site samples. In addition to being the largest tooth, L.H. 3 also shows the greatest lingual crenulation.

c) Wear: The only worn Laetoli $\underline{C}$, L.H. 5, is more heavily worn than the Hadar counterparts A.L.200-1a and 400-lb. All of these canines share a planar facet on the mesial occlusal edge, while the Hadar specimens show a distally facing dentin exposure on the distal occlusal edge instead of the more lingually facing exposure seen on L.H. 5. It is likely that the exposure of the Hadar specimens would have come to face more lingually as wear progressed, but more fossils are needed from both sites to verify this possibility.

Comment: Wolpoff's (1980) assertion that Afar canines are less projecting is not supported by our comparisons. Although L.H. 3 appears to be the largest tooth in the sample, on the basis of MD and BL dimensions, the massive A.L.333-1 $\underline{C}$ would surely have been a more impressive specimen in its combination of a great BL dimension with what must have been a very tall, projecting crown before it was broken. Variation at both sites is about equivalent (L.H. 5 vs. L.H. 3 for Laetoli and A.L.400-1a vs. A.L.333-1 for Hadar). This variation is greater than that seen in *A. africanus* (Sts 52a; TM 1512) but is in keeping with that expressed by the fuller sample of *A. robustus* from Swartkrans (SK 845; SK 94).

4) Upper third premolar: Five Laetoli and four Hadar individuals.

a) Size: There is near identity in MD dimensions, with the exception of one large Laetoli individual (L.H. 6) and one small Hadar specimen (A.L.199-1). In BL dimensions the two large Laetoli specimens fall 1.0 mm or less above the largest Hadar specimen, while the diminutive A.L.199-1 crown is again smallest. The crown modules reflect these differences. Shape indices show complete overlap of ranges.

b) Crown morphology: The Laetoli sample shows a more pronounced mesiobuccal ridge and groove than seen in the Hadar P^3s and also shows slightly more mesiobuccal extension of the enamel line and hence less symmetry of the buccal crown face. The sample ranges do overlap in the latter features, however (A.L.200-1a; Garusi I). The occlusal crown outline appears to separate the two site samples, but the apparent asymmetry at Laetoli is confined to relatively unworn individuals, interproximal attrition having reduced the asymmetry in the four worn Hadar counterparts.

c) Wear: The only comparably worn Hadar P^3 (A.L.199-1) shows less disparity between buccal and lingual cusp wear than the L.H. 5 and L.H. 25 specimens.

Comment: An equivalent amount of total variation in mesiobuccal groove depth and mesiobuccal enamel line projection is seen in *A. africanus* (TM 1511; Sts 52a). The addition of a small individual (of A.L.199-1 size) to the Laetoli sample would bring complete overlap in ranges. A larger sample of worn teeth is required to address the significance of the relatively heavily worn buccal cusp of the A.L. 199-1 P^3, but similar variation is encountered in modern human populations. The predominance of the buccal crown half in cusp area and MD length in the Laetoli sample is, however—on the basis of the known samples—a discrete difference between the site samples. Its biological significance remains unclear, however, because similar differences are also observed between *A. africanus* individuals (Sts 52a; Stw 73).

5) Upper fourth premolar: Five Laetoli and five Hadar individuals.

a) Size: In MD dimension, a single Laetoli P^4 exceeds the Hadar range by 0.2 mm. There is complete overlap between the samples in BL dimensions, shape indices, and crown modules, largely due to the fact that only one Laetoli specimen has a measurable BL dimension. In overall size, it is evident that the major difference between the site samples lies with the absence of an A.L.199-1-sized individual from Laetoli.

b) Crown morphology: No significant differences are evident.

c) Wear: No differences, where comparable.

Comment: Overall size variation in the combined Hadar/Laetoli sample is equivalent to or less than that seen in *A. africanus* (Sts 47; Sts 30) and Swartkrans *A. robustus* (SK 48; SK 845).

6) Upper first molar: Six Laetoli and five Hadar individuals.

a) Size: There is only minimal size overlap in site samples for MD and BL diameters, as reflected in the crown module, where there is no overlap. These results stem largely from the fact that two of the smaller Laetoli individuals (L.H. 5 and L.H. 22b) are too broken to measure. In shape index, the Hadar sample lies fully within the Laetoli range.

b) Crown morphology: The large lingual cingulum of L.H. 17 is not present on any other specimen from Hadar or Laetoli. In crown outline, three of the four Laetoli specimens show relatively MD-abbreviated buccal crown halves, but the samples overlap in this character.

c) Wear: No differences.

Comment: In overall size the site samples show little overlap for reasons described above. The difference between large (L.H. 3) and small (A.L.199-1) ends of the range for the combined site sample is, however, only slightly greater than that seen in *A. africanus* (Sts 28; MLD 6) and *A. robustus* (SK 52; SK 47).

7) Upper second molar: Four Laetoli and two Hadar individuals.

a) Size: In MD diameter and crown module, the three measurable, nearly identical Laetoli specimens are encompassed in the Hadar range. In BL diameter they lie just above the high end of this specimen range. The shape index differences between the two site samples are not significant.

b) Crown morphology: No differences.

c) Wear: Similar where comparable.

Comment: In contrast to the M^1, M^2 crown dimensions show a total level of variation *less* than that seen in *A. africanus* (Sts 56; Stw 6) or *A. robustus* (SK 48; SK 13).

8) Upper third molar: Five Hadar and four Laetoli individuals.

a) Size: Laetoli lies in the Hadar range of BL dimensions, and crown modules overlap considerably. Less overlap is evident in MD dimensions between the two site samples. Shape indices show no overlap, the Laetoli teeth being relatively broad.

b) Crown morphology: No differences.

c) Wear: Similar where comparable.

Comment: The L.H. 5, L.H. 8, and Garusi II molars represent individuals equivalent in size to the small Hadar individual A.L.199-1. This combines with the fact that the large L.H. 3 and L.H. 6 M^3s are not available to produce an effect opposite that described above for M^1. At the M^3 position, the Laetoli specimens fall at the *small* end of the total pooled sample range. Total variation at Hadar (A.L.333x-1 compared to A.L.161-40) is not matched at Laetoli because no M^3 from a large individual is yet available. When the total pooled variation is compared to other related taxa, similar variation is observed in crown size and shape (*A. africanus*: Sts 28, Stw 43; *A. robustus*: SK 836, SK 48).

B. Lower Permanent Dentition:

1) Lower central incisor: Two Laetoli and four Hadar individuals.

a) Size: BL measures are tightly clustered and the apparent MD outlier value is from L.H. 2, an unerupted and hence unworn crown. The Hadar sample would have more closely approximated the Laetoli value if the former site's I_1's were unworn. These comments apply as well to the shape index and crown module values.

b) Crown morphology: No differences.

c) Wear: Not comparable.

2) Lower lateral incisor: Two incomplete Laetoli crowns and four Hadar crowns.

a) Size: The low MD value from Laetoli is from a very worn crown (L.H. 14) and is not an accurate indication of the nearly complete size overlap *between* and wide range of variation *within* the two site samples. Only half of the very large L.H. 3 I_2 remains.

b) Crown morphology: No differences.

c) Wear: Incisal bevel and dentine exposure are similar in the comparable specimens L.H. 14 and A.L. 198-18.

Comment: The small samples of *A. africanus* and *A. robustus* do not match the wide range of size variation seen at *either* Hadar or Laetoli. The differences between the latter two sites are insignificant.

3) Lower canine: Four Laetoli and nine Hadar individuals.

a) Size: The MD elongate L.H. 3 crown is similar to this individual's upper counterpart canine (see point A-3 under "The Permanent Dentition," above) in surpassing all values known for the Hadar sample. This exception aside, the Laetoli BL values, crown modules, and shape indices are all accommodated in the Hadar range.

b) Crown morphology: No differences.

c) Wear: The M 18773 $\overline{C}$, with its large distal dentin exposure, represents a very aged individual. The wear is thus more extreme in degree but no different in functional kind ($C/\overline{C}$ clash) from that seen in Hadar specimens A.L.198-1 and 128-1. On the other hand, some Laetoli specimens match the pattern where apical wear predominates (compare L.H. 14 and A.L.333w-58).

Comment: The great degree of size variation in MD, BL, and Ht at Hadar and Laetoli is not excessive when compared to that seen in *A. africanus* (Sts 3; Sts 51b) or *A. robustus* (SK 876; SK 96).

4) Lower third premolar: Five Laetoli and thirteen Hadar individuals.

a) Size: As described in Materials and Methods above, two sets of measures, standard and minimum/maximum, were taken on each crown. The Laetoli sample includes only one relatively small individual and hence appears large in BL and MD dimensions compared to the Hadar sample, with L.H. 3, 4, 14, and 24 all clustering at or slightly above the Hadar range. Crown modules reflect this fact, while the shape indices show no significant differences between the site samples, except for the outlying L.H. 3 shape index that results from a skewed, BL-elongate crown (associated, interestingly, with a MD-elongate $\overline{C}$ crown).

b) Crown morphology: The full range of metaconid expression (from virtually absent to present) as well as crown profile asymmetry (fairly round to very elliptical) is seen in both site samples. Most of the Hadar specimens are heavily worn, so that occlusal crenulation like that seen in L.H. 3 is obliterated. No other characters consistently differentiate the two site samples. While White et al. (1981) count six of 15 *A. afarensis* P_3s as lacking metaconids, only one of the five Laetoli individuals (L.H. 24) lacks this lingual cusp on its worn occlusal surface.

c) Wear: Where comparable, there is no difference between the site samples. The unique wear pattern on the right L.H. 14 P_3 (but not the left) is probably the result of malocclusion. Wolpoff (1980) describes the Afar P_3 as fully incorporated into the grinding process and contrasts this with the Laetoli condition, where he perceives a functionally different pattern. His mischaracterization of the Laetoli sample seems to be based on the maloccluded L.H. 14 right P_3. All of the other Laetoli P_3 crowns display wear similar to that seen at Hadar (see Wolpoff and Russell, 1981, and White, 1981b, for views on L.H. 14 and its significance).

Comment: The absence of very small (female) individuals combined with the occurrence of large (male) individuals (L.H. 3, 4, 14, 26) at Laetoli results in mean crown size differences between the sites. However, recovery of a smaller individual from Laetoli will substantially change this apparent difference (see "The Deciduous Dentition," comment above and more discussion below). Comparing variation in the pooled Hadar/Laetoli sample to variation seen in *A. africanus* (Sts 24; MLD 2) and *A. robustus* (SK 96; SK 858) shows that similar degrees of variation characterize the latter two taxa.

5) Lower fourth premolar: Three Laetoli and twelve Hadar individuals.

a) Size: The Laetoli specimens are contained within the Hadar range for both MD and BL dimensions, but they cluster near the top of that range, resulting in crown modules for L.H. 3 and L.H. 14 that slightly exceed the Hadar range (by 0.25 mm). Laetoli shape indices also fall within the Hadar range.

b) Crown morphology: The BL-elongate, smaller crowns at Hadar (A.L.198-1, 288-1) are not matched by the Laetoli sample in shape or size. Most Hadar specimens are worn so that the kind of crenulation so apparent on L.H. 3 is not evident. No other significant, site-differentiating characters are evident.

c) Wear: No differences, where comparable.

Comment: The comments about the large Laetoli average compared to the more sex-balanced Hadar sample average made above also apply to the P_4. When the total range of variation between A.L.277-1 and A.L.128-23 is compared to that seen in *A. africanus* (Stw 5; Stw 14) or Swartkrans *A. robustus* (SK 7; SK 34), it is evident that the Hadar range is slightly larger. The significance of this difference will only be revealed by increased samples.

6) Lower first molar: Five Laetoli and fourteen Hadar individuals.

a) Size: The Laetoli specimens are accommodated within the Hadar range of MD and crown module values but lie, as with P_4, at the upper end of that range. For BL dimensions, one Laetoli specimen exceeds the Hadar range by 0.4 mm. In crown shape index, the Laetoli sample is centered in the Hadar range.

b) Crown morphology: Most of the Hadar teeth are well worn and hence lack the intensive occlusal crenulation seen in L.H. 2 and L.H. 3. No other significant features sort the site samples.

c) Wear: Where comparable, wear is similar, but the two worn Laetoli molars show somewhat greater relative wear on the metaconid apex than do their Hadar counterparts (compare L.H. 4, 16 with A.L.333w-60, -12), forming a more elevated, sharper lingual occlusal edge on the latter specimens.

Comment: The Laetoli sample, as with P_3 and P_4, again lacks the very small individuals on the order of A.L.128-23 or 288-1. Whether Laetoli (L.H. 3) or Hadar (A.L.145-35) is taken as the maximum value when compared to A.L.128, however, the combined Hadar/Laetoli site sample range in M_1 size and morphology is about equivalent to the range seen for Sterkfontein *A. africanus* (Sts 18; Sts 24) or for *A. robustus* from Swartkrans (SK 6; SK 1588).

7) Lower second molar: Four Laetoli and eighteen Hadar individuals.

a) Size: For MD and BL values, as well as for crown modules and shape indices, the Laetoli specimens are all accommodated in the Hadar range and are more centered in that range than was the case for P_3 to M_1.

As with P_3 to M_1, no Laetoli specimen is yet known to reach the small size of the A.L.128-23 or 288-1 specimens.

b) Crown morphology: No significant differences are evident, but it should be noted that no Laetoli specimen preserves substantial unworn relief.

c) Wear: The sharp crest forming the buccal metaconid ridge is very pronounced in L.H. 23 but not seen to this degree on any other specimen (the closest approach being A.L.145-35, a less worn specimen). No other significant differences distinguish the site samples.

Comment: The impressive, encompassing crown size variation at Hadar (A.L.188-1, A.L.128-23) is somewhat greater than that seen in *A. africanus* (MLD 2; Sts 9) or *A. robustus* (SK 34; SK 5).

8) Lower third molar: Two Laetoli and eleven Hadar individuals.

a) Size: In MD dimensions and crown modules the two Laetoli specimens encompass the entire Hadar range. Both BL values and shape indices for the Laetoli M_3s fall into the Hadar range.

b) Crown morphology: No Laetoli M_3 displays the round profile of A.L.333w-60 and no Hadar M_3 has the skewed crown outline seen in L.H. 15. Besides these minor differences, no features are seen to sort the site samples.

c) Wear: No significant differences.

Comment: The variation seen in crown size between A.L. 188-1 and A.L. 288-1 is slightly less than that seen at Sterkfontein (TM 1518; Sts 52b) or Swartkrans (SK 851; SK 840). As for the upper third molar, the M_3 comparison between sites shows that small individuals (presumably female) are indeed present at Laetoli as well as at Hadar.

TOOTH ROOTS

Ward et al. (1982; and in Johanson et al., 1982) has undertaken the most comprehensive comparisons of Laetoli and Hadar subocclusal anatomy. About the Hadar maxillary premolar roots Ward concludes: "This condition is identical to that present in the Garusi maxillary fragment," and about the mandibular molars he states: "An identical pattern is observed on all mandibular molars recovered from the Laetolil Beds in Tanzania" (Ward et al., 1982, p. 384).

The bilobed buccal P^4 root seen on Garusi I and L.H. 5 suggest fused roots for this tooth at Laetoli. Ward (1982) describes a similar condition for A.L.199-1 and A.L.200-1a. Lower premolar roots do not distinguish between the two site samples. Ward's research is still underway, and the deciduous tooth roots have yet to be considered. It is evident, however, that no significant differences in permanent tooth roots differentiate the Hadar and Laetoli samples. For the deciduous dentition the only feature of note is the absence at Hadar of the deep buccal groove on the dc′ root seen on one L.H. 3/6 dc′ and on L.H. 21.

SUPERNUMERARY TEETH

The only known instance in either site sample is the LM_4 on A.L.198-1. No Laetoli third molar shows an interproximal contact facet to indicate supernumerary teeth.

DIASTEMATA

Three of four Laetoli $\overline{C}$s and P_3s show evidence of a diastema, while six of sixteen Hadar specimens show such a gap. Wolpoff's (1980) contention that mandibular diastemata represent a "main distinction" between Afar and Laetoli hominids is simply unsubstantiated. In the maxillary dentition, the one relevant Laetoli specimen (L.H. 5) shows evidence for open maxillary diastemata in the adult dentition, while one of five Hadar specimens also displays this feature. Larger samples are clearly needed to determine whether the somewhat higher incidence of diastemata at Laetoli reflects biological significance or mere sampling error.

RELATIVE TOOTH PROPORTIONS

Visual inspection of the associated Laetoli teeth in maxillary (L.H. 3, 5, 6, 8, 21, 22) and mandibular (L.H. 2, 3, 4, 14) rows or sections of rows show that tooth size proportions among the Laetoli hominids as currently known are very similar to proportions among associated Hadar teeth. Thus, the very large L.H. 3 individual of Laetoli has the largest incisors, canines, premolars, and first molars along its tooth row, while the markedly smaller L.H. 5 individual shows smaller crowns at each position.

Not one of three Laetoli specimens where the $\overline{C}/P_3$ ratios can be measured shows the condition seen in two of the three Hadar specimens, where the ratio is greater than 1.0. The A.L.400-1a ratio, however, closely matches the L.H. 14 ratio. All three Laetoli specimens with P_3 and P_4 show a P_3/P_4 ratio of greater than 1.0, while only A.L.333-1 of the eight available Hadar specimens shows a ratio of 1.0. The others have smaller ratios (P_4 modules greater than P_3). Besides these minor differences, relative tooth proportions are closely similar in the two samples for both permanent and deciduous dentitions.

DISCUSSION

The comparisons presented here focus exclusively on biologically significant differences between the Hadar and Laetoli dental samples. No attempt has been made to balance the presentation by referring to shared characteristics that link the two site samples. Previous publications (Johanson et al., 1978; Johanson and White, 1979; White et al., 1981) have dealt with the most significant of the many characters that teeth from the two sites share. In fact, so many are shared that monographic treatment would be required to elaborate on each.

The metric data shown in Figures 1 through 6 provide clear comparisons between teeth from the two Pliocene sites. The only "trend" seen in the MD, BL, and crown module comparisons is the canine through first molar size dominance of the Laetoli sample. This trend is upset at the second and third molar positions, where small Laetoli individuals are represented. This finding is in accordance with Tobias's (1980a) study with restricted samples and with Blumenberg and Lloyd's (1983) findings of statistically significant differences between Hadar and Laetoli at only the M^1, P_3, P_4, and M_1 positions. The differences documented here and by Blumenberg and Lloyd are likely to be biologically significant differences between the two site samples. Several large Laetoli individuals, L.H. 2, 3, and 4, account for many of the high values. These three presumably male individuals lack measurable M_2 or M_3 crowns. Thus, the biological significance of the relatively large Laetoli size values for some crowns is probably sex-related: small female individuals are not yet well represented in the Laetoli sample *at these crown positions*. That such individuals do exist at Laetoli is fully documented, for example, at the *third* molar position. Unfortunately, these are *not* the same individuals whose measurable premolars and first molars are preserved; if these smaller individuals had associated premolars and first molars, the site sample range overlap would have been more extensive.

CONCLUSIONS

This comparison of Hadar and Laetoli teeth may be brought to bear on two separate but related issues. First, do intersite dental differences indicate, in the words of Tobias (1981a, p. 47), that " . . . the tying of

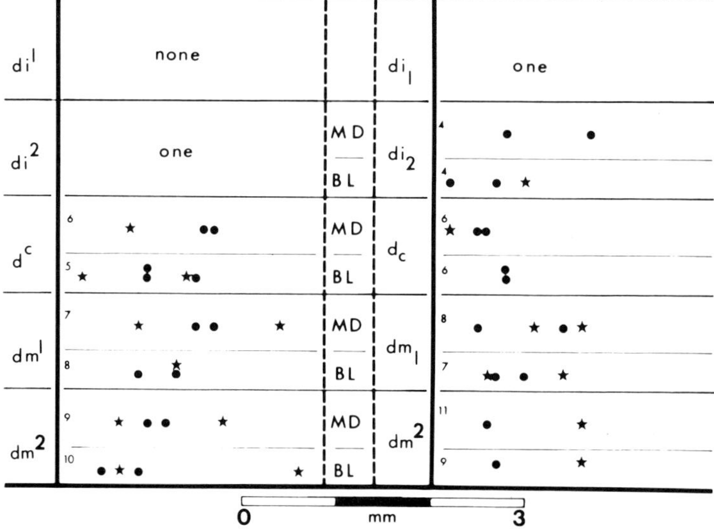

Fig. 1. Deciduous crown dimensions for Laetoli and Hadar hominids. The stars represent Laetoli and the circles Hadar tooth measurements. The scale is shown and the numerical value to the left of each distribution coincides with the dark vertical line to its immediate left. Measurements are provided according to procedures described in the text.

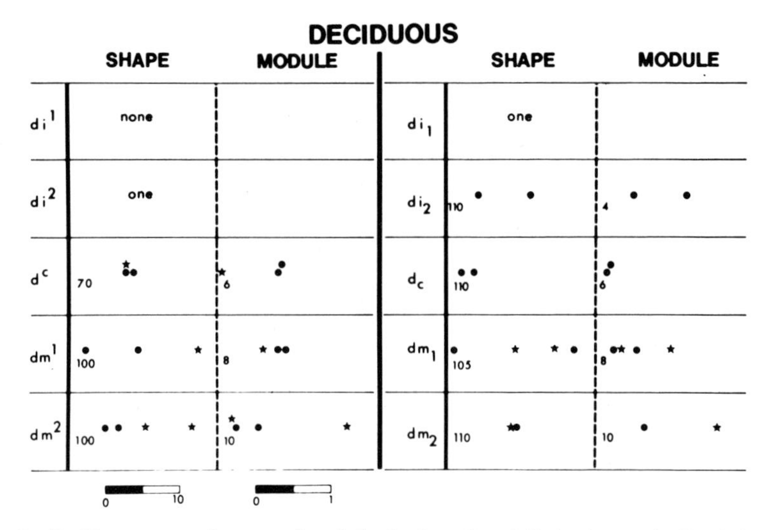

Fig. 2. Deciduous crown shapes and modules for Laetoli and Hadar hominids. Symbols and scales as in Figure 1.

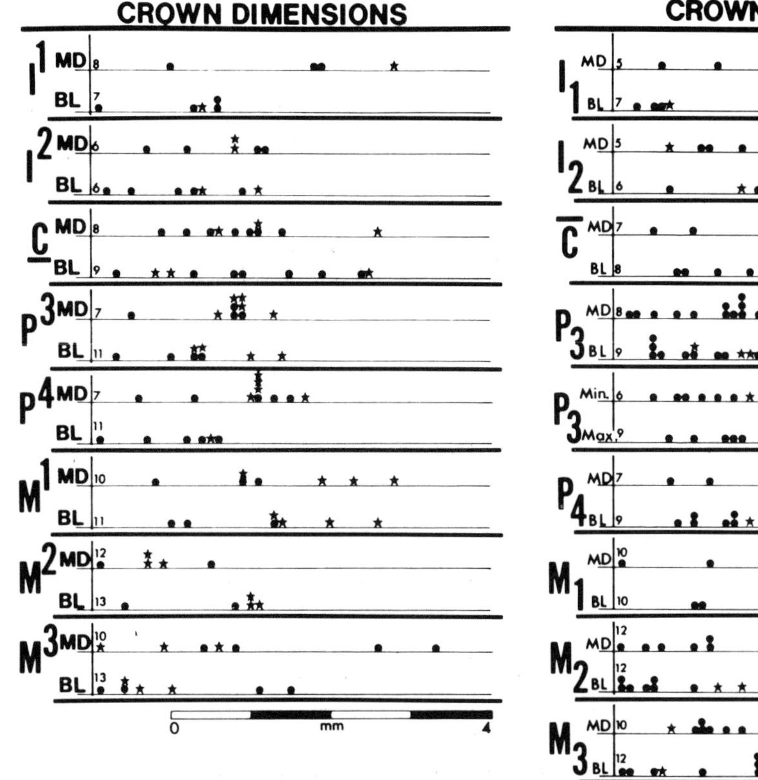

Fig. 3. Permanent upper crown dimensions for Laetoli and Hadar hominids. Symbols and scale as in Figure 1.

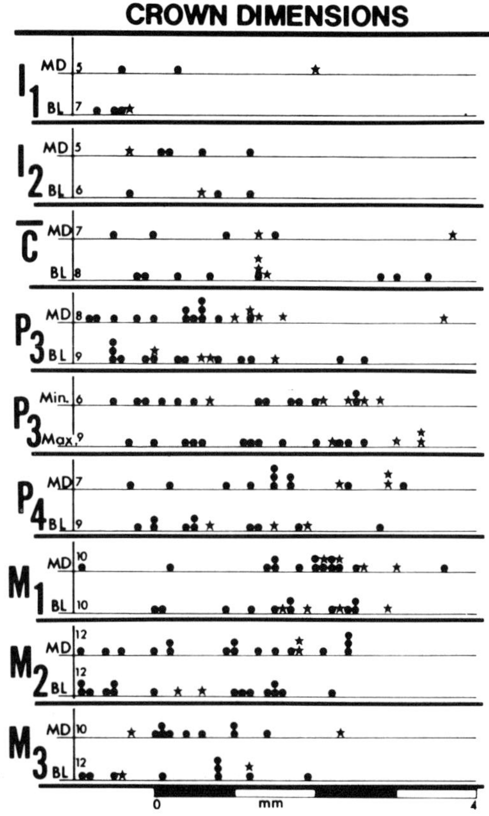

Fig. 4. Permanent lower crown dimensions for Laetoli and Hadar hominids. Symbols and scale as in Figure 1. Note that two sets of P_3 measurements, standard (above) and minimum–maximum (below) are provided.

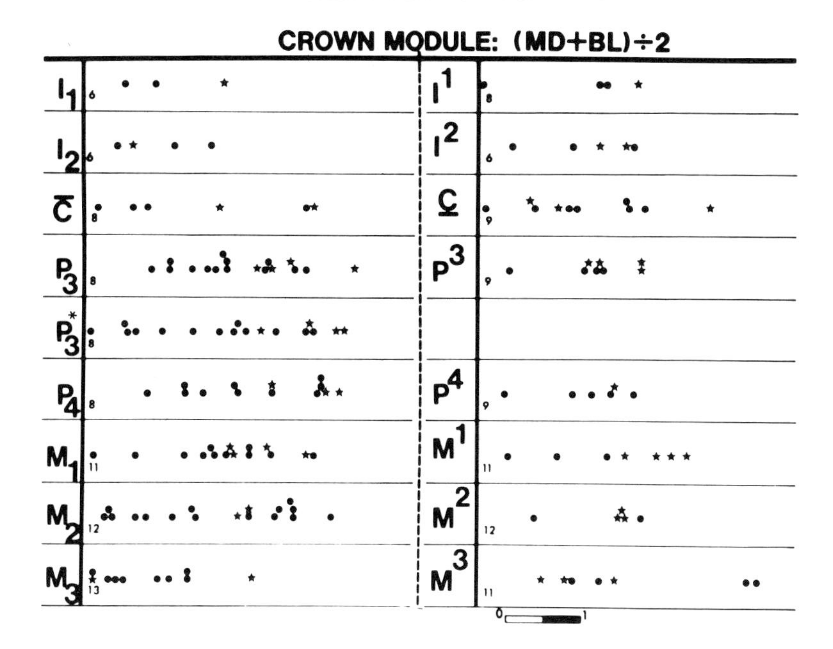

Fig. 5. Crown modules for permanent Laetoli and Hadar hominid teeth. Symbols and scale as in Figure 1. Note that two sets of P_3 measurements, standard (above) and minimum-maximum (below) are provided.

the name '*A. afarensis*' to the Laetoli fossils is manifestly inappropriate . . . "? In other words, is the pooling of the two site samples justifiable in systematic practice?

The second issue concerns the analysis of evolution among mid-Pliocene Hominidae. The oft-stated suggestion that a "young" date for Hadar and/or an "old" date for Makapansgat is relevant to the taxonomic status of *A. afarensis* is patently inappropriate. Morphology alone, as Weidenreich appreciated, must be the only consideration in such decisions. However, a recent rekindling of concern with evolutionary mode and tempo does make the Laetoli/Hadar comparison significant. Laetoli dates to between 3.6 and 3.8 million years (m.y.). The upper hominid-bearing portion of the Hadar Formation has been dated to around 2.9 m.y., while the dating of the base of this formation is uncertain. Some geophysical and biochronological evidence suggests a ca. 3.7 m.y. age for the stratigraphically oldest Hadar hominids (Aronson et al., 1983 and references therein; White et al., 1985), while other evidence suggests that the age of these Sidi Hakoma hominids rests closer to 3.3 m.y. (Brown, 1982, 1983; Brown et al., 1985). Given this chronology, it is appropriate to assess any differences between the Hadar and Laetoli site samples for possible evidence of evolutionary change.

The present analysis of the dentition confirms the strong similarities between the Hadar and Laetoli site samples. This is not a novel conclusion—the fact that so many workers (20 of 37 authors in the 128 publications on Hadar and Laetoli fossils reviewed in preparing this paper) have followed Johanson and myself in assigning Laetoli and Hadar hominids to a single hominid taxon shows the utility of this pooling. The dental element-by-element comparison above reinforces this pooling, as surprisingly few metric or morphological characters are found to differentiate the site samples. The total range of variation that results from a pooling of Hadar and Laetoli teeth is fully consistent with ranges universally documented for other early and modern hominoid taxa (a point that the 17 authors or author-teams attributing Hadar hominids to two or more taxa have not noted, even though it was evident in Johanson and White's Figure 8 (1979); see also Kimbel et al., 1985). We are currently engaged in de-

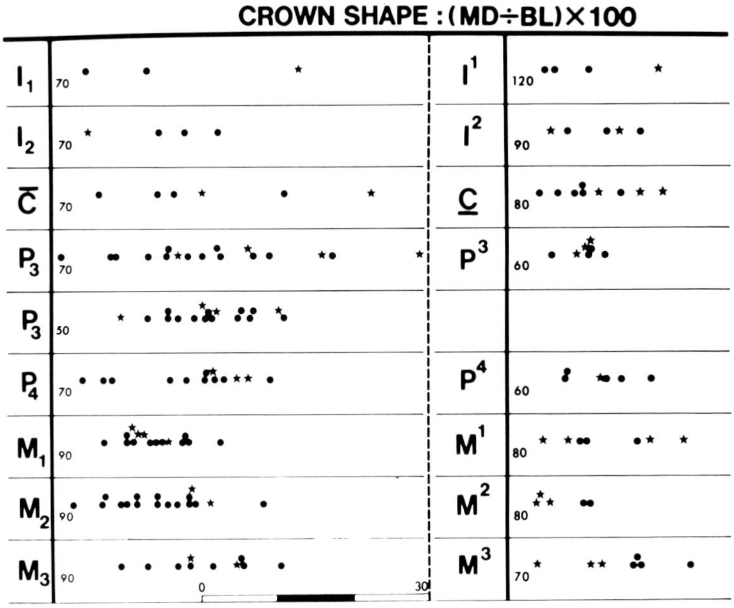

CROWN SHAPE : (MD÷BL)X100

Fig. 6. Crown shape indices for permanent Laetoli and Hadar hominid teeth. Symbols and scale as in Figure 1. Note that two sets of P_3 measurements, standard (above) and minimum–maximum (below) are provided.

tailed comparisons between early hominid taxa in an attempt to discern evolutionary changes in intraspecific variation as a corollary of sexual dimorphism. Much work remains to be done, but it is clear that a pooling of Hadar and Laetoli hominids has not led to an excessive amount of intraspecific variation. In fact, this detailed dental analysis indicates that the two site samples are strongly unified by a shared, unique (for hominids) series of plesiomorphous features. Given the parallel resemblances in the cranial (maxillary and mandibular) and postcranial (ulna, femur, phalanges) bony skeleton represented by individuals in both site samples, it appears that a species level distinction between the hominids from the two sites is entirely unwarranted.

Only future collection at Hadar, Laetoli, and other localities will further test the hypothesis of conspecificity. Further fieldwork may one day provide evidence requiring subspecies to be designated for *A. afarensis*. Much enlarged samples will be required before this possibility should be put into taxonomic practice. The designation of L.H. 4 as a holotype serves to reinforce what we had intended—the continuity between the site samples. We believe that the unification of the eastern African mid-Pliocene hominids under the nomen *A. afarensis* is a sound systematic position.

The issue of early hominid evolutionary rate, like that of conspecificity of Laetoli and Hadar hominid fossils, must remain open as evidence accumulates. The conspecificity issue is fairly easy to resolve on the basis of the present samples, but the evolutionary rate issue remains a more difficult problem, because of dating and sample size deficiencies. The apparent lack of change *within* the Hadar Formation and the strong resemblances between Hadar and Laetoli led Johanson and White (1979) to suggest stasis within the taxon *A. afarensis*. Other authors (Wolpoff, 1980) have approached the data and found what they have suggested as a more primitive status for the older Laetoli sample. Some features could arguably be made to fit a predisposition for progressive change within *A. afarensis* (see sections on canines, diastemata, P^3s above). On the other hand, the opposite relationship can also be found. For example, the youngest Hadar specimen (A.L.288-1) displays a unicuspid P_3, while only one of five Laetoli specimens can (arguably) be said to lack a P_3 metaconid.

Whatever the ultimate resolution of the evolutionary rate issue, it is evident that the presence of very similar hominids at different times in very different ecological settings may be related to a broad niche—to a stable but flexible adaptation that formed the foundations of further hominid diversification and evolution.

ACKNOWLEDGMENTS

I thank Mary Leakey for asking me to describe the Laetoli fossils; these descriptions formed the basis for my subsequent interpretive work. Don Johanson kindly made available the Hadar dental collections. Phillip Tobias, C.K. Brain, and E. Vrba assisted in my comparative study of the South African early hominids. Thanks go to the American Museum of Natural History, especially to Eric Delson, for providing me the opportunity to participate in "Ancestors." Gen Suwa assisted in the production of this paper, while E. Delson, W. Kimbel, and B. Schmucker provided critical comments. Much of the research reported was supported by the National Science Foundation (grant BNS 79 06849).

LITERATURE CITED

Aronson, JL, Walter, RC, and Taieb, M (1983) Correlation of Tulu Bor Tuff at Koobi Fora with the Sidi Hakoma Tuff at Hadar. Nature *306:*209–210.

Blumenberg, B, and Lloyd, AT (1983) *Australopithecus* and the origin of the genus *Homo*: Aspects of biometry and systematics with accompanying catalog of tooth metric data. Biosystems *16:*127–167.

Brown, FH (1982) Tulu Bor Tuff at Koobi Fora correlated with the Sidi Hakoma Tuff at Hadar. Nature *300:*631–633.

Brown, FH (1983) Correlation of Tulu Bor Tuff at Koobi Fora with the Sidi Hakoma Tuff at Hadar. Nature *306:*210.

Brown, FH, McDougall, I, Davies, T, and Maier, R (1985) An integrated Plio-Pleistocene chronology for the Turkana Basin. In E Delson (ed): Ancestors: The Hard Evidence. New York: Alan R. Liss, Inc., pp. 82–90.

Clarke, RJ (1977) The cranium of the Swartkrans hominid, SK 847 and its relevance to human origins. Unpublished Ph.D. Thesis, University of Witwatersrand, Johannesburg.

Coppens, Y (1983) Le Singe, l'Afrique et l'Homme. Paris: Fayard.

Cronin, JE, Boaz, NT, Stringer, CB, and Rak, Y (1981) Tempo and mode in hominid evolution. Nature *292:*113–122.

Day, MH (1982) "Lucy" jilted? Nature *300:*574.

Day, MH, Leakey, MD, and Olson, TR (1980) On the status of *Australopithecus afarensis*. Science *207:*1102–1103.

Delson, E, Eldredge, N, and Tattersall, I (1977) Reconstruction of hominid phylogeny: A testable framework based on cladistic analysis. J. Hum. Evol. *6:*263–278.

Eldredge, N, and Tattersall, I (1982) The Myths of Human Evolution. New York: Columbia University Press.

Grine, FE (1981) Trophic differences between "gracile" and "robust" australopithecines: A scanning electron microscope analysis of occlusal events. S. Afr. J. Sci. *77:*203–230.

Harris, JM (1985) Age and paleoecology of the Upper Laetolil Beds, Laetoli, Tanzania. In E. Delson (ed): Ancestors: The Hard Evidence. New York: Alan R. Liss, Inc., pp. 178–183.

Johanson, DC, Taieb, M, and Coppens, Y (1982) Pliocene hominids from the Hadar Formation, Ethiopia (1973–1977): Stratigraphic, chronologic, and palaeoenvironmental contexts, with notes on hominid morphology and systematics. Am. J. Phys. Anthropol. *57:*373–402.

Johanson, DC, and White, TD (1979) A systematic assessment of early African hominids. Science *203:* 321–330.

Johanson, DC, White, TD, and Coppens, Y (1978) A new species of the genus *Australopithecus* (Primates: Hominidae) from the Pliocene of eastern Africa. Kirtlandia *28:*1–14.

Johanson, DC, White, TD, and Coppens, Y (1982) Dental remains from the Hadar Formation, Ethiopia: 1974–1977 collections. Am. J. Phys. Anthropol. *57:*545–603.

Kennedy, G (1979) Difficulties in the definition of new hominid species. Nature *278:*400–401.

Kimbel, WH, White, TD, and Johanson, DC (1985) Craniodental morphology of the hominids from Hadar and Laetoli: Evidence of "*Paranthropus*" and *Homo* in the mid-Pliocene of Eastern Africa? In E Delson (ed): Ancestors: The Hard Evidence. New York: Alan R. Liss, Inc., pp. 120–137.

Leakey, MD (1981) Tracks and tools. Phil. Trans. Roy. Soc. Lond. *B292:*95–102.

Leakey, MD, and Hay, RL (1979) Pliocene footprints in the Laetolil Beds at Laetoli, northern Tanzania. Nature *278:*317–323.

Leakey, MD, Hay, RL, Curtis, G, Drake, RE, Jackes, MK, and White, TD (1976) Fossil hominids from the Laetolil Beds. Nature *262:*460–466.

Mayr, E (1969) Principles of Systematic Zoology. New York: McGraw Hill.

Mayr, E (1982) Reflections on human paleontology. In F Spencer (ed): A History of American Physical Anthropology 1930–1980. New York: Academic, pp. 231–237.

Olson, TR (1981) Basicranial morphology of the extant hominoids and Pliocene hominids: The new material from the Hadar Formation, Ethiopia, and its significance in early human evolution and taxonomy. In C Stringer (ed): Aspects of Human Evolution. London: Taylor and Francis, pp. 99–128.

Olson, TR (1985) Cranial morphology and systematics of the Hadar Formation hominids and "*Australopithecus" africanus*. In E Delson (ed): Ancestors: The Hard Evidence. New York: Alan R. Liss, Inc., pp. 102–119.

Protsch, RR (1981) The palaeoanthropological finds of the Pliocene and Pleistocene. In H Müller-Beck (ed): Die Archäologischen und Anthropologischen Ergebnisse der Kohl-Larsen-Expeditionen in Nord-Tanzania 1933–1939. Tubingen: Verlag Archaeologica Venatoria, Vol. 3.

Remane, A (1951) Die zähne des *Meganthropus africanus*. Zeit. Morphol. Anthropol. *42:*311–329.

Remane, A (1954) Structure and relationships of *Meganthropus africanus*. Am. J. Phys. Anthropol. *12:* 123–126.

Robinson, JT (1953) *Meganthropus*, australopithecines and hominids. Am. J. Phys. Anthropol. *11:*1–38.

Robinson, JT (1955) Further remarks on the relationship between "*Meganthropus*" and australopithecines. Am. J. Phys. Anthropol. *13:*429–446.

Robinson, JT (1956) The dentition of the Australopithecinae. Transv. Mus. Mem. *9:*1–179.

Senyürek, M (1955) A note on the teeth of *Meganthropus africanus* Weinert from Tanganyika Territory. Belleten *73:*1–55.

Simpson, GG (1961) Principles of Animal Taxonomy. New York: Columbia University Press.

Tobias, PV (1978a) Position et rôle des australopithecinés dans la phylogenèse humaine, avec étude particulière de *Homo habilis* et des théories controversées avancées a propos des premiers hominidés fossiles de Hadar et de Laetolil. In Les Origines Humaines et les Epoques de l'Intelligence. Paris: Masson, pp. 38–77.

Tobias, PV (1978b) The earliest Transvaal members of the genus *Homo* with another look at some problems of hominid taxonomy and systematics. Z. Morphol. Anthropol. *69:*225–265.

Tobias, PV (1980a) "*Australopithecus afarensis*" and *A. africanus*: Critique and an alternative hypothesis. Palaeont. Afr. *23:*1–17.

Tobias, PV (1980b) A survey and synthesis of the African hominids of the late Tertiary and early Quaternary periods. In LK Königsson (ed): Current Argument on Early Man. Oxford: Pergamon Press, pp. 1–12.

Tobias, PV (1981a) The emergence of man in Africa and beyond. Phil. Trans. Roy. Soc. Lond. *B292:*43–56.

Tobias, PV (1981b) The anatomy of hominization. In: Eleventh International Congress of Anatomy: Advances in the Morphology of Cells and Tissues. New York: Alan R. Liss, Inc., pp. 101–110.

Tuttle, R (1981) Evolution of hominid bipedalism and prehensile capabilities. Phil. Trans. Roy. Soc. Lond. *B292:*89–94.

Ward, S, Johanson, DC, and Coppens, Y (1982) Subocclusal morphology and alveolar process relationships of hominid gnathic elements from the Hadar Formation: 1974–1977 collections. Am. J. Phys. Anthropol. *57:*605–630.

Weidenreich, F (1939) The classification of fossil hominids and their relations to each other, with special reference to *Sinanthropus pekinensis*. Bull. Geol. Soc. China *19(1):*64–75.

White, TD (1977) New fossil hominids from Laetolil, Tanzania. Am. J. Phys. Anthropol. *46:*197–229.

White, TD (1980) Additional fossil hominids from Laetoli, Tanzania: 1976–1979 specimens. Am. J. Phys. *53:*487–504.

White, TD (1981a) Primitive hominid canine from Tanzania. Science *213:*348–349.

White, TD (1981b) On the evidence for "anterior dental cutting" in Laetoli hominids. Am. J. Phys. Anthropol. *54:*107–108.

White, TD, Johanson, DC, and Kimbel, WH (1981) *Australopithecus africanus*: Its phyletic position reconsidered. S. Afr. J. Sci. *77:*445–470.

White, TD, Moore, RV, and Suwa, G (1985) Hadar biostratigraphy and hominid evolution. J. Vert. Paleontol. (in press).

White, TD, and Suwa, G (in press) Hominid footprints at Laetoli: Facts and interpretations. In DC Johanson (ed): The Origins of Human Locomotion. San Diego: Academic Press.

Wolpoff, MH (1980) Paleoanthropology. New York: A.A. Knopf.

Wolpoff, MH, and Russell, M (1981) Anterior dental cutting at Laetolil. Am. J. Phys. Anthropol. *55:*223–224.

Ancestors: The Hard Evidence, pages 153–167

Australopithecine Evolution: The Deciduous Dental Evidence

Frederick E. Grine
Department of Anthropology, State University of New York, Stony Brook, New York 11794

ABSTRACT Examination of the deciduous teeth of specimens comprising various australopithecine samples has revealed a number of traits that differ between them. An analysis of the distributions of deciduous crown variants amongst these samples suggests that the known australopithecine specimens are attributable to five species: *A. afarensis, A. africanus, A. robustus, A. crassidens,* and *A. boisei.* A testable hypothesis of the phylogenetic (cladistic) relationships of these taxa is proposed from an analysis of synapomorphy of deciduous dental characters. According to this evidence, *A. robustus* appears to have been phylogenetically intermediate between *A. africanus* and *A. crassidens. A. boisei* probably evolved from *A. crassidens,* or an *A. crassidens*-like ancestor.

INTRODUCTION

The recent paleoanthropological literature, including several papers in the present volume, attests to the level of debate concerning the recognition of australopithecine species and the systematic relationships of these taxa. As a first step to considering the phylogenetic relationships of fossil hominid taxa, it is necessary to determine which of the specimens under consideration may be attributed to a single species. All hypotheses concerning systematic relationships and all scenarios about ancestry and descent and the course of evolutionary events within a group's phylogeny are dependent upon the allocation of individual specimens to phena, or taxa.

The deciduous dentition forms only part of the total morphological package that must be assessed before any reasonably secure taxonomic distinctions can be drawn among australopithecine samples, and while the deciduous dental sample for African Plio-Pleistocene Hominidae is of limited size, the milk teeth have been of singular pertinence to the study of hominid evolution since *Australopithecus* was introduced some 60 years ago by Raymond Dart (1925). Indeed, deciduous teeth have played a critical role, not only in the definition of many of the proposed australopithecine taxa but also in debates over their phylogeny (Broom, 1929, 1946, 1950; Broom and Robinson, 1950, 1952; Le Gros Clark, 1952; Robinson, 1954, 1956; Ashton et al., 1957; Tobias, 1967; von Koenigswald, 1967; Wolpoff, 1974; Howell, 1978; White et al., 1981).

In this paper the morphological characters of the milk teeth of specimens comprising various australopithecine samples are examined, in an attempt to elucidate certain questions pertaining to both the number and the phylogenetic relationships of the species represented by these elements.

AUSTRALOPITHECINE SAMPLES CIRCUMSCRIBED

The first Plio-Pleistocene hominid specimen to be recovered in Africa was described briefly by Dart (1925), who designated it the holotype of a new genus and species, *Australopithecus africanus.* Broom (1936) attributed the initial finds from Sterkfontein to a second species of *Australopithecus, A. transvaalensis,* although subsequently he (1938) transferred this species to the genus *Plesian-*

thropus. For his part, Dart (1948) termed the hominid specimens from Makapansgat *Australopithecus prometheus*.

While Broom (1946, 1950) considered the Taung and Sterkfontein fossils to be generically distinct, and those from Makapansgat to be referrable to a separate subfamily, Robinson (1954, 1956) argued that the specimens from these three sites could be accommodated within a single species. He proposed also that the Taung specimen be differentiated at a subspecific level from the combined Sterkfontein–Makapansgat sample on the basis of a few differences in details of dental morphology. Subsequent studies of the Sterkfontein and Makapansgat fossils have revealed several apparent differences between them (Tobias, 1967; Aguirre, 1970; Sperber, 1973; Rak, 1983), but most authorities are agreed that they represent but a single taxon (Clarke, 1977; White, 1977; Howell, 1978; Tobias, 1978a, 1980; White et al., 1981; Rak, 1983). On the other hand, whereas most workers do not recognize any taxonomic distinction between the Sterkfontein-Makapansgat sample and the Taung specimen, Tobias (1973, 1978c) has suggested that the latter does not, in fact, belong to the species represented by the hominids from Makapansgat and Sterkfontein Member 4.

In the present analysis, then, the deciduous specimens from Sterkfontein Member 4 and Makapansgat Member 3 are pooled to comprise a single sample containing 12 specimens with one or more milk teeth, nearly twice as many were available to Robinson (1956) when he undertook his meticulous dental study. The Taung specimen is treated independently.

Broom (1938) attributed the hominid remains from Kromdraai to a distinct genus and species, *Paranthropus robustus*, while the Swartkrans australopithecines were considered by him (1949) to represent a second species of *Paranthropus*, *P. crassidens*. Robinson (1954, 1956), on the other hand, argued that these remains could be accommodated in a single species (*P. robustus*), and that the differences between them were ascribable to subspecific differentiation.

Notwithstanding the polemics over the generic distinctiveness of *Paranthropus* (e.g., Mayr, 1950; Le Gros Clark, 1964; Pilbeam and Simons, 1965; Tobias, 1967; Robinson, 1968; Clarke, 1977; Olson, 1978), while some workers are of the opinion that all of the South African australopithecines are conspecific (Brace, 1973; Campbell, 1974; Wolpoff, 1974, 1980), an overwhelming body of morphological evidence suggests to most that the forms from Kromdraai and Swartkrans are at least specifically distinct from the Sterkfontein and Makapansgat australopithecines (Tobias, 1967, 1978a; Bilsborough, 1972; Wood and Stack, 1980; White et al., 1981; Rak, 1983). Although contemporary opinion seems to hold that the Kromdraai and Swartkrans australopithecine specimens are not separable at any taxonomic level, Howell (1978) has proposed that a species-specific distinction between them is warranted. I have elsewhere (Grine, 1981b, 1982) alluded to differences between these samples that are suggestive of some level of taxonomic distinction.

Thus, in the present analysis, the Kromdraai and Swartkrans australopithecine specimens are grouped separately for comparative purposes. The Kromdraai sample comprises four specimens possessing one or more milk teeth, and the Swartkrans sample consists of some 15 such specimens. In both instances these samples are somewhat larger than those available to previous investigators.

In 1959 a massively constructed hominid cranium was discovered in Olduvai Gorge, and this specimen was attributed by L.S.B. Leakey (1959) to a novel genus and species, *Zinjanthropus boisei*. Tobias (1967), in his thorough comparative study of the Olduvai fossil, argued that it should be recognized as representing a distinct species of *Australopithecus*, *A. boisei*. Numerous additional specimens attributable to this taxon have been recovered subsequently from Beds I and II of Olduvai Gorge, the Humbu and Chemoigut Formations, Member D through G and Member L of the Shungura Formation, and from the Lower and especially the Upper and Ileret Members of the Koobi Fora Formation (see Howell (1978) and Grine (1981b) for reviews of these discoveries).

Tobias has suggested that these eastern African specimens, together with those from Kromdraai and Swartkrans, might be regarded as representing "a superspecies, *A. robustus*, comprising two semispecies, *A. robustus* and *A. boisei*" (1975, p. 293). Others (e.g., Clarke, 1977; Bilsborough, 1978) have implied an even closer relationship amongst these forms by viewing them as comprising two subspecies of *A. robustus*, while Rak (1983) has argued that *A. boisei* is a distinct species because it possesses fundamental structures of facial architecture that are substantially different from those evinced by *A. robustus*.

Among the numerous specimens from diverse eastern African localities that have been attributed to *A. boisei*, seven comprise one or more deciduous elements. This sample includes three specimens (OH 3, OH 30, and OH 32) from Bed II of Olduvai Gorge (Tobias, 1985), a juvenile mandible (KNM-ER 1477) from the Upper Member of the Koobi Fora Formation (Wood, 1978), an isolated, incomplete upper molar (L 338x-32) from Member E of the Shungura Formation (Howell and Coppens, 1976), and two isolated lower molars (L 64-2 and L 704-2) from Member D of the Shungura Formation (Grine, 1981b, 1984).

The mid-1970s witnessed significant fossil hominid discoveries at Hadar, Ethiopia, and Laetoli, Tanzania. Between 1973 and 1977, field work in the Afar Depression resulted in the recovery of numerous specimens, representing at least 35 individuals, from the Sidi Hakoma, Denen Dora, and Kada Hadar Members of the Hadar Formation (Johanson et al., 1982). The remains of some 22 individuals have been discovered in the Laetoli Beds between 1974 and 1979, adding appreciably to the previously available "Garusi" sample (White et al., 1981).

Preliminary announcements of the Hadar fossils suggested that they represented at least three hominid taxa (Taieb et al., 1976). Several elements (e.g., A.L.166-9 and A.L.211-1) were observed to resemble *A. robustus* (or *A. boisei*) specimens, some (e.g., A.L.128-1 and A.L.129-1) were noted to correspond to *A. africanus* homologues, while others (eg, A.L.199-1 and A.L., 200-1) were held to resemble specimens assigned to the genus *Homo* (Taieb et al., 1974, 1975, 1976; Johanson and Taieb, 1976). At the same time, the Laetoli sample was provisionally interpreted as representing early evidence of the genus *Homo* (Leakey et al., 1976).

Subsequent analyses of these samples, however, prompted Johanson et al. (1978) to argue that all known specimens from both localities could be accommodated in a single species, *Australopithecus afarensis*, characterized by a suite of primitive features that serves to distinguish it from other early hominid taxa. This assignment has been questioned by several workers, although there is a notable lack of consensus amongst the critics as to the proper taxonomic attribution of these specimens. Tobias (1978b, 1980) and Boaz (1979, 1983) have argued that the Laetoli and Hadar samples are referable to *A. africanus*, although Tobias has ascribed them to separate subspecies. On the other hand, several workers have opined that these samples are variously divisible into at least two discrete taxa: *Homo* and "late *Ramapithecus*" (Leakey, 1976), *Homo* and *Paranthropus or Australopithecus* (Olson, 1981, 1985; Senut and Tardieu, 1985), or even an unspecified hominid species and *Sivapithecus sivalensis* (Ferguson, 1983).

Although many of these and other such arguments have been cogently dismissed (White et al., 1981; White, 1985; Kimbel et al., 1985), in the present study, the specimens possessing deciduous teeth from Hadar and Laetoli are treated independently. The Hadar sample comprises 13 such specimens, all of which derive from a single locality (333) in the Denen Dora Member of the Hadar Formation, while six juvenile specimens make up the sample from the Laetolil Beds.

CHARACTER VARIABILITY AND POLARITY

Among the metrical and non-metrical features of the deciduous teeth, many show some degree of variability within the small early hominid samples that are available. Characters that display high degrees of intraspecific (intrademic) variability are commonly considered to be of low "phyletic valence" (Robinson, 1960a) or "weight" (Mayr, 1969) in systematic analyses. Conversely, characters of low phenotypic variability are usually accorded high weight in such studies (Farris, 1966). Thus, in the present study, only those traits that display little or no variability within (but vary among) samples are utilized in the reconstruction of systematic relationships.

Once a suite of analytically useful characters has been identified, the erection of hypotheses on plesiomorph–apomorph polarity becomes the most important aspect of any reconstruction of phylogenetic relationships. The most widely used method of polarity determination is that based on out-group comparisons (Hennig, 1966; Wiley, 1981). With regard to the analysis of fossil taxa (such as are represented by the australopithecines), however, there is some debate over the choice of an appropriate out-group: fossil taxa might be treated as terminal sisters of extant species, or comparisons might be limited to relevant fossil groups (Farris, 1976; Patterson and Rosen, 1977). Despite these apparent difficulties, comparisons of the

Laetoli specimens with both extinct and extant hominoid taxa have revealed the Laetoli specimens to possess an array of dental features that would seem to be primitive for the Hominidae (*sensu stricto*) (de Bonis et al., 1981; White et al., 1981). As a working hypothesis, then, the character states evinced by the Laetoli deciduous crowns are considered to be primitive for the australopithecines.

Some workers persist in the belief that if a certain character state is possessed by a geologically older taxon, and an alternate state is shown by a stratigraphically later species of the same monophyletic group, then the state evinced by the older must represent the primitive condition (e.g., Harper, 1976). However since the actual biochronological range of a species cannot be assumed to coincide with its observed stratigraphic (geochronological) range, geochronological precedence cannot be used to determine character state polarity (Schaeffer et al., 1972). While not sufficient in and of itself for polarity determination, the criterion of geochronological precedence does maintain a striking degree of paleontological correlation, and it does not abrogate the working hypothesis used here. The Laetoli specimens, dated to between 3.59 and 3.77 million years (m.y.) (Leakey et al., 1976), are, by all recent accounts, older than those comprising the other australopithecine deciduous samples (Brown et al., 1985; Harris, 1985).

DISTRIBUTION OF DECIDUOUS DENTAL CHARACTERS AMONG AUSTRALOPITHECINE SAMPLES

As noted above, only those deciduous dental features that show little or no variability within samples are useful in the reconstruction of the phylogenetic relationships of the australopithecine taxa represented by these samples. Characters that display intrasample variability, but that appear to display differences in "central tendency" among samples are discussed, however, inasmuch as they comprise part of the overall dental patterns that serve to characterize individual samples or groups of samples.

Deciduous Incisors

The only australopithecine di^1s presently available are represented by an isolated, incomplete tooth from Sterkfontein and the damaged antimeres of a single individual from Swartkrans. The only observable difference between these specimens pertains to the somewhat stronger linguocervical and distocervical bevel evinced by the incisal wear facet on the Sterkfontein crown, but this tooth is also somewhat more heavily worn than the Swartkrans homologues.

An isolated crown from the Hadar Formation represents the only early hominid di^2 known at present, apart from the severely damaged crown of the Rdi^2 from Taung. Originally, the latter preserved "an appreciable part of the crown still intact," according to Broom (1946, p.37), who noted that "it has indications of its having had a somewhat shovel type of crown." The lingual face of the Hadar homologue is flat, showing absolutely no development of the median or either marginal ridge.

Australopithecine di_1s are limited to one specimen from Sterkfontein, the damaged antimeres of a single individual from Swartkrans, and an isolated crown from Hadar. There are no noteworthy differences among these teeth.

Mandibular lateral incisors are somewhat better represented, being known for the Hadar, Sterkfontein, Kromdraai, and Swartkrans samples. The distal portion of the incisal margin of the di_2 is strongly bevelled in the Sterkfontein specimen and in one out of the two Hadar specimens; the other Hadar incisor shows a moderate distal bevel. In the Kromdraai specimen and in one out of two Swartkrans specimens the incisal edge of the di_2 is horizontally disposed, while the second Swartkrans crown possesses a slight downward bevel over the distal part of the margin.

Deciduous Canines

Maxillary milk canines are known only for the Taung specimen, two Laetoli and three Hadar specimens, and two individuals attributed to the *A. boisei* sample.[1] The Laetoli, Hadar, and Taung dc^1s are similar to one another in that the median lingual ridge is vertically disposed to slightly convex mesially, whereas on the *A. boisei* homologues this ridge is robustly convex mesially. The Laetoli, Hadar, and Taung dc^1s are rather more symmetrical in buccal outline than the *A. boisei* crowns; in the former the mesial

[1]Tobias (1985) follows Wallace (1978) in referring to the milk canine of Swartkrans specimen SK 839 as a Ldc^1. This tooth is interpreted by the present author as a Rdc_1 (Grine, 1981a).

and distal apical edges diverge at similarly steep angles from the horizontal, while on *A. boisei* dc^1s the mesial edge is nearly horizontal.

Mandibular deciduous canines are better represented, being known for the Taung specimen, one Sterkfontein specimen, two Kromdraai specimens, four Swartkrans specimens, two specimens each from Laetoli and Hadar, and one individual attributed to the *A. boisei* sample. Several features of these crowns appear to be sufficiently discrete to warrant an analysis of their distributions among these small samples. These characters are as follows:

Distal apical edge. The distal apical edge (character 1, Fig. 1A) of the dc_1 is elongate and vertically disposed in the Laetoli and Hadar specimens. On the other hand, the Sterkfontein, Taung, Kromdraai, Swartkrans, and *A. boisei* dc_1s are similar to one another but differ from the Laetoli and Hadar homologues in that the distal apical edge is both shorter and more shallowly (= obliquely) inclined.

Relative mesiodistal crown size. The Laetoli, Hadar, Sterkfontein and Taung specimens are similar to one another and to constituents of extant hominoid samples in the relative mesiodistal proportions (character 2, Fig. 1A) of the dc_1 and dm_1, compared to the corresponding diameter of the dm_2. On the other hand, the Kromdraai, Swartkrans, and *A. boisei* samples differ notably in possessing relatively smaller canines and relatively larger first deciduous molars. Relative reduction in dc_1 size would appear to be a unique character shared by the Kromdraai, Swartkrans, and *A. boisei* samples (Grine, 1984). This relative reduction in size applies also to these samples for known dc^1s and for the upper and lower permanent canines (Robinson, 1954, 1956; Tobias, 1967; Wood and Stack, 1980; Grine, 1982), and it may attest to functionally significant differences in canine usage (Grine, 1981b).

Lingual cingulum disposition. On the Laetoli, Hadar, Sterkfontein, Taung, and Kromdraai dc_1s the lingual cingulum (character 3, Fig. 1A) rises to a slightly higher

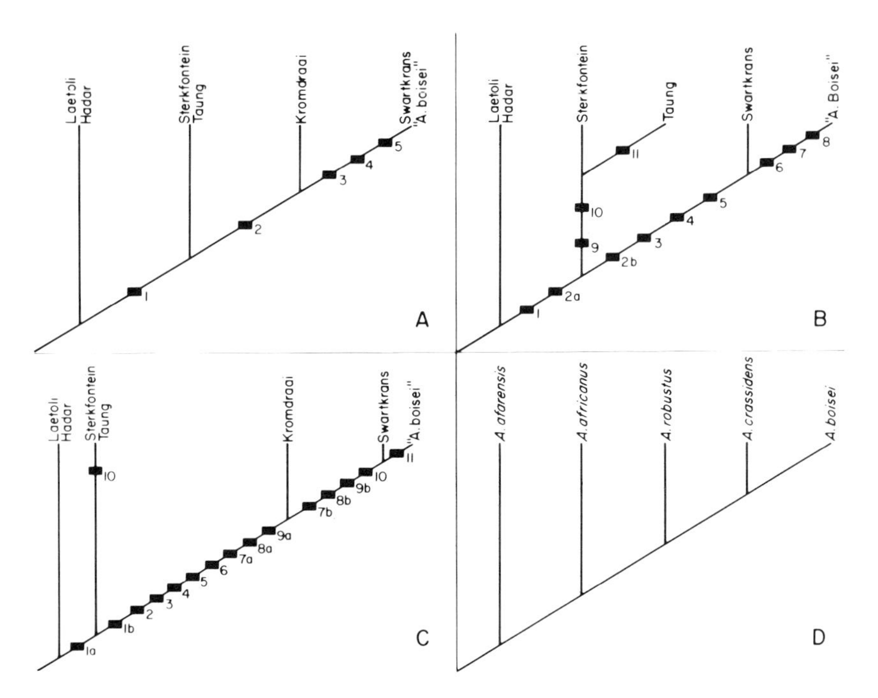

Fig. 1. A–C, distribution of character states for mandibular deciduous canines (A), maxillary deciduous molars (B), and mandibular deciduous molars (C) among australopithecine samples. Note that "A. boisei" in quotes reflects the use of a taxonomic term to denote a phenon assemblage and does not imply doubt as to its validity or specific distinction. The individual characters are discussed in the text. D, hypothesized phylogenetic relationships of australopithecine taxa based upon deciduous dental characters.

level mesially than is attained distally. In Swartkrans and *A. boisei* specimens, on the other hand, the cingulum is markedly higher mesially than distally.

Height of mesial crown convexity. The maximum mesial convexity of the dc_1 (character 4, Fig. 1A) is situated near mid-crown height in the Laetoli, Hadar, Sterkfontein, Taung, and Kromdraai specimens, whereas on Swartkrans and *A. boisei* homologues the maximum convexity of the mesial surface closely approximates the level of the apex.

Position of apex. The apex of the dc_1 (character 5, Fig. 1A) is centrally positioned in the Laetoli, Hadar, Sterkfontein, Taung, and Kromdraai specimens, while the tip is displaced mesially on Swartkrans and *A. boisei* homologues.

In addition to these five features, several characters of the mandibular canines that serve to differentiate among samples or groups of samples might be considered in terms of their contribution to the total morphological pattern of each sample. Thus, while the degree of lingual surface relief and the development of the distobuccal groove and distal stylid are variable within samples, the Kromdraai dc_1s are noteworthy for the very weak expression of these features. Also, in comparison to other australopithecine examples, the mesiobuccal groove of the dc_1 is very shallow in Kromdraai specimens, and in the Swartkrans specimens it is so feeble as to be almost imperceptible. Finally, the specimen from Sterkfontein appears to be unique in the possession of a distinct mesial stylid. However, whilst this stylid is evident on the Ldc_1, its antimere, which is more heavily worn, evinces no indication of this feature. Although a distinct mesial stylid would appear to be lacking on other australopithecine dc_1s, a number of these are worn such that its presence might have been obliterated by attrition.

Maxillary Deciduous Molars

Although upper milk molars are unknown from Kromdraai, specimens of both the dm^1 and dm^2 are represented in all other australopithecine samples. A number of traits exhibited by these crowns appear to be sufficiently discrete to warrant an analysis of their distributions among known samples. These characters are as follows:

Distal marginal ridge of dm^1. In Sterkfontein, Taung, Swartkrans, and *A. boisei* specimens the distal marginal ridge of the dm^1 (character 1, Fig. 1B) is of moderate to substantial thickness. This crest is notably thinner on Laetoli and Hadar homologues.

Lingual bevel of protocone. This feature (character 2, Fig. 1B) evinces three different states amongst the australopithecine samples, and these are interpreted here as representing a morphocline, which is presumed to reflect the probable pathway of character state transformation. In the Laetoli and Hadar specimens the lingual surface of the protocone on both the dm^1 and dm^2 is strongly bevelled. On the Sterkfontein and Taung molars these surfaces are moderately bevelled, showing a degree of inflation not present on Laetoli and Hadar crowns (character 2a, Fig. 1B). By comparison, in the Swartkrans and *A. boisei* specimens these surfaces are so inflated as to be nearly horizontal (character 2b, Fig. 1B). Inasmuch as these differences relate to the resultant angle of inclination displayed by the Phase I facets on the protocones of the upper molars, they would appear to be functionally significant (Grine, 1981b).

Sizes of paracone and metacone on dm^1. On the Laetoli, Hadar, Sterkfontein, and Taung dm^1s the paracone is notably larger than the metacone (character 3, Fig. 1B). In Swartkrans and *A. boisei* specimens, on the other hand, the metacone shows marked enlargement relative to the size of the paracone, the disparity between the sizes of these cusps being minimal to altogether absent.

Tuberculum molare of dm^1. In the Laetoli, Hadar, Sterkfontein, and Taung specimens the *tuberculum molare* (character 4, Fig. 1B) is moderate to well-developed, being protuberant and convex both buccally and cervically. On the Swartkrans and *A. boisei* dm^1s, however, this feature is but exiguously developed, if present at all.

Distal marginal ridge of dm^1. The buccal end of the distal marginal ridge is both low and narrow on the Laetoli, Sterkfontein, and Taung dm^2s, (character 5, Fig. 1B), and in one out of two Hadar specimens; in the second Hadar individual this ridge, while being buccally narrow, attains a moderate height. By contrast, in all five Swartkrans specimens and in all three *A. boisei* specimens the buccal end of the distal marginal ridge is notably better developed, being thick and attaining an occlusally high confluence with the metacone.

Accessory lingual cuspule on dm². In the Laetoli, Hadar, Sterkfontein, Taung, and Swartkrans specimens the lingual groove between the protocone and hypocone of the dm^2 follows an uninterrupted course. In both *A. boisei* specimens, on the other hand, this fissure is interrupted by the presence of an accessory cuspule (character 6, Fig. 1B) of variable size. The presence of this feature, which appears to be unique to *A. boisei* specimens, was the source of debate over the interpretation of the OH 3 molar (Leakey, 1958; Robinson, 1960b).

Occlusal outline of dm². The Laetoli, Hadar, Sterkfontein, Taung, and Swartkrans dm^2s have a nearly quadrangular occlusal outline (character 7, Fig. 1B), while *A. boisei* homologues are asymmetric in the mesiobuccal projection of the paracone.

Crown shape index value of dm². The buccolingual (BL) diameter of the dm^1 exceeds the corresponding mesiodistal (MD) diameter for known Laetoli, Hadar, Sterkfontein, Taung, and Swartkrans specimens. The crown shape index values (character 8, Fig. 1B) for these specimens therefore fall above 100%, ranging between 116.7 (L.H. 3) and 101.8 (Sts 2). In both *A. boisei* specimens, on the other hand, the MD diameter is greater than the corresponding BL diameter, resulting in crown shape index values that fall below 100% (92.1 and 98.0).

Buccal depression on dm¹. The Sterkfontein and Taung specimens appear to be unique in that the buccal groove of the dm^1 (character 9, Fig. 1B) is set in a distinct V-shaped depression. Laetoli, Hadar, Swartkrans, and *A. boisei* homologues lack this feature.

Mesiobuccal groove on dm¹. In the Laetoli, Hadar, Swartkrans, and *A. boisei* specimens the mesial fissure on the buccal face of the dm^1 (character 10, Fig. 1B) is only weakly defined, whereas the Sterkfontein and Taung specimens possess a very distinct mesiobuccal groove.

Size of hypocone on dm². The hypocone (character 11, Fig. 1B) is moderately large to well-developed on Laetoli, Hadar, Sterkfontein, Swartkrans, and *A. boisei* dm^2s, and the relative development of this cusp does not differ appreciably amongst these samples. The Taung specimen, however, appears to be unique amongst australopithecines in that the hypocone is noticeably reduced on the dm^2 (and M^1), imparting a degree of curvature to the distolingual corner of the crown not expressed on other australopithecine homologues.

In addition to the characters detailed above, there are several other features of the upper milk molars that are somewhat variable within samples, but which tend to differentiate among samples or groups of samples. Thus, the distal trigon crest of the dm^1 tends to be obliquely oriented at about 45° to the MD crown axis in the Laetoli and Hadar specimens, while on other australopithecine homologues this crest follows a more transverse course. The epicrista of the dm^2 tends to be incised or markedly constricted in the Laetoli and Hadar specimens, whereas it is complete on other australopithecine molars. The principal buccal groove of the dm^2 lies in a broad, U-shaped depression in both Laetoli and one out of two Hadar specimens, whereas this feature is lacking from all other australopithecine homologues, with the possible exception of one Sterkfontein individual, where an oblique enamel ridge distal to the buccal groove may represent an homologous development.

The mesial fissure of the trigon basin of the dm^2 tends to be better developed than the distal limb in the Laetoli, Hadar, Sterkfontein, and Taung specimens, while in Swartkrans and *A. boisei* specimens the distal limb tends to be equivalent to or stronger than the mesial fissure. In the Laetoli, Hadar, Sterkfontein, and Taung specimens there is a distinct tendency for the buccal limb of the fovea posterior (= talon basin) of the dm^2 to be notably dominant over the lingual limb, whereas the lingual fissure tends to be equal to or stronger than the buccal limb in Swartkrans and *A. boisei* specimens.

While the Laetoli, Hadar, Sterkfontein, and Taung dm^2s lack an accessory distobuccal fissure, this groove is evinced by all three *A. boisei* specimens and by one out of four Swartkrans specimens.

Finally, in the Laetoli, Sterkfontein, and Taung specimens the buccal limb of the trigon basin cuts a deep, V-shaped notch at the occlusobuccal margin of the dm^2 between the paracone and metacone. On the other hand, in all five Swartkrans specimens and on two out of three *A. boisei* dm^2s bulbous crests from the paracone and metacone join to form a high occlusobuccal wall. In the third *A. boisei* specimen, however, the occlusobuccal margin is deeply albeit narrowly incised. Both Hadar specimens possess an enamel wall at the occlusobuccal margin, though in neither individual does this crest assume the proportions evinced by the Swartkrans and *A. boisei* crowns.

Mandibular Deciduous Molars

Examples of both lower milk molars are known for all australopithecine samples. The following features of these teeth appear to be sufficiently discrete to warrant an analysis of their distributions:

Position of the protoconid on dm$_2$. Three different states of this feature (character 1, Fig. 1C) are evinced amongst the early hominid samples, and these differing states are interpreted here along a morphocline, which presumably reflects the pathway of transformation. In the Laetoli and Hadar specimens the protoconid is set strongly mesial to the level of the metaconid. On the Sterkfontein-Makapansgat and Taung dm$_2$s the protoconid is also set mesial to the level of the metaconid, but the difference between these apical levels is not marked as on the Laetoli and Hadar homologues (character 1a, Fig. 1C). In comparison to the foregoing, the Kromdraai, Swartkrans, and *A. boisei* specimens exhibit what would appear to be a further derived condition (character 1b, Fig. 1C), with the protoconid and metaconid apices approximating the same buccolingual plane.

Position of the protoconid on dm$_1$. On the Laetoli, Hadar, Sterkfontein, and Taung dm$_1$s the protoconid (character 2, Fig. 1C) is set mesial to the level of the metaconid, whereas these cusps are aligned more transversely on the Kromdraai, Swartkrans, and *A. boisei* homologues.

Proportional MD length of dm$_1$ trigonid. The trigonid of the dm$_1$ (character 3, Fig. 1C) is mesiodistally dominant over the talonid in the Laetoli, Hadar, Sterkfontein, and Taung specimens, whereas in the Kromdraai, Swartkrans, and *A. boisei* specimens the talonid is relatively longer, being usually dominant over, or at least equivalent to, the length of the trigonid.

Depth of buccal groove on dm$_1$. The principal buccal groove (i.e., that between the protoconid and hypoconid) of the dm$_1$ (character 4, Fig. 1C) is shallow to moderate in the Laetoli, Hadar, Sterkfontein, and Taung specimens, while it is represented by a notably deeper, distinct fissure on the Kromdraai, Swartkrans, and *A. boisei* homologues.

Mesial marginal ridge of dm$_1$. The lingual end of the mesial marginal ridge (character 5, Fig. 1C) is low on Laetoli, Hadar, Sterkfontein, and Taung dm$_1$s. In the Kromdraai, Swartkrans, and *A. boisei* specimens, on the other hand, the lingual end of the mesial marginal ridge is high. Moreover, this crest is separated from the metaconid, usually by a distinct fissure, in the Laetoli, Hadar, Sterkfontein, and Taung specimens, whereas it serves to completely enclose the fovea anterior (= trigonid basin) on Kromdraai, Swartkrans, and *A. boisei* homologues.

Fovea anterior of dm$_1$. The fovea anterior, or trigonid basin, of the dm$_1$ (character 6, Fig. 1C) is skewed to the lingual side of the median MD crown axis in the Laetoli, Hadar, Sterkfontein, and Taung specimens, whereas it is positioned symmetrically on Kromdraai, Swartkrans, and *A. boisei* crowns. Similarly, while this fovea is represented by a triradiate or broad triangular basin in the Laetoli, Hadar, and Taung dm$_1$s, and in two out of three Sterkfontein specimens, in all known Kromdraai, Swartkrans, and *A. boisei* specimens this fossid is represented by a single fissure of predominantly buccolingual orientation.

Tuberculum molare of dm$_1$. This feature shows (character 7, Fig. 1C) shows varying degrees of development among the australopithecine samples, and these differing states are interpreted here in terms of a transformation morphocline that presumably reflects the pathway of character change. The *tuberculum molare* is moderate to well-developed, being protuberantly convex both buccally and cervically on the Laetoli, Hadar, Sterkfontein, and Taung dm$_1$s. By comparison, the *tuberculum molare* shows weaker development in the Kromdraai specimens; on these crowns it is less protuberant, and only slightly convex cervically and buccally (character 7a, Fig. 1C). In three out of six Swartkrans specimens and in one out of two *A. boisei* specimens the *tuberculum molare* is represented by a slight downward extension of the protoconidal cervical line immediately mesial to the hypoconidal margin, while the other Swartkrans and *A. boisei* dm$_1$s lack any development of this trait. Thus, this feature, which displays reduction in the Kromdraai specimens, is further reduced or altogether absent from Swartkrans and *A. boisei* specimens (character 7b, Fig. 1C).

Cuspal height disparity on dm_1. This character (character 8, Fig. 1C) also displays varying degrees of expression amongst the early hominid samples, and these different configurations are interpreted in terms of a transformation morphocline. The protoconid and metaconid are substantially higher than the hypoconid and entoconid on the faintly worn Laetoli and Hadar dm_1s, and a notable disparity between the heights of these cusps is evident also on the moderately worn Laetoli, Sterkfontein, and Taung crowns. On the unworn and faintly worn Kromdraai homologues the disparity in cuspal heights is similar to that shown by the moderately worn Sterkfontein and Taung molars. Thus, compared to the Laetoli, Hadar, Sterkfontein, and Taung specimens, those from Kromdraai display a reduction in the disparity between the heights of the trigonid and talonid cusps; on the latter molars the talonid cusps appear to have increased in their relative heights (character 8a, Fig. 1C). By contrast, in the Swartkrans and *A. boisei* specimens the trigonid and talonid cusps closely approximate the same elevation, and this lack of disparity is manifest on very slightly to moderately worn crowns comprising both samples. Thus, the relative heights of the talonid cusps on the dm_1 show an increase in the Swartkrans and *A. boisei* specimens (character 8b, Fig. 1C) over the condition possessed by the Kromdraai specimens.

Cuspal size on dm_1. This feature (character 9, Fig. 1C), which also displays different degrees of expression among the australopithecine samples, is considered here in terms of a morphocline. While the composition of this feature is probably related to two others discussed above (position of the protoconid on the dm_1 and cuspal height disparity on the dm_1), because its distribution does not correspond precisely to those of the other characters, it is treated independently. Thus, in the Laetoli, Hadar, Sterkfontein, and Taung specimens the protoconid is the largest cusp on the dm_1. In the Kromdraai specimens, on the other the metaconid is equivalent in size to the protoconid (character 9a, Fig. 1C). The Swartkrans and *A. boisei* specimens appear to display a further derived condition in the cuspal size relationships of the dm_1 (character 9b, Fig. 1C), where the metaconid is the largest cusp and the hypoconid also is larger than the protoconid.

Mesioconulid on dm_1. A distinct mesioconulid (character 10, Fig. 1C) (i.e., a separate cuspulid situated mesial to the protoconid) is present on the Sterkfontein and Taung dm_1s as well as on the Swartkrans and *A. boisei* molars. The Laetoli, Hadar, and Kromdraai specimens, on the other hand, lack this feature. While the size of the mesioconulid varies somewhat within the Sterkfontein and Swartkrans samples, the presence or absence of this feature does not vary within any one sample.

Length of buccal groove on dm_1. In the Laetoli, Hadar, Sterkfontein, Taung, Kromdraai, and Swartkrans specimens the principal buccal groove of the dm_1, (character 11, Fig. 1C) despite its difference in depth among these samples, terminates about half way down the buccal face; i.e., well above the cervical margin. *A. boisei* specimens, on the other hand, appear to be unique in that the buccal fissure continues downward to incise the cervical enamel margin between the protoconid and hypoconid.

In addition to the features discussed above, there are a number of other lower molar traits that, while being somewhat variable, tend to differentiate between samples or groups of samples. Thus, the fovea anterior of the dm_2 is subdivided in the Laetoli, Hadar, Sterkfontein-Makapansgat, Taung, and Swartkrans specimens, and in two out of three Kromdraai specimens, while it is represented by a single transverse fissure in the *A. boisei* specimens. Similarly, on the Laetoli, Hadar, Sterkfontein-Makapansgat, Taung, and Swartkrans dm_2s, and in two out three Kromdraai specimens, the primary occlusal fissures form a symmetrical Y, whereas on *A. boisei* homologues the mesiobuccal and lingual occlusal fissures are arranged in a nearly straight transverse line. (The Kromdraai specimen that shows this transverse alignment of the distal borders of the metaconid and protoconid also possesses a *tuberculum intermedium,* a feature lacking from the *A. boisei* homologues.) In a word, the structure of the fovea anterior and the arrangements of the primary occlusal fissures on the dm_2 tend to distinguish *A. boisei* specimens from those of other samples.

Among those samples in which the fovea anterior of the dm_2 is subdivided, the distal trigonid crest is better developed than the accessory trigonid crest in the Laetoli, Hadar, Sterkfontein-Makapansgat, and Taung specimens. In the Kromdraai and Swartkrans specimens, on the other hand, the accessory trigonid crest tends to be stronger. Moreover, in these latter specimens the distal fossid of the fovea anterior tends to course transversely, while in the others it tends to follow an oblique (mesiolingual–distobuccal) course.

While the size of hypoconulid of the dm_1 is variable within samples, it tends to be better developed in the Kromdraai, Swartkrans, and *A. boisei* specimens than in those from Laetoli, Hadar, Sterkfontein, and Taung. At the same time, while the presence and expression of the *tuberculum sextum* on the dm_2 is variable within samples, there would seem to be a tendency for it to be better developed in the Taung, Swartkrans, and *A. boisei* specimens than in the specimens comprising the other australopithecine samples.

IMPLICATIONS FOR AUSTRALOPITHECINE TAXONOMY AND PHYLOGENY

The present study has disclosed a number of deciduous dental traits that differ between individual samples and groups of samples. The specimens comprising the *A. boisei*, Swartkrans, and Kromdraai samples share a substantial suite of synapomorphies that clearly distinguish them from the Laetoli, Hadar, Taung, and Sterkfontein-Makapansgat specimens. A variety of features also serve to differentiate among the Kromdraai, Swartkrans, and *A. boisei* samples.

On the other hand, while the teeth comprising the Laetoli and Hadar samples differ in a few traits that are known to be variable within other samples, no discrete morphological character serves to differentiate the Laetoli from the Hadar specimens. Similarly, although the Taung specimen differs from Strekfontein specimens in the size of the hypocone on its dm^2 and in several other (variable) features, the Taung and Sterkfontein individuals also share several traits not found in other australopithecine representatives. Suggestions that the Laetoli and Hadar fossils be considered subspecifically distinct, and the Taung specimen is taxonomically distinct from the australopithecines of Sterkfontein and Makapansgat are not supported by morphological evidence from the deciduous dentition.

The specimens attributed to the *A. boisei* and Swartkrans samples share a number of synapomorphies that serve to distinguish them from Kromdraai individuals, and the eastern African fossils differ from those from Swartkrans in several apparent autapomorphies (Fig. 1B,C). Together with these apparently autapomorphic character states, there are several other features in which the specimens attributed to *A. boisei* tend to differ from the Swartkrans individuals. The combination of these various traits imparts a unique overall morphological pattern to the deciduous dentitions of these eastern African specimens. Thus, the deciduous dental evidence appears to be in keeping with the view that the distinction at the species level be retained for *A. boisei*: a distinction clearly defined by Tobias (1967), and more recently argued for by Howell (1978) and Rak (1983).

While the Kromdraai and Swartkrans (and *A. boisei*) specimens share a suite of features that clearly differentiate them from the Sterkfontein-Makapansgat and Taung homologues, there are a number of deciduous dental characters in which the Kromdraai and Swartkrans specimens differ (Fig. 1A,C). In several instances these differences are related to synapomorphic states shared by the Swartkrans and *A. boisei* specimens, in which the states evinced by the Kromdraai specimens are the same as those possessed by the Sterkfontein-Makapansgat and Taung specimens. In several non-metrical characters the states revealed by the Kromdraai specimens appear to be intermediate along a morphocline between those shown by the Sterkfontein-Makapansgat and Taung specimens, on the one hand, and those possessed by the Swartkrans specimens, on the other. Moreover, in a variety of metrical features of the lower molars, the Kromdraai values tend to fall between those for the Sterkfontein, Makapansgat, and Taung samples, and those for the Swartkrans sample (Grine, 1982, 1984). In addition, the distal stylid and distobuccal groove of the dc_1, are better developed, and the dc_1 evinces greater lingual surface relief in Swartkrans specimens than in Kromdraai specimens, and functionally relevant differences in cuspal facet angulation between the Kromdraai and Swartkrans lower deciduous molars have been noted (Grine, 1981b).

These deciduous dental differences, together with several features of some permanent teeth (Grine, 1982), support Howell's (1978) proposal that distinction at the species level be retained for *A. robustus* and *A. crassidens*. To subsume these specimens in a single taxon, or to relegate the differences between them to a subspecific distinction, serves only to mask the evolutionary significance of the Kromdraai australopithecines.

As noted above, there is little morphological evidence to support a taxonomic distinction among the Taung, Sterkfontein, and Makapansgat specimens, nor are there any significant differences between the deciduous specimens from Laetoli and Hadar. On the other hand, the Sterkfontein, Makapansgat, and Taung (*A. africanus*) specimens differ from the Laetoli and Hadar fossils in several discrete characters. In some cases the states possessed by *A. africanus* specimens are shared also by *A. robustus, A. crassidens,* and *A. boisei* specimens, while in some instances the states evinced by *A. africanus* specimens appear to be intermediate along a morphocline between the states possessed by the Laetoli and Hadar specimens and those shown by *A. robustus, A. crassidens,* and *A. boisei* individuals. *A. africanus* specimens also possess several apparently unique features, and they also tend to differ from Laetoli and Hadar specimens in a variety of somewhat more variable dental traits.

The combination of these differing traits serves to impart recognizably distinct overall deciduous dental patterns to the *A. africanus* sample and to the sample comprising both the Laetoli and Hadar specimens. The deciduous dental evidence appears to support the opinion (Johanson et al., 1978; White et al., 1981) that these two groups of specimens be accorded separate specific status.To subsume *A. afarensis* in *A. africanus* masks the evolutionary and functionally significant differences between these samples (White et al., 1981; Rak, 1983).

Recently it has been argued that, because of its occipito-temporal morphology, the A.L.333-45 cranium from Hadar represents a member of the *Paranthropus* clade (Olson 1981, 1985; Falk and Conroy, 1983, with reference to the venous drainage pattern of this specimen), and that the nasal bone outline of the juvenile specimen from the same locality, A.L.333-105, attests to its paranthropine affinities (Olson, 1985). Moreover, Olson (1981) has claimed that the dental specimens associated with the A.L.333-45 cranium—this includes the entire deciduous sample from Hadar—as well as the dental specimens from Laetoli compromise members of *Paranthropus* because they possess "a series of derived characters which ally them with this lineage" (p. 101). Nonetheless, according to Olson (1981) the primitiveness of the Laetoli and Hadar teeth easily differentiates them from the Kromdraai, Swartkrans, and *A.* (= *Paranthropus*) *boisei* homologues.

As observed above, however, the only feature in which the Hadar (but not Laetoli) deciduous specimens come to resemble paranthropine (*A. crassidens* and *A. boisei*) specimens is in the "walled" configuration of the occlusobuccal margin of the dm^2, and this feature is known to be variable within the small *A. boisei* sample. Rather, *A. robustus, A. crassidens,* and *A. boisei* (paranthropine) specimens share a multitude of synapomorphies that clearly distinguish them from Hadar and Laetoli individuals. Indeed, in a cladistic context, the "walled" occlusobuccal margin of the dm^2 possessed by Hadar and paranthropine specimens constitutes homoplasy. It does not, therefore, suggest any special phylogenetic relationship between these specimens. In a word, if paranthropine elements do occur in either the Hadar Formation or Laetoli Beds, they are not represented by specimens possessing deciduous teeth. (Proposals that certain characters of the Hadar cranial remains attest to their paranthropine affinity (Olson, 1981; Falk and Conroy, 1983) have been shown to be specious (Kimbel, 1984; Kimbel et al., 1985).

The hypotheses proposed here concerning synapomorphy and the phylogenetic relationships of australopithecine species are summarized in the cladogram depicted in Figure 1D. A cladogram, while depicting hypotheses of phylogenetic relationship, is not an evolutionary tree. Evolutionary trees infer specific sequences of ancestry and descent that are not reflected in a cladogram. Once a cladogram has been constructed, however, phylogenetic trees may then be proposed and evaluated. Attempts to reverse this process, that is, to draw cladograms to corroborate statements of ancestry and descent, have resulted in some rather profound inconsistencies (cf. Figure 10A in Johanson and White (1979) and Figure 4 in White et al. (1981)).

Delson (1977) has aptly observed that, in terms of morphology, the proposition of ancestry is rendered more plausible if the presumptive ancestor is intermediate along a morphocline between the putative descendant and an earlier ancestral condition. With regard to the australopithecines, since *A. afarensis* specimens appear to lack any significant autapomorphies in the deciduous dentition, it seems reasonable to suggest that *A. africanus* evolved from *A. afarensis,* or an *A. afarensis*-like ancestor (White et al., 1981). Similarly, the deciduous dental evidence is in keeping with the conclusions reached by numerous workers, whose morphological studies have led them to postulate *A. africanus* or an *A. africanus*-like form as having been ancestral to the lineage comprising *A. boisei* and the Kromdraai and Swartkrans forms. It must be stressed, however, that this particular proposition has no bearing on questions pertaining to the relationship between *A. africanus* and the genus *Homo (sensu stricto)*.

At the same time, while *A. robustus* specimens share a number of synapomorphies with *A. crassidens* and *A. boisei*, the Kromdraai specimens are morphologically intermediate in a variety of dental traits between *A. africanus*, on the one hand, and *A. crassidens*, on the other. This evidence strongly suggests *A. robustus* to have been ancestral to *A. crassidens*; it does not appear to be in keeping with the evolutionary sequence proposed by Robinson (1956) and von Koenigswald (1967) among others, for the Swartkrans and Kromdraai australopithecines.

Finally, while the Swartkrans and *A. boisei* specimens share a number of derived features not shared by *A. robustus*, *A. boisei* teeth show several apparent autapomorphies not possessed by *A. crassidens*. On the other hand, Swartkrans specimens do not display any apomorphic characters that would preclude *A. crassidens* from having been ancestral to *A. boisei*. Together with features of craniofacial morphology (Rak, 1983), the deciduous dental evidence suggests that *A. boisei* evolved from *A. crassidens* or an *A. crassidens*-like ancestor. This evidence certainly appears to contradict the obverse—i.e., the common, albeit facile notion that because the earlier *A. boisei* specimens are (apparently) geochronologically older than either the Kromdraai or Swartkrans fossils, *A. robustus* (= Kromdraai + Swartkrans) evolved from *A. boisei* (Boaz, 1979).

In closing, it is well to reiterate that the deciduous dentition forms only part of the totality of dental and skeletal morphology that must be assessed before any secure taxonomic distinctions can be drawn amongst the Plio-Pleistocene fossils. However, assessment of taxonomic (e.g., species) distinctiveness and determination of systematic relationships are not the same thing. Thus, whereas Le Gros Clark's (1964) concept of total morphological pattern is relevant in assessing species distinctiveness, only analysis of synapomorphy of the individual characters comprising this pattern is germane to determining the phylogenetic relationships of a fossil species. The phylogenetic hypotheses proposed here require critical evaluation through cladistic analyses of other relevant dental and skeletal morphologies.

SUMMARY AND CONCLUSIONS

This analysis of australopithecine deciduous dentitions has resulted in the formulation of proposals concerning the number of species currently represented in the Plio-Pleistocene fossil record, and in the formulation of hypotheses pertaining to the phylogenetic relationships of these taxa.

Nearly 60 australopithecine specimens in possession of one or more deciduous teeth were attributed to seven site-specific or phenetic samples. The groups delineated comprised the Laetoli, Hadar, Sterkfontein + Makapansgat, Taung, Kromdraai, and Swartkrans samples, in addition to specimens attributed to an *A. boisei* sample. Morphological attributes of the milk teeth in these samples suggest that five separate taxa (species) are represented.

The deciduous dental evidence supports the proposal by Johanson et al. (1978) that the Hadar and Laetoli fossils are attributable to a single species (*A. afarensis*), which is distinct from South African *A. africanus*. The odontological evidence also indicates a taxonomic distinction among the specimens from Sterkfontein, Makapansgat, and Taung to be unwarranted. The Sterkfontein and Makapansgat fossils are referable to the hypodigm of *A. africanus.*

Howell's (1978) proposal that distinction at the species level be retained for *A. robustus* and *A. crassidens* is most strongly supported by the multitudinous features in which the Kromdraai and Swartkrans australopithecine dentitions differ (Grine, 1981b, 1982, 1984). In addition, several apparent autapomorphies possessed by known *A. boisei* specimens tend to corroborate recent arguments that it represents a valid species (Howell, 1978; Rak, 1983), as clearly defined by Tobias (1967).

Analysis of synapomorphy in deciduous dental characters has led to the proposal of hypotheses on the phylogenetic (i.e., systematic) relationships of these australopithecine taxa. Based upon these cladistic relationships, it is argued that *A. africanus* or an *A. africanus*-like form was likely ancestral to the so-called "robust" australopithecine lineage (i.e., the clade comprising *A. robustus, A. crassidens,* and *A. boisei*). This proposition, however, has no bearing on questions pertaining to the relationship between *A. africanus* and the genus *Homo* (*sensu stricto*).

A number of the dental traits evinced by *A. robustus* (Kromdraai) specimens are intermediate between the character states possessed by *A. africanus* and *A. crassidens* specimens, while *A. crassidens* and *A. boisei* teeth display a variety of synapomorphies not shared by *A. robustus* homologues. This evidence suggests *A. robustus* to have been ancestral to *A. crassidens*. Similarly, it is argued that *A. boisei* evolved from *A. crassidens* or an *A. crassidens*-like ancestor.

The deciduous dentition forms only part of the total package of dental and skeletal morphology that must be assessed before truly secure taxonomic distinctions can be drawn amongst assemblages of Plio-Pleistocene fossils. The phylogenetic hypotheses posited here require evaluation through analyses of synapomorphy for the other dental and skeletal characters comprising the available paleoanatomical evidence.

ACKNOWLEDGMENTS

I would like to express my appreciation to Drs. Eric Delson and Ian Tattersall for inviting me to participate in the symposium "Paleoanthropology: the Hard Evidence," and to the American Museum of Natural History for its generous hospitality.

I thank Professor P.V. Tobias for his extraordinarily thorough and invaluable comments, suggestions, and criticisms of drafts of my doctoral thesis, a partial distillation of which comprises this paper. I also thank Drs. C.K. Brain, F.C. Howell, H.E.H. Paterson, E.S. Vrba, T.D. White, and B.A. Wood for discussions on various aspects of this work. Responsibility for the conclusions expressed here, however, remains my own.

This research was supported through generous grants from the L.S.B. Leakey Foundation, the Ernest Oppenheimer Memorial Trust, the Institute for the Study of Man in Africa, The C.S.I.R. of South Africa, and the Senate Research Committee, the S.L. Sive Memorial Travelling Fellowship, the Medical Referee's Fund, and the Bernard Price Institute for Palaeontological Research, University of the Witwatersrand.

LITERATURE CITED

Aguirre, E (1970) Identification de *"Paranthropus"* en Makapansgat. Cronica del XI Cong. Nacional de Arqueologia, Merida, 1969, pp. 98–124.

Ashton, EH, Healey, MJR, and Lipton, S (1957) The descriptive use of discriminant functions in physical anthropology. Proc. Roy. Soc. Lond. (B) *146:* 555–572.

Bilsborough, A (1972) Anagenesis in hominid evolution. Man *7:* 481–483.

Bilsborough, A (1978) Some aspects of mosaic evolution in hominids. In DJ Chivers and KA Joysey (eds): Recent Advances in Primatology, Vol. 3, Evolution. London: Academic, pp. 335–350.

Boaz, NT (1979) Hominid evolution in eastern Africa during the Pliocene and early Pleistocene. Ann. Rev. Anthropol. *8:* 71–85.

Boaz, NT (1983) Morphological trends and phylogenetic relationships from Middle Miocene hominoids to Late Pliocene hominids. In RL Ciochon and RS Corruccini (eds): New Interpretations of Ape and Human Ancestry. New York: Plenum, pp. 705–720.

Brace, CL (1973) Sexual dimorphism in human evolution. Yrbk. Phys. Anthropol. *16:*50–68.

Broom, R (1929) Note on the milk dentition of *Australopithecus.* Proc. Zool. Soc. Lond. *1928:*85–88.

Broom, R (1936) A new fossil anthropoid skull from South Africa. Nature *138:*486–488.

Broom, R (1938) The Pleistocene anthropoid apes of South Africa. Nature *142:*377–379.

Broom, R (1946) The occurrence and general structure of the South African ape-men. In R Broom and GWH Schepers: The South African Fossil Ape-Men, the Australopithecinae. Mem. Transvaal. Mus. *2:*1–272.

Broom, R (1949) Another type of fossil ape-man (*Paranthropus crassidens*). Nature *163:*57.

Broom, R (1950) The genera and species of the South African fossil ape-men. Am. J. Phys. Anthropol. *8:*1–13.

Broom, R, and Robinson, JT (1950) Further evidence of the structure of the Sterkfontein ape-man, *Plesianthropus.* In R Broom, JT Robinson and GWH Schepers: Sterkfontein Ape-Man, *Plesianthropus.* Mem. Transvaal. Mus. *4:*1–117.

Broom, R, and Robinson, JT (1952) Swartkrans ape-man *Paranthropus crassidens.* Mem. Transvall. Mus. *6:*1–123.

Brown, FH, McDougall, I, Davies, T, and Maier, R (1985) An integrated chronology for the Turkana Basin. In E. Delson (ed): Ancestors: The Hard Evidence. New York: Alan R. Liss, Inc. pp. 82–90.

Campbell, BG (1974) A new taxonomy of fossil man. Yrbk. Phys. Anthropol. *17:*195–201.

Clarke, RJ (1977) The cranium of Swartkrans hominid SK 847, and its relevance to human origins. Unpublished Ph.D. Thesis, University of the Witwatersrand, Johannesburg.

Dart, RA (1925) *Australopithecus africanus:* The Man-Ape of South Africa. Nature *115:*195–199.

Dart, RA (1948) The Makapansgat proto-human, *Australopithecus prometheus.* Am. J. Phys. Anthropol. *6:*259–283.

de Bonis, L, Johanson, DC, Melentis, J, and White, TD (1981) Variations métriques de la denture chez les Hominidés primitifs: Comparison entre *Australopithecus afarensis* et *Ouranopithecus macedoniensis.*

C.R. Acad. Sci. Paris, D, *292:*373–376.

Delson, E (1977) Catarrhine phylogeny and classification: Principles, methods and comments. J. Hum. Evol. *6:*433–459.

Falk, D, and Conroy, GC (1983) The cranial venous sinus system in *Australopithecus afarensis.* Nature *306:*779–781.

Farris, JS (1966) Estimation of conservatism of characters by consistency within biological populations. Evolution *20:*587–591.

Farris, JS (1976) Phylogenetic classification of fossils with recent species. Syst. Zool. *25:*271–282.

Ferguson, WW (1983) An alternative interpretation of *Australopithecus afarensis* fossil material. Primates *24:*397–409.

Grine, FE (1981a) Description of some juvenile hominid specimens from Swartkrans, Transvaal. Ann. S. Afr. Mus. *86:*43–71.

Grine, FE (1981b) Trophic differences between "gracile" and "robust" australopithecines: A scanning electron microscope analysis of occlusal events. S. Afr. J. Sci. *77:*203–230.

Grine, FE (1982) A new juvenile hominid (Mammalia: Primates) from Member 3, Kromdraai Formation, Transvaal, South Africa. Ann. Transvaal. Mus. *33:*165–239.

Grine, FE (1984) The deciduous dentition of the Kalahari San, the South African Negro and the South African Plio-Pleistocene hominids. Unpublished Ph.D. Thesis, University of the Witwatersrand, Johannesburg.

Harper, CW (1976) Phylogenetic inference in paleontology. J. Paleont. *50:*180–193.

Harris, JM (1985) Age and paleoecology of the Upper Laetolil beds, Laetoli, Tanzania. In E. Delson (ed): Ancestors: The Hard Evidence. New York: Alan R. Liss, Inc., pp. 76–81.

Hennig, W (1966) Phylogenetic Systematics. Urbana: University of Illinois Press.

Howell, FC (1978) Hominidae. In VJ Maglio and HBS Cooke (eds): Evolution of African Mammals. Cambridge, Massachusetts: Harvard University Press, pp. 154–248.

Howell, FC, and Coppens, Y (1976) An overview of the Hominidae from the Omo Succession, Ethiopia. In Y Coppens, FC Howell, GL Isaac, and REF Leakey (eds): Earliest Man and Environments in the Lake Rudolf Basin. Chicago: University of Chicago Press, pp. 522–532.

Johanson, DC, and Taieb, M (1976) Plio-Pleistocene hominid discoveries in Hadar, Ethiopia. Nature *260:*293–297.

Johanson, DC, Taieb, M, and Coppens, Y (1982) Pliocene hominids from the Hadar Formation, Ethiopia (1973–1977): Stratigraphic, chronologic and palaeoenvironmental contexts, with notes on hominid morphology and systematics. Am. J. Phys. Anthropol. *57:*373–402.

Johanson, DC, White, TD, and Coppens, Y (1978) A new species of the genus *Australopithecus* (Primates: Hominidae) from the Pliocene of eastern Africa. Kirtlandia *28:*1–14.

Kimbel, WH (1984) Variation in the pattern of cranial venous sinuses and hominid phylogeny. Am. J. Phys. Anthropol. *63:*243–263.

Kimbel, WH, White, TD, and Johanson, DC (1985) Craniodental Morphology of the Hominids From Hadar and Laetoli: Evidence of "*Paranthropus*" and *Homo* in the Mid-Pliocene of Eastern Africa? In E. Delson (ed): Ancestors: The Hard Evidence. New York: Alan R. Liss, Inc., pp. 120–137.

Leakey, LSB (1958) Recent discoveries at Olduvai Gorge, Tanganyika. Nature *181:*1099–1103.

Leakey, LSB (1959) A new fossil skull from Olduvai. Nature *184:*491–493.

Leakey, MD, Hay, RL, Curtis, GH, Drake, RE, Jackes, MK, and White, TD (1976) Fossil hominids from the Laetoli Beds, Tanzania. Nature *262:*460–466.

Leakey, REF (1976) Hominids in Africa. Am. Sci. *64:*174–178.

Le Gros, Clark WE (1952) Hominid characters of the australopithecine dentition. J. Roy. Anthrop. Inst. *80:*37–54.

Le Gros, Clark WE (1964) The Fossil Evidence for Human Evolution. An Introduction to the Study of Paleoanthropology, 2nd ed. Chicago: University of Chicago Press.

Mayr, E (1950) Taxonomic categories in fossil hominids. Cold Spring Harbor Symp. Quant. Biol. *15:*108–118.

Mayr, E. (1969) Principles of Systematic Zoology. New York: McGraw-Hill.

Olson, TR (1978) Hominid phylogenetics and the existence of *Homo* in Member I of the Swartkrans Formation, South Africa. J. Hum. Evol. *7:*159–178.

Olson, TR (1981) Basicranial morphology of the extant hominoids and Pliocene hominids: The new material from the Hadar Formation, Ethiopia, and its significance in early human evolution and taxonomy. In CB Stringer (ed): Aspects of Human Evolution. London: Taylor and Francis, pp. 99–128.

Olson, TR (1985) Cranial morphology of the Hadar Formation hominids and "*Australopithecus*" *africanus.* In E. Delson (ed): Ancestors: The Hard Evidence. New York: Alan R. Liss, Inc., pp 102–119.

Patterson, C, and Rosen, DE (1977) Review of icthyodectiform and other Mesozoic teleost fishes and the theory and practice of classifying fossils. Bull. Am. Mus. Nat. Hist. *158:*83–172.

Pilbeam, DR, and Simons, EL (1965) Some problems of hominid classification. Am. Sci. *53:*237–259.

Rak, Y (1983) The Australopithecine Face. New York: Academic Press.

Robinson, JT (1954) The genera and species of the Australopithecinae. Am. J. Phys. Anthropol. *12:*181–200.

Robinson, JT (1956) The dentition of the Australopithecinae. Mem. Transvaal. Mus. *9:*1–179.

Robinson, JT (1960a) The affinities of the new Olduvai australopithecine. Nature *186:*456–458.

Robinson, JT (1960b) An alternative interpretation of the supposed giant deciduous hominid tooth from Olduvai. Nature *185:*407–408.

Robinson, JT (1968) The origin and adaptive radiation of the australopithecines. In G Kurth (ed): Evolution und Hominization, 2nd ed. Stuttgart: Gustav Fischer, pp. 150–175.

Schaeffer, B, Hecht, MK, and Eldredge, N (1972) Phylogeny and paleontology. In T Dobzhansky, MK Hecht, and WC Steere (eds): Evolutionary Biology, Vol. 6, pp. 31–46.

Senut, B, and Tardieu, C (1985) Functional Aspects of Plio-Pleistocene Hominid Limb Bones: Implications for Taxonomy and Phylogeny. In E. Delson (ed): Ancestors: The Hard Evidence. New York: Alan R. Liss, Inc., pp. 193–201.

Sperber, GH (1973) Morphology of the cheek teeth of early South African hominids. Unpublished Ph.D. Thesis, University of the Witwatersrand,

Johannesburg.

Taieb, M, Johanson, DC, Coppens, Y, Bonnefille, R, and Kalb, J (1974) Découverte d'hominidés dans les series Plio-Pléistocènes d'Hadar (Bassin de l' Awash; Afar, Éthiopie). C.R. Acad. Sci . Paris, D, *279:*735–738.

Taieb, M, Coppens, Y, Johanson, DC, and Bonnefille, R (1975) Hominidés de l'Afar Central, Éthiopie (site d'Hadar, campagne 1973). Bull. Mem. Soc. Anthrop. Paris *13:*117–124.

Taieb, M, Johanson, DC, Coppens, Y, and Aronson, JL (1976) Geological and palaeontological background of Hadar hominid site, Afar, Ethiopia. Nature *260:*289–293.

Tobias, PV (1967) The Cranium and Maxillary Dentition of *Australopithecus (Zinjanthropus) boisei.* Olduvai Gorge, Vol. 2. Cambridge: Cambridge University Press.

Tobias, PV (1973) Implications of the new age estimates of the early South African hominids. Nature *246:* 79–83.

Tobias, PV (1975) New African evidence on the dating and the phylogeny of the Plio-Pleistocene Hominidae. Trans. Roy. Soc. New Zealand, Bull. *13:*289–296.

Tobias, PV (1978a) The place of *Australopithecus africanus* in hominid evolution. In DJ Chivers and KA Joysey (eds): Recent Advances in Primatology, Vol. 3, Evolution. London: Academic, pp. 373–394.

Tobias PV (1978b) Position et rôle des australopithecines dans la phylogenèse humaine, avec étude particulière de *Homo habilis* et des théories controversées avancées à propos des premiers hominidés fossiles de Hadar et de Laetolil. In Les Origines Humaines et les Epoques de l'Intelligence. Paris: Masson, pp. 38–77.

Tobias, PV (1978c) The South African australopithecines in time and hominid phylogeny, with special reference to the dating and affinities of the Taung skill. In CJ Jolly (ed): Early Hominids of Africa. London: Duckworth, pp. 45–84.

Tobias, PV (1980) "*Australopithecus afarensis*" and *A. africanus*: Critique and an alternative hypothesis. Paleont. Afr. *23:*1–17.

Tobias, PV (1985) Olduvai Gorge, Vol. 4. Cambridge: Cambridge University Press. (in press).

von Koenigswald, GHR (1967) Evolutionary trends in the deciduous molars of the Hominidae. J. Dent. Res. *46:*779–786.

Wallace, JA (1978) Evolutionary trends in the early hominid dentition. In CJ Jolly (ed): Early Hominids of Africa. London: Duckworth, pp. 285–310.

White, TD (1977) The anterior corpus of early African Hominidae: Functional significance of shape and size. Unpublished Ph.D. Thesis, University of Michigan, Ann Arbor.

White, TD (1985) The hominids of Hadar and Laetoli: An element-by-element comparison of the dental samples. In E Delson (ed): Ancestors: The Hard Evidence. New York: Alan R. Liss, Inc. pp. 138–152.

White, TD, Johanson, DC, and Kimbel, WH (1981) *Austrapithecus africanus:* Its phyletic position reconsidered. S. Afr. J. Sci. *77:*445–470.

Wiley, EO (1981) Phylogenetics. The Theory and Practice of Phylogenetic Systematics. New York: Wiley.

Wolpoff, MH (1974) The evidence for two australopithecine lineages in South Africa. Yrbk. Phys. Anthropol. *17:*113–139.

Wolpoff, MH (1980) Paleoanthropology. New York: Knopf.

Wood, BA (1978) Classification and phylogeny of East African hominids. In DJ Chivers and KA Joysey (eds): Recent Advances in Primatology, Vol. 3, Evolution. London: Academic, pp. 351–372.

Wood, BA, and Stack, CG (1980) Does allometry explain the differences between "gracile" and "robust" australopithecines? Am. J. Phys. Anthropol. *52:*55–62.

Ancestors: The Hard Evidence, pages 168–170

Systematic and Functional Implications of the Facial Morphology of *Australopithecus* and Early *Homo*

Yoel Rak

Department of Anatomy and Anthropology, Sackler Faculty of Medicine, Tel-Aviv University, Tel-Aviv, Israel

ABSTRACT The morphology of the masticatory system bears heavily on decisions regarding the taxonomic and phylogenetic status of *Australopithecus africanus*. Five morphological features have been selected in order to demonstrate the place that *A. africanus* occupies on the robust line.

The advancement of the masseteric origin, and hence also of the lateral part of the infraorbital region in the transverse plane of both the *A. africanus* and *Australopithecus robustus* faces, results in the concave, or "dished," facial profile, which, along with the sagittally oriented zygomatic arch, forms the sharply protruding zygomatic prominence. In the sagittal plane the infraorbital region of *A. africanus* and *A. robustus* slopes anteroinferiorly, thus modifying the canine fossa. The zygomaticoalveolar crest in *A. robustus* and *A. africanus* is straight and ascends steeply from the maxillary body toward the root of the zygomatic arch, thereby maximizing the vertical component of the occlusal load through its long axis. The formation of anterior pillars, located adjacent to the pyriform aperture, is regarded as a response to increased occlusal load on the anterior part of the palate; this increased load is implied by the molarization of the premolars. The pillars are found in both *A. africanus* and *A. robustus*. Retraction of the palate in the robust australopithecines causes the flat nasoalveolar clivus seen in *A. africanus* (part of the nasoalveolar triangular frame) to appear as a gutter in *A. robustus*. The affinity between *A. africanus* and *A. robustus* indicated by these morphological traits is confirmed by their absence in *Homo* and many other primates.

INTRODUCTION

The fact that *Australopithecus robustus* exhibits a highly specialized masticatory system seems to be agreed upon by most of those who have studied the unique morphology of the robust clade. It is precisely this extreme specialization—clearly the manifestation of simple biomechanical principles—that renders the interpretation of the masticatory-related morphology a relatively straightforward task. Many of these explainable features can also be recognized in the face of *Australopithecus africanus*, especially when it is contrasted with another hominid to which it is often compared, *Homo habilis* (though such comparisons are usually intended to demonstrate similarities!). Hence, the morphology of the masticatory system bears heavily on attempts to clarify the taxonomic and phylogenetic status of *A. africanus*. *Homo habilis* emerges as an undisputed representative of an evolutionary lineage parallel to that of the robust australopithecines. Its generalized masticatory system and the primitive topography of its face have been preserved in principle all along the *Homo* line to this day. The approach that aspires to reach a taxonomic decision regarding *A. africanus* by identifying the similarities between it and *A. robustus* ought to be preferred over that based on an arbitrary comparison between *A. africanus* and *Homo*, because only in the first approach are the facial differences examined in the context of their functional significance.

Among the many metric and morphological features expressing the specialization of the *A. robustus* masticatory system and shared by *A. africanus*, five that demonstrate *A. africanus*'s position on the robust lineage have been singled out for this discussion.

THE DISHED FACE AND THE ZYGOMATIC PROMINENCE (THE ADVANCEMENT OF THE PERIPHERAL FACE IN THE TRANSVERSE PLANE)

The biomechanical advantage achieved by the advancement of the muscles operating upon the mandibular lever is generally recognized. The extreme advancement of the site of origin of the masseter in the robust australopithecines is manifested in the change of orientation of the infraorbital region in two planes, the transverse and the sagittal. In the transverse plane, the advancement results in the migration of the peripheral section of the infraorbital region to a more anterior location than that of the medial part of the face, and thus the transverse profile of the face becomes concave: the so-called "dished face" appearance. The point of junction between the concave frontal aspect of the face and the lateral aspect (i.e., the anterior root of the zygomatic arch and the arch itself, which are both oriented in a parasagittal plane) protrudes dramatically as a sharp corner, the *zygomatic prominence*. The farther the lateral part of the facial contour advances, the deeper the dishing will become, and the more the zygomatic prominence will protrude. This prominence is also readily observed in the face of *A. africanus*, where the same conditions bring about its formation: 1) the advancement of the lateral part of the infraorbital region and 2) its junction with a sagittally oriented zygomatic arch. The presence of the prominence, therefore, represents a most significant reorganization of the face; it is easily identified in specimens Sts-5 and Sts-71 and was undoubtedly present Sts-17. Not only can the dished face be seen on these fossils but it also can be inferred from the remains constituting specimens MLD-6 and MLD-9, which would in turn indicate the presence of a zygomatic prominence.

This particular specialization is also brought to light through examination of the ontogeny of the *A. africanus* face. As in many other highly specialized primates, comparison of the young and the adult dramatizes the specific specialization of the latter (see, for example, Schultz, 1960; Cramer, 1977). The reorganization of the infraorbital region such that its lateral part protrudes more than its medial section stands out clearly in the comparison between the transverse contours of the face seen in the Taung juvenile and in Sts-5. An even more dramatic difference is revealed when the transverse facial profile of Taung is viewed against that of Sts-71. Comparable contours in a juvenile and an adult chimpanzee show no significant ontogenetic change. In *Homo* and many other primates, the transverse contour of the face retreats continuously from the protruding lateral nasal margin in a posterolateral direction. The transition between the frontal aspect of the face and the zygomatic arch is smooth and imperceptible.

THE INCLINATION OF THE INFRAORBITAL REGION IN THE SAGITTAL PLANE AND THE FORMATION OF THE MAXILLARY FOSSULA

The advancement of the site of the masseteric origin also manifests itself in changes undergone by the infraorbital plate in the sagittal plane. The infraorbital region slopes anteroinferiorly from the inferior margins of the orbit. Therefore, the zygomaticoalveolar crest in *A. robustus* is located substantially anterior to the plane of the orbital openings. This configuration also characterizes *A. africanus*, though not in such an extreme fashion. The bone surface in side view thus slopes parallel to the straight anterior profile of the face, as evidenced by specimens Sts-5, Sts-71, Sts-17, and undoubtedly also MLD-6. This is in sharp contrast to the bone inclination in *Homo* and in many other primates, where the infraorbital region slopes inferoposteriorly, forming, in side view, an angle with the anterior profile of the face. In the more medial section, the unique inclination of the infraorbital plate in *A. robustus* brings about an interesting and significant modification: just a minute vestige of what had constituted the canine fossa in the more generalized face now remains in the form of the *maxillary fossula*. *A. africanus*, in which the advancement of the infraorbital plate is not as great, offers us a view of an intermediate state in the formation of the fossula—the *maxillary furrow*, a modified canine fossa consisting of only a long, confined groove. Transitional stages between the furrow and the fossula can be seen on some of the *A. robustus* and *A. africanus* specimens.

THE ZYGOMATICOALVEOLAR CRESTS (THE LATERAL PILLARS)

The zygomaticoalveolar crest in both *A. robustus* and *A. africanus* is straight and ascends steeply from the body of the maxilla toward the root of the zygomatic arch. In contrast, the primitive face, as in *Homo* and other primates, exhibits a crest the lateral portion of which runs almost horizontally and at times even descends inferolaterally, forming a distinct notch.

This straightening of the crests may be viewed as a response to extreme stress, as they are situated like a beam between a strong, massive masseter and a heavy occlusal load. The relatively upright posture of the lateral pillars maximizes the vertical component of the masseteric force and the occlusal load through their long axis.

THE ANTERIOR PILLARS

The most dramatic evidence of the affinity between *A. africanus* and the robust australopithecines consists of the *anterior pillars*, found in the face of both *A. africanus* and *A. robustus*. The presence of these pillars adjacent to the pyriform aperture causes the nasal margins to be rounded in a transverse cross-section. This is in contrast to the paper-thin lateral nasal margins that are the medial termination of thin plates, as seen in the primitive face of *Homo*. The appearance of the anterior pillars can be viewed as related to the molarization of relatively anterior teeth, the premolars. The occlusal load that can be inferred from the molarization at the front of the palate necessitates structural modification of the facial skeleton. Hence, the anterior pillars are regarded here as a response to this load. The lower part of these massive pillars that descend from the glabellar region of the frontal bone supports the front of the palate.

It is true that the premolars of *A. africanus* indicate just an early stage of the molarization process. However, another factor for support increases. Clearly, then, the advanced molarization combined with the retracted palate in *A. robustus* results in the must be taken into account: the degree of protrusion of the palate anterior to the more peripheral facial frame, a protrusion that is seen to be greater in *A. africanus* than in *A. robustus* (Rak, 1983). As the palate extends farther forward, its anterior portion's need same need for pillars as in *A. africanus*, where the upper jaw protrudes considerably, but the degree of molarization is still modest. Support for this functional interpretation of the anterior pillars can be drawn from the fact that the relative rigidity of these structures enables them to resist attrition better than other parts of the face. This can be seen when specimens of both *A. africanus* (such as Stw-13) and *A. robustus* (for example, SK-46) are examined taphonomically.

THE RETRACTION OF THE PALATE AND THE FORMATION OF THE NASOALVEOLAR GUTTER IN *A. ROBUSTUS*

In *A. africanus* the anterior pillars and the flat nasoalveolar clivus, which resembles a ribbon stretched between the lower third of the pillars, join together to form a distinct morphological unit named the *nasoalveolar triangular frame*. The deep maxillary furrow that extends all along the lateral aspect of the pillars helps accentuate this frame even more dramatically. The retraction of the dental arcade seen in the more robust australopithecines leads to the transformation of the nasoalveolar clivus as it appears in the nasoalveolar triangular frame of *A. africanus* into the nasoalveolar gutter of *A. robustus*. That is to say, the more upright pillars lateral to the clivus continue to support the front of the palate, while the rest of the upper jaw, including the medial part of the clivus, retreats. This retraction achieves the same biomechanical effect as the advancement of those parts of the face that serve as the origin of the masticatory muscles. Indeed, these two antagonistic shifts, which also manifest themselves metrically (Rak, 1983), constitute the major determinants of australopithecine facial topography. The absence of the pillars on the one hand, and the presence of a nasoalveolar clivus that is convex in both the sagittal and the transverse planes on the other, characterize *Homo* in general.[1]

CONCLUSIONS

These traits unite with a large complex of morphological features to lead us to the conclusion that the face of *Homo* is primitive in the most basic sense. Hence, *A. africanus*, with all the modifications it is seen to share with the robust clade, can no longer be accorded its traditional role of perfect common ancestor to the two parallel branches.

LITERATURE CITED

Cramer, DL (1977) Craniofacial morphology of *Pan paniscus*. Contrib. Primatol. *10:*1–64

Rak, Y (1983) The Australopithecine Face. New York: Academic.

Schultz, AH (1960) Age changes and variability in the skulls and teeth of the Central American monkeys *Alouatta, Cebus* and *Ateles*. Proc. Zool. Soc. London *133:*337–390.

[1]Some early representatives of *Homo* do exhibit a flat nasoalveolar clivus. Nevertheless, the smooth transition to the more lateral part of the alveolar process, combined with the lack of anterior pillars, still categorically distinguishes this region from the nasoalveolar triangular frame of *A. africanus*.

Ancestors: The Hard Evidence, pages 171–177

Australopithecus and Early *Homo* in Southern Africa

Ronald J. Clarke
Palaeoanthropology Research Group, Department of Anatomy, University of the Witwatersrand Medical School, Parktown 2193 Johannesburg, South Africa

ABSTRACT Disagreement over interpretation of the fossil hominid record stems from a lack of clear understanding of what is meant by *Australopithecus*. I suggest that the generic name *Paranthropus* be retained for the hominids commonly called *Australopithecus robustus* and *Australopithecus boisei*. Clarke has shown SK 847 from Swartkrans to be distinct from both *A. africanus* and *Paranthropus* and has assigned it to the earliest form of *Homo erectus*. Clues to the ancestry of *Homo erectus* and *Paranthropus* are provided at the Sterkfontein site. *H. habilis* cranium STw 53, reconstructed by Clarke, is virtually identical to the Olduvai H. 24 cranium of *H. habilis*. These crania provide a logical morphological link between *A. africanus* and early *Homo erectus*. Variety in cranial structure in Sterkfontein *A. africanus* could be due to individual, sexual, temporal, or specific variation. The hominid sample from Sterkfontein Member 4 probably embraces a long period of time, and variation should be expected. The presence of an early form of *H. habilis* plus the absence of *Paranthropus* in the Member 5 breccia suggests it is of an earlier date than the neighbouring site of Swartkrans. The ancestry of *Paranthropus* is not yet clear. Although it possibly evolved from *A. africanus*, there are no intermediate forms known, and it could have evolved from another perhaps unknown species of *Australopithecus*.

INTRODUCTION

Although the name "*Australopithecus*" has since the time of its first usage by Dart in 1925 been entrenched in anthropological literature and in the minds of anthropologists, we are far from understanding exactly what we mean by the name. The symposium upon which this volume is based was entitled "Paleoanthropology: The Hard Evidence." Perhaps it should have been "The Hard-to-Understand Evidence." The fossils are certainly hard facts, but what they are "evidence" for depends entirely on the investigator. Thus we have fossils that are "evidence" of early *Homo* for some but of late *Australopithecus africanus* for others. We have "evidence" of a distinct genus called *Paranthropus* for some that is "evidence" of a flat-faced species of *Australopithecus* for others. The old Broom left us in 1951 with a legacy of fossil hominids that represented three distinct genera: *Australopithecus*, *Paranthropus*, and *Homo*. Clearly we need a new broom to sweep away the cobwebs of confusion and leave us with a clear understanding of what we mean by these terms.

PARANTHROPUS

Some years ago (Clarke 1977), I suggested that it was valid to retain a generic distinction between *Paranthropus* and *Australopithecus*. The following diagnosis summarizes my results:

Paranthropus is a genus of the family Hominidae characterized by the following mainly apomorphous characters in the cranium, compared to other genera within the family:

1) A brain that is on the average larger than that of *Australopithecus*, yet not as large as that of *Homo*.
2) Formation of a slightly concave, low forehead with a frontal trigone delimited laterally by posteriorly-converging temporal crests.
3) Presence of a flattened "rib" of bone across each supra-orbital margin.
4) A glabella that is situated at a lower level than the supraorbital margin.
5) Formation of a central facial hollow associated with a completely flat nasal skeleton, and a cheek region that is situated anterior to the plane of the piriform aperture.
6) Naso-alveolar clivus sloping smoothly into the floor of the nasal cavity.
7) Small incisive canals that open into the horizontal surface of the nasal floor.
8) Great enlargement of premolars relative to the molars and canines.
9) Great enlargement of molars and massiveness of tooth-bearing bone.
10) Anterior teeth small when compared to premolars and molars.
11) A tendency for the maxillary canine and incisor sockets to be situated in an almost straight line across the front of the palate.
12) Formation on the naso-alveolar clivus of prominent ridges marking the central incisor sockets but concavities marking the lateral incisor sockets.
13) Cusps of cheek teeth low and bulbous and situated closer to the centre of the crown than in other hominid genera.
14) Formation of flat occlusal wear surfaces to the cheek teeth, accompanied by smoothly rounded borders between the occlusal surfaces and the sides of the crowns of the cheek teeth.
15) Virtually completely molarised $dm_{\bar{1}}$ with anterior fovea centrally situated and with complete margin.
16) Great increase in the size of the masticatory musculature and attachments relative to the size of the skull.
17) Temporal fossa capacious and mediolaterally expanded.
18) Formation of a broad gutter on the superior surface of the posterior root of the zygoma.
19) A tendency for the palate to be shallow anteriorly and deep posteriorly.
20) Formation of either a marked pit or a groove across the zygomaticomaxillary suture of the cheek region—at least in the South African *Paranthropus*.

With so many clearly defined diagnostic features, it is easier to distinguish *Paranthropus* from *Australopithecus africanus* than it is to distinguish *Australopithecus africanus* from early *Homo*. The apparent reason for this is that the earliest known representatives of *Paranthropus* already present their distinct suite of characters and had probably been separate from *Australopithecus africanus* for a longer time than had early *Homo*. The earliest *Homo*, in the form of *Homo habilis* (e.g., Olduvai H. 24), is so morphologically close to *Australopithecus africanus* that some researchers find it difficult to separate them, as Boaz (1983) has observed. I would stress that if it is valid to place *Homo habilis* in a genus distinct from *Australopithecus*, it is far more justifiable to separate *Paranthropus* from *Australopithecus*.

AUSTRALOPITHECUS

If we maintain *Paranthropus* as a distinct genus, what then is *Australopithecus*? The type specimen of *A. africanus* from Taung is that of a child, but it exhibits the characteristic morphology seen in the adults of the species from Sterkfontein and Makapansgat between ca. 2.5–3.2 million years (m.y.) ago. *Australopithecus africanus* can best be distinguished from *Paranthropus* and *Homo* through its paucity of apomorphous characters. In other words, *A. africanus* lacks the specialised characters just listed for *Paranthropus*. It differs from *Homo* and *Paranthropus* in its small brain size (between 420–490 cm^3) and in having the incisor-bearing portion of the premaxilla projecting as a distinct entity anterior to the canines. In *A. africanus* the premaxillary suture can still be seen in two individuals (Taung and MLD 6) running lateral to the lower nasal margin and indicating that, as in the apes, the incisor-bearing surface of the face was composed only of premaxilla. By contrast, in the *Homo habilis* specimen STw 53, the premaxillary suture can be seen lateral to the inferior margin of the nasal bones and then curving medially into the superior nasal cavity.

Thus, in this specimen the incisor-bearing surface of the face is composed of maxillary bone that had overgrown the premaxilla in the embryo. Such overgrowth of the facial element of the premaxilla by the maxilla is a characteristic of *Homo* (Callender, 1869; Woo, 1949; Wood Jones, 1948) and seems to be associated with the evolutionary retraction of the incisor region beneath the nose.

A second, even more ape-like species of *Australopithecus* is *A. afarensis*, which occurs at Laetoli and Hadar in East Africa between 3.7 and 3 m.y. ago. So far this species has not been found in South Africa, but it seems reasonable to accept *A. afarensis* as the immediate evolutionary antecedent of *A. africanus* (Johanson and White, 1979).

HOMO

It was 35 years ago that Broom and Robinson discovered at Swartkrans a fragment of mandibular corpus (SK 45) that they claimed to be "of a type of early man which seems to us to be very near to that of *Homo*." This mandible fragment from the same breccia that was yielding remains of the ape-man *Paranthropus* demonstrated the contemporaneity of the two forms of hominid. They also recovered a maxilla fragment (SK 80) that Robinson (1953) considered to belong to the same type of man as the SK 45 mandible fragment. Another 20 years were to pass before I recognized that a large portion of the face and temporal (SK 847) fitted on to SK 80 (Clarke et al., 1970). Subsequently (Clarke 1977), I demonstrated that this cranium was that of an early *Homo* contemporary with *Paranthropus*. I concluded that the cranium probably should be regarded as belonging to a "sapient-like" subspecies of *Homo erectus* but suggested that until a bigger and more complete sample of early *Homo* cranial remains was available it would be better to consider SK 847 as *Homo* sp. indet. In 1977, after I had just completed my study on SK 847, Alan Walker brought to Johannesburg the cast of a recently discovered early *Homo erectus* from East Lake Turkana (KNM-ER 3733). Walker (e.g., 1976) had been unconvinced that SK 847 was a representative of early *Homo*, but when I pointed out to him the striking resemblance between SK 847 and KNM-ER 3733, he was then convinced. Together we took a series of comparative photographs of SK 847 and KNM-ER 3733. Later, Walker (1981) published these photographs, but I was disappointed to note that there was no reference to my publications of SK 847 nor to my suggestion of the resemblance of SK 847 to KNM-ER 3733.

Although I had found some similarities between SK 847 and *Homo habilis*, there were also strong resemblances to *Homo erectus*. The discovery of KNM-ER 3733 has demonstrated that Clarke and Howell (1972) were probably correct in their contention that SK 847 showed indications of a larger brain than in the australopithecines. Although one cannot say whether 847 had a brain size like that of *Homo habilis* or early *Homo erectus*, the *erectus*-like morphology of the frontal bone (which is not seen in any of the *Homo habilis* crania) plus the remarkable overall similarity to 3733 convinces me that 847 must now be classified as an early *Homo erectus*. Morphologically and temporally the early *Homo erectus* represented by 847 and 3733 bridges the gap between *Homo habilis* (as represented by O.H. 24, STw 53, and KNM-ER 1470) and classic *Homo erectus* (as represented by *Homo erectus pekinensis*).

It has been adequately demonstrated that early *Homo erectus* in the form of SK 847 and KNM-ER 3733 was a contemporary of *Paranthropus* about 1.5 m.y. ago in both South and East Africa (Clarke 1977; Leakey and Walker 1976). The question to be answered now is: What was the ancestry of these two diverse forms of hominid—*Homo erectus* with its small cheek teeth and large brain, and *Paranthropus* with its large cheek teeth and small brain?

HOMO HABILIS AT STERKFONTEIN

For possible clues, we must turn to an earlier site than Swartkrans (i.e., Sterkfontein) a mere kilometre away on the other side of the road. This cave site is famous for the quantities of fossils attributed to the taxon *Australopithecus africanus*, found in the Member 4 breccia dated to about 2.5–3.0 m.y. ago. The Member 5 breccia above this yielded in 1976 fragments of a cranium STw 53 classified as *Homo habilis* (Hughes and Tobias, 1977; Tobias, 1978). It has also yielded crude stone tools (Robinson, 1957, 1962; Mason, 1957, 1961, 1962; Stiles, 1979; Stiles and Partridge, 1979).

I have just made a preliminary reconstruction of STw 53 (Fig. 1) based solely on anatomical considerations and symmetry. When I had completed the reconstruction, I found

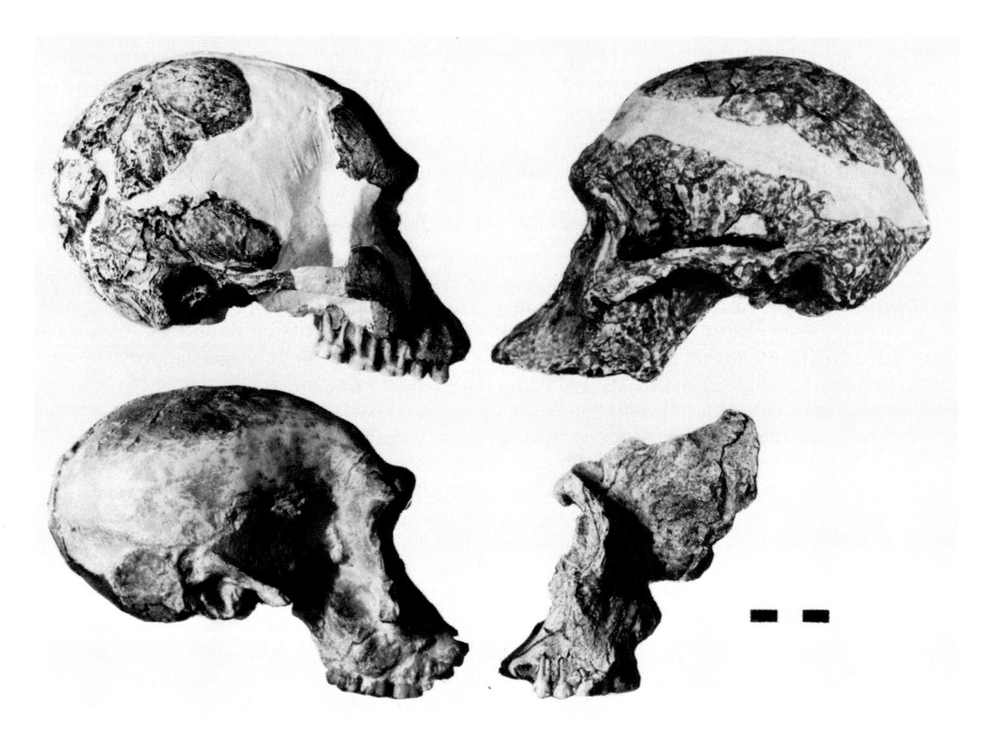

Fig. 1. Lateral views of STw 53 (top left), Sts-5 (top right), O.H. 24 (bottom left), and Sts-17 (bottom right). (All are casts.) Scale bar in cm.

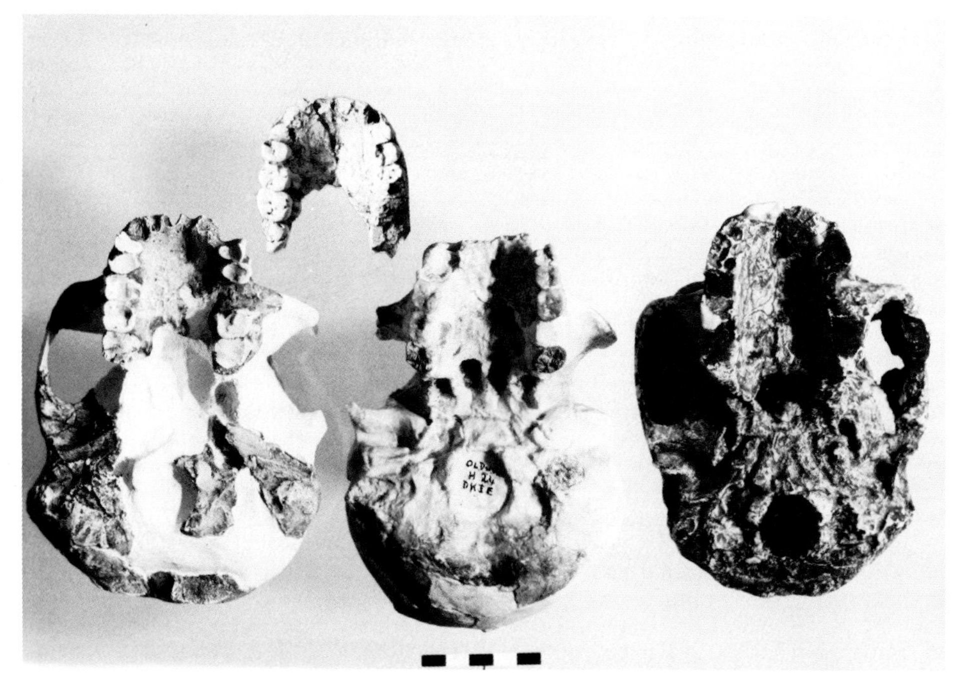

Fig. 2. Basal view from left to right of STw 53, Sts-17 palate, O.H. 24, and Sts-5. (All are casts.) Note the short, narrow brain case of Sts-5 and the posterior retraction of the palate in STw 53 and O.H. 24. Scale bar in cm.

that it looked almost identical to the cranium O.H. 24, a *Homo habilis* from Bed I of Olduvai Gorge. Such a similarity had been observed by Tobias (1978) on the frontal alone.

When STw 53 and O.H. 24 are compared to the *A. africanus* cranium Sts-5, there are striking differences, in that Sts-5 has a narrow cranial base, a foramen magnum situated further posteriorly, and a very prognathic muzzle. Sts-5, however, although the most complete, is not the only cranium of *A. africanus* from Sterkfontein. There are in particular two others that are somewhat different. One is Sts-17, which has rather small though buccolingually wide cheek teeth, and in facial structure is more similar than Sts 5 to O.H. 24 and STw 53 (Figs. 1,2). The other is Sts 71, which has large cheek teeth, a more rugged facial morphology, and little prognathism.

It is clear that there is a range of cranial morphology represented among the fossil sample attributed to *A. africanus* from the more lightly structured cranium of Sts-17 to the more rugged Sts-71. Similarly, among the mandibles and teeth from Sterkfontein Member 4 there are large and small varieties. This variation could be explained through individual or sexual differences or through temporal change during deposition of the Member 4 breccia. Sts-17 was recovered from the top of the Member 4 deposit excavated by Broom (Broom et al., 1950), so it would appear to have been temporally close to the Member 5 breccia that yielded the morphologically similar STw 53.

Several questions spring to mind when one notes this similarity: 1) Is Sts-17 really an *A. africanus*, or is it a *Homo habilis*? 2) Did Broom recover Sts-17 from the top of Member 4 or the bottom of Member 5? 3) Could one expect to find *H. habilis* in Member 4 anyway? 4) Does *A. africanus* occur in Member 5? 5) Is the great variability in molar tooth size seen in the Sterkfontein Member 4 assemblage due to sexual, individual, temporal, or specific variability? 6) Could there be two species of *Australopithecus* in Member 4? 7) Finally, did *A. africanus* necessarily disappear once *H. habilis* had appeared? We should beware of making assumptions when dealing with material blasted out of limestone caves, or indeed with material excavated from such stratigraphically complicated situations.

In view of the proximity of Sterkfontein to Swartkrans, one would expect that if Member 5 were only 1.5 m.y. old, it would contain remains of *Homo erectus* and *Paranthropus*. Both of those hominids were represented in the Swartkrans deposits of 1.5 m.y. ago. If Sterkfontein Member 5 were of the same age as the Swartkrans *Paranthropus* breccia, then the animal species responsible for the Swartkrans *Paranthropus* and *Homo erectus* accumulation would undoubtedly have accumulated remains of *Paranthropus* and possibly *Homo erectus* in the Sterkfontein Member 5 deposit. Although *Homo erectus* is rare at Swartkrans, *Paranthropus* is common, but so far there is nothing that can be assigned to *Paranthropus* from the hominid sample at Sterkfontein Member 5. Thus, the presence of an early form of *Homo habilis* (not present at Swartkrans), plus the absence of *Paranthropus*, suggests that Sterkfontein Member 5 is earlier than Swartkrans, i.e., earlier than 1.5 m.y.

Just as deposition in lake basins of East Africa covered a long span of time, so did deposition in the dolomite caves of South Africa. It is therefore erroneous to expect all fossil hominids from one geological member to be of similar morphology and geological age. Probably the sample of *Australopithecus africanus* from Member 4 of Sterkfontein includes specimens widely separated in geological age and stage of evolutionary development. Similarly, in Member 5 the stone tools may not necessarily be of one age or industry. Stiles and Partridge (1979) concluded that the presence of early *Homo*, the character of the bovid fauna (after Vrba, 1975), and the stone artefact assemblage all indicated an age for Member 5 in the range of 1.5 $\pm$ 0.3 m.y. The presence of early *Homo habilis* (STw 53), so similar to the 1.9 m.y. O.H. 24, would suggest an age closer to 2.0 m.y.

Although Mason (1962) and Stiles (1979) both considered the Sterkfontein artifacts to belong to the early Acheulean, Mary Leakey (1970) believed them to be Developed Oldowan. Some crude handaxes were found in the overburden, but aside from one protohandaxe, it is my opinion that the rest of the (mainly undescribed) tools so far known from excavation of the *in situ* deposits could well belong to a facies of the Oldowan. This is also the industry associated with the Olduvai *Homo habilis* (O.H. 24). These tools could thus be of earlier age than the handaxes found in the overburden. Volman (in press) has independently reached the same conclusion. Furthermore, the artefacts from Member 5 are not necessarily as old as the *Homo habilis*. The sample from one cave member can accumulate over a long period of time, and not necessarily in sequential order from bottom to top.

Australopithecus africanus so far has only been found in the three South African cave sites of Taung, Sterkfontein, and Makapansgat. Claims for the presence of this species in East Africa are not, in my view, well founded, as certain East African specimens attributed to *A. africanus* do not show its typical morphology but rather can be attributed to *Homo habilis*. Such cranial specimens are O.H. 24, KNM-ER 1813, and O.H. 13, all of which have small, buccolingually narrow cheek teeth and are temporally much later than *Australopithecus africanus*. The strong similarity between *Homo habilis* crania such as O.H. 24 and STw 53 and lightly structured *A. africanus* crania such as Sts-17 and Sts-52 would seem to suggest that *H. habilis* could well have evolved from a population of lightly-structured *A. africanus*. If one accepts that man evolved from a broadly chimpanzee-like ancestor in the Miocene, through the moderately chimp-like *Australopithecus afarensis* of 3.5 m.y. ago, and then *Australopithecus africanus* of 2.5 m.y. ago, then O.H. 24 and STw 53 at 2 m.y. ago do seem to present the next logical step in the evolution towards early *Homo erectus*. Reduction in cheek teeth size, retraction of the palate under the nose, and increase in brain size are the major evolutionary trends.

EVOLUTION OF ROBUST AUSTRALOPITHECINES

The ancestry of *Paranthropus* is not so obvious. It is possible to distinguish *Paranthropus* from *A. africanus* in most areas of cranial and dental morphology (Clarke, 1977). The earliest *Paranthropus* crania already show the flat frontal, flat face, and massive grinding cheek teeth that are characteristic of the genus. There are some specimens of *A. africanus* from Sterkfontein and Makapansgat that have massive mandibles and large cheek teeth (e.g., STw 14 and MLD 2), but they do not resemble *Paranthropus*. So far, there is nothing that one can readily say is an ancestor to *Paranthropus* or an intermediate between *A. africanus* and *Paranthropus*. We should not assume that such an ancestor to *Paranthropus* would necessarily occur in any of the known South African sites. *Paranthropus* was a widely spread and highly specialised form of hominid. It probably evolved from a species of *Australopithecus*, but not necessarily from *A. africanus* of South Africa. It could have even had an earlier origin from *A. afarensis*, or it might have first developed in East Africa from another, as-yet unrecognised species of *Australopithecus*. Although no fossils that occur in East Africa can be assigned with certainty to *A. africanus*, the presence of an *Australopithecus* species comparable to *A. africanus* may be suggested by teeth found at Omo in Ethiopia, dated between 3 and 2.5 m.y. ago (Howell and Coppens, 1976). *Paranthropus* first occurs in the Omo succession at least 2.1 m.y. ago (Howell, 1982, p. 99) but does not appear in South Africa until about 1.5 m.y. ago (Brain, 1982). On present indications, we can thus tentatively say that *Paranthropus* seems to have been present at an earlier date in East Africa than in South Africa. The fact that no earlier form of *Paranthropus* morphologically intermediate between *A. africanus* and *Paranthropus* has been found in Sterkfontein Member 5 could be seen as a tentative argument against *Paranthropus* having evolved from the South African *A. africanus*.

Johanson and White (1979) and Johanson et al. (1981) have argued that *A. africanus* should be classified as already on the specialised lineage towards *Paranthropus*. This view is supported by the excellent, detailed analysis by Rak (1983) of the structure of the australopithecine face. If we accept this view, then we are still left with the question of why *Homo habilis* is so morphologically similar to certain *A. africanus* specimens, while all *Paranthropus* specimens are so morphologically distinct from all *A. africanus* specimens. That *Paranthropus* evolved from a species of *Australopithecus* is a reasonable conclusion to be drawn from the available fossil record and from morphological comparisons, but exactly how, when, where, and why that evolution took place is, I believe, still open to speculation.

CONCLUSIONS

In conclusion and with reference to the symposium title, "Paleoanthropology: The Hard Evidence," I would request that we rid our minds and our writings of this misguided concept of "evidence." After all, "evidence" is what is seen and not how it is interpreted. We can all see a bone and know it is a bone, but what it is "evidence" for depends upon one's interpretation. For example, Robinson deserves great credit for having seen the SK 80 maxilla and interpreted its morphology as that of *Homo*. He made the error, however, of assuming that nearly all other Swartkrans hominids were *Paranthropus* and thus classified the SK 847 cranium (which I joined 20 years later to SK 80) as *Paranthropus*. In other words, he unknowingly utilised two parts of one individual as "evidence" of two different genera! Similarly, Broom and Robinson (1952) assumed that SK 27 was "evidence" of a large-brained *Paranthropus* with abnormally large canines. It wasn't. It was a small-brained *Homo* with normally sized canines. The 1470 cranium from East Lake Turkana was accepted by Richard Leakey as "evidence" for *Homo* at 2.9 m.y. ago. It wasn't. It was evidence of erroneous dating. All of us are open to being misled by our own preconceptions or readiness to accept dates, geological interpretations, or taxonomic assessments without question or qualification.

What we need is not "evidence" but more hard facts cautiously and wisely interpreted with a willingness to amicably discuss our colleagues' differing viewpoints. It is pleasant to recall that such friendly discussion around the original fossils was a highlight of the study sessions that preceded the "Ancestors" symposium on which this volume is based.

LITERATURE CITED

Boaz, NT (1983) Morphological trends and phylogenetic relationships from Middle Miocene hominoids to Late Pliocene hominids. In RL Ciochon and RS Corruccini (eds): New Interpretations of Ape and Human Ancestry. New York: Plenum, pp. 705–720.

Brain, CK (1982) The Swartkrans site: Stratigraphy of the fossil hominids and a reconstruction of the environment of early *Homo*. Prétirage, 1er Congrès Internat Paléontol Humaine. Nice: CNRS, 676–706.

Broom, R, and Robinson, JT (1952) Swartkrans ape-man, *Paranthropus crassidens*. Transvaal Mus. Mem. *6*:1–123.

Broom, R., Robinson, JT, and Schepers, GWH (1950) Sterkfontein ape-man, *Plesianthropus*. Transvaal Mus. Mem. *4*:1–117.

Callender, GW (1869) The formation and early growth of the bones of the human face. Phil. Trans. *159*:163–172.

Clarke, RJ (1977) The Cranium of the Swartkrans Hominid, SK 847, and its Relevance to Human Origins. Unpublished Ph.D. thesis, University of the Witwatersrand.

Clarke, RJ, Howell, FC, and Brain, CK (1970) More evidence of an advanced hominid at Swartkrans. Nature *225*:1219–1222.

Clarke, RJ, and Howell, FC (1972) Affinities of the Swartkrans 847 hominid cranium. Am. J. Phys. Anthropol. *37*:319–336.

Dart, RA (1925) *Australopithecus africanus*: The man-ape of South Africa. Nature *115*:195–199.

Howell, FC (1982) Origins and evolution of the African Hominidae. In JD Clark (ed): The Cambridge History of Africa, Vol 1, Cambridge: Cambridge Univ. Press, pp. 70–156.

Howell, FC, and Coppens, Y (1976) An overview of Hominidae from the Omo succession, Ethiopia. In Y Coppens, FC Howell, G Isaac, and REF Leakey (eds): Earliest Man and Environments in the Lake Rudolf Basin. Chicago: Univ. of Chicago Press, pp. 522–532.

Hughes, AR, and Tobias, PV (1977) A fossil skull probably of the genus *Homo* from Sterkfontein, Transvaal. Nature *265*:310–312.

Johanson, D, and White, T (1979) A systematic assessment of early African hominids. Science *202*:321–330.

Johanson, D, White, T, and Kimbel, W (1981) *Australopithecus africanus:* Its phyletic position reconsidered. S. Afr. J. Sci. *77*:445–470.

Leakey, M (1970) Stone artefacts from Swartkrans. Nature *225*:1222–1225.

Leakey, REF, and Walker AC (1976) *Australopithecus, Homo erectus* and the single species hypothesis. Nature *261*:572–574.

Mason, R (1957) Occurrence of stone artefacts with *Australopithecus* at Sterkfontein, Part 2. Nature *180*:521–524.

Mason, R (1961) The earliest tool-makers in South Africa. S. Afr. J. Sci. *57*:13–16.

Mason, R (1962) The Sterkfontein stone artefacts and their maker. S. Afr. Archaeol. Bull. *17*:109–125.

Mason, R (1976) The earliest artefact assemblages in South Africa. Prétirage, 9th Congrés Union Internationale des Sciences Préhistoriques et Protohistoriques, Nice: Colloque V, 140–156.

Rak, Y (1983) The Australopithecine Face. New York: Academic Press.

Robinson, JT (1953) *Telanthropus* and its phylogenetic significance. Am. J. Phys. Anthropol. *11*:445–501.

Robinson, JT (1957) Occurrence of stone artefacts with *Australopithecus* at Sterkfontein, Part 1, Nature *180*:521–524.

Robinson, JT (1962) Australopithecines and artefacts at Sterkfontein, Part 1: Sterkfontein stratigraphy and the significance of the extension site. S. Afr. Archaeol. Bull. *17*:87–107.

Stiles, D (1979) Recent archaeological findings at the Sterkfontein site. Nature *277*:381–382.

Stiles, DN, and Partridge TC (1979) Results of recent archaeological and palaeoenvironmental studies at the Sterkfontein extension site. S. Afr. J. Sci. *75*:346–352.

Tobias, PV (1978) The earliest Transvaal members of the genus *Homo* with another look at some problems of hominid taxonomy and systematics. Zeit. Morphol. Anthropol. *69*:225–265.

Volman, TP (1984) Early prehistory of southern Africa. In RG Klein (ed): Southern African Prehistory and Paleoenvironments. Rotterdam: Balkema, pp. 169–220.

Vrba, ES (1975) Some evidence of chronology and palaeoecology of Sterkfontein, Swartkrans and Kromdraai from the fossil Bovidae. Nature *254*:301–304.

Walker, A (1976) Remains attributable to *Australopithecus* in the East Rudolf succession. In Y Coppens, FC Howell, GL Isaac, and RE Leakey (eds): Earliest Man and Environments in the Lake Rudolf Basin. Chicago: University of Chicago Press, pp. 484–489.

Walker, A (1981) The Koobi Fora hominids and their bearing on the origins of the genus *Homo*. In BA Sigmon and J Cybulski (eds): *Homo erectus*, Papers in Honor of Davidson Black. Toronto: University of Toronto Press, pp. 193–215.

Woo, JK (1949) Ossification and growth of the human maxilla, premaxilla and palate bone. Anat. Rec. *105*:737–761.

Wood Jones, F (1948) Hallmarks of Mankind. London: Baillière, Tindall and Cox.

Ancestors: The Hard Evidence, pages 178–183

Implications of Postcanine Megadontia for the Origin of *Homo*

Henry M. McHenry
Department of Anthropology, University of California, Davis, California 95616

ABSTRACT This report builds on the work of many, but particularly two recent papers by Johanson and White (1979) and White et al. (1981). From these studies, two alternative phylogenies agree best with available fossil evidence: 1) *Australopithecus afarensis* is the common ancestor of two lineages, one leading to *Homo habilis, H. erectus*, and *H. sapiens* and the other leading to *A. africanus* and *A. robustus* + *boisei*; 2) *A. afarensis* is ancestral to *A. africanus*, and *A. africanus* is the common ancestor of the *Homo* lineage and the *A. robustus* + *boisei* lineage.

The most compelling reason to support the first alternative is the observation that *A. africanus* appears to be too specialized in the direction of *A. robustus* and *boisei* to be considered ancestral to *H. habilis*, particularly in features related to postcanine megadontia. On the other hand, the second alternative receives support from the fact that *A. africanus* and *A. robustus* + *boisei* share a suite of derived features with *H. habilis* that are absent in *A. afarensis* (McHenry, 1984; Skelton et al., 1984; Wolpoff, 1982). These features may have evolved through parallel evolution, but it is more parsimonious to assume their presence in a common ancestor. If *A. africanus* is taken as the common ancestor of *A. robustus* + *boisei* and *H. habilis*, then the evolutionary transition from *Australopithecus* to *Homo* involved rapid reduction in features relating to postcanine megadontia. Although this may be regarded as an unparsimonious evolutionary reversal, there is some evidence that *H. habilis* did originate from an australopithecine characterized by postcanine megadontia and associated features.

INTRODUCTION

My topic concerns the origin of *Homo*: which species of *Australopithecus* gave rise to the first member of our own genus, *Homo habilis*? By most accounts there are two known species of *Australopithecus* that might fill this role: *A. afarensis* and *A. africanus*.[1] There is, however, an apparent contradiction between two lines of evidence on the issue of the immediate ancestor of *Homo habilis*. On the one hand, *A. afarensis* appears to fill this role, because *A. africanus* is specialized in the direction of the robust australopithecines (Johanson and White, 1979; White et al., 1981). On the other hand, *A. africanus* and *A. robustus* + *boisei* share a suite of derived features with *H. habilis* that

[1]The validity of the species *A. afarensis* is the subject of much debate (Anonymous, 1976; Boaz, 1983; Day et al., 1980; Leakey, 1981; Leakey and Walker, 1980; Tobias, 1978, 1980a,b). There is also some doubt about whether the Hadar samples represent a species that gave rise to *A. africanus* (Falk, 1983, Falk and Conroy, 1983; Olson, 1978, 1981, 1985; although see Kimbel, 1984 and Kimbel et al., 1985). In this paper I follow the classification used by Johanson and White (1979) and White et al. (1981), since the forceful case that they make for the species validity in the latter publication has not received an equally thorough rebuttal. I test the two phylogenies that they believe are the most likely (B and C of their Fig. 18).

is absent in *A. afarensis* (McHenry, 1984; Skelton et al., 1984; Wolpoff, 1982). These shared derived traits support the view that *A. africanus* is the immediate ancestor of *H. habilis*. The purposes of this paper are to explore this contradiction and to suggest a resolution.

SPECIALIZATIONS OF *A. AFRICANUS*

Australopithecus africanus is excluded by some from direct ancestry of *H. habilis* by the apparent fact that it is specialized in the direction of *A. robustus* + *boisei*. Johanson and White (1979, p. 328) list four of these specializations: "stronger molarization of the premolars, increased relative size of the postcanine dentition, increased buttressing of the mandibular corpus, and increased robustness of the corpus itself." White et al. (1981) add several more traits that could be included in this list, although they do not explicitly do so. Almost all of these are closely related to the hypertrophy of the posterior dentition: larger postcanine teeth are functionally associated with increased mandibular buttressing, more strongly developed muscle attachment areas on the cranium, a shift forward of the center of action of the temporalis muscle, midfacial buttressing (Rak, 1983), and many other features.

Is it true that *A. afarensis* has relatively smaller postcanine dentition than *A. africanus*? From the context of Johanson and White's (1979) assertion, it is apparent that they meant postcanine tooth size relative to that of anterior teeth, which is the conventional comparison. A biologically more meaningful comparison is between postcanine size and body size, but the latter parameter is very difficult to estimate in fragmentary fossils. I sidestepped the problem of body size estimation in a preliminary study (McHenry, 1983) by comparing cheek-tooth size to postcranial size in associated skeletons. When compared to extant great apes and man, the one associated skeleton representing *A. afarensis* (A.L.288-1) has cheek teeth 2.8 times larger than expected from the size of its humerus, whereas the representatives of *A. africanus* (Sts 7) and *A. robustus* (TM 1517) have cheek teeth about twice the size predicted from their humeri. Relative postcanine size measured in this way reveals the opposite relationship from that described by Johanson and White (1979): *A. afarensis* is the most megadont of the australopithecines.

A subsequent study (McHenry, 1984) showed that this result is due to the fact that the relationship between postcanine area and humeral size in A.L.288-1 is atypical of *A. afarensis*. When the absolute postcanine area is taken, the average for *A. afarensis* (757 mm^2) is smaller than the average for *A. africanus* (856 mm^2). The only way *A. afarensis* could be more megadont than *A. africanus* relative to body weight is for the former to weigh much less than the latter. The known postcranial samples do not support this. Indeed, my own estimation of body weight (McHenry, 1982) shows *A. afarensis* to weigh more than *A. africanus* (37 kg and 35 kg, respectively).[2] Compared to extant great apes and humans, the cheek teeth of *A. afarensis* are 1.7 times larger than expected from estimated body weight, compared to 1.9–2.0 times for both *A. africanus* and *A. robustus* and 2.3 times for *A. boisei*. *Homo habilis* is 1.4 times larger. Living species deviate no more than 17% from the regression line.

From these studies two findings are important: 1) all species of *Australopithecus* are characterized by postcanine megadontia with cheek teeth averaging twice the size expected for their estimated body weight, when compared with extant great apes and humans, and 2) relative cheek tooth size tends to increase through time from *A. afarensis* to *A boisei*. The increase in relative cheek tooth size through time shows positive allometry (1.26) that is higher than expected from geometric similarity (0.67) or from metabolic rate (0.75). This implies that the species are not merely "scaled variants of the same animal" (Pilbeam and Gould, 1974, p. 892), but instead show morphological adaptations independent of body size increase.

THE PRIMITIVENESS OF *A. AFARENSIS*

From the evidence presented above, it is apparent that *A. afarensis* is more similar to

[2]These estimates of body weight (37 kg for *A. afarensis* and 35 kg for *A. africanus*) are based on the midpoints between the small and large morphs at Hadar and Sterkfontein. While these estimates have a range of possible error, I do not believe available evidence supports the view that *A. africanus* is significantly larger than *A. afarensis*. The sample of postcrania of *A. africanus* is small (reports on less than two dozen individuals have been published), but it is unlikely that sampling bias resulted in a collection where more dentitions of large bodied individuals were preserved, but more postcranial specimens of small-bodied individuals survived. There are two distinct postcranial size morphs of *A. africanus*, perhaps representing female with the diminutive Sts 14, Sts 65, TM 1526 and male with the larger Sts 34, TM 1513, Sts 73, and Sts 7, for example. The more complete distal femur of this larger morph, TM 1513, is 13% smaller than the largest complete distal femur of *A. afarensis* (A.L. 333-4), based on seven measurements.

H. habilis in features relating to postcanine megadontia than is *A. africanus*. In absolute size of the cheek teeth, *A. afarensis* is almost identical to *H. habilis* (757 mm^2 and 759 mm^2, respectively). However, the latter two species are not very much alike in many other ways. *Australopithecus afarensis* is exceptionally primitive in many traits, and in most of these respects *A. africanus* and *A. robustus + boisei* more closely resemble *H. habilis*. In fact, *H. habilis* and the later species of *Australopithecus* share a suite of derived traits not present in *A. afarensis*. This implies that the immediate ancestor of *H. habilis* also shared these traits.

Table 1 lists these derived traits shared by *H. habilis*, *A. africanus*, and *A. robustus + boisei*, which differentiate them from the primitive *A. afarensis*. These traits are described in White et al. (1981) to show the distinctiveness of *A. afarensis* from other species of *Australopithecus*. In several of the cranial, facial, and mandibular traits and in all of the dental traits I have interpreted their descriptions and made original observations on *H. habilis* which they did not provide. In some cases, this is a potential source of error because my interpretation of what a "strong basal lingual tubercle" is, for example, may not be the same as theirs. As they point out, these are not always clear-cut diagnostic traits but often represent tendencies. They are not all independent pieces of information, either, since there is a great deal of intercorrelation among sets of traits.

Despite these qualifications and the possibility that some of the descriptions listed in Table 1 may need revision, the list is impressively long and involves several functionally independent units. If *A. afarensis* is the immediate ancestor of *H. habilis*, then a great deal of parallel evolution must have occurred. From this point of view it is clearly more parsimonious to assume that the hominid species that immediately predates the appearance of *Homo* (*A. africanus*) is also the immediate common ancestor of *H. habilis* and *A. robustus + boisei*.

RESOLUTION

There are at least three possible resolutions to the contradiction described above: 1) none of the known species of *Australopithecus* represents the immediate ancestor of *H. habilis*; 2) *A. afarensis* is the immediate ancestor of *H. habilis*, and the features shared by *H. habilis* and the later species of *Australopithecus* evolved independently; or 3) *A. africanus* is the immediate ancestor of *H. habilis*, implying that *H. habilis* evolved from a form with hypermasticatory complex associated with postcanine megadontia.

The first resolution (i.e., the immediate ancestor of *H. habilis* is unknown) is certainly possible and is championed by some (e.g., Leakey, 1981). It is difficult to prove or disprove since it relies on negative evidence. One consideration makes it less likely: it implies that the 52 traits listed in Table 1 evolved independently in the *Homo* and later *Australopithecus* lineages. The transition from *A. afarensis* to *A. africanus*, for example, involved canine reduction, an increase in the metaconid of P_3, the reduction in the diastema, an increase in basicranial kyphosis, reduction in facial prognathism, and so on. These same modifications are seen in *H. habilis*, which therefore implies that they arose from a common evolutionary source.

This liability also weakens the second resolution: if *A. afarensis* is the immediate ancestor of *H. habilis*, then the traits shared by *H. habilis* and later australopithecines must have evolved independently. This could be true of some of the traits, of course. Encephalization in *H. habilis* might move the center of action of the temporalis forward, whereas selection for the maximization of biting power moved it forward in *A. africanus*. But to explain all of the traits shared by *H. habilis* and later australopithecines and not by *A. afarensis* as the result of parallel evolution is not parsimonious.

The third alternative has its strengths: *A. africanus* immediately predates *H. habilis*, and *A. africanus* shares the derived features with *H. habilis* listed in Table 1. But its masticatory apparatus appears to be specialized in the direction of the robust australopithecines. Could *H. habilis* have evolved from a form with the hypermasticatory complex associated with enlarged check teeth? This would imply that the reduced masticatory apparatus characteristic of *H. habilis* is not a retention of the primitive condition but is a specialization characteristic of the genus *Homo*. Dental reduction is certainly a characteristic feature of the *Homo* lineage through time. The vector of relative postcanine tooth size from *H. habilis* to *H. sapiens* shows very strong negative allometry. An evolutionary transition from *A. africanus* to *H. habilis* would require similar negative allometry, although slightly stronger.

TABLE 1. Traits shared by early Homo *and* Australopithecus africanus *and* A. robustus + boisei *but not by* A. afarensis[1]

Calvarium	1. No compound temporal/nuchal crest in females
	2. Divergence of temporal lines is high (well above lambda)
	3. Absence of strong asterionic notch
	4. Nuchal plane not steeply inclined
	5. Absence of transverse curvature of nuchal plane
	6. Relatively deep mandibular fossa
	7. Strong and steep articular eminence
	8. Tympanics cone-shaped in sagittal cross-section
	9. Tympanics directed nearly vertically
	10. Medium to weak inflection of mastoid process beneath cranial base
	11. More pronounced basicranial kyphosis
	12. Anterior fibers of temporalis muscle expanded relative to posterior fibers
Face	13. Premaxilla less expanded
	14. Flatter and straighter nasoalveolar clivus
	15. I^2 roots situated medial to nasal aperture margins
	16. Reduced C/P^3 juga
	17. Reduced or absent canine fossa
	18. Reduced or absent transverse buttress from superior part of juga to zygomatic arch
	19. Wider palate
	20. Arched palate that is less shallow
	21. Maxillary tooth row diverge posteriorly
Mandible	22. Reduced or absent hollow above and behind mental foramen
	23. Mental foramen tends to be above mid corpus
	24. More vertical anterior corpus
	25. No lateral convexity of tooth row
Dentition	26. Maxillary and mandibular diastema less frequent
	27. Deciduous canine less tall, wider, and less pointed
	28. I^1 incisal edge medium to short in length
	29. Strong basal lingual tubercle on I^2 usually lacking
	30. Fewer than seven mammelons on I_1
	31. Upper canine tends to be more symmetric in labial view
	32. Upper canine projects less above occlusal surface
	33. Upper canine tends to have apical wear only
	34. Lingual ridge on lower canine less pronounced
	35. Mesial occlusal edge of lower canine shorter, more horizontal, and less straight
	36. Distal cingulum of lower canine set at a relatively higher crown position, which limits the distal occlusal edge of the tooth
	37. Absence of flattened distal occlusal edge of lower canine
	38. Lower canine tends to have exclusively apical wear
	39. P_3 occlusal crown outline tends to be rounder
	40. P_3 bicuspid in both males and females with metaconid more strongly developed
	41. Anterior fovea commonly closed in P_3
	42. Talonid of P_3 not as low or narrow
	43. P_3 distal marginal ridge more commonly cuspidate
	44. Less frequent mesiobuccal extension of P_3 buccal enamel line
	45. Occlusal wear of P_3 tends to keep pace with other postcanine teeth
	46. Distolingual crown corner of P_3 tends to be less abbreviated
	47. Occlusal outline of P_4 tends to be more rounded and symmetrical
	48. Mesial and distal marginal ridges of P^4 tend to be larger relative to buccal cusp length
	49. Upper molars tend to have accessory cusps
	50. Buccal cusps of the maxillary molars tend to wear down as quickly as the lingual cusps
	51. Medium to weak mesial appression of the M_1 and M_2 hypoconulid
	52. Breadth of talonid and trigonid equal in M_2

[1]Traits are those discussed by White et al. (1981).

Some of the characteristics closely related to postcanine megadontia indicate that *H. habilis* might well be derived from a form "specialized" in the direction of the robust australopithecines. The form of the lower anterior premolar is a good example. Johanson and White (1979) point out that *A. africanus* appears to be specialized in the direction of *A robustus + boisei* in having a tendency towards molarization of this tooth. This implies that the form of this tooth in *H. habilis* most likely derived from the form characteristic of *A. afarensis*. The P_3 in *A. afarensis* is usually unicuspid, and if the metaconid is present, it is weakly developed. In *H. habilis*, the metaconid is always well developed, and in at least one specimen (KNM-ER 1802) the tooth is more molarized than it is in some specimens of *A. africanus*. One explanation for the strong P_3 metaconid and the occasional molarization of the tooth in *H. habilis* is that it evolved from a form characterized by the degree of postcanine megadontia and associated molarization of premolars typical of *A. africanus*.

Australopithecus africanus may not fit the role of immediate ancestor to *H. habilis* perfectly, of course, but the evidence presented here indicates that it is closer than any other known species. It does not fit perfectly with available evidence from the subnasal morphology, for example (Ward and Kimbel, 1983). The nasoalveolar clivus ends at the inferior nasal margin in *A. africanus* (except in MLD 9) but it extends into the nasal cavity of *A. afarensis* and *H. sapiens*. On the other hand, the morphology of *H. sapiens* is so very specialized that the way in which the clivus enters the nasal cavity is quite different looking from *A. afarensis*. One can imagine a scenario in which the reorganization of the face from *A. afarensis* to *H. habilis* involved retreat and reentry of the clivus out and back into the nasal cavity in a different arrangement.

CONCLUSIONS

This paper explores the consequences of two hominid phylogenies proposed by Johanson and White (1979) and White et al. (1981), one in which *A. afarensis* is the immediate common ancestor of *H. habilis* and the other in which *A. afarensis* gives rise to *A. africanus*, which in turn gives rise to both *H. habilis* and *A robustus + boisei*. The strength of the first phylogeny derives from the apparent fact that *A. africanus* is too specialized in features relating to postcanine megadontia to be an ancestor of *H. habilis*. I conclude, however, that the strengths of the second phylogeny outweigh this apparent liability: *A. africanus* immediately predates the appearance of *H. habilis* and *A robustus + boisei*, and it shares with these species a suite of derived features that is absent in *A. afarensis*. In this view, *H. habilis* evolved from a species with a complex of traits associated with well developed postcanine megadontia. There is some preliminary evidence that this did occur.

ACKNOWLEDGMENTS

For the privilege of examining the original fossil material used in this study and for many other kindnesses I thank C.K. Brain, E. Vrba, and the staff of the Transvaal Museum, Pretoria; P.V. Tobias, A. Hughes, and the staff of the Department of Anatomy, University of the Witwatersrand, Johannesburg; R.E. Leakey, M.D. Leakey, the late L.S.B. Leakey, L. Jacobs, and the staff of the National Museums of Kenya, Nairobi; D.C. Johanson and the staff of the Cleveland Museum of Natural History; and Tadesse Terfa, Mammo Tessema, Woldesenbet Abomssa, and the staff of the National Museum of Ethiopia, Addis Ababa. I am grateful to R.R. Skelton and G.M. Drawhorn for many of the ideas expressed herein. I thank L.J. McHenry for invaluable assistance. Partial funding for this research was provided by the Committee on Research, University of California, Davis.

LITERATURE CITED

Anonymous (1976) Hominid remains from Hadar, Ethiopia. Nature *260*:389.

Boaz, NT (1983) Morphological trends and phylogenetic relationships from Middle Miocene hominoids to Late Pliocene hominids. In RL Ciochon and RS Corruccini (eds): New Interpretations of Ape and Human Ancestry. New York: Plenum, pp. 705–720.

Day, MH, Leakey, MD, and Olson, TR (1980) On the status of *Australopithecus afarensis*. Science *207*:1102–1103.

Falk, D (1983) Cerebral cortices of East African early hominids. Science *221*:1072–1074.

Falk, D, and Conroy, GC (1983) The cranial venous sinus system in *Australopithecus afarensis*. Nature *306*:779–781.

Johanson, DC, and White, TD (1979) A systematic assessment of early African hominids. Science *202*:321–330.

Kimbel, WH (1984) Variation in the pattern of cranial

venous sinuses and homind phylogeny. Am. J. Phys. Anthropol. *63*:243–263.

Kimbel, WH, White, TD, and Johanson, DC (1985) Craniodental morphology of the hominids from Hadar and Laetoli: Evidence of *Paranthropus* and *Homo* in the mid-Pliocene of eastern Africa? In E Delson (ed): Ancestors: The Hard Evidence. New York: Alan R. Liss, Inc., pp 120–137.

Leakey, RE, and Walker, A (1980) On the status of *Australopithecus afarensis*. Science *207*:1103.

Leakey, RE (1981) The Making of Mankind. New York: Dutton.

McHenry, HM (1982) The pattern of human evolution: Studies on bipedalism, mastication, and encephalization. Ann. Rev. Anthropol. *11*:151–173.

McHenry, HM (1983) Relative size of the cheek teeth in Australopithecus. Am. J. Phys. Anthropol. *60*:224.

McHenry, HM (1984) Relative cheek tooth size in *Australopithecus*. Am. J. Phys. Anthropol. *64*:297–306.

Olson, TR (1978) Hominid phylogenetics and the existence of *Homo* in Member I of the Swartkrans Formation, South Africa. J. Hum. Evol. *7*:159–178.

Olson, TR (1981) Basicranial morphology of the extant hominoids and Pliocene hominids: The new material from the Hadar Formation, Ethiopia, and its significance in early human evolution and taxonomy. In CB Stringer (ed): Aspects of Human Evolution. London: Taylor and Francis, pp. 99–128.

Olson, TR (1985) Cranial morphology of the Hadar Formation hominids and "*Australopithecus*" *africanus*. In E. Delson (ed): Ancestors: The Hard Evidence. New York: Alan R. Liss, Inc., pp 102–119.

Pilbeam, D, and Gould, SJ (1974) Size and scaling in human evolution. Science *186*:892–901.

Rak, Y (1983) The Australopithecine Face. New York: Academic.

Skelton, RR, McHenry, HM, and Drawhorn, GM (1984) Phylogenetic analysis of Plio-Pleistocene hominids. Am. J. Phys. Anthropol. *63*:219.

Tobias, PV (1978) The place of *Australopithecus africanus* in hominid evolution. In DJ Chivers and KA Joysey (eds): Recent Advances in Primatology, Vol. 3, Evolution. London: Academic, pp. 373–394.

Tobias, PV (1980a) "*Australopithecus afarensis*" and *A. africanus*: A critique and an alternative hypothesis. Palaeontol. Africana *23*:1–17.

Tobias, PV (1980b) A study and synthesis of the African hominids of the late Tertiary and Early Quaternary Periods. In LK Königsson (ed): Current Arguments on Early Man. Oxford: Pergamon, pp. 86–113.

Ward, SC, and Kimbel, WH (1983) Subnasal alveolar morphology and the systematic position of *Sivapithecus*. Am. J. Phys. Anthropol. *61*:157–171.

White, TD, Johanson, DC, and Kimbel, WH (1981) *Australopithecus africanus*: Its phyletic position reconsidered. S. Afr. J. Sci. *77*:445–470.

Wolpoff, MH (1982) Australopithecines: The unwanted ancestors. In KJ Reichs (ed): Hominid Origins. Washington, D.C.: University Press of America, pp. 109–126.

Ancestors: The Hard Evidence, pages 184–192

Locomotor Adaptations in the Hadar Hominids

Randall L. Susman, Jack T. Stern, Jr., and William L. Jungers
Department of Anatomical Sciences, Health Sciences Center, State University of New York, Stony Brook, New York 11794

ABSTRACT Numerous studies of the locomotor skeleton of the Hadar hominids have revealed traits indicative of both arboreal climbing/suspension *and* terrestrial bipedalism. These earliest known hominids must have devoted part of their activities to feeding, sleeping, and/or predator avoidance in trees, while also spending time on the ground where they moved bipedally. In this paper we offer a synopsis of new data on morphology of the foot and on aspects of body proportions that further strengthen the argument for arboreality in the Hadar hominids. These same data provide additional evidence that the mode of bipedality practiced at Hadar differed from that documented in modern humans.

Consideration of the ecology at Hadar, in conjunction with modern primate models, also supports the notion of arboreality in these earliest australopithecines. We speculate that selection for terrestrial bipedality may have intensified through the Plio-Pleistocene as forests and woodland patches shrunk and the need arose to move increasingly longer distances on the ground.

INTRODUCTION

Most published accounts of the locomotor anatomy of the Hadar hominids have suggested that they climbed (Senut, 1978, 1980; Senut and Le Floch, 1981; Tardieu, 1979, 1981, 1983; Tuttle, 1981; Jungers, 1982; Stern and Susman, 1983; Susman, 1983; Jungers and Stern, 1983; Marzke, 1983; Cook et al., 1983, Schmid, 1983) and/or that these earliest known hominids, while bipedal, engaged in a mode of bipedalism different from that practiced by modern humans (Tardieu, 1979, 1981, 1983; Jungers, 1982; Jungers and Stern, 1983; Stern and Susman, 1983). Others who have studied the Hadar material have indicated a "primitiveness" in the postcranial morphology but have been reluctant to attribute functional significance to specific anatomical differences between *A. afarensis* and later hominids (Tuttle, 1981; McHenry, 1983; Bush, 1980, Bush et al., 1982; Gomberg and Latimer, 1984). In our own earlier studies of the hand, shoulder, hip, knee, and foot (Stern and Susman, 1983) and studies of biomechanically important aspects of unique body proportions in *A. afarensis* (Jungers, 1982; Jungers and Stern, 1983) we offered the opinion that the Hadar hominids engaged in a significant amount of climbing but that when they were on the ground they moved bipedally—albeit in a manner unique to this morphologically primitive grade.

In this paper we will summarize certain aspects of our earlier discussions and elaborate upon a number of "non-anatomical" considerations of the habitus of *A. afarensis; viz.*, that diminutive (30+ kg), small-canined, non-tool-using hominids are not likely to have been fully terrestrial.[1] To suggest

[1]Because of space limitations in this volume, a more detailed consideration of these and other points will be found in Susman et al., 1984.

otherwise contradicts the lessons of behavioral ecology of all the living higher primates. Neither can paleoecological evidence be used to falsify the hypothesis that the Hadar hominids made extensive use of the trees.

MORPHOLOGICAL ANALYSES

Limb Proportions

At a value of approximately 85, the humero-femoral index of Lucy (and presumably her male counterparts, because this index is not sexually dimorphic in hominoids) is intermediate between the values that characterize humans of small stature (pygmies, $\bar{X}$ = 73.7) and small-bodied African pongids (bonobos, $\bar{X}$ = 97.8). It has been demonstrated that this unique ratio of fore- to hindlimb bony elements is due to a relatively short hindlimb in A.L. 288-1 comparable to that seen in apes of similar body size (Jungers, 1982; Jungers and Stern, 1983) (Fig.1). Although such proportions would not necessarily endow Lucy with a capability for climbing equivalent to modern chimpanzees, such an inter-limb ratio would clearly facilitate climbing in comparison to modern humans. Considerable competence in climbing can be preserved in a relatively large-bodied primate like *A. afarensis* even if relative forelimb length has been reduced to modern proportions, provided that hindlimb length remains relatively even shorter (Cartmill, 1974; Jungers, 1978, 1984, 1985).

The relatively short hindlimb of the Hadar hominids also has important implications for the nature of bipedality practiced 3–4 million years ago. Absolute and relative stride length would necessarily have been shorter in Lucy than in comparably-sized humans (e.g., pygmies), unless of course her bipedality was completely unlike that observed in living *Homo sapiens* (Jungers, 1982; Jungers and Stern, 1983).This implies that energetic costs per distance moved were probably greater in Lucy and maximum velocity of bipedal gait was probably lower (again, compared to humans of comparable body size). These inferences are reinforced by the consideration of Lucy's long feet (especially the forefoot) relative to hindlimb length (Jungers and Stern, 1983; Susman et al., 1984).

Foot

Stern and Susman (1983) compared A.L.333-115 pedal phalanx length to A.L.333-3 femoral head diameter in order to assess relative length of the toes in the Hadar species. The results indicate that the fossils occupy an intermediate position compared to bonobos and humans (Fig. 2). The second proximal phalanx of the A.L.333-115 foot is shorter than either the third or fourth. To our knowledge this condition does not occur in humans (Martin and Saller, 1959) but typifies apes. A consideration of the one pedal proximal phalanx (A.L.288-1y) of Lucy (as being either from the second or fourth ray) shows that the length of A.L. 288-1y relative to femoral head diameter is nearly the same as in the A.L.333 specimens (Fig. 2). It is also possible to compare the A.L.288-1y specimen directly to femoral length. Such a comparison confirms that, regardless of the ray to which this phalanx is assigned, its relative length is far outside the normal range of variation seen in modern humans, including pygmies. Finally, White and Suwa (in press) have reconstructed Lucy's foot with the result that the relative length of her toes (expressed as a proportion of the length of the foot from the calaneal tuberosity to the tip of the third metatarsal) is virtually halfway between that of gorillas and that of modern humans and about 45–50% longer than the toes of a scaled-down human. Our own analyis of foot length in Lucy (Susman et al., 1984), based in part on correlations between talus length and foot length in hominoids as documented by White and Suwa (in press), indicates that skeletal foot length (and especially forefoot length) relative to hindlimb length is very long indeed, and outside the normal range of variation in modern humans. This relationship alone implies kinematic and biomechanical differences in the bipedal gait of Lucy in comparison to that typical of adult modern humans (Jungers and Stern, 1983).

Although the toes of *A. afarensis* were long compared to modern humans, still they were short compared to living apes. The question arises as to whether or not the Hadar hominids could use them effectively in climbing. One answer may come from a comparison of their length to the fingers of young human children. Comparing the estimates of Lucy's toe length presented by White and Suwa (in press) with radiologic measurements of young children (Garn et al., 1972) demonstrates that the small individuals of the fossil species had toes about as long as the fingers of 2-year-old humans. The toes of the

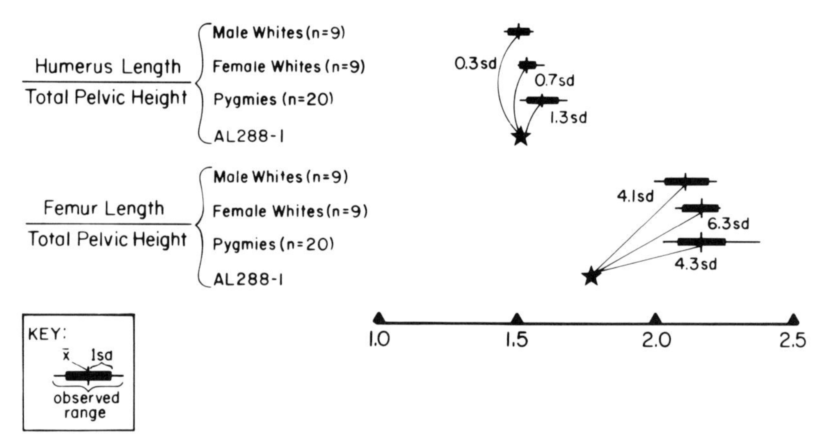

Fig. 1. Relative limb lengths of modern humans (Americans of European descent and pygmies) and "Lucy." Total pelvic height has been used as the size variable here because it was demonstrated to be significantly correlated and virtually isometric with body mass in a sample of modern humans (Jungers and Stern, 1983). Relative humerus length of Lucy is well within the normal range of variation exhibited by modern humans, but relative femur length is very low and far outside observed ranges. Lucy's distance from each mean of the modern humans is indicated in standard deviation units.

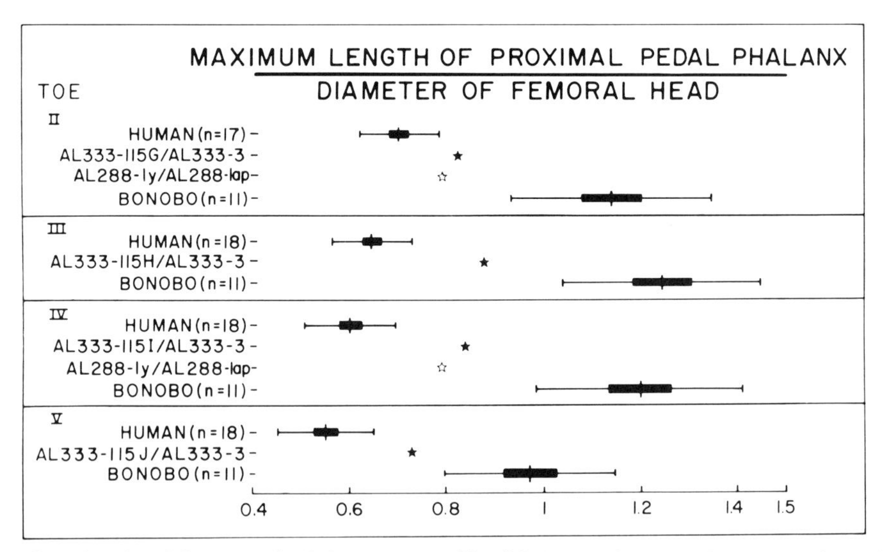

Fig. 2. Length of pedal proximal phalanges expressed as a proportion of femoral head diameter. The vertical line = the mean, the black bar = 95% confidence limits of the mean, and the horizontal line spans the 95% fiducial limits of the population in each species. The A.L.333-115 foot not only possesses long phalanges compared to modern humans, but it also possesses a second proximal phalanx that is shorter than the third—an ape-like trait.

larger individuals from Hadar can be estimated to be more than 40% longer than those of Lucy (White and Suwa, in press). Thus, if there were no anatomical restrictions to toe flexion in *A. afarensis*, at the very least the small individuals should have been able to grab with their toes as well as 2-year-old children grab with their fingers. The large Hadar individuals probably could use their toes for simple grasping as effectively as considerably older human children use their fingers. If the pedal flexor musculature of the fossil species were highly developed, the strength of toe grip may have well exceeded the strength of hand grip in young humans.

Even though the Hadar toe bones are relatively long, if they were also straight this would have to be considered an indication that the Hadar hominids did not use their toes in climbing. Thus, we performed an analysis that sought to quantify pedal phalangeal curvature (Stern and Susman, 1983). Using the reciprocal of the radius of curvature as our measure, we found that the fossil pedal proximal phalanges were more curved than in the average bonobo. But we neglected to consider a potential relationship between length and curvature. When this is done, it is seen that, for arboreal species, as phalangeal length increases, radius of curvature actually decreases; however, the included angle is length-independent and is the most informative way of comparing species (for a full discussion see Susman et al., 1984). Figure 3 is a graph of included angle for pedal proximal phalanges from several species of nonhuman primates, humans, and the Hadar fossils. This graph shows that chimpanzees and bonobos have pedal proximal phalanges of nearly identical curvature, and they have the most curved toe bones of any ape plotted in Figure 3. (However, to place matters in perspective, the single specimen of orangutan that we measured was characterized by an average included angle for the lateral four pedal proximal phalanges of 84.3°.) Figure 3 also shows that gorilla proximal toe bones are significantly less curved than those of *Pan*, and that human pedal phalanges are nearly straight. The 95% fiducial limits for humans do not overlap those for the African apes. With the exception of the proximal phalanx of the fifth toe, the Hadar fossils lie far outside the human 95% fiducial limits and closest to the means for chimpanzees or bonobos. Stern and Susman (1983) noted that the proximal phalanx of the fifth toe from Hadar is comparatively straight, given the curvature of the other toe bones. Figure 3 shows that this particular proximal phalanx lies at the extreme end of the human 95% fiducial limits. Therefore, it does suggest the beginnings of evolution toward the human condition. Figure 3 also includes data on the proximal pedal phalanx (A.L.288-1y) from Lucy, which Stern and Susman (1983) neglected to analyze. Quantitative analysis demonstrates that it is as curved as the A.L.333-115 specimens.

Because baboons have proximal phalanges of approximately the same absolute length as those from the A.L.333-115 foot, we thought a comparison of these highly terrestrial primates to the fossils might be revealing. Indeed, Figure 3 shows that the third pedal proximal phalanges of baboons are relatively straight and that the corresponding fossil (A.L.333-115H) is far removed from the baboon distribution. This is not due to a pecularity of Old World monkeys, since two proximal pedal phalanges from third rays of the highly arboreal genus *Nasalis* (Hornaday, 1929; Davis, 1962; Kern, 1964; Kawabe and Mano, 1972) were characterized by included angles of 32.8° and 35.8°, well within the distributions of African apes.

It has also been suggested (Latimer, personal communication) that the relatively straight toe bones of gibbons indicate that curvature is not correlated highly with arboreality. The fact is that the toe bones of *Hylobates* are indeed curved (as we expect in a highly arboreal animal), and the fact that they are the least curved of all the apes might suggest some influence of body weight on degree of curvature. On the other hand, woolly monkeys, which are comparable in size to gibbons, have toe bones considerably more curved that those of *Hylobates* (Fig. 3). It may be that during climbing gibbons use their feet in a manner different from other climbing primates, a possiblity alluded to by Stern and Susman (1981). Regardless, the inclusion of data on gibbons confirms, not dispels, the link between curvature and arboreality indicated by all other data.

Not only are the Hadar pedal phalanges both exceptionally long and curved, but there are other, non-metric criteria to support our contention that the foot of *A. afarensis* was unlike that of later hominids in function. One such morphological feature of *afarensis* that sets it apart from humans is the lack of dorsoplantar expansion at the base of proximal phalanges II–V (A.L.333-115G-J). Rather, the A.L.333-115 phalanges reveal a basal morphology similar to that of the apes (Fig. 4). The functional significance of this similarity is appreciated by reference to the work of Preuschoft (1970), who has suggested that the unique morphology of the plantar aspect of the base of human pedal proximal phalanges in related to the pres-

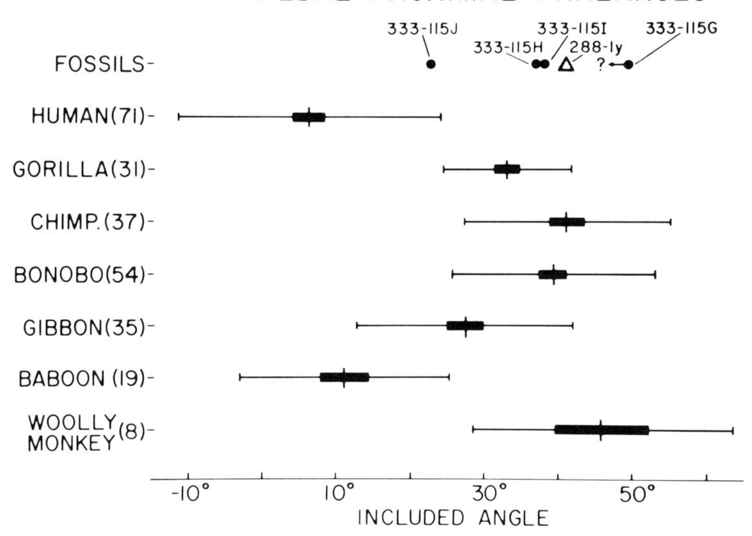

Fig. 3. Included angles of pedal proximal phalanges. The same descriptive statistics are indicated here as in Figure 2 for the included angles in each species. The Hadar phalanges are most closely allied to chimpanzees and bonobos (pygmy chimpanzees) with respect to phalangeal curvature. The arrow for A.L.333-115G represents the inexact value for this specimen due to breakage and distortion. The baboon is a mixed group of *Papio* spp., in Ray III only.

ence of a well-developed plantar aponeurosis in humans. Preuschoft noted that during the terminal phase of toe-off the "windlass" of the human foot (Hicks, 1953) is driven, tightening the plantar aponeurosis and increasing the dorsoplantar bending moments at the base of the proximal phalanges. Susman (1983) and others (e.g., Loth, 1908; Sokoloff, 1971) have noted that great apes lack a well-developed plantar aponeurosis and lack a toe-off mechanism such as occurs in humans. The implication is that the Hadar species had an incompletely developed toe-off mechanism.

A point to be emphasized, however, is that the bases of the A.L.333–115 pedal proximal phalanges also display a human-like trait wherein they are excavated on their proximo-dorsal margins (see best in lateral view, Fig. 4). Correlated with this trait is the presence of extended dorsal articular surfaces on the metatarsal heads. There is general agreement that these aspects of the structure of the fossil metacarpophalangeal joints are to be interpreted as indicating an enhanced range of dorsiflexion compared to that of apes. These traits are indicative of bipedality, since the acme of their expression is seen in human (versus pongid) feet and can be related to the expected dorsiflexion of the toes at the termination of stance phase during bipedal progression.

One very important point regarding the determination of the range of dorsi- and plantarflexion at the metatarsophalangeal joints of A.L.333-115 has been overlooked in previous considerations of the pedal anatomy of *A. afarensis.* In animals in which the closed-packed position of a metatarso- or metacarpophalangeal joint is in *flexion,* the plantar aspect of the metatarsal or metacarpal head is wide (medio-lateral dimension), and there is an obvious narrowing of the head near its dorsal margin. This configuration characterizes fingers and toes of all the apes and humans with two notable exceptions: 1) the principal *weight bearing* fingers of the African "knuckle-walkers," in whom the metacarpal heads are widest on their dorsal aspect (Susman, 1979), and 2) the heads of metatarsals I–IV in humans, which are also widened dorsally. The African apes load their fingers (principally rays III and IV) with the metacarpophalangeal joints in dorsiflexion—this is the closed-packed position—thus the articular surfaces are expanded medio-laterally on their dorsal aspects. Human metatarsal heads I–IV are also

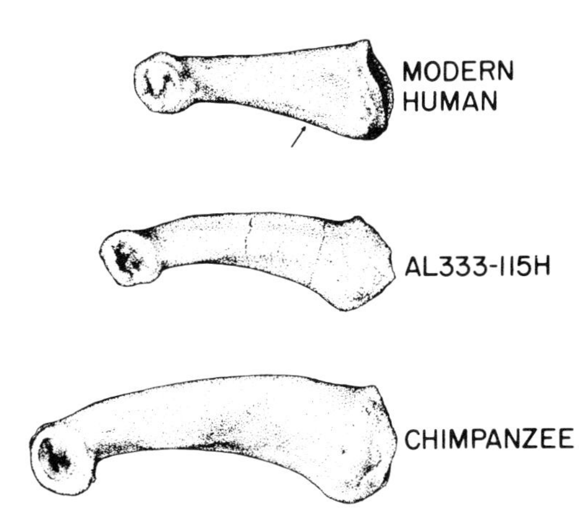

Fig. 4. Proximal phalanges (ray III) in lateral view. Similarities between the fossil (A.L.333-115H) and chimpanzee are particularly evident in 1) degree of curvature and 2) the parallel plantar and dorsal surfaces. The dorso-plantarly expanded base of the pedal proximal phalanges (II–IV) in humans (arrow) may be related to their way of toeing-off during bipedal walking. This type of toe-off is not observed in apes, nor was it present in the Hadar hominids (see text).

expanded dorsally (especially compared to those of apes, Fig. 5), indicating the enhancement of dorsiflexion at toe-off and a reduced emphasis on toe flexion in human locomotion. The A.L.333-115 metatarsals are *not* medio-laterally expanded on their dorsal aspects, as are those of humans; rather, they resemble pongids with wide plantar expansions of the metatarsal heads (Fig. 5). This indicates an emphasis on flexion of the toes exceeding that in modern humans.

While *A. afarensis* walked bipedally on the ground, there is little question that the stability of its first through fourth (and perhaps fifth) metatarsophalangeal joints at toe-off was less than in later hominids and modern humans. The toes of *afarensis* were undoubtedly more mobile (especially medio-laterally) than those of modern humans, and this ancient hominid had not entirely sacrificed toe flexion, a capability which would have been maintained for arboreal activities.

EVOLUTIONARY, PRIMATOLOGICAL, AND ECOLOGICAL PERSPECTIVES

Many people find it intuitively difficult to accept the existence of an animal that spent a significant amount of time in the trees without being as adept as an ape and that also lived on the ground without being as quick and agile on two legs as are humans. We find it equally difficult to conceive of such a stage having been averted by the human lineage. Rose (1984) has dealt with this issue, and presented a convincing argument that "If bipedalism is seen within the context of changing positional repertoires then far from having been abrupt and complete, its evolution can be seen as having taken place over a considerable time span and to have been closely linked with changing patterns of food acquisition behavior." The crucial question may reduce to whether or not this time span was considerable enough to be picked up in the fossil record.

We wonder why there is such reluctance to accept the idea that a behaviorally and structurally intermediate human ancestor could exist for a geologically significant period of time. Would the same resistance be offered to the suggestion that the fossil record might reveal a stage during which the ancestors of whales were neither as mobile on the ground as they had been, nor as adept in the water as they would become (Gingerich et al., 1983)? Furthermore, there exist such intermediately adapted animals today. The chimpanzee spends a great portion of its time in the trees, but is not as highly adapted for movement in this milieu as is the orangutan. In the wild, chimpanzees also spend a substantial portion of their time on the ground, as quadrupeds, but their manner of quadrupedalism is more costly than that of almost any other ground dwelling quadrupedal mammal (Taylor et al., 1982). Compromise to increase the range of habitat use seems to be the best explanation for this apparently "stable" adaptation in chimpanzees.

It probably was not necessary for the earliest biped to have been as adept on two legs as are humans. There is compelling evidence that hunting of savanna game played no part in the subsistence pattern of the earliest hominids (Lovejoy, 1981; Isaac, 1982, in press; Potts, 1984; Walker, cited by Lewin, 1983). They may have walked on two legs for the purpose of passing between trees or clumps of trees (Rodman and McHenry, 1980), or for brief ventures into the grassland to obtain food. Such bouts of bipedalism may not have been of longer duration than those of which some extant nonhuman primates are capable. But even if they were, the Hadar hominids had undergone changes

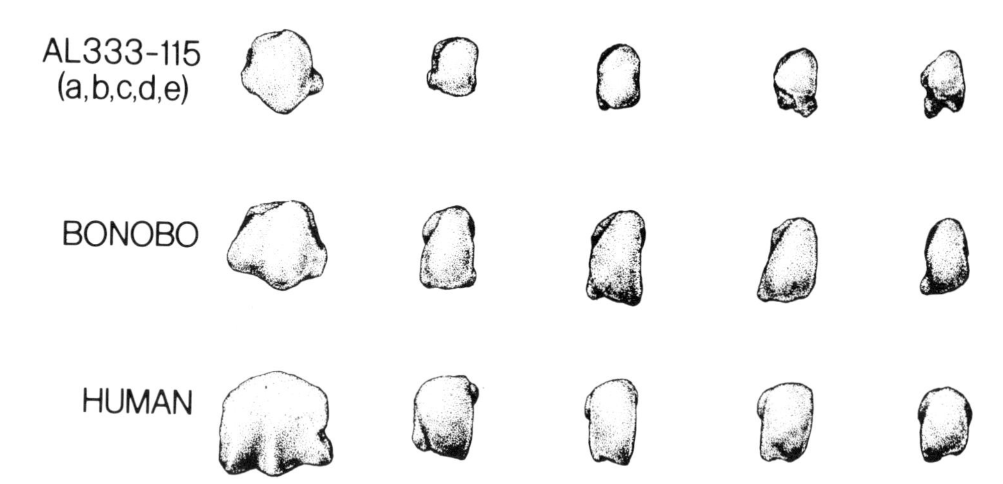

Fig. 5. Distal view of Metatarsal Heads I–V (from left to right). Note the narrowing of the dorsal aspects of the metatarsal heads in the ape *and in A.L.333–115.* The metatarsal heads in humans, on the other hand, remain medio-laterally broad on their dorsal aspects, a feature that accommodates the close-packing of the metatarsophalangeal joints (especially in rays I–IV) during toe-off.

to enable the trunk to be more readily balanced on the femoral heads, and such changes alone would have been sufficient to increase the bidpedal range of the animal well beyond that of extant nonhuman primates. There is no need to expect adaptations for especially low cost and high speed of bipedal progression in the earliest hominids.

The final point we wish to make regards non-morphological evidence for our suggestion that *Australopithecus afarensis* was a more skillful climber while at the same time a less adept biped than later hominids. This evidence we regard as ancillary, but it lends "outside" support to the anatomic evidence and allows the possibility for falsifying our hypothesis with extra-morphological data.

The first issue is that of the ecological status of *A. afarensis; viz.,* how might 30-kg hominids, such as "Lucy" and her sisters, have survived in a completely terrestrial niche—one that occasioned "full-blown bipedality." We feel, based on the extensive literature on free-ranging primates, that creatures such as represented by A.L.288-1 could not have survived full-time on the ground. Today, all primates from common chimpanzees (which range from 27 kg up to 70 kg) to vervet monkeys and baboons (which range from less than 3 kg to over 40 kg) are obliged at least to sleep in trees (or on rocky cliff-faces). They all feed in trees, and it has been noted by fieldworkers ever since the pioneering studies of Washburn and DeVore (1961) that sleeping trees and food trees are principal factors limiting group ranges and ranging patterns. It should be pointed out that *A. afarensis* possessed neither the formidable canine teeth of baboons nor the body size and/or canines of chimpanzees. How could this ancient hominid have fared better on the ground than these extant primates, which must rely on the trees? The only way, it seems to us, is for *afarensis* to have made and used some sort of tools, or perhaps to have lived in highly organized social groups such as those of modern-day terrestrial primates. But the evidence of tool use is lacking from the time ranges in question at Hadar, and the model for the socioecology of *A. afarensis* presented by those most familiar with the paleontological and paleoecological evidence (Lovejoy, 1981) claims that *afarensis* most probably lived as monogamous pair-bonded adults with their offspring. We are aware of the many counter arguments against this scenario, but within such a framework, it is inconceivable to us that, if the larger male *A. afarensis* were off foraging during the day, leaving the diminutive females with their offspring to fend for themselves, that the latter would have survived without recourse to the trees. Indeed, no living primate does so. Even comparatively large female gorillas (85 kg) are highly dependent on the trees for sleeping. Thus, it seems that the only possibilities of survival, especially if the Hadar hominids were monogamous, would have rested in their being at least part-time arborealists.

Finally, we would like to consider briefly the issue of the habitats at Hadar and Laetoli. Of course it would weaken any suggestion of arboreality for *afarensis* if it could be shown that these sites, at the times in question, were devoid of trees. It has been stated by a number of experts that at Hadar there was a mosaic of forest, woodland, and more open habitats. The forests in the vicinity of the Hadar localities, based on microfauna and pollen analysis, were likely similar to forests of the Ethiopian highlands today and the Kakamega Forest of Kenya (F.C. Howell, personal communication). There were arboreal colobine monkeys at Hadar and Laetoli (Szalay and Delson, 1979) and additional elements of the fauna that indicate forested conditions. Laetoli was, by numerous accounts, drier than Hadar (Harris, 1985), but even the driest reaches (including deserts) inhabited by primates today have at least some trees (e.g., Kummer, 1968; McGrew et al., 1981). These may be along rivers or perennial water courses or concentrated near waterholes or ephemeral lakes. It may be that these areas, however small, were where the small-bodied hominids lived. It may have been these areas, shrinking through time in relative proportion to the overall habitat, that the hominids sought out for food and as a haven from predators. It must be kept in mind that no matter what the sedimentology, palynology, macrobotanical, or faunal remains tell us about the paleoenvironment, they cannot place the hominids in any *one* component of a mosaic habitat (Oxnard, 1983). Any number of habitats were available to the hominids at Hadar and Laetoli, and those hominids most probably ranged widely in both areas over a variety of habitats.

Bipedalism may have been a response to the need to travel to and among the ever shrinking forested areas. As time went on, increasing body size, tool use, and social cohesiveness would have eventually freed hominids from reliance on trees. With *Homo erectus* we encounter hominids that are large, robust, and with sufficiently sophisticated culture (including refined stone tools, fire, etc.) to allow for "full-time" terrestriality. We do not, however, feel that at the *afarensis* (or even *Homo habilis*) grade hominids were of sufficient size or cultural capacity for full-time terrestrial life.

ACKNOWLEDGMENTS

We are grateful to the National Science Foundation for support of our research on primate locomotion (grants BNS 81-19664, BNS 82-17635, BNS 83-11206, and BNS 83-18013). We thank the following persons for generously allowing us access to fossils and skeletal material in their care: D.C. Johanson, W.H. Kimbel, M.D. Leakey, R.E.F. Leakey, G. Musser, B. Senut, C. Stringer, and D.F.E. Thys Van den Audenaerde. We also thank Susan Larson, John Fleagle, Norman Creel, David Krause, and Fred Grine for many lunch-times of animated discussion.

LITERATURE CITED

Bush, ME (1980) The thumb of *Australopithecus afarensis*. Am. J. Phys. Anthropol *52*:210.

Bush, ME, Lovejoy, CO, Johanson, DC, and Coppens, Y (1982) Hominid carpal, metacarpal and phalangeal bones recovered from the Hadar Formation: 1974–1977 collections. Am. J. Phys. Anthropol. *57*:651–667.

Cartmill, M (1974) Pads and claws in arboreal locomotion. In FA Jenkins, Jr. (ed): Primate Locomotion. New York: Academic, pp. 45–83.

Cook, DC, Buikstra, JE, DeRousseau, CJ, and Johanson, DC (1983) Vertebral pathology in the Afar australopithecines. Am. J. Phys. Anthropol. *60*:83–102.

Davis, DD (1962) Mammals of the lowland rain forest of north Borneo. Bull. Natl. Mus. Singapore *31*:1–129.

Garn, SM, Hertzog, KP, Poznanski, AK, and Nagy, JM (1972) Metacarpophalangeal length in the evaluation of skeletal malformation. Radiology *105*:375–381.

Gingerich, PD, Wells, NA, Russell, DE, and Ibrahim Shah, SM (1983) Origin of whales in epicontinental remnant seas: New evidence from the Early Eocene of Pakistan. Science *220*:403–406.

Gomberg, DN, and Latimer, B (1984) Observations on the transverse tarsal joint of *A. afarensis*, and some comments on the interpretation of behavior from morphology. Am. J. Phys. Anthropol. *63*:164.

Harris, JM (1985) Age and paleoecology of the upper Laetolil beds, Laetoli, Tanzania. In E Delson (ed): Ancestors: The Hard Evidence. New York: Alan R. Liss, Inc., pp. 76–81.

Hicks, JH (1953) The mechanics of the foot. II. The plantar aponeurosis and the arch. J. Anat., Lond. *88*:25–30.

Hornaday, WT (1929) Two Years in the Jungle, 10th ed. New York: Scribner.

Isaac, GL (1982) Aspects of human evolution. In DS Bendall (ed): Essays on Evolution: A Darwin Centenary Volume. Cambridge: Cambridge University Press.

Isaac, GL (in press) The archeology of human origins: Studies of the Lower Pleistocene in East Africa, 1971–1981. In F Wendorf (ed): Advances in World Archeology, Vol. 3. New York: Academic.

Jungers, WL (1978) The functional significance of skeletal allometry in *Megaladapis* in comparison to living prosimians. Am. J. Phys. Anthropol. *19:*303–314.

Jungers, WL (1982) Lucy's limbs: Skeletal allometry and locomotion in *Australopithecus afarensis*. Nature *297:*676–678.

Jungers, WL (1984) Scaling of the hominoid locomotor skeleton with special reference to the lesser apes. In DJ Chivers, H Preuschoft, W Brockelman, and N Creel (eds): The Lesser Apes: Evolutionary and Behavioural Biology. Edinburgh: Edinburgh University Press, pp. 146–169.

Jungers, WL (1985) Body size and scaling of limb proportions in primates. In WL Jungers (ed): Size and Scaling in Primate Biology. New York: Plenum, pp. 345–381.

Kawabe, M, and Mano, T (1972) Ecology and behavior of the wild proboscis monkey, *Nasalis larvatus* (Wurmb), in Sabah, Malaysia. Primates *13:*213–228.

Kern, JA (1964) Observations on the habits of the proboscis monkey, *Nasalis larvatus* (Wurmb), made in the Brunei Bay area, Borneo. Zoologica, N.Y. *49:*183–192.

Kummer, H (1968) Social Organization of Hamadryas Baboons. A Field Study. Chicago: University of Chicago Press.

Lewin, R (1983) Studying humans as animals. Science *220:*1141.

Loth, E (1908) Die Aponeurosis plantaris in der Primatenreich. Gegenb. Morph. Jahrb. *38:*194–322.

Lovejoy, CO (1981) The origin of man. Science *211:*341–350.

Marzke, MW (1983) Joint function and grips of the *Australopithecus afarensis* hand, with special reference to the region of the capitate. J. Hum. Evol. *12:*197–211.

McGrew, WC, Baldwin, PJ, and Tutin, GEG (1981) Chimpanzees in a hot, dry and open habitat: Mt. Asserik, Senegal, West Africa. J. Hum. Evol. *10:*227–244.

McHenry, HM (1983) The capitate of *Australopithecus afarensis* and *A. africanus*. Am. J. Phys. Anthropol. *62:*187–198.

Oxnard, CE (1983) The Order of Man. New Haven: Yale University Press.

Potts, R (1984) Hominid Hunters? Problems of Identifying the Earliest Hunter/Gatherers. In R Foley (ed): Human Evolution and Community Ecology. Prehistoric Human Adaptation in Biological Perspective. London: Academic, pp. 129–166.

Preuschoft, H (1970) Functional anatomy of the lower extremity. The Chimpanzee, Vol. 3. Basel: Karger, pp. 221–294.

Rodman, PS, and McHenry, HM (1980) Bioenergetics and the origin of hominid bipedalism. Am. J. Phys. Anthropol. *52:*103–106.

Rose, MD (1984) Food acquisition and the evolution of positional behavior: The case of bipedalism. In DJ Chivers, BA Wood, and A Bilsborough (eds): Food Acquisition and Processing in Primates. London: Plenum.

Schmid, P (1983) Eine Rekonstruktion des Skelettes von A. L. 288–1 (Hadar) und deren Konsequenzen. Folia Primatol. *40:*283–306.

Senut, B (1978) Contribution a l'étude de l'humerus et de ses articulations chez les Hominides du Plio-Pleistocene. Thèse 3ème cycle, University of Paris VI.

Senut, B (1980) New data on the humerus and its joints in Plio-Pleistocene hominids. Colleg. Antropol. *4:*87–93.

Senut, B, and LeFloch, P (1981) Divergence des piliers de la palette humerale chez les Primates hominoides. C.R. Acad. Sci. Paris, Ser. II *292:*757–760.

Sokoloff, S (1971) The muscular anatomy of the chimpanzee foot. Gegenb. Morph. Jahrb. *119:*86–125.

Stern, JT, Jr, and Susman, RL (1983) The locomotor anatomy of *Australopithecus afarensis*. Am. J. Phys. Anthropol. *60:*279–317.

Susman, RL (1979) Comparative and functional morphology of hominoid fingers. Am. J. Phys. Anthropol. *50:*215–236.

Susman, RL (1983) Evolution of the human foot: Evidence from Plio-Pleistocene hominids. Foot and Ankle *3:*365–376.

Susman, RL, Stern, JT, Jr, and Jungers, WL (1984) Arboreality and bipedality in the Hadar hominids. Folia Primatol, *43:*113–156.

Szalay, F, and Delson, E (1979) Evolutionary History of the Primates. New York: Academic.

Tardieu, C (1979) Analyse morpho-functionelle de l'articulation du genou chez les primates. Application aux hominides fossiles. Thèse, Université Pierre et Marie Curie, Science Naturelles, Paris.

Tardieu, C (1981) Morpho-functional analysis of the articular surfaces of the knee-joint in primates. In AB Chiarelli and RS Corrucini (eds): Primate Evolutionary Biology. Berlin: Springer-Verlag, pp. 68–80.

Tardieu, C (1983) L'Articulation du Genou. Analyse morphofonctionnelle chez les Primates. Application aux Hominides fossiles. Paris: CNRS.

Taylor, CR, Heglund, NC, and Maloiy, CMO (1982) Energetics and mechanics of terrestrial locomotion. I. Metabolic energy consumption as a function of speed and body size in birds and mammals. J. Exp. Biol. *97:*1–21.

Tuttle, RH (1981) Evolution of hominid bipedalism and prehensile capabilities. Phil. Trans. R. Soc. Lond., B *292:*89–94.

Washburn, SL, and DeVore, I (1961) The social life of baboons. Sci. Am. *204:*62–71.

White, TD, and Suwa, G (in press) Hominid footprints at Laetoli: Facts and interpretations. In DC Johanson (ed): The Origins of Human Locomotion. San Diego, Academic.

Ancestors: The Hard Evidence, pages 193–201

Functional Aspects of Plio-Pleistocene Hominid Limb Bones: Implications for Taxonomy and Phylogeny

Brigitte Senut and Christine Tardieu

Laboratoire d'Anthropologie du Muséum National d'Histoire Naturelle et Laboratoire Associé 49 au C.N.R.S., Musée de l'Homme, 75116 Paris (B.S.) and Equipe de Recherche au C.N.R.S. 06-0246 "Locomotion Animale" associée au Muséum National d'Histoire Naturelle, Laboratoire d'Anatomie Comparée, 75005 Paris (C.T.), France

ABSTRACT This paper is concerned with the study of two intermediate limb joints (elbow and knee) in Plio-Pleistocene hominids. From comparative osteology and dissection we assess their morpho-functional, taxonomic, and phylogenetic implications. Numerous catarrhine and platyrrhine monkeys have been studied. Laxity of the elbow joint would be combined with stability of the knee joint in early *Homo*, whereas stability of the elbow joint would be combined with laxity of the knee joint in early *Australopithecus*. That would mean that we could assess taxonomy from limb bones; in that case, we might have two different taxa in "*Australopithecus afarensis*" at Hadar (Ethiopia). Moreover, early hominids (e.g., *Australopithecus*) were probably not exclusively terrestrial bipeds but retained a propensity for climbing or suspension. In the australopithecines that we isolate in Hadar (presenting clear derived human postcranial features), the chimp-like elbow joint morphology (probably convergently derived with *Hylobates, Pan,* and *Gorilla*, and not primitive) does not permit us to consider this hominid as a direct human ancestor. But its knee joint features (commonly inherited with non-human hominoids) could represent a good ancestral pattern to later hominids.

RÉSUMÉ Cet article concerne l'étude des articulations intermédiaires du coude et du genou chez les Hominidés plio-pléistocènes. A partir de l'ostéologie comparée, pratiquée sur un large échantillon de primates catarhiniens et platyrhiniens, et de plusieurs dissections, nous proposons une explication fonctionnelle des caractères sélectionnés, dégageons les implications taxonomiques de ces résultats, et, à la lumière des fossiles antérieurs au Plio-Pléistocène, présentons les apports phylogénétiques. Il ressort de ce travail que, chez les *Homo* anciens, la laxité de l'articulation du coude aurait été associée à la stabilité de l'articulation du genou, tandis que chez les australopithèques anciens, la stabilité de l'articulation du coude aurait été associée à la laxité de celle du genou. Ces derniers n'étaient probablement pas des bipèdes terrestres exclusifs, mais avaient sans doute conservé une capacité au grimper et à la suspension arboricoles. Nous suggérons, à titre d'hypothèse, la présence de deux genres différents,

---

Authors' names appear in alphabetical order; there is no junior author.

Homo et *Australopithecus* dans le matériel fossile récolté à Hadar (Ethiopie) et rassemblé par certains sous la seule espèce *Australopithecus afarensis* et dont le squelette postcrânien présente des caractères humains reconnus par tous. Chez les australopithèques que nous reconnaissons à Hadar, les caractères de l'articulation du coude décrits comme proches de ceux du chimpanzé (probablement dérivés de façon convergente chez *Hylobates, Pan,* et *Gorilla* et non pas primitifs) conduisent à éloigner ces hominidés de notre ancestralité directe. Par contre chez ces mêmes australopithèques éthiopiens, les caractères de l'articulation du genou, décrits comme proches du chimpanzé (probablement hérités de façon commune chez tous les hominoïdes non humains), pourraient représenter un bon schéma ancestral pour les hominidés ultérieurs.

INTRODUCTION

This paper deals with studies realized by Tardieu on the hind limb and Senut on the forelimb of hominids[1] from the Plio-Pleistocene. We will mainly focus on sites from East Africa: East Turkana and Kanapoi in Kenya; and the Omo Valley, Hadar, and Melka Kunturé in Ethiopia. That Plio-Pleistocene hominids were bipedal is clear from the vast literature on these forms.

Among the more imporant previous works bearing on this topic are those by: Broom and Robinson (1949); Broom et al. (1950); Prost (1967); Sigmon (1969); Genet-Varcin (1969); Heiple and Lovejoy (1971); Leakey et al. (1971); Lovejoy (1972); Robinson (1972); Lovejoy et al. (1973); Napier (1974); Ciochon and Corruccini (1975); Oxnard (1975 a,b); Day et al. (1976); Johanson and Coppens (1976); McHenry and Corruccini (1978); Day (1978); Wood (1978); Tardieu (1979, 1983); Berge (1980); Berge and Ponge (1983); Tuttle (1981); Vancata (1982); Lamy and Nossant (1983); Schmid (1983); Stern and Susman (1983); and Susman et al. (1985). Yet the numerous chimpanzee-like features pointed out by some of these authors and others (Senut, 1978, 1981b; Susman and Stern, 1979; Feldesman, 1982; McHenry, 1983; Marzke, 1983; Ferguson, 1983) would suggest a different type of bipedality in certain hominids, especially in *Australopithecus afarensis* (Johanson, White, and Coppens, 1978), which was probably not completely isolated from an arboreal environment. In this paper, we assemble our results on the intermediate joint of the upper and lower limbs and discuss the functional significance of the features and their implication for taxonomy and phylogeny.

ELBOW JOINT (B.S.)

All of the East African Plio-Pleistocene distal humeri have been studied: KNM-ER 739, KNM-ER 740, KNM-ER 1504, and KNM-ER 3735 from East Turkana (Kenya); KNM-KP 271 from Kanapoi (Kenya); Gomboré IB 7594 from Melka Kunturé (Ethiopia); A.L.288-1M, A.L.288-1S, A.L.137-48A, A.L. 322-1, and A.L. 333W-29 from Hadar (Ethiopia).

Morphology

In the humerus, as is usual in long bones, the most interesting areas are the extremities, in this case the distal humeral end, which permits one to recognize two distinct patterns among the fossil hominids.

In the first group, the lateral anterior trochlear crest is poorly developed and does not separate the articular area into two subequal parts. The lateral border of this crest is neither long nor strongly angulated. The *epicondylus lateralis* projects weakly and is located almost or at the same level as the most superior point of the *capitulum humeri*, which is distally poorly developed and truncated medio-laterally. The medial and lateral borders of the shaft converge superiorly. Its anterior border is strongly salient anteriorly.

This group includes Gomboré IB 7594 and KNM-KP 271 (in which no diaphyseal feature could be observed). A.L.333W-29 could also fit this pattern; the specimen is very weathered, but the weak projection of the *epicondylus lateralis* is clear, as well as its low position relative to the *capitulum humeri*, the moderately developed lateral anterior crest, and the salience of the anterior border.

The second group is generally characterized by an anterio-posterior flattening of the shaft as well as of the *epicondylus lateralis* and the higher setting of the *epicondylus*

[1] In this paper, we restrict the term hominidae to *Australopithecus* and *Homo*.

lateralis relative to the *capitulum humeri*, which is strongly distally developed. This second group includes all other known specimens: KNM-ER 739, KNM-ER 740, KNM-ER 1504, KNM-ER 3735, A.L. 137-48A, A.L.288-1M, A.L.288-1S, and A.L.322-1. Within this second group, two subgroups can be recognized.

The first subgroup (KNM-ER 739, KNM-ER 1504, and KNM-ER 3735) exhibits an articular surface not very much different from the modern human one, with a lateral anterior trochlear crest poorly salient. The *epicondylus lateralis* is strongly laterally salient and positioned high relative to the *capitulum humeri*. The *fossa olecrani* is shallow, and the minimal width of the two humeral pillars is almost equal.

In the second subgroup, we note the presence of an anterior lateral trochlear crest that is strongly developed and separates the articular surface into two sub-equal parts (we may speak of two spools stuck together). The two borders of the keels are equal in length and in angulation with reference to the most salient point. The *epicondylus lateralis* is less salient than in the first subgroup, and the lateral margin of the shaft is rectilinear. In this subgroup are found all the specimens from Hadar, which in overall morphology are more gracile than the previous ones.

Comparisons

The fossils have been compared with representatives of all extant hominoids: more than 100 modern humans from all over the world, 40 *Pan troglodytes*, 12 *Pan paniscus*, 60 *Gorilla gorilla*, 40 *Pongo pygmaeus*, and 76 *Hylobates*. All the non-human primates were wild and adult. The first group of fossils exhibits features that cannot be distinguished from those of humans: lateral anterior trochlear crest poorly developed and *epicondylus lateralis* poorly salient and located almost at the same level as the *capitulum humeri*, which is distally weakly extended (Fig. 1).

The second group presents traits intermediate between man and chimpanzee: the flattening of the shaft and of the *epicondylus lateralis* is found in the non-human hominoids and the salience of the *epicondylus lateralis* is clearly more African ape-like (Fig. 1). The first subgroup exhibits an articular surface more like *Homo* and *Pongo* compared to the more *Pan*- and *Gorilla*-like second one, with a *trochlea humeri* that resembles two reels stuck together, as confirmed by virtual sectioning of the distal humerus (Senut, 1981a).

Function

The African ape-like features pointed out in the gracile forms could suggest a particular form of locomotion using the upper limbs, such as suspension or climbing, as also suggested by Stern and Susman (1983). For example, the lateral anterior trochlear crest present in "Lucy" and its relatives is strongly salient in chimpanzees and gibbons. In these primates, it prevents dislocation of the joint during climbing or suspension.

Among modern primates, only *Pan, Hylobates*, and *Gorilla* exhibit a clearly distally extended *capitulum humeri* that allows a longer articular traverse for the radius and perhaps a better extension of the forearm on the arm. The salience of the *epicondylus lateralis* reflects the strength of the extensor muscles of the hand and the digits, and its high position would give a longer lever arm for these muscles.

The elbow joint is not stabilized for a quadrupedal terrestrial form of walking, as it does not show a strong lateral crest to the *fossa olecrani*, as is the case in *Pan, Gorilla*, or *Papio*, for example. The early hominids, then, were probably neither quadrupeds nor, by extension, knuckle-walkers. The combination of several of these features found in some early hominids suggests a certain degree of tree-dwelling, with the most gracile form apparently better adapted to arboreality.

Taxonomy

The group exhibiting modern hominid features may represent early *Homo* (KNM-KP 271, Gomboré IB 7594, A.L.333W-29). The second group, presenting some *Pan*-like features, may be allocated to *Australopithecus*: one subgroup would be more robust (KNM-ER 739, KNM-ER 740, KNM-ER 1504, KNM-ER 3735) and the other more gracile (Hadar specimens). We do not know any distal humeri of *Australopithecus africanus*, but it is impossible, at the moment, to isolate *A. africanus* from *A. afarensis* on the basis of the proximal humerus. We have, thus, a challenge: either it is impossible to recognize specific levels from the limb bones, or the species *africanus* is synonymous with *afarensis* as proposed by Tobias (1980). However, we think that we could isolate two different groups at Hadar, and these differences would suggest the presence of two hominid genera in the Ethiopian Pliocene.

Could the two groups simply suggest sexual dimorphism? After examination of hundreds of modern non-human, male and female wild specimens, we do not think so. Among extant hominoids, we can easily (except for Hylobatidae) differentiate the males from the females on the basis of bone robusticity, but sexual dimorphism does not seem to affect the morphology of the epiphysis. In Plio-Pleistocene hominids, articular areas seem different enough to be allocated to distinct forms—genera or species. Moreover, from East Turkana there are several distal humeri of various sizes exhibiting exactly the same morphological pattern (KNM-ER 739, KNM-ER 1504, KNM-ER 3735). Alternatively, could the two groups simply represent two species of the same genus? The differences observed here are larger than between *Pan troglodytes* and *Pan paniscus* or between *Hylobates lar* and *H. concolor*, for example. Thus, we suggest that the two Hadar groups may belong to the different genera *Homo* (A.L.333W-29) and *Australopithecus* (A.L.137-48A., A.L.288-1M, A.L.288-1S, A.L.322-1).

Phylogeny

The African ape-like features observed in the *Australopithecus afarensis* forelimb are often considered as primitive. This would suggest that their ancestors possessed them. But *Pan* and *Gorilla* are probably not primitive animals in several respects. After examination of several Oligocene and Miocene primates, we do not think that these traits are ancestral; on the contrary, they are probably convergently derived ones expressed in African great apes and Hylobatidae, as well as some Plio-Pleistocene hominids (Senut, 1982).

For instance, if we look at the morphology of the articular area in modern *Homo*, we see a weakly salient anterior lateral trochlear crest without symmetry in the *condylus humeri*. This crest is strongly salient in non-human hominoids, dividing the joint almost in two equal parts, but it is weak in platyrrhine monkeys and almost nonexistent in some cercopithecoids. A weak development is generally the case in early strepsirhines (Szalay and Dagosto, 1980; Novacek, 1980) and in Oligocene primates from the Fayum in Egypt (Fleagle, 1983; Bown et al., 1982; Fleagle and Simons, 1978, 1982, 1983; Conroy, 1976). In most Miocene East African hominoids, such as *Proconsul africanus* (KNM-RU 2036) or *Dendropithecus macinnesi* (KNM-RU 2097) (Le Gros Clark and Leakey, 1951; Le Gros Clark and Thomas, 1951; Andrews and Simons, 1977; Napier and Davis, 1959; Morbeck, 1975, 1976; Rose, 1983; McHenry and Corruccini, 1975; Senut, 1982, in press; Walker and Pickford, 1983), the crest is better expressed than in the older fossils (as is generally the case in modern hominoids), but never exhibits the "double-trochlea" pattern seen in non-human hominoids and *Australopithecus afarensis*. If the latter pattern is primitive, we must assume (following Johanson and White, 1979, who consider Lucy primitive and ancestral to *Homo*) that it appeared in the Plio-Pleistocene and then disappeared. This seems incompatible with the parsimony rule and its important functional meaning.

KNEE JOINT (C.T.)

The fossils studied include: KNM-ER 1481 A and B, KNM-ER 1472, KNM-ER 1476 B, and KNM-ER 1500 A and B from East Turkana (Kenya); A.L. 129-1A and -1B, A.L.288-1AP and -1AQ, A.L.333-4, and A.L.333x-26 from Hadar (Ethiopia).

Morphology

Our osteological study of the knee joint in fossil and extant primates leads us to recognize two different groups among Plio-Pleistocene hominids. This distinction is based mainly on the following traits (Tardieu, 1982, 1983): 1) the proportions of the femoral distal epiphysis, which schematically resembles a square in humans and a rectangle in pongids (in the former, we see an antero-posterior lengthening of the articular surfaces and in the latter a medio-lateral broadening of the epiphysis); 2) The shape of the intercondylar notch, which is higher than wide in humans, following the "square" morphology, and wider than high in pongids, following the "rectangular" morphology; 3) the symmetry of the epiphysis about the parasagittal plane passing through the middle of the *trochlea femoris*: a slight asymmetry, due to the curvature of the anterior segment of the internal condyle (used in the terminal rotation mechanism) is present in humans; a marked asymmetry with preponderance of the internal side over the external one is seen in pongids (the internal condyle does not show any curvature in its anterior segment, but a strong and continuous divergence); and 4) the close fitting of the tibial spines in the intercondylar femoral notch in modern humans and their loose fit in pongids.

The human-like group includes the fossils: KNM-ER 1481 A and B, KNM-ER 1472, A.L.333-4, and A.L.333x-26.[2] The pongid-like group includes the fossils A.L.129-1a and b, A.L.288-1AP and -1AQ, and probably KNM-ER 1500 A and B.

Contribution of dissection

The last feature concerning the fitting of the spines in the intercondylar notch has been supplemented by dissection of the knee menisci in one human, one chimpanzee, and one gorilla. Figure 2 shows that the morphology of the external meniscus is significantly different: in the chimpanzee and gorilla, it is ring-shaped with one tibial insertion, anterior to the external spine; in humans, it is crescent-shaped with two tibial insertions—one anterior to the external spine and one posterior; the latter is unique among primates.

Consequently the dry tibia presents two areas of insertion for the external meniscus in modern man. The posterior border of the external plateau is long, discontinuous, and notched by the posterior insertion of the meniscus. There is only one area of insertion, anterior to the external spine, in the chimpanzee and the gorilla. The posterior border of the plateau is short, continuous, and abrupt. In KNM-ER 1481B, the morphology is clearly human, but in A.L.129-1b, A.L.288-1AQ, and A.L.333x-26, the morphology is clearly chimpanzee-like (Fig. 3).

[2]These two last controversial fossils are placed at the lower limit of the variation of the human group on the basis of previous features.

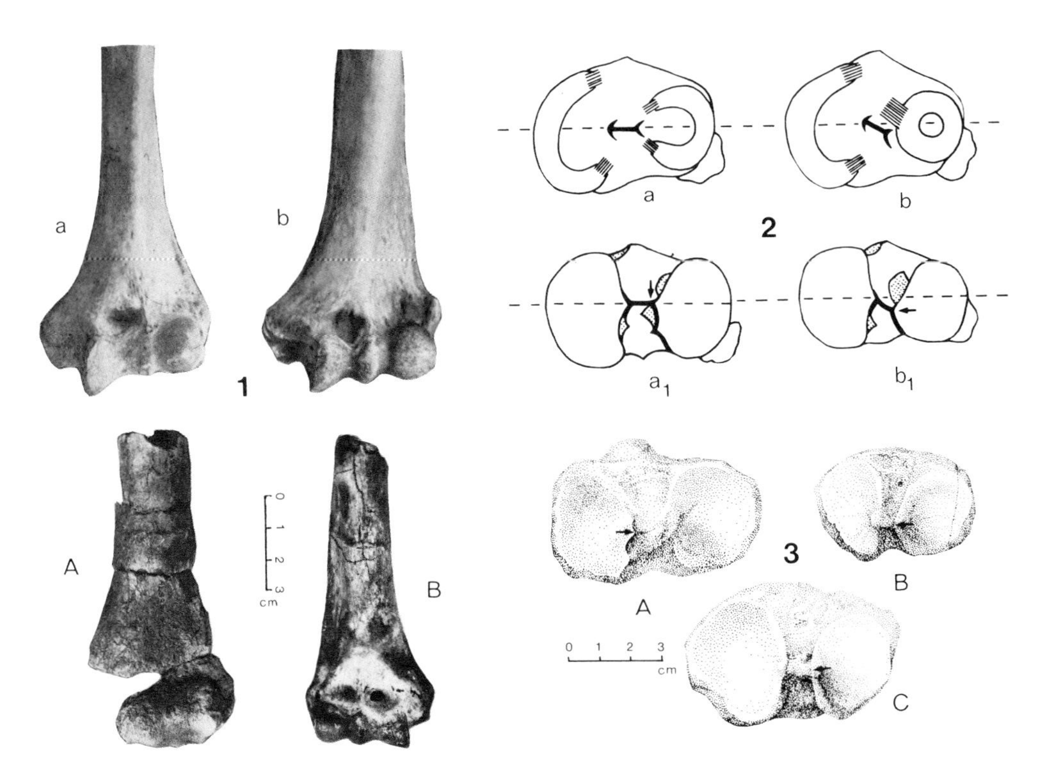

Fig. 1. Anterior view of the distal humerus in two extant hominoids and two Hadar fossils, probably representing the taxa *Homo* and *Australopithecus*. a) Modern human (left side); b) Chimpanzee (left); A) A.L.333W;29 (left), more *Homo*-like; B) A.L.288-1M (right), more *Pan*-like.

Fig. 2. Proximal view of right tibia and fibula in modern human (a and a_1) and African pongid (b and b_1) showing orientation of the two knee menisci. Top row: dissection view of shape and attachments (thin parallel lines) of menisci; heavy line in center of tibia represents intercondylar eminence terminating in medial and lateral spines. Bottom row: dry-bone view showing areas of meniscal attachment (stippling), outline of tibial plateau (heavy line—note the notch of the posterior border of the external plateau in a_1) and summit of external spine (tip of arrow).

Fig. 3. Proximal view of three Plio-Pleistocene hominid tibiae: A) KNM-ER 1481B (left side); B) A.L.129-1B (right); C) A.L. 333x-26 (right). The arrow points to the summit of the external spine.

Function

To complete this study, we took angular measurements of knee rotation in a human and a chimpanzee. After implantion of rods on the femur and tibia, degrees of internal and external rotation of the tibia about the femur at different degrees of flexion–extension were measured (Tardieu, in press). It appears that there is a strong stability of the knee in man, especially in extension, and a greater knee laxity in the chimp.

Thus results from osteology, dissection, and measurements are closely functionally linked. In humans the close fitting of the knee joint and the double insertion of the external meniscus limit the amplitude of rotation. In the chimp, its loose fit and the single external meniscal insertion permit greater amplitude of rotation. The mode of insertion of the meniscus determines its mobility, which thus determines the amplitude of rotation.

Considering now the Plio-Pleistocene hominids, these results suggest that two locomotor groups exhibiting distinct types of bipedality could have been present. In the first one, the stability of the knee joint is similar to modern man. In the second one, however, the femoro-tibial joint shows a significantly stronger rotation, closer to the chimpanzee. This consequent laxity of the knee indicates that the leg and the foot could be placed on the substrate in a much freer fashion than in man and that the mobility and prehensility of the foot were complemented at the knee.

Taxonomy

Thus, on the basis of the meniscal features, two groups equivalent to those separated by osteological features can be distinguished among Plio-Pleistocene hominids. One is fully modern and called *Homo*, including KNM-ER 1481 A and B, KNM-ER 1476 B, KNM-ER 1472, A.L.333-4, and A.L.333x-26.[3] The other, more chimpanzee-like one, is attributed to *Australopithecus*, including A.L.129-1a and -1b and A.L.288-1AP and -1AQ (*Australopithecus afarensis*, in our definition) and probably KNM-ER 1500 A and B.

Phylogeny

A detailed phylogenetic interpretation of the knee joint's osteological features was presented in a previous paper on Miocene and Plio-Pleistocene fossils (Tardieu, 1982), based also on Ford's (1980) study of Paleocene and Eocene primate femora. With the new contribution of meniscal features, it can be summarized as follows:

1) A square and symmetrical distal femoral epiphysis might be a primitive primate feature, whereas the rectangular and asymmetrical epiphysis would be derived and would have appeared several times as a result of convergence in some Strepsirhini, Platyrrhini and Catarrhini. However, the rectangular-asymmetrical morphology might have been inherited in common by hylobatids, pongids, and *Australopithecus afarensis* (as defined in this paper). At the base of the hominoid line, a slight medio-lateral broadening of the epiphysis would have appeared, resulting in a slightly rectangular but symmetrical epiphysis, illustrated by the Middle Miocene "*Paidopithex rhenanus*" (Germany). Then a slight asymmetry with preponderance of the internal side would have been associated with the rectangular shape, as present in the Middle Miocene *Pliopithecus vindobonensis* (Czechoslovakia), in extant hylobatids, pongids, and in *Australopithecus afarensis*. Medio-lateral broadening and asymmetry of the epiphysis are strongly emphasized in extant pongids. In modern and fossil *Homo*, a derived antero-posterior lengthening of the articular surfaces results in a square-shaped epiphysis, as a convergence with the primitive condition. The curvature of the anterior segment of the internal condyle would also be a derived human feature, present in the *Homo*-like fossil group.

2) It is not possible at the moment to determine the primitive and derived conditions of the fitting of the tibial spines in the intercondylar notch in Primates. The combined influence of weight and locomotor factors on this feature requires further investigation. We can only suggest that the close fit of the two articular elements is a derived human feature, present in the *Homo*-like group, whereas other hominoids display a looser fit, as in *Australopithecus afarensis* (of this paper).

3) We suggest that in hominid evolution the knee joint has evolved from having a single insertion of the external meniscus to a double one. The fossils A.L.129-1b, A.L.288-1AQ, and A.L.333x-26 are thus more primitive than KNM-ER 1481B and KNM-ER 1476 B for meniscal insertion. However, the single insertion of the external meniscus is not only a primitive hominid feature, but

[3]The taxonomic attribution of the composite knee joint (A.L.333-4 and A.L. 333x-26) that shows a mosaic pattern is explained below in the section on phylogeny.

a primitive primate feature (Tardieu, in press). Its common presence in A.L.333x-26, AL.L.129-1b, and A.L.288-1AQ thus does not constitute an acceptable argument for placing these fossils in the same species as published by Johanson et al. (1978): it is a shared primitive feature, not a shared derived one. We think that the controversial composite knee joint fossil, A.L.333-4 and A.L.333x-26, shows a mosaic pattern with derived shape features of the *Homo* line combined with a primitive meniscal insertion.

CONCLUSIONS

From these results, it appears that in early *Homo* the laxity of the elbow joint was combined with the solidity of the knee joint. In early *Australopithecus*, the opposite is seen: the laxity of the knee joint combined with the solidity of the elbow joint. That means that suspension or climbing could have been important for *Australopithecus* (and especially *A. afarensis*) and that its bipedalism was probably different from ours. Thus, these hominids were probably not exclusively terrestrial bipeds as previously proposed (Lovejoy et al., 1982; Johanson et al., 1982). This paper suggests, moreover, that just over 3 million years ago two types of hominid were present in Ethiopia. This was also suggested by Tuttle (1983), Olson (1981), Schmid (1983), and Coppens (1981, 1983), who proposed to name some of the chimp-like *Australopithecus afarensis* "Pre-*Australopithecus*."

Susman et al. (1985) have suggested that the two forms recognized at Hadar could represent different adaptations for male and female *Australopithecus*. The former could have been more terrestrial and the latter more arboreal. But they point out that it is difficult to be certain. For us, the two forms might represent two genera: one could be tentatively allocated to a modern type called *Homo* and the other to *Australopithecus*.

ACKNOWLEDGMENTS

For allowing access to the original fossils, we deeply thank Y. Coppens (Collège de France, Paris), D.C. Johanson (Cleveland Museum of Natural History), and R.E.F. Leakey (National Museums of Kenya). Extant primates were studied with the help of J. Anthony and F. Jouffroy (Laboratoire d'Anatomie Comparée du Muséum National d'Histoire Naturelle, Paris), J. Biegert and P. Schmid (Anthropologisches Institut der Universität Zürich), Y. Coppens (Collège de France, Paris), P. Napier (Department of Zoology, British Museum (Natural History), London), C. Smeenk (Department of Zoology, Rijksmuseum van Natuurlijke Historie, Leiden), and D. Thys van den Audenaerde (Department of Zoology, Musée Royal de l'Afrique Centrale, Tervuren). We gratefully acknowledge the skillful assistance of P. Mars (drawings) and A. Gordon and D. Ponsard (photographs). This work was supported by R.C.P. 292 and L.A. 49 funding from C.N.R.S. (through Professor Yves Coppens) and by the Laboratoire d'Anthropologie du Muséum National d'Histoire Naturelle.

LITERATURE CITED

Andrews, P, and Simons, EL (1977) A new African Miocene gibbon-like genus *Dendropithecus* (Hominoidea, Primates) with distinctive postcranial adaptations: Its significance to the origin of Hylobatidae. Folia Primatol. *28:*161–168.

Berge, C (1980) Biométrie du bassin des primates. Application aux primates fossiles de Madagascar et aux anciens Hominidés. Thèse Doctorat 3ème cycle (Anthropologie Biologique) de l'Université Paris VII.

Berge, C, and Ponge, J-F (1983) Les caractéristiques du bassin des Australopithèques (*A. robustus, A. africanus* et *A. afarensis*) sont-elles liées à une bipédie de type humain? Bull. Mém. Soc. Anthropol. Paris, *10:*335–354.

Bown, TM, Kraus, MJ, Wing, SL, Fleagle, JG, Tiffney, BH, Simons, EL, and Vondra, CF (1982) The Fayum primate forest revisited. J. Hum Evol. *11:*603–632.

Broom, RL, and Robinson, JT (1949) The lower end of the femur of *Plesianthropus*. Ann. Transvaal Mus. *21:*181–182.

Broom, RL, Robinson, JT, and Schepers, GWT (1950) Sterkfontein ape-man *Plesianthropus*. Transvaal Mus. Mem. *4*, 1–17.

Ciochon, RL, and Corruccini, RS (1975) Morphometric analysis of platyrrhine femora with taxonomic implication and notes on two fossil forms. J. Hum Evol. *4:*193–217.

Conroy, GC (1976) Primate postcranial remains from the Oligocene of Egypt. Contrib. Primatol. *8:*1–134.

Coppens, Y (1981) Le cerveau des hommes fossiles. C.R. Acad. Sci. Paris, *292*, Suppl. Vie Acad:3–24.

Coppens, Y (1983) Les plus anciens fossiles d'Hominidés. Pontif. Acad. Scient. Scripta Varia, *50:*1–9.

Day, MH (1978) Functional interpretations of the morphology of postcranial remains of Early African hominids. In CJ Jolly (ed): Early Hominids of Africa. London: Duckworth, pp. 311–345.

Day, MH, Leakey, REF, Walker, AC, and Wood BA (1976) New hominids from East Turkana, Kenya. Am. J. Phys. Anthropol. *45:*369–436.

Feldesman, MR (1982) Morphometric analysis of the distal humerus of some Cenozoic catarrhines: The late divergence hypothesis revisited. Am. J. Phys. Anthropol. *59:*73–95.

Ferguson, WW (1983) An alternative interpretation of *Australopithecus afarensis* fossil material. Primates *24:*397–409.

Fleagle, JG (1983) Locomotor adaptations of Oligocene and Miocene hominoids and their phyletic implications. In RL Ciochon and RS Corruccini (eds): New Interpretations of Ape and Human Ancestry. New York: Plenum, pp. 301–324.

Fleagle, JG, and Simons, EL (1978) Humeral morphology of the earliest apes. Nature *276:*705–707.

Fleagle, JG, and Simons, EL (1982) The humerus of *Aegyptopithecus zeuxis*: A primitive anthropoid. Am. J. Phys. Anthropol. *59:*175–193.

Fleagle, JG, and Simons, EL (1983) Skeletal remains of *Propliopithecus chirobates* from the Egyptian Oligocene. Folia Primatol. *39:*161–177.

Ford, SM (1980) Phylogenetic relationships of the Platyrrhini: The evidence of the femur. In RL Ciochon and AB Chiarelli (eds): Evolutionary Biology of the New World Monkeys and Continental Drift. New York: Plenum, pp. 317–331.

Genet-Varcin, E (1969) Structure et comportement des Australopithèques d'après certains os post-crâniens. Ann. Paléontol. Paris, *LV:*139–148.

Heiple, KG, and Lovejoy, CO (1971) The distal femoral anatomy of *Australopithecus*. Am. J. Phys. Anthropol. *35:*75–84.

Johanson, DC, and Coppens, Y (1976) A preliminary anatomical diagnosis of the first Plio/Pleistocene discoveries in the Central Afar, Ethiopia. Am. J. Phys. Anthropol. *45:*217–223.

Johanson, DC, Lovejoy, CO, Kimbel, WH, White, TD, Ward, SC, Bush, ME, Latimer, BM, and Coppens, Y (1982) Morphology of the Pliocene Hadar hominid skeleton (A.L. 288-1) from Hadar Formation, Ethiopia. Am. J. Phys. Anthropol. *57:*403–451.

Johanson, DC, and White TD (1979) A systematic assessment of early African hominids. Science *203:*321–330.

Johanson, DC, White, TD, and Coppens, Y (1978) A new species of the genus *Australopithecus* (Primates: Hominidae) from the Pliocene of Africa. Kirtlandia *28:*1–14.

Lamy, P, and Nossant, P (1983) Considérations sur le pied de quelques Hominidés fossiles du Plio-Pléistocène d'Afrique. C.R. Acad. Sci. Paris, *296*, III:621–624.

Leakey, REF, Mungai, JM, and Walker, AC (1971) New Australopithecines from East Rudolf, Kenya. Am. J. Phys. Anthropol. *35:*175–186.

Le Gros Clark, WE, and Thomas, DP (1951) The Miocene Hominiodea of East Africa. Foss. Mamm. Afr. *1:*1–117.

Le Gros Clark, WE, and Thomas, DP (1951) Associated jaws and limb bones of *Limnopithecus macinnesi.* Foss. Mamm. Afr. *3:*1–27.

Lewin, R (1983) Were Lucy's feet made for walking? Science *220:*700–702.

Lovejoy, CO (1972) The biomechanics of stride and their bearing on the gait of *Australopithecus*. Comm. 55rd Ann. Meet. Amer. Asc. Phys. Anthropol.

Lovejoy, CO, Heiple, KG, and Burstein, AH (1973) The gait of *Australopithecus*. Am. J. Phys. Anthropol. *38:*757–780.

Lovejoy, CO, Johanson, DC, and Coppens, Y (1982) Hominid lower limb bones recovered from the Hadar Formation: 1974–1977 collections. Am. J. Phys. Anthropol. *57:*679–700.

Marzke, MW (1983) Joint function and grips of the *Australopithecus afarensis* hand, with special reference to the region of the capitate. J. Hum. Evol. *12:*197–211.

McHenry, HM (1983) The capitate of *Australopithecus afarensis* and *Australopithecus africanus*. Am. J. Phys. Anthropol. *62:*187–198.

McHenry, HM, and Corruccini, RS (1975) Distal humerus in hominoid evolution. Folia Primatol. *23:*227–244.

McHenry, HM, and Corruccini, RS (1978) The femur in early human evolution. Am. J. Phys. Anthropol. *49:*473–488.

Morbeck, ME (1975) *Dryopithecus africanus* forelimb. J. Hum. Evol. *4:*39–46.

Morbeck, ME (1976) Problems in reconstruction of fossil anatomy and locomotor behavior: the *Dryopithecus africanus* elbow complex. J. Hum. Evol. *5:*223–233.

Napier, JR (1974) The antiquity of human walking. Readings from Sci. Amer. Biol. Anthropol., pp. 36–46.

Napier, JR, and Davis, PR (1959) The forelimb skeleton and associated remains of *Proconsul africanus*. Foss. Mamm. Afr. *16:*1–69.

Novacek, MJ (1980) Cranioskeletal features in tupaiids and selected Eutheria as phylogenetic evidence. In WP Luckett (ed): Comparative Biology and Evolutionary Relationships of the Tree-Shrews. New York: Plenum, pp. 35–93.

Olson, TR (1981) Basicranial morphology of the extant hominoids and Pliocene hominids: The new material from the Hadar Formation, Ethiopia and its significance in early human evolution and taxonomy. In CB Stringer (ed): Aspects of Human Evolution. London: Taylor and Francis, pp. 99–128.

Oxnard, CE (1975a) Uniqueness and Diversity in Human Evolution: Morphometric Studies of Australopithecines. Chicago: University of Chicago Press.

Oxnard, CE (1975b) The place of Australopithecines in human evolution: Grounds for doubt? Nature *258:*389–395.

Prost, JH (1967) Bipedalism of man and gibbon compared using estimates of joint motion. Am. J. Phys. Anthropol. *26:*135–148.

Robinson, JT (1972) Early Hominid Posture and Locomotion. Chicago: University of Chicago Press.

Rose, MD (1983) Miocene hominoid postcranial morphology: Monkey-like, ape-like, neither, or both? In RL Ciochon and RS Corruccini (eds): New Interpretation of Ape and Human Ancestry. New York: Plenum, pp. 405–417.

Schmid, P (1983) Eine Rekonstruktion des Skelettes von A.L. 288-1 (Hadar) und deren Konsequenzen. Folia Primatol. *40:*283–306.

Senut, B (1978) Contribution à l'étude de l'humérus et de ses articulations chez les Hominidés du Plio-Pléistocène. Thèse Doctorat 3ème cycle, Paris: Université Pierre et Marie Curie.

Senut, B (1981a) Humeral outlines in some hominoid primates and in Plio-Pleistocene hominids. Am. J. Phys. Anthropol. *56:*275–282.

Senut, B (1981b) L'humérus et ses articulations chez les Hominidés plio-pléistocènes. Cah. Paléoanthropol. Paris: C.N.R.S.

Senut, B (1982) Réflexions sur la brachiation et l'origine des Hominidés à la lumière des Hominoïdes miocènes et des Hominidés plio-pléistocènes. Géobios, Mém. Spéc. *6:*335–344.

Senut, B (in press) Nouvelles données sur l'évolution du coude chez les primates hominoïdes. Round Table "Morphogenetik und Evolution," Xanthi, 1983.

Sigmon, BA (1969) Anatomical structure and locomotor habit in Anthropoidea with special reference to the evolution of erect bipedality in man. Unpublished Ph.D. thesis, Madison: University of Wisconsin.

Stern, JT, and Susman, RL (1983) The locomotor anatomy of *Australopithecus afarensis*. Am. J. Phys. Anthropol. *60:*279–317.

Susman, RL, and Stern, JT (1979) Telemetered electromyography of *flexor digitorum profundus* and *flexor digitorum superficialis* in *Pan troglodytes* and implications for interpretation of the OH 7 hand. Am. J. Phys. Anthropol. *50:*565–574.

Susman, RL, Stern, JT, and Jungers, WL (1985) Locomotor adaptations of the Hadar hominids. In E Delson (ed): Ancestors: The Hard Evidence. New York: Alan R. Liss, Inc., pp. 184–192.

Szalay, FS, and Dagosto, M (1980) Locomotor adaptations as reflected in the humerus of Paleogene primates. Folia Primatol. *34*:1–45.

Tardieu, C (1979) Aspects biomécaniques de l'articulation du genou chez les primates. Bull. Soc. Anat. Paris *4*:66–86.

Tardieu, C (1982) Caractères plésiomorphes et apomorphes de l'articulation du genou chez les primates hominoïdes. Géobios, Mém. Spéc. *6*:321–334.

Tardieu, C (1983) Analyse morpho-fonctionnelle de l'articulation du genou chez les primates et les Hominidés fossiles. Cah. Paléoanthropol. Paris: C.N.R.S.

Tardieu, C (1984) Evolutionary pattern of the external meniscus in Primates. Abstract Comm. Xth Cong. Intern. Primatol. Soc. July, 22–27, Nairobi. Intern. J. Primatol. *5*:(4):385.

Tardieu, C (in press) Knee-joint in three hominoid primates: Osteology, meniscal dissection, rotation measurements. Application to Plio-Pleistocene hominids. Evolutionary implication. IXth Cong. Intern. Primatol. Soc. August 1982, Atlanta.

Tobias, PVT (1980) "*Australopithecus afarensis*" and *A. africanus*: Critique and alternative hypothesis. Palaeont. Afr. *23*:1–17.

Tuttle, R (1983) cited in R Lewin, Were Lucy's feet made for walking? Science *220*:700–702.

Vancata, V (1982) Chimpanzee locomotion and implications for the origin of hominid bipedality. Anthropos *21*:41–45.

Walker, AC, and Pickford, M (1983) New postcranial fossils of *Proconsul africanus* and *Proconsul nyanzae*. In RL Ciochon and RS Corruccini (eds): New Interpretations of Ape and Human Ancestry. New York: Plenum, pp. 325–351.

Wood, BA (1978) An analysis of early hominid postcranial material: Principles and method. In CJ Jolly (ed): Early Hominids of Africa. London: Duckworth, pp. 347–360.

Ancestors: The Hard Evidence, pages 202–205

Early and Early Middle Pleistocene Hominids From Asia and Africa

Milford H. Wolpoff and Abel Nkini

Department of Anthropology, University of Michigan, Ann Arbor, Michigan (M.H.W.), and Department of Anthropology, Frankfurt Universität, Frankfurt-am-Main, West Germany (A.N.)

Homo erectus may well be the most interesting and the most important of the fossil hominids. It is the species of the genus *Homo* immediately preceding ourselves, the first hominid species to show a recognizably human adaptive pattern, the first hominid to successfully inhabit regions outside of Africa, and the first hominid to, as Louis Leakey put it, "make tools according to a set and regular pattern." While *Homo erectus* is obviously not modern in form, with regard to the two things that have changed most over the course of human evolution, the species ranges of variation for tooth size and brain size largely fall within the modern range. Moreover, the species shows geographic distinctions that parallel those of *Homo sapiens*, and which in the view of some workers may represent the evolutionary precursors of the modern variations.

The papers presented during this session touched on some of these topics, and the subsequent discussion reflected the areas where the interest of the audience was particularly heightened. This discussion session focused on the earliest remains of *Homo erectus*, from Africa and from East and Southeast Asia. The session was fortunate to feature many of the researchers who discovered and subsequently studied the original materials. Six papers were presented. They aroused considerable interest and the subsequent discussions were at times vigorous.

None of the papers was directly concerned with the origin of *Homo erectus*. Bernard Wood focused on the earliest materials from Lake Turkana in Kenya, dealing mainly with the taxonomic relations of the various East African specimens included in the genus *Homo*. Wood recognized three species among the earliest representatives of this genus. When asked how these are related, he responded that they evolved from a common ancestor not unlike *Australopithecus africanus*, but declined to provide further details. Wood also avoided naming the three species, admitting difficulties in priority because of a previous publication by Groves and Mazak (1975). He contended that it is too early in the study to attempt addressing these problems. Subsequent questions reflected the interest of the audience in the details of his proposal, and the questions became quite specific. Wood argued that the ER 1470 male and the ER 1813 female were in different species. This prompted the question of what the male of the species represented by the ER 1813 female would look like. Wood responded that it would be more robust, and in particular differ from the ER 1470 male in (unnamed) details of facial form and frontal morphology.

The other paper specifically dealing with African *Homo erectus* was presented by Wolfgang Maier. His analysis was based on a new study of the O.H. 9 hominid in collaboration with Abel Nkini. Maier contended that this specimen shows very little basicranial flexion. This claim engendered considerable discussion during the study periods, and the problem of whether distortion of the cranial base underlay Maier's observations was raised persistently. Thus, this question was presented during the symposium session, in the context of how the lack of flexion in the OH 9 cranial base could be reconciled with the very flexed cranial bases of other more complete *Homo erectus* crania such as ER 3733 and Sangiran 17. Maier responded that he had not seen midsagittal diagrams of these specimens, and was not convinced by their casts that a comparison with O.H. 9 demonstrated that the Olduvai specimen was distorted in the region. When asked about the seemingly peculiar gyral alignment of the frontal region evident in his diagrams of O.H. 9, Maier responded that the region had not yet been studied in detail, and that the diagram in question reflected how the endocast appeared to the artist who made the drawing.

The papers concerning East and Southeast Asia also evoked considerable discussion, although the questions were informative rather than argumentative. John de Vos was questioned about why the Indonesian *erectus* was found with *Pongo* while the African hominids were never found with pongids. De Vos suggested that the same ecological association actually applied in Indonesia, and that the problem was with the "evidence" for the association. None of the *Pongo* remains from either Indonesia or China date to the Early Pleistocene, in his view, and while there may be associated remains at Punung, the fact that it is a fissure deposit makes the ecological interpretation difficult. Jens Franzen also responded, adding that fossil associations do not necessarily reflect living associations and pointing to the fact that the *Pongo* remains are all dental, as evidence that they might have been transported further before fossilization.

Wu Rukang was asked about the evidence for fire at the Zhoukoudian cave. He indicated that most of the ash layers in the cave were very thick, and are believed to be natural occurrences. There are some pits with ashes and fish bones, but these have proved difficult to interpret. Wu was also questioned about whether the animal remains at Zhoukoudian were the result of human activities. He responded by pointing out that both humans and hyenas were known to occupy the cave at various times.

De Vos, Franzen, and Wu were jointly asked a question about the dating of the earliest hominids in East and Southeast Asia. There was surprising agreement about an approximate 700,000 year age for the earliest *Homo erectus* occupations in both China and Indonesia.

Finally, the session ended with a discussion of Philip Rightmire's paper. In his presentation he established a trend for increasing brain size within *Homo erectus*. Rightmire was asked about the magnitude of the correlation of brain size against time in his regression, and whether the observed trend took possible body size changes into account. He responded that the correlation coefficient was 0.4, and that, if anything, body weight decreased over the time period concerned, so that it could not account for the brain size increase.

The subtitle of this symposium was "The Hard Evidence," and it would be fair to say that a considerable amount of hard evidence was presented in this session on Early and Middle Pleistocene *Homo erectus*. Yet it is interesting to observe what this evidence was used *for*. In particular, with the exception of Rightmire's paper, it was *not* used for an examination of evolutionary issues. Most of the information imparted was taxonomic and/or descriptive. The questions concerning the descriptive presentations focused on clarifications of the data offered, while those concerned with the taxonomic discussions focused on phylogenetic relationships and the taxonomy of individual specimens.

The tenor of the questions also differed, with the taxonomic questions engendering the most heated discussions. Perhaps this reflects a general level of disbelief about the presentations, but it is more likely that this tone reflects the fact that the important problems of human evolution can only rarely be solved by individual specimens. Evolution is a description of what happens to populations over time, and in the fossil record it is the *sample* and not the *individual* that comes closest to accurately reflecting this process. This is because the characteristics of a sample more closely estimate the parameters of the population than the characteristics of an individual ever can, and ultimately the evolutionary process accounts for changes among populations and does not account for differences among specific individuals.

Obviously, samples are made up of individuals, and it is important for the remains of these individuals to be described. When these descriptions are comparative and when they involve an analysis of function, they can be helpful in understanding the populations that the individuals once lived in, but the characteristics of populations can no more be derived directly from the description of an individual than the characteristics of a house can be derived directly from the description of a brick with which it is built. Admittedly, there are numerous examples in which an individual is all that remains of a sample, but this is hardly the case for the

Homo erectus remains that are the objects of this session. Thus, for instance, whether the morphology of the O.H. 9 cranial base was actually as Maier described or whether it is a consequence of crushing as many other observers suggested is ultimately irrelevant in view of the fact that this base is both morphologically unique and functionally inexplicable in the context of a *Homo erectus* sample for which a number of cranial bases are known. Maier's work is important, and this issue clearly must be resolved. The point is that it is an issue about O.H. 9 and not about the taxon *Homo erectus* and its evolution, and will remain such at least until the taphonomic issue and the problem of its functional interpretation can be satisfactorily resolved.

The focus on taxonomy is also unfortunate, since ultimately the taxonomy is the *consequence* of evolutionary studies. It was right and proper for Wood to claim that it was too early in his study for evolutionary interpretations, but, if so, it was probably too early for the taxonomy as well. In paleoanthropology, taxonomy has grown to be a red herring of gigantic proportions (?*Gigantourithroculpea*).

There is a complex interrelation between evolutionary analysis and the identification of biologically valid samples to compare. This is a different issue than that of the taxonomy of the samples, because the taxonomy is not simply based on the identification of biologically distinct samples. It must also reflect the phylogeny of these samples, and phylogeny is an outgrowth of an evolutionary analysis. In the American Museum of Natural History—the North American capital of Popperian science as represented by cladistics—it is surely not improper to argue that science advances through progressive refutations of hypotheses. In particular, the identification of a biologically valid sample should be the consequence of a failure to refute the hypothesis that the sample is comprised of more than one taxon. Subsequent analysis of such samples eventually results in a hypothesis of phylogeny that cannot (at the moment) be refuted, and this in turn provides the basis for taxonomy. As long as this process is muddled and confused in paleoanthropology, taxonomic debates will continue to be time and energy consuming, without end, and without resolution.

How did this state of affairs come to be? Within this symposium it was almost certainly a consequence of the exclusive focus on "the hard evidence." Popper wrote, after all, that without theory there are no data. If the superstructure of theory is lifted away, there is no answer to the question of what "the hard evidence" is evidence for.

Rightmire did present an evolutionary account of changes within the *Homo erectus* sample. This was significant because it represents a clarification of his *Paleobiology* paper (Rightmire, 1981), which has been widely quoted as a demonstration of stasis within a hominid species. Rightmire's point, which was well taken, addressed rates of change rather than the issue of stasis vs. evolution. He contended that the rate of change for cranial capacity accelerated after *Homo erectus*, and that this rate was more rapid in early (i.e., Middle Pleistocene) *Homo sapiens* (although the rate was much less rapid in the Late Pleistocene *Homo sapiens* sample). Presumably this will help remove *Homo erectus* from the punctuationalism debate, and one can hope that subsequent research will address the more interesting question of why these rates differ.

This session, and the papers presented within it, serve to outline a number of critical and as yet unresolved problems and/or disagreements surrounding the understanding of *Homo erectus*. It is evident that the origin of the species will be an important research focus. There appears little agreement on *how* the species originated. This would have been even more problematic had African archaeology and the interpretation of the Oldowan sites been brought into the picture, because these data are relevant to the issue of *why* the species originated, and their interpretation has been the subject of many recent publications. The question of which australopithecine species *H. erectus* originated from was clearly a focus of discussion, and this was even further complicated by the recurrent suggestions that the species may have evolved in Asia and was intrusive to Africa.

Adaptation was not addressed at all, but one of its consequences, the spread of *Homo erectus*, did receive attention. After all, *Homo erectus* is far better represented outside of Africa than it is within Africa. There has been real progress in the important geological issues of time and sequence for the hominid remains in both East Asia and Indonesia, as reflected in this symposium. This provides a much better basis for answering the question of what happened and why, and these should increasingly become objects of research in the region. The fact is that within Asia itself there are marked morphological differences between the northern and southern ends of the *Homo erectus* range, and there are equally dramatic contrasts in environment, climate, and archaeology (for instance, archaeological remains have yet to be firmly associated with the Middle Pleistocene deposits in Indonesia). This makes Asia an ideal testing ground for a number of conflicting evolutionary hypotheses.

The fate of *Homo erectus* is a final problem of interest. Wu contended in his presentation that at least in China the *erectus* populations were immediately ancestral to later populations of archaic *Homo sapiens* and ultimately ancestral to the modern Chinese. If this is also the case in other regions it has important implications for two related issues: the nature of the *Homo erectus–Homo sapiens* distinction and the origin of living populations (i.e., of modern geographic variation).

The first issue stems from the problem of how regionally distinct populations of *Homo erectus* can all evolve into regionally distinct populations of *Homo sapiens*. Solutions offered thus far include claims of parallelism, of the overwhelming importance of the common cultural adaptation, and of a rapidly spreading macromutation. Alternatively, it has been suggested that there is nothing to explain, because there is no basis for the species distinction between *Homo erectus* and *Homo sapiens*. This is becoming an important research area.

The second issue follows from a potential contradiction between two models. Modern populations cannot both show evolutionary continuity with preceding populations of *Homo erectus* in varying regions of the world and also be the result of a rapid replacement of these earlier populations by a small source population of modern *Homo sapiens* originating in a single area. Thus, an understanding of *Homo erectus* impinges directly on an understanding of ourselves and of our immediate ancestry.

In sum, the papers and discussion presented in this symposium were informative, interesting, and in some cases provocative. Their most important role, however, may best be seen as a benchmark of the progress and prospects for subsequent research in this often overlooked area of human evolution—the natural history of *Homo erectus*.

LITERATURE CITED

Groves, CP, and Mazak, U (1975) An approach to the taxonomy of the Hominidae: Gracile Villafranchian hominids of Africa. Casopis Mineral. Geol. *20:*225–247.

Rightmire, GP (1981) Patterns in the evolution of *Homo erectus*. Paleobiol. *7:*241–246.

Ancestors: The Hard Evidence, pages 206–214

Early *Homo* in Kenya, and Its Systematic Relationships

Bernard Wood

Department of Anatomy and Biology as Applied to Medicine, The Middlesex Hospital Medical School, London, W1P 6DB, England

ABSTRACT The diagnostic features of the cranial vault, face, and base of *A. africanus* are reviewed and classified according to their distribution within fossil and extant hominids and hominoids. Two crania from Koobi Fora, Kenya, KNM-ER 1470 and 1813, are compared with *A. africanus*, and it is concluded that they are significantly different from the "gracile" australopithecines. The possible systematic relationships of the two specimens are examined, and it is suggested that differences in size and shape between the two crania may merit their assignment to separate taxa. The differences between either KNM-ER 1470 or KNM-ER 1813 and KNM-ER 3733 and 3883 are equally profound, thus suggesting the working hypothesis that there may be three "non-australopithecine" taxa at Koobi Fora.

INTRODUCTION

Twenty years have elapsed since the publication of the paper that claimed sufficient evidence to justify establishing a new species of *Homo, Homo habilis* (Leakey et al., 1964). In the two decades following the announcement of *H. habilis*, research at Olduvai and at other East African sites has led to the discovery of similar, if not like, remains (Leakey et al., 1971; Coppens, 1981; Leakey et al., 1978; Howell, 1978). Yet despite the apparent abundance of evidence, it must be said that our understanding of the early (pre-*Homo erectus*) phase of the evolution of our own genus is still woefully poor.

Reasons for this are not hard to find. The samples available for particularly important anatomical regions are still small. Also, while efforts have been made to document the comparative context of variation in specific anatomical areas of the early hominid skeleton (e.g., White, 1977; Dean, 1982; Rak, 1983; Wood and Abbott, 1983; Wood et al., 1983), many important regions that are comparatively well represented in the early hominid fossil record have not been so intensively studied. A further hindrance has been the lack of adequate metrical and morphological information about *Australopithecus africanus*, for, whether it proves to be the most likely candidate for a common hominid ancestor (Tobias, 1973, 1981), or not (Johanson and White, 1979; Rak, 1983), it is still the taxon most likely to be confused with early representatives of *Homo*, and any candidates for inclusion in such taxa have first to be shown to be significantly different from *A. africanus*.

This short paper does not claim to put right this deficiency, but it will present a summary of the features of the vault, face, and cranial base of *A. africanus* with comments about their likely significance, whether symplesiomorphic for hominids, apomorphic for *Australopithecus* (i.e., "gracile" plus "robust"), or autapomorphic for *A. africanus*. It will then compare, area by area, two crania from Koobi Fora, Kenya, which are candidates for inclusion in early *Homo* (KNM-ER 1470 and 1813), with the observed morphological and metrical features of the *A. africanus* hypodigm. The significance of any differences will then be discussed in relation to the likely systematic relationships of the Koobi Fora specimens.

The reference hypodigm of *A. africanus* is taken to include the hominid collections from Taung (Dart, 1925), Sterkfontein Member 4

(Partridge, 1978; but see Wilkinson, 1983), and Makapansgat Members 3 and 4 (Partridge, 1979). The paper will concentrate on the appearance of the adult form, and the metrical data referred to will be published in more detail in a monograph by Wood and in a series of papers that are being prepared by Bilsborough and Wood.

A SUMMARY OF THE CRANIAL MORPHOLOGY OF *AUSTRALOPITHECUS AFRICANUS*

Vault

The overall shape of the cranial vault in *A. africanus* is remarkably primitive. However, the increased cranial capacity with respect to extant pongids of similar skull length (i.e., glabella–opisthocranion) is reflected in a vault that both is higher and has a flatter sagittal parietal border than is the case in the more peramorphic pongids. These differences are reflected in significantly different values for the chord/arc indices of the individual vault bones, as well as in Le Gros Clark's (1964) supraorbital height index and indices that relate auricular height to cranial length and width. These and other differences in cranial shape are less marked when the fossil crania are compared to the more paedomorphic crania of *Pan paniscus* (Cramer, 1977; Zihlman et al., 1978). Compared to the pongids, the emphasis of the temporal markings in *A. africanus* is more anteriorward, and the area for nuchal muscle attachment is reduced. The anterior emphasis of the temporal attachment is apparently less in *A. afarensis* (White et al., 1981), but it is even more marked in the "robust" australopithecines. The result is that *A. africanus* does not usually have a compound nuchal crest. The reduction in the nuchal area, and the relative expansion of the squamous part of the occipital, combine to produce a relatively longer occipital upper scale than is seen in the extant pongids.

Face

The face of *A. africanus* is short and narrow, particularly in the upper part. A prominent glabella is apparently plesiomorphic for the ape/hominid clade, but the narrow upper face is likely to be a derived hominid feature. The frontal morphology varies in the degree of development of a sulcus between the orbital margin and the frontal squame. Except for the most paedomorphic *Pan paniscus* crania, extant pongids have a flat or arched supraorbital bar that is separated from the anterior end of the temporal line by a marked lateral depression. Hominids lose this feature as the face moves beneath the neurocranium or the cranial cavity extends forwards over the face. The degree of development of the supraorbital bar varies among *A. africanus* crania, but a reduction relative to the apes (especially at its lateral end) is a common feature and is likely to be primitive for the hominid clade. The vault bones of *A. africanus* are thin, like those of the extant apes, suggesting that this condition is primitive for the African ape/hominid clade.

The nasal bones are relatively straight and are wider below than above. The mid-face is broader than the upper face (*contra* most of the extant apes), but it is particularly noteworthy that the degree of midline projection is similar in *A. africanus* and chimpanzee skulls of the same overall size; Clarke (1977, p. 282) notes that this is due to the retention in *A. africanus* of a forward facing frontal process of the maxilla. However, the malar region is more forward facing than in the African apes, and its lower border lies ahead of the upper; the resulting inclination of the malar surface is derived with respect to the African apes. The origin of the maxillary process of the maxilla is also higher and set more anterior with respect to the palate than it is in the apes. The degree of prognathism of *A. africanus* varies from the highly prognathic Sts-5 to the more orthognathic Sts-71. The nasoalveolar clivus is flat, not rounded as in the apes, and is separated from the nasal floor by a sill of variable height and pitch. The lateral nasal walls are better defined in *A. africanus* than in the apes, and the more vertical orientation of the plane of the nasal aperture and the separation of the nasal cavity from the face are clearly likely to be derived hominid features.

Rak has made the most detailed study of the facial structure of the australopithecines to date and concludes that "almost every aspect of the *A. africanus* face indicates that the species as a whole was already embarked on the evolutionary course leading to the robust australopithecines" (Rak, 1983, p. 120). Ward and Kimbel (1983) come to similar conclusions on the basis of the subnasal alveolar morphology. However, to be set against this are the evident general similarities in facial architecture between *A. africanus* and similar sized extant pongids, with the major difference being the size of the maxillary canine roots to be accommodated within the facial skeleton. There is also the evidence (from the same parts of the face, as well as other features) that has led other authors to conclude that there are marked differences between the faces of the "gracile" and "robust" australopithecines (Robinson, 1954; Clarke, 1977). Observations on other fossil hominid crania (including those specimens that are to be reviewed) also suggest that some forward migration of the zygomatic process of the maxilla may be a primitive trait of hominids and not a character confined to the "robust" australopithecine clade.

Base

The overall shape of the cranial base of the "gracile" australopithecines is pongid-like (Du Brul, 1977; Dean and Wood, 1981, 1982; Laitman and Heimbuch, 1982). The basicranial axis is relatively "unflexed" and indeed most closely corresponds to the degree of cranial base flexion in stage 5 (i.e., adult) *Gorilla* for Sts 5, and to stage 4 *Gorilla*, or stage 5 *Pan*, for MLD 37/38 (Laitman and Heimbuch, 1982). The base of *A. africanus* is still relatively elongated and narrow, as it is in the pongids, but the carotid foramina are relatively further apart. Apparently derived features of the early hominid cranial base are the more anterior situation of the foramen magnum (so that its anterior border lies at, or in front of, the bitympanic line) and shortening of the tympanic plates. However, the "gracile" australopithecine cranial base has retained pongid-like petrous bones, which are long and sagittally oriented, and the foramen magnum faces downwards and backwards (i.e., in a pongid direction, in Sts-5). Likewise, the elongated bony ridges for the attachment of longus capitis on the underside of the basiocciput and the expression of the eustachian process in Sts-5 and MLD 37/38 are likely to be retained features of the great ape/hominid clade (Dean, 1982). Clarke has also commented on the similarity between the relationship of the sigmoid sinus and the petrous pyramid in pongids and in Sts-5. He noted that, whereas in these forms the sigmoid sinus lies behind the sloping posterior surface of the petrous pyramid, in later hominids it is situated beneath the more vertically oriented posterior aspect of the petrous (Clarke, 1977, p. 261). The projected position of porion in relation to the glabella–opisthocranion line in Sts-5 is also little changed from the position in the great apes, though in Sts-71 the posterior skull length index is higher, and thus more similar to values in later *Homo*.

The mastoid process in the "gracile" australopithecines is poorly defined, but whereas in the pongids the small, nipple-like process is dwarfed by the anterior extension of the compound nuchal crest, the process is nearer to the lateral border of the cranium in the "gracile" australopithecines. Deepening of the mandibular fossa and a tendency towards a more vertical orientation of the tympanic plate are clearly hominid features that are present, if weakly expressed, in the "gracile" australopithecines.

The features that have been presented above are summarised in Table 1. The letter after the description indicates how specific that character state is likely to be; i.e., whether it is synapomorphic for the African ape and hominid clade, for hominids, or for the "robust" australopithecine clade. If the interpretation of the polarities are correct, it is clear that most of the character states do not indicate that the "gracile" australopithecines have any special relationship with the "robust" australopithecines. The possible exceptions are the facial features referred to by Rak (1983) and Ward and Kimbel (1983). It is noteworthy that none of the characters listed in Table 1 are considered to be autapomorphic for *A. africanus*.

INDIVIDUAL SPECIMENS COMPARED TO *A. AFRICANUS*

KNM-ER 1813

The cranial vault of KNM-ER 1813 is similar in thickness, shape, and size to that of *A. africanus*. Comparison of the estimated cranial capacity of KNM-ER 1813 (510 ml) and the average of the estimates for six *A. africanus* crania (442 ml) (Holloway, 1976) provides the best estimate of relative neurocranial volumes. The length of the occipital margin of the parietal is greater in KNM-ER 1813 than in *A. africanus*, but the sagittal length of the parietal is shorter than that of the "gracile" australopithecines. However, these differences are probably at least partly due to presence of accessory ossicles at the apex of the occipital. The major differences between KNM-ER 1813 and *A. africanus* relate to the shape of the parietal and the occipital, and the latter's contribution to the sagittal arc of the vault. The relatively long coronal border of the parietal results in

TABLE 1. Characteristics of Australopithecus africanus *with an indication of their specificity*

Characteristic	Specificity
Vault	
Thin bones	C[1]
Greatest width across supramastoid crest	C
Increased height and roundness	H[2]
Decreased emphasis on posterior attachment of temporalis and apparent lack of a compound nuchal crest	H
Reduction in nuchal area associated with a relatively larger occipital upper scale	H
Face	
Short and narrow in the upper part	H
Prominent glabella	C
Reduction of supraorbital bar, particularly at the lateral end	H
Relatively straight and inferiorly broad nasal bones	C
Midface broader than upper face	R[3] or ?H
Little midline projection in the midface	C
Frontal process of the maxilla forward facing	C
Malar region more forward facing and protracted than in apes	R or ?H
Reduced alveolar prognathism	H
Nasal sill and sharp lateral nasal border	H
Straight sagittal profile to nasoalveolar clivus	R or ?H
Anterior pillars and maxillary furrows	C or R
Zygomatic tubercles	R
Base	
Relatively long and narrow	C
Anterior border of the foramen magnum migrated up to, and beyond, the bitympanic line	H
Increased bicarotid foramen breadth	H
Reduced length of the tympanic	H
Relative position of porion	C
Slope of foramen magnum	C
Slope of posterior surface of the petrous in the posterior cranial fossa	C
Mastoid larger and more laterally situated	?H
Basioccipital morphology and eustachian process	C
Deeper mandibular fossa	H

[1]C = Primitive for African ape/hominid clade.
[2]H = Derived for hominid clade.
[3]R = Derived for "robust" australopithecine clade.

a shape index (88) that is a good deal lower than the *A. africanus* mean of 112. Even if the Wormian bones are not counted as part of the occipital, the upper scale of the occipital is relatively larger than that of *A. africanus*, and the occipital/parietal sagittal arc index value is 119, compared to an *A. africanus* average value of 83 $\pm$ 3 (n = 3). If the Wormian bones are included with the occipital, then the discrepancy is even more marked. The occipital is also more rounded, with a modest central torus.

When the face of KNM-ER 1813 is compared to the range of preserved *A. africanus* facial remains, the differences are mainly in terms of facial proportions and the morphology of the frontal, nasal, and malar regions. The upper face is broader than the midface, whereas in *A. africanus* the reverse is the case; the relatively broader upper face may be a retained shared feature with the great apes, and would thus suggest that the condition in *A. africanus* may indeed be derived for the "robust" clade (but see below that this is also the condition for KNM-ER 1470). The face of KNM-ER 1813 is also shorter and less prognathic and has a vertical malar process. Glabella, instead of being on a rounded projection, is situated in a depression between the arched, but slight, supraorbital tori. The curvature of the tori and the central depression between them are reflected in a transverse frontal torus chord/arc index of 85, which compares with index values of 92 and 91 for Sts-5 and -71. The lateral border of the nose is crest-like and everted, and the frontal process of the maxilla faces antero-laterally.

The base of KNM-ER 1813 is poorly preserved, but sufficient landmarks can be reconstructed to compare it metrically with *A. africanus* (Dean and Wood, 1982) and estimate the relative position of porion and the slope of the foramen magnum. The foreshortening of the skull base, the relative widening of the sphenoid, and particularly the more coronal angulation of the petrous axes are all ways in which KNM-ER 1813 differs from *A. africanus* and is more like later *Homo*. Likewise, the value of the porion position index of KNM-ER 1813 (42) is more *Homo*-like than the values of 28 and 33 for Sts-5 and Sts-71, respectively, and the foramen magnum is estimated to slope antero-inferiorly, with an angle of 13°, whereas in Sts-5 it has a negative angle of 20° (Tobias, 1967).

KNM-ER 1470

The overall shape and thickness of the cranial vault of KNM-ER 1470 is similar to that in *A. africanus*, but the major differences are its larger size and the preferential increase in the length of the upper scale of the occipital. Otherwise, in terms of relative postorbital constriction and the position of maximum cranial breadth, KNM-ER 1470 is close to the *A. africanus* condition. The exception in the frontal area is glabella, which is situated in a shallow depression, *contra* the condition in *A. africanus*. The difference in size of the vault components is illustrated

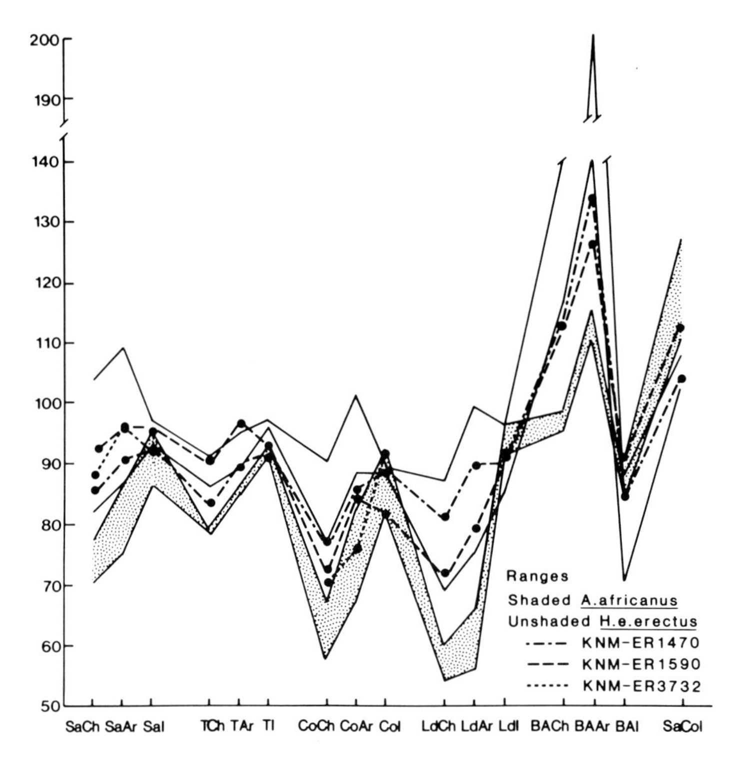

Fig. 1. Diagram to show the comparative size of the parietal bones in australopithecines and other early hominids. The three individual specimens are tentatively identified as *Homo habilis*. The y-axis is in mm for chord and arc measures and percent for indices. The x-axis represents measures (Ch = chord, Ar = arc) and indices (I = 100 Ch/Ar) for the four parietal borders (Sa = sagittal, T = temporal, Co = coronal, Ld = lambdoid), the bregma-asterion dimension (BA), and the sagittal-coronal chord index (SaCoI = 100 SaCh/CoCh).

with respect to the parietal (Fig. 1). It can be seen that the parietal of KNM-ER 1470 is within the *H. erectus* range for the dimensions of the sagittal and lambdoid borders but below it for the coronal and temporal dimensions. However, the shape index of the parietal falls within the *H. erectus* range.

The face of KNM-ER 1470 resembles that of *A. africanus* in being broader in the middle part and in sharing supraorbital features, but otherwise it is distinguished from it by being longer and a good deal less projecting in the midline in the lower and middle parts of the lateral face. The tall, relatively anteriorly-situated and forward sloping malar region superficially resembles this area in the "robust" australopithecines, but the anterior migration of the zygomatic process of the maxilla has not proceeded so far in KNM-ER 1470. The similarity in lateral profile between KNM-ER 1470 and the "robust" crania is due more to the lack of central facial projection in KNM-ER 1470 than to an exaggeratedly anterior situation of the malar region (Bilsborough and Wood, submitted).

The cranial base of KNM-ER 1470 is poorly preserved, and what has survived has undergone plastic deformation (Walker, 1981). Nonetheless, it is possible to use the preserved margins of vascular and neural foramina and the position of the petrous to estimate some measurements (Dean and Wood, 1982). It is then apparent that the cranial base of KNM-ER 1470 is both expanded laterally and foreshortened. The orientation of the petrous can also be estimated, and, if the available measurements and angles are any guide, then the base of KNM-ER 1470 was unlike that of *A. africanus* and metrically very similar to that of three East African *Homo* crania and calvariae attributed to *H. erectus* (O.H. 9, KNM-ER 3733, and 3883). Computation of the porion position index is complicated by distortion of the skull, but nonetheless, if the value of posterior skull length is taken to be that of the apparently undistorted left side, then the index of 38 compares with values of 28 and 33 for Sts-5 and Sts-71, respectively.

PHENETIC COMPARISONS WITH *A. AFRICANUS*: RESULTS AND IMPLICATIONS

KNM-ER 1813

The phenetic comparisons of three regions of the cranium of KNM-ER 1813 with *A. africanus* have resulted in a list of character differences that are summarised in Table 2. To be set against these differences are evident similarities in overall size and cranial vault shape, and the choice facing the hominid taxonomist is clear. If the latter phenetic similarities are emphasised, then a "grade" classification within *Australopithecus* (with specific or sub-specific distinction from *A. africanus*) is indicated. However, if the differences in Table 2 are recognised, then a case can be made that KNM-ER 1813 has acquired a sufficiently comprehensive set of character states (shared in common with later *Homo* taxa) to justify its exclusion from *A. africanus* and inclusion in *Homo*. While recognising the arguments in favour of the former arrangement, I nonetheless believe the balance of the evidence is in favour of excluding KNM-ER 1813 from *A. africanus*. Walker (1976) has elsewhere emphasised that the relative size relationship between the neurocranium and the splanchnocranium is an important part of the *Homo*/*Australopithecus* distinction. While KNM-ER 1813 has the small neurocranium of *Australopithecus*, its splanchnocranium is also small, resulting in facial-cranial measurement indices that are within the *Homo* range (e.g., nasion–alveolare/glabella–opisthocranion = 44%).

TABLE 2. Character differences between KNM-ER 1813 and the *A. africanus* hypodigm with an indication of their specificity

KNM-ER 1813	Specificity
Vault	
More dominant upper occipital scale	ES[1]
Greater occipital contribution to sagittal arc	ES
Increased parietal lambda/asterion chord and arc	ES
Smaller parietal sagittal-coronal arc index	E[2]
Broader postorbitally	H[3] or ?ES
Face	
Supraorbital bar arched, rounded, and torus-like	ES
Glabella in central depression	ES
Relatively broader upper face	C[4]
More orthognathic	ES
Vertical malar processes	H
Everted nasal margins	ES
Frontal process of the maxilla more laterally facing	ES
Base	
Foreshortened	ES
Wider sphenoid	ES
More coronal petrous axes	ES or R[5]
More anteriorly situated porion	ES
Foramen magnum facing antero-inferiorly	ES

[1]ES = Shared with *H. erectus* and *H. sapiens*.
[2]E = Shared with *H. erectus* only.
[3]H = Derived for hominid clade.
[4]C = Primitive for African ape/hominid clade.
[5]R = Derived for "robust" australopithecine clade.

KNM-ER 1470

The overall size difference between *A. africanus* and KNM-ER 1470 is large, but Walker (1976) has already drawn attention to the ways in which KNM-ER 1470 resembles australopithecine crania, particularly in its overall proportions and vault shape. To be set against this evidence is the context of these similarities and of the observed differences from *A. africanus*, which are listed in Table 3. When the similarities with *A. africanus* are examined, it is clear that they relate to features that are primitive for the African ape/hominid clade, and in isolation they are thus not relevant to deciding systematic relationships within that clade. In contrast, when the differences listed in Table 3 are examined, it is clear that for the majority of characters that differ between *A. africanus* and KNM-ER 1470 the condition in the latter is likely to be derived with respect to *Homo*, with some features interpreted as resembling the derived condition for the "robust" australopithecine clade. Thus, when the difference in overall size is taken together with the character differences outlined in Table 3, the evidence is strongly in favour of recognising a taxonomic distinction between KNM-ER 1470 and *A. africanus*.

Relationship Between KNM-ER 1470 and 1813

Before the detailed implications of the observed differences between the two fossil crania and *A. africanus* can be assessed, the crania themselves have to be compared to judge whether they are likely to belong to the same phenon grouping. Put simply, when examining the systematic relationships of KNM-ER 1470 and 1813, are we dealing with one taxon, or two?

TABLE 3. Character differences between KNM-ER 1470 and the *A. africanus* hypodigm with an indication of their specificity

KNM-ER 1470	Specificity
Vault	
Larger overall	ES[1]
More dominant upper occipital scale	ES
Greater occipital contribution to sagittal arc	ES
Smaller parietal sagittal-coronal arc index	E[2]
Face	
Longer overall	E or R[3]
More orthognathic	ES
Glabella in central depresssion	ES
Everted nasal margins	ES
Deep malar region	R
More anterior and forward facing malar region	R
Base	
Foreshortened	ES
Wider sphenoid	ES
More coronal petrous axes	ES or R
More anteriorly situated porion	ES

[1]ES = shared with *H. erectus* and *H. sapiens*.
[2]E = shared with *H. erectus* only.
[3]R = Derived for "robust" australopithecine clade.

The null hypothesis is that the differences between them are intraspecific ones, and in favour of this interpretation are the evident similarities in overall cranial shape, for despite differences in calvarial measurements, indices derived from them are similar in the two crania. There are also similarities in the proportions of the occipital bone. However, the similarities in overall vault shape are most likely primitive retentions, and thus do not imply any particularly close relationships between the two crania. Likewise, the occipital similarities may be common to all taxa traditionally regarded as hominine and thus have no specificity beyond that of defining the *Homo* clade.

The major differences in shape are in the face, particularly in the supraorbital and malar regions with the latter the more profound. Indices that relate the positions of the zygomatic process of the maxilla and other facial landmarks (Bilsborough and Wood, submitted) show clear differences between the protracted forward sloping and deep malar region of KNM-ER 1470 and the smaller excavated and backward sloping process of KNM-ER 1813. Some context for the shape differences between the facial regions of the two crania is provided by the matrix of D^2 distances between early hominid crania (Bilsborough and Wood, submitted). The D^2 distance between KNM-ER 1470 and 1813 is 11.6, which compares with an *A. africanus*/KNM-ER 1813 distance of 7.7, an *A. boisei*/KNM-ER 1470 distance of 11.0, and a Neanderthal/Modern *Homo sapiens* distance of 9.5.

The significance of the size difference between the cranial vault of KNM-ER 1470 and 1813 is most conveniently investigated by using the estimated cranial capacities, both of which were scored as relatively reliable estimates (Holloway, 1976). It is possible to assess the likelihood of the two crania coming from the same population by assuming a value for C.V. and then computing the likely distribution based on the mean of the two values (KNM-ER 1470 = 770; KNM-ER 1813 = 510; $\bar{X}$ = 640). A guide to values of C.V. can be taken from the parameters of cranial capacities for extant and fossil samples, and available evidence suggests that a C.V. = 10 would be a reasonable estimate for a hominid fossil taxon. Such an assumption results in a 2 S.D. range of $\pm$ 128; i.e., 512 to 768. The cranial capacity estimates of the two fossil crania thus lie almost exactly at these admittedly inexactly derived distribution ranges, which should include approximately 95% of the population. The chance of one of the two crania being part of the population is thus 5%, with the chance of both crania being included correspondingly small (i.e., 0.25%). There is also evidence that KNM-ER 1470 is not at the upper size limit of its phenon grouping. When the parietal dimensions of KNM-ER 1590 and 3732 are compared to those of KNM-ER 1470 (Fig. 1), it is evident that these specimens are similar in overall size, with indications that the intact cranium of KNM-ER 1590 may have been larger than KNM-ER 1470. If we assume that these two crania belong to the same phenon grouping as KNM-ER 1470, then it is even less likely that specimens with cranial capacity values of 510 and more than 770 are sampled from the same taxon.

These statistical tests are simplistic, but nonetheless, when taken with the evidence of the facial differences, they suggest that KNM-ER 1470 and 1813 should not be uncritically regarded as belonging to the same taxon. Walker and Leakey (1978) also concluded that the two crania are taxonomically distinct, but likened KNM-ER 1813 to *Australopithecus africanus* rather than to *Homo*. However, it should be noted that other careful studies have concluded that they are conspecific (e.g., Howell, 1978), and Tobias (1983) has interpreted the cranial capacity data (but his own, not Holloway's) in terms of one species with a percentage sexual dimorphism of 85%, and thus within the range of such dimorphism in *Homo sapiens* populations.

CONCLUSIONS

Clearly this paper can be no more than an exploration of the systematic relationships of two fossil hominid crania. Its phenetic conclusions are based on three areas only and are hypotheses that need to be tested against a wider range of anatomical features. The exercise in phylogenetic analysis is also tentative, but the possibilities it raises are tantalizing. The phenetic differences between crania such as KNM-ER 3733 and 3883 and either KNM-ER 1470 or KNM-ER 1813 are more impressive than those between the two latter crania, and if the taxonomic hypotheses referred to above are substantiated, then this would lead to the conclusion that there are at least three non-australopithecine taxa in East Africa in the Early Pleistocene.

The task of testing these hypotheses, and deducing their broader systematic and narrower taxonomic implications, is in hand.

ACKNOWLEDGMENTS

The author is grateful to Andrew Chamberlain for contributing to the ideas expressed in this paper and to Paula Smith for typing the manuscript. The research incorporated in this paper was supported by the N.E.R.C., The Royal Society, and The Nuffield Foundation. The author is indebted to the Government of Kenya and the Director and Trustees of the National Museums of Kenya for permission to make a detailed study of specimens in their care.

LITERATURE CITED

Bilsborough, A, and Wood, BA (submitted) Cranial morphometry of early hominids. I: Facial region.

Clarke, RJ (1977) The cranium of the Swartkrans hominid SK 847 and its relevance to human origins. Unpublished Ph.D. Thesis. Johannesburg: University of the Witwatersrand.

Coppens, Y (1981) The difference between *Australopithecus* and *Homo*: Preliminary conclusions from the Omo Research Expedition's studies. In L-K Königsson (ed): Current Argument on Early Man. Oxford: Pergamon, pp. 86–113.

Cramer, DL (1977) Craniofacial morphology of *Pan paniscus*: A morphometric and evolutionary appraisal. Contrib. Primatol. *10:*1–64.

Dart, RA (1925) *Australopithecus africanus*, the man-ape of South Africa. Nature *115:*195–199.

Dean, MC (1982) The comparative anatomy of the hominoid cranial base. Unpublished Ph.D. Thesis, University of London.

Dean, MC, and Wood, BA (1981) Metrical analysis of the basicranium of extant hominoids and *Australopithecus*. Am. J. Phys. Anthropol. *54:*63–71.

Dean, MC, and Wood, BA (1982) Basicranial anatomy of Plio-Pleistocene hominids from East and South Africa. Am. J. Phys. Anthropol. *59:*157–174.

Du Brul, EL (1977) Early hominid feeding mechanisms. Am. J. Phys. Anthropol. *47:*305–320.

Holloway, RL (1976) Some problems of hominid brain endocast reconstruction, allometry and neural organisation. In PV Tobias and Y Coppens (eds): Prétirage, Les Plus Anciens Hominidés. Paris: C.N.R.S., pp. 69–119.

Howell, FC (1978) Hominidae. In VJ Maglio and HBS Cooke (eds): Evolution of African Mammals. Cambridge, Massachusetts: Harvard University Press, pp. 154–248.

Johanson, DC, and White, TD (1979) A systematic assessment of early African hominids. Science *202:*321–330.

Laitman, JT, and Heimbuch, RC (1982) The basicranium of Plio-Pleistocene hominids as an indicator of their upper respiratory systems. Am. J. Phys. Anthropol. *59:*323–343.

Leakey, LSB, Tobias, PV, and Napier, JR (1964) A new species of the genus *Homo* from Olduvai Gorge. Nature *202:*7–9.

Leakey, MD, Clarke, RJ, and Leakey, LSB (1971) New hominid skull from Bed I, Olduvai Gorge, Tanzania. Nature *232:*308–312.

Leakey, RE, Leakey, MG, and Behrensmeyer, AK (1978) The Hominid Catalogue. In MG Leakey and RE Leakey (eds): Koobi Fora Research Project. Vol. 1: The Fossil Hominids and an Introduction to Their Context, 1968–1974. Oxford: Clarendon Press, pp. 86–182.

Le Gros, Clark, WE (1964) The fossil evidence for human evolution: An introduction to the study of palaeoanthropology, 2nd Ed. Chicago: University of Chicago Press.

Partridge, TC (1978) Re-appraisal of lithostratigraphy of Sterkfontein hominid site. Nature *275:*282–287.

Partridge, TC (1979) Re-appraisal of lithostratigraphy of Makapansgat Limeworks hominid site. Nature *279:*484–488.

Rak, Y (1983) The Australopithecine Face. New York: Academic Press.

Robinson, JT (1954) The genera and species of the Australopithecinae. Am. J. Phys. Anthropol. *12:*181–200.

Tobias, PV (1967) Olduvai Gorge, Volume 2. The cranium and maxillary dentition of *Australopithecus (Zinjanthropus) boisei*. Cambridge: Cambridge University Press.

Tobias, PV (1973) Implications of the new age estimates of the early South African hominida. Nature *246:*79–83.

Tobias, PV (1981) A survey and synthesis of the African hominids of the late Tertiary and early Quaternary periods. In L-K Königsson (ed): Current Argument on Early Man. Oxford: Pergamon, pp. 86–113.

Tobias, PV (1983) Recent advances in the evolution of the hominids with especial reference to brain and speech. Pont. Acad. Sci. Scr. Varia *50:*85–140.

Walker, A (1976) Remains attributable to *Australopithecus* in the East Rudolf succession. In Y Coppens, F

Clark Howell, GLl Issac, and REF Leakey (eds): Earliest Man and Environments in the Lake Rudolf Basin. Chicago: University of Chicago Press, pp. 484–489.

Walker, A (1981) The Koobi Fora hominids and their bearing on the origins of the genus *Homo*. In BA Sigmon and JS Cybulski (eds): *Homo erectus*: Papers in Honour of Davidson Black. Toronto: University of Toronto Press, pp. 193–215.

Walker, A, and Leakey, REF (1978) The hominids of East Turkana. Sci. Am. 44–56.

Ward, SC, and Kimbel, WH (1983) Subnasal alveolar morphology and the systematic position of *Sivapithecus*. Am. J. Phys. Anthropol. *61:*157–171.

White, TD (1977) The anterior mandibular corpus of Early African Hominidae: Functional significance of size and shape. Unpublished Ph.D. Thesis, University of Michigan.

White, TD, Johanson, DC, and Kimbel, WH (1981) *Australopithecus africanus*: Its phyletic position reconsidered. S. Afr. J. Sci. *77:*445–470.

Wilkinson, MJ (1983) Geomorphic perspectives on the Sterkfontein australopithecine breccias. J. Arch. Sci. *10:*515–529.

Wood, BA, and Abbott, SA (1983) Analysis of the dental morphology of Plio-Pleistocene hominids. I. Mandibular molars—crown area measurements and morphological traits. J. Anat. *136:*197–219.

Wood, BA, Abbott, SA, and Graham, SA (1983) Analysis of the dental morphology of Plio-Pleistocene hominids. II. Mandibular molars—study of cusp areas, fissure pattern and cross-sectional shape of the crown. J. Anat. *137:*287–314.

Zihlman, AL, Cronin, JE, Cramer, DL, and Sarich, VM (1978) Pygmy chimpanzee as a possible prototype for the common ancestor of humans, chimpanzees and gorillas. Nature *275:*744–745.

Ancestors: The Hard Evidence, pages 215–220

Faunal Stratigraphy and Correlation of the Indonesian Hominid Sites

J. de Vos
Rijksmuseum van Natuurlijke Historie, 2300 RA Leiden, The Netherlands

ABSTRACT Sondaar (1984) has suggested a sequence of mammalian faunas from Java including (from oldest to youngest): Satir, Ci Saat, Trinil H.K., Kedung Brubus, Ngandong, Punung, and Wajak. The first few assemblages are species-poor and reflect island isolation, while Kedung Brubus documents the greatest interchange with the mainland. The Trinil H.K., Kedung Brubus, and perhaps Ngandong and Wajak faunas inhabited dry, open woodland during glacial phases, while Punung was a warm and humid forest (interglacial). The Trinil site yielded *Homo erectus* (skullcap, femur and P_3) and *Meganthropus* (M^2, M^3), while a more progressive hominid is known from Kedung Brubus. At Sangiran, the oldest human fossils (*P. dubius* and *P. robustus*) are of Trinil H.K. or Ci Saat age, skull II (*H. erectus*) and the *Meganthropus* type are equivalent to Trinil H.K., and several younger skulls correspond to the Kedung Brubus fauna. Teeth from Punung, here referred to *H. sapiens*, are between Ngandong and Wajak in age. The Mo(d)jokerto child skull is of uncertain provenance and not well associated with faunal remains, although *Leptobos* suggests a Kedung Brubus age and paleomagnetism a date younger than 0.73 million years (m.y.). Morphologically, it seems similar to the Ngandong hominid taxon.

INTRODUCTION

Sondaar (1984) proposes a faunal stratigraphy for Java which I shall summarize here (Table 1, first column):

The Satir Fauna

Characterized by poverty in species, only *Mastodon*, *Hexaprotodon*, and cervids are known so far from this level. The *Mastodon* and *Hexaprotodon* are both archaic and may point to a relatively old age. The unbalanced character of the fauna points to island conditions.

The Ci Saat Fauna

Though poorly known, it can be said that *Stegodon*, *Hexaprotodon*, and cervids are common, and felids are present, while bovids are rare. The fauna is still poor in species, which might point to isolated conditions, though the presence of large felids shows that the mainland could not be far off.

The Trinil H.K. Fauna

This fauna is still relatively poor in species, though much more balanced than the preceeding faunas. This fauna contains endemic taxa, and large bovids are abundantly represented. The relatively low number of species points to an isolated position and little faunal interchange with the mainland. The high number of large bovids implies a drier biotope, and the environment is considered to have been an open woodland. This might point to a glacial interval.

The Kedung Brubus Fauna

New arrivals of *Epileptobos*, *Tapirus*, *Elephas*, *Hyaena*, and *Rhinoceros kendengindicus* are evidenced The fauna bears a more mainland stamp, reflecting maximal faunal interchange with the mainland. This fauna also suggests an open woodland during an ice age.

The Ngandong Fauna

This fauna is still insufficiently studied. The faunal list of von Koenigswald (1934) resembles in some aspects that of Kedung Brubus. The environment was probably an open woodland with a drier climate than exists nowadays.

The Punung Fauna

The assemblage is quite different from the preceding. *Stegodon* and *Elephas hysudrindicus* are absent, while *Elephas maximus* might be considered as a new migrant. In this fauna we find for the first time the extant species *Sus vittatus*. Typical for this fauna is the high quantity of *Pongo*, also probably a new migrant, as well as *Hylobates* and *Ursus malayanus*. On the basis of the high number of *Pongo* and *Hylobates* fossils, this fauna is interpreted as living in a humid forest, while the total fauna suggests an interglacial period.

The Wajak (Formerly Wadjak) Fauna

Elephas is not known from this fauna, and it probably documents the last occurrence of *Tapirus* on Java. This fauna also bears an open woodland character and might be considered younger in age than that of Punung; it probably flourished during a colder period than at present in Java.

Fossil hominids are known from several localities in this scheme (Table 1). The aim of this paper is to make some remarks upon these fossil hominids.

TRINIL HAUPT-KNOCHENSCHICHT (= TRINIL H.K.)

Nine hominid fossils are known from this locality (Trinil 1–9 of Oakley et al., 1975—all numbers of Javan specimens follow this catalogue). The skullcap (Tr. 2) is the holotype of *Homo erectus* (see Dubois, 1894).

Since the publication of the find of femur I (Tr. 3) there has been continual debate concerning the classification of this fossil because of its relatively "modern" morphology. Dubois (1894) stated that the femur and the skullcap were found in the same stratum, and he believed that the skullcap and the femur belonged to one individual. As the morphology of the femur is rather modern, colleagues first doubted that the femur and the skullcap belonged to one individual and later doubted that they originated from the same stratum. For the suggestion of Dubois (1894) that they belonged to one individual, there is no basis. However, there is no reason to doubt that they originated from the same stratum.

First of all, there are in the files of the Dubois Collection two letters from Kriele and de Winter, who excavated for Dubois, which tell us something concerning the provenance. In the first letter, dated August 31, 1892, Kriele and de Winter wrote (my translations from the Dutch): "That bone aforementioned is found at the same side of the river where in the past that skull from that chimpanzee is found. . . The bone is found somewhere about at the depth of the chimpanzee skull at the same level of the former water level, separated from each other about 12 meters." In the second letter, dated September 1892, we can read: ". . . that bone is found at the same side as where

TABLE 1. Biostratigraphy of Java in relation to fossil hominids

Faunal succession	Hominid Localities						
	Sangiran[1]	Trinil H.K.	Kedung Brubus	Mojokerto	Ngandong	Punung	Wajak
Recent							
Wajak							Wadjak 1&2
Punung						*Homo* sp.	
Ngandong				↑	Ng. 1–14		
				Mo 1			
Kedung Brubus	S3, S12, S17[2]		KB1, femur	↓			
Trinil H.K.	S2, S6	Tr. 1–9					
Tr.HK or Ci S.	S1, S4, S5						
Ci Saat							
Satir							

[1]Hominid positions follow Sondaar (1984).
[2]Hominid catalog numbers follow Oakley et al. (1975).

the skull is found, also about the same depth, equal with the former lower water level, separated from each other 12 meters." Secondly, examination of the fluorine content by Bergman and Karsten (1952) and chemical analysis of the fluorine content and the radioactivity counts for uranium-family elements by Day and Molleson (1973) apparently indicated the contemporaneity of the skullcap and the femora with the Trinil fauna and did not suggest any marked difference in stratigraphic age. Although the debate will go on, based on the letters from Kriele and de Winter and the physico-chemical analyses, I see no reason to doubt the provenance of the skullcap and femur from the "Haupt-knochenschicht." I still consider the femur, although the morphology is modern, as belonging to *Homo erectus*. The left P_3 (Tr. 5), according to von Koenigswald (1967), is undoubtedly from *Homo erectus*.

The right M^3 (Tr.1) and the left M^2 (Tr. 4) were attributed by Dubois (1894, 1924) to *Pithecanthropus erectus* (= *Homo erectus*). Miller (1923) suggested that these molars were those of fossil orang-utan; this opinion was doubted by Weinert (1928), but since then it has been accepted by Gregory and Hellman (1939), Weidenreich (1937), and von Koenigswald (1940). In dealing with an extensive collection of prehistoric teeth of *Pongo pygmaeus*, Hooijer (1948) confirmed Miller's identification and agreed with the suggestion of von Koenigswald (1940) that the second molar must be a third molar and the third molar must be a fourth one. However, according to von Koenigswald (1967), these molars are undoubtedly from *Meganthropus*. I support this idea, because *Pongo* does not fit in the Trinil H.K. fauna, which suggests a more open woodland, while for *Pongo* one should expect a more humid forest. Moreover, it is difficult to accept—although it happens sometimes that *Pongo* has one—that we are dealing here with an M^4.

We may conclude that it seems that in the Trinil H.K. fauna there are two hominids; *viz.*, *Homo erectus* and *Meganthropus*.

KEDUNG BRUBUS

There are two fossils known from Kedung Brubus: mandible A (K.B. 1) and a fragment of femur. The mandible was first described by Dubois (1891) as *Homo* sp. indet. Thirty-three years later Dubois (1924) described it as belonging to *Pithecanthropus erectus* (=*Homo erectus*). According to Tobias (1966), who reexamined the Kedung Brubus mandible, the specimen was from a juvenile; he concluded that "... tempting as it may be to attribute the fragment to *Homo erectus* or even *Homo ?erectus*, the morphology is within the limits possible for *Homo sapiens neanderthalensis*." He thus concluded that the specimen must be identified as *Homo* sp. indet.

The femur, which was probably found at Kedung Brubus, was attributed by Dubois (1935) to *Pithecanthropus*. Day and Molleson (1973), who reexamined this fossil, found a pit close to the linea aspera. They stated that if this pit is for muscle attachment, the femur could not be identified as a hominine, hominid, or other primate. They further reported that paleontologists who have examined this bone (including Dr. Hooijer, then of Leiden) have suggested that it might belong to one of the large carnivores known from the Middle Pleistocene faunas of this area or that it could be a suid femoral fragment. However, in fact, it is neither from a large carnivore nor from a suid. Lucien Jourdan (Marseille; personal communication) suggested that the pit was caused by chipping and that the bone is human.

Based on the mandible, we may conclude that in the Kedung Brubus fauna there is a hominid that seems more progressive than the ones from Trinil.

SANGIRAN

Based on the work of Matsu'ura (1982), Sondaar (1984) suggested the following faunal indication for three groups of Sangiran hominids (see Table 1): 1) those from the upper part of the black clays; i.e., S4 (= *Pithecanthropus* IV, "*P. robustus*"), S1 (= *Pithecanthropus* B), and S5 (= *Pithecanthropus dubius*) equate to a Ci Saat or Trinil H.K. fauna; 2) those from the "Grenzbank", namely S2 (= *Pithecanthropus* II) and S6 (= *Meganthropus* holotype) equate to the Trinil H.K. fauna; and 3) hominids from the so-called Kabuh Fm. between the middle and lower tuff, e.g., S17 (= *Pithecanthropus* VIII), S12 (= *Pithecanthropus* VII), and S3 (= *Pithecanthropus* III) with a Kedung Brubus faunal age.

According to Holloway (1981), the brain endocasts of S17 (1,059 ml) and S12 (1,004 ml) are larger than those of S2 (813 ml) and S4 (908 ml). The suggestions of Sondaar (1984) and Holloway (1981) results fit in with what we found for Trinil H.K. and Kedung Brubus themselves. At Trinil H.K. we have skull I and *Meganthropus*, the same association we find for the suggested faunal correlation of the Sangiran specimens S2, which resembles skull I, and S6, the *Meganthropus* holotype. At Kedung Brubus, we suggested the presence of a more progressive hominid than the Trinil ones, the same as we find for the suggested faunal indication of the Sangiran specimens S17 and S12.

NGANDONG, PUNUNG, AND WAJAK

From the Ngandong locality come the remains of 14 hominid specimens (Ng. 1–14, the so-called Solo Man), which are often considered more progressive than *Homo erectus* from Trinil.

Badoux (1959) reported two upper incisors, an upper canine, a lower canine, and an upper molar among the Punung fossils, which are all too small to belong to the dentition of an orang-utan. He further stated that the von Koenigswald collection contained an upper molar from the same locality, which may provisionally be identified as belonging to *Homo* (cf. *Pithecanthropus*). This identification was probably given because von Koenigswald (1940) considered the Punung fauna as a Trinil fauna. In 1975 von Koenigswald considered the Punung fauna as post-Trinil and stated that a few isolated teeth indicate the presence of man. Recently (de Vos, 1983), I considered the Punung fauna much younger, between Ngandong and Wajak, and I now suggest that the human remains are of *Homo sapiens*.

From Wajak there is a rather complete skull (Wajak 1), skull fragments (Wajak 2), some teeth, and some postcranial skeletal elements. There is general agreement that they belong to *Homo sapiens sapiens*.

MODJOKERTO

As there has always been a debate concerning the age and status of *Homo modjokertensis*, the child of Modjokerto, it is interesting to reconsider some data concerning this fossil.

The Find Circumstances

The skull was found in February, 1936. Von Koenigswald (1936a) announced the find in a newspaper dated March 28, 1936. He stated, concerning the circumstances of the find, that (my translations from the Dutch): "Unfortunately this skull is a surface find and rather eroded." Concerning the geological layer from which this skull came he stated: "Geologically it is of importance, that our find is coming from a layer which must be, concerning the faunal elements, older than the layer from which come the well-known fossils from Trinil." In a newspaper article, Dubois (1936a) doubted both that the skull was from a *Pithecanthropus* and that it originated from a layer older than the Trinil one. In a later newspaper article, von Koenigswald (1936b) stated that it was not a surface find and quoted Ir. J. Duifjes, who stated: "I can positively assure Prof. Dubois that at the find of the new skull a mixing with bones found on the surface is absolutely excluded. It was excavated by our collector Andojo from a hole of 1 m depth in hard conglomeratic sandstone, which with certainty belongs to the fossil bone level of Modjokerto." Von Koenigswald (1936c) also stated that it was great luck that the fossil was found, because the layer was not rich. In 1956 (and 1975), von Koenigswald wrote: "The skull is that of an infant, very thin and only fourteen centimeters long. It was found in a shallow excavation by Andojo, one of the Indonesian assistants working for Ir. Duifjes, who mapped the area. By accident, this excavation was the only one in that region. Andojo had observed the skull of a *Leptobos* partly exposed by erosion and he excavated the fossil; just underneath it, at a depth of less than one meter, he discovered the human skull."

The Position of the Locality

According to von Koenigswald (1936c) the find-spot is 0.5 km from Soembertengah; according to von Koenigswald (1940) it is about 12 km east-northeast of Modjokerto; according to Sartono (1982) its position is some 4 km from Perning, a village about 14 km east-northeast of Modjokerto; according to Oakley et al. (1975), it lies 3 km north of Perning and 15 km northeast of Modjokerto.

Morphology of the Modjokerto Skull

Concerning the morphology, von Koenigswald (1936a) stated that: ". . . the skull of the child is rather high and broad at the rear. The frontal is relatively low, but not flat. Of the strong bony wall above the eye, so typical for *Pithecanthropus* and *Sinanthropus*, there is still nothing to see . . . . Although more than 4 cm shorter than the skullcap found by Dubois, the height plotted against the largest length of the skull is still in both cases the same. This was one of the reasons to consider this small skull as a skull of *Pithecanthropus*." Later, von Koenigswald (1936c, p. 1007) wrote: 'Die Unterschiede sind aber immerhin doch so gross, dass eine Zugehörigkeit des Schädels von Soembertengah zum *Pithecanthropus* nur vermutet, aber beim geringfügigen Material vorläufig nicht bewiesen werden kann. Aus diesen Gründen scheint es besser, unserem Schädel einen besonderen Namen zugeben; ich schlage hierfür den Namen *Homo modjokertensis* vor. Den Namen *Homo* verdient er durch seine ausgesprochene und ins Auge fallende Menschenähnlichkeit, *modjokertensis* nach dem bekannten Ort Modjokerto, der in der Nähe der Fundstelle liegt." In 1975 von Koenigswald explained why he called it *Homo*, stating: "Although it is not directly comparable to the Trinil skull-cap, because of the differences in individual age and fauna, we felt that we could give a separate name to this apparently new *Pithecanthropus*. We had been asked to publish the first description in Holland. In my original manuscript the name *Pithecanthropus modjokertensis* had to be changed to *Homo modjokertensis* because at that time (1936) Professor Dubois regarded his *Pithecanthropus* as an anthropoid and thus no apparently human skull could be called *Pithecanthropus*." Dubois (1936b) and Weidenreich (1939) went further and included the skull of Modjokerto in *Homo soloensis*, on morphological grounds. However, Weidenreich (1939) dated the Modjokerto fossil to the Early Pleistocene on geological grounds. Sartono et al. (1981) concluded on geological and faunal grounds that *Homo modjokertensis* must be included in *Pithecanthropus erectus soloensis* (= *Homo soloensis*).

The Accompanying Fauna and Radiometric Age

Von Koenigswald (1956, 1975) stated that Andojo had found the skull below the skull of a *Leptobos*. In the concept of von Koenigswald (1934), *Leptobos* is a guide fossil of his Jetis (Djetis) fauna. Sartono et al. (1981) reported fossils from the area where the skull was found, including *Hippopotamus namadicus*, "*Buffelus bubalus* ? var. *sondaicus fossilis*," and *Sus brachygnathus*, which they considered characteristic for the Middle Pleistocene Trinil and Ngandong fauna. De Vos et al. (1982) considered the Jetis fauna in the concept of von Koenigswald as a Kedung Brubus fauna.

An age of 1.9 ± 0.4 m.y. was reported by Jacob and Curtis (1971). However, according to Pope (1983), this date has little bearing on the age of either Sangiran 8 or Perning 1 (= child of Modjokerto). According to Sartono (1982), *Homo modjokertensis* (the child of Modjokerto) should be younger than 0.73 m.y.

Thus, for the Modjokerto child's skull, we may conclude that: 1) the find circumstances are not yet clear; 2) the position of the locality is not exactly known; 3) the morphology points to *Homo soloensis*; 4) the fauna is not found in context with the skull; however, *Leptobos* points to a Kedung Brubus fauna; and 5) the age is probably younger than 700,000 years. From this I suggest that the specimen probably should be associated with the Kedung Brubus or Ngandong fauna.

CONCLUSIONS

The faunal succession proposed by Sondaar (1984) is supported by the hominid fossils, which become more progressive from Trinil H.K. to Wajak. The overall indication for the child of Modjokerto is a Kedung Brubus or Ngandong faunal equivalent.

ACKNOWLEDGMENTS

The author is grateful to Dr. P.Y. Sondaar and Dr. J.J.M. Leinders for discussions and to Mr. R. van Zelst for his help in many ways.

LITERATURE CITED

Badoux, DM (1959) Fossil mammals from two fissure deposits at Punung (Java). Utrecht, 1–151.

Bergman, RAM, and Karsten, P (1952) The fluorine content of *Pithecanthropus* and other specimens from the Trinil fauna. Verh. Kon. Ned. Akad. Wetensch. B *55*:150–152.

Day, MH, and Molleson, TI (1973) The Trinil femora. In M Day (ed): Hum. Evol., Symp. Soc. Study Hum. Biol. *11*:127–154.

de Vos, J (1983) The *Pongo* faunas from Java and Sumatra and their significance for biostratigraphical and paleo-ecological interpretations. Proc. Kon. Akad. Wetensch., B *86*:417–425.

de Vos, J, Sartono, S, Hardja-Sasmita, S, and Sondaar, PY (1982) The fauna from Trinil, type locality of *Homo erectus*: A reinterpretation. Geol. Mijnbouw *61*:207–211. [See also discussions and reply in Geol. Mijnbouw *62*:329–343.]

Dubois, E (Anonymous) (1891) Palaeontologische onderzoekingen op Java. Verslag van het mijnwezen 1890 *(4)*:14–15.

Dubois, E (1894) *Pithecanthropus erectus*, eine Menschenaehnliche Uebergangsform aus Java. Batavia: Landersdruckerei.

Dubois, E (1924) Over de voornaamste onderscheidende eigenschappen van den schedel en de hersenen, onderkaak en het gebit van *Pithecanthropus erectus*. Verslag gew. Verg. Wis- en Natuurk. Afd. Kon. Ak. Wetensch. Amsterdam *33:*135–148.

Dubois, E (1935) The sixth (fifth new) femur of *Pithecanthropus erectus*. Proc. Kon. Ned. Ak. Wetensch. *38:*850–852.

Dubois, E (1936a) Nieuwe *Pithecanthropus* ontdekt? Algemeen Handelsblad, Zaterday 18 April 1936, Avondblad, 5de blad, p. 10.

Dubois, E (1936b) Racial identity of *Homo soloensis* Oppenoorth (including *Homo modjokertensis* von Koenigswald) and *Sinanthropus pekinensis* Davidson Black. Proc. Kon. Ak. Wetensch. *39:*1180–1185.

Gregory, WK, and Hellman, M (1939) The South African fossil man-apes and the origin of the human dentition. J. Am. Dent. Assoc. *26:*558–564.

Holloway, RL (1981) The Indonesian *Homo erectus* brain endocasts revisited. Am. J. Phys. Anthropol. *55:*503–521.

Hooijer, DA (1948) Prehistoric teeth of man and of the orang-utan from central Sumatra, with notes on the fossil orang-utan from Java and southern China. Zool. Med. Mus. Leiden *29:*175–301.

Jacob, T, and Curtis, GH (1971) Preliminary potassium-argon dating of early man in Java. Contr. Univ. Calif. Archeol. Res. Facil. *12:*50.

Matsu'ura, S (1982) A chronological framing for the Sangiran Hominids. Bull. Nat. Sci. Mus. Tokyo D. *8:*1–53.

Miller, GS (1923) Notes on the casts of the *Pithecanthropus* molars. Bull. Am. Mus. Nat. Hist. *48:*527–530.

Oakley, KP, Campbell, BG, and Molleson, TI (1975) Catalogue of Fossil Hominids. Part III: Americas, Asia, Australasia. London: Trustees of the British Museum (Natural History).

Pope, GG (1983) Evidence on the age of the Asian Hominidae. Proc. Natl. Acad. Sci. USA *80:*4988–4992.

Sartono, S (1982) Characteristics and chronology of early men in Java. Congrès Int. Pal. Hum. 1er Congrès: 491–541.

Sartono, S, Seman, F, Astadieredia, KAS, Sukendarmono, M, and Djubiantono, T (1981) The age of *Homo modjokertensis*. Mod. Quat. Res. SE Asia *6:*111–119.

Sondaar, PY (1984) Faunal evolution and the mammalian biostratigraphy of Java. Cour. Forsch.-Inst. Senckenberg *69:*219–235.

Tobias, P (1966) A re-examination of the Kedung Brubus mandible. Zool. Med. Mus. Leiden *41:*307–320.

von Koenigswald, GHR (1934) Zur Stratigraphie des javanischen Pleistocän. De Ing. in Ned. Indië *1:*185–201.

von Koenigswald, GHR (1936a) Een nieuwe *Pithecanthropus* ontdekt. Algemeen Indisch Dagblad, Zaterdag 28 Maart 1936. Derde Blad.

von Koenigswald, GHR (1936b) *Pithecanthropus erectus*, Antwoord dr. Von Koenigswald. Algemeen Handelsblad, 7 mei 1936.

von Koenigswald, GHR (1936c) Erste Mitteilung über einen fossilen Hominiden aus dem Altpleistocän Ostjavas. Proc. Kon. Ned. Akad. Wetensch. *39:*1000–1009.

von Koenigswald, GHR (1940) Neue *Pithecanthropus*-Funde 1936–1938. Ein Beitrag zur Kenntnis der Praehominiden. Wet. Med. Dienst Mijnb. Ned. Indië *28:*1–205.

von Koenigswald, GHR (1956) Speurtocht in de prehistorie; ontmoetingen met onze voorouders. Amsterdam: Spieghel.

von Koenigswald, GHR (1967) De *Pithecanthropus*-kiezen uit de collectie Dubois. Verslag gew. Verg. Afd. Nat Kon. Akad. v. Wetensch., Amsterdam *76:*42–45.

von Koenigswald, GHR (1975) Early man in Java: Catalogue and Problems. In RH Tuttle (ed): Paleoanthropology. The Hague: Mouton, pp. 303–309.

Weidenreich, F (1937) The dentition of *Sinanthropus pekinensis*: A comparative odontography of the Hominids. Pal. Sinica, n.,s., D *1:*1–180.

Weidenreich, F (1939) The classification of fossil hominids and their relations to each other, with special reference to *Sinanthropus pekinensis*. Bull. Geol. Soc. China *19:*64–75.

Weinert, H (1928) *Pithecanthropus erectus*. Zeitschr. Anat. Entw. Gesch. *87:*429–547.

Ancestors: The Hard Evidence, pages 221–226

What is "*Pithecanthropus dubius* Koenigswald, 1950"?

Jens Lorenz Franzen
Forschungsinstitut Senckenberg, D 6000 Frankfurt am Main 1, Federal Republic of Germany

ABSTRACT *Pithecanthropus dubius* Koenigswald, 1950 is the oldest hominoid known up to now from Java. It is dated at about 1.4 million years (m.y.) as a minimum. The taxon has been regarded by various authors as either an orang-like anthropoid or an early member of *Homo erectus ("Pithecanthropus")* or a female of *Meganthropus palaeojavanicus* Koenigswald, 1945. Recent comparisons of the type specimen and of mandible C with various hominoids have led to the conclusion that *Pithecanthropus dubius* is definitely not an anthropoid but an early hominid with australopithecine affinities.

ZUSAMMENFASSUNG

Pithecanthropus dubius Koenigswald, 1950 ist bis heute der älteste Hominoide von Java. Sein Alter beträgt wenigstens 1.4 Mio. Jahre. Verschiedene Autoren haben das Taxon entweder als Orang-artigen Anthropoiden oder als einen frühen *Homo erectus ("Pithecanthropus")* oder als ein weibliches Individuum von *Meganthropus palaeojavanicus* Koenigswald, 1945 betrachtet. Neuerdings durchgeführte Vergleiche des Holotypus und von Mandibel C mit verschiedenen Hominoiden haben zu dem Schluß geführt, daß *Pithecanthropus dubius* definitiv kein Anthropoide, sondern ein früher Hominide ist, der Beziehungen zu den Australopithecinen aufweist.

INTRODUCTION

Pithecanthropus dubius, as it was named by the late Professor von Koenigswald in 1950 still belongs among the most enigmatic fossil hominoids. The type of the taxon is a mandible fragment of the right side with M_2 and a heavily worn M_1, the roots of P_4, and some alveolae of the anterior dentition. A remarkable feature is the robustness of the ramus horizontalis, more massive than in the so-called mandible B, which was ascribed by von Koenigswald (1950) to *Pithecanthropus* or, as it is now called, *Homo erectus modjokertensis*. On the other hand, the type mandible of *Pithecanthropus dubius* is distinctly more gracile than that of *Meganthropus palaeojavanicus* Koenigswald (in Weidenreich), 1944 (see Fig. 1).

THE FIRST SPECIMEN: MANDIBLE, 1939

The mandible of *Pithecanthropus dubius* was discovered by Javanese collectors in the spring of 1939 within the dome of Sangiran. From von Koenigswald (1968, p. 102) it can be deduced that the specimen seems to have come from a locality south of Tegapati. Like most of the fossils from Sangiran, it had been collected from the surface, but Professor von Koenigswald told me he was sure that it had come from the lower part of the Pucangan Formation or, as he called it, the Djetis layers.

The specimen was published for the first time by Franz Weidenreich (1945, pp. 15–16, 56–62, Pls. 8–9), who wrote (1945, p. 15): "When I received the cast in 1940, I identified the fragment, not as a portion of a hom-

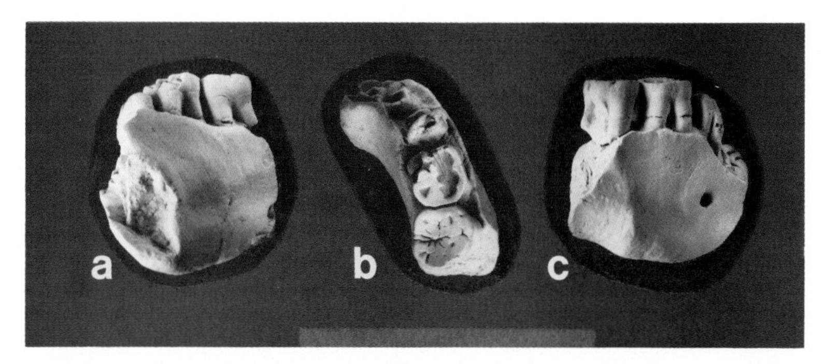

Fig. 1. "*Pithecanthropus dubius* Koenigswald 1950," Holotype, Sangiran 5, fragment of right mandible with M_{2-1} (P_4). a, lingual, b, occlusal, and c, buccal. Whitened with ammonium chloride. a–c ×0.5. Photo: Senckenberg Museum, E. Pantak.

inid mandible but as one from an anthropoid, possibly an orang-utan, isolated teeth of which very commonly occur in the Trinil horizon." Von Koenigswald did not accept this view. In a letter dated February 3, 1941, he wrote to Weidenreich: "It certainly does not belong to *Pithecanthropus* (symphysis) nor to *Simia* (dental arch). Perhaps it is a kind of ancestral pithecus." Later in 1941, having discovered the mandible of *Meganthropus*, von Koenigswald concluded that both mandibles should belong to the same taxon, that of 1939 representing a female, and that of 1941 a male of *Meganthropus*. It is interesting to observe that at that time he regarded *Meganthropus* as being "perhaps related to *Australopithecus*," as it is reported in a letter of W.C.B. Koolhoven, then director of the Geological Survey of the Netherlands Indies, to Franz Weidenreich dated January 15, 1942. It is well known that von Koenigswald later abandoned this idea in opposition to Robinson (von Koenigswald, 1968, p. 106).

After the Second World War, having had the opportunity to study his discoveries from Java together with Weidenreich at the American Museum of Natural History, von Koenigswald changed his view. In 1950 he published a new interpretation of the 1939 mandible, for which he now created a new taxon, *Pithecanthropus dubius*. Obviously decisive was the fact that the "molars show a very curious highly specialized pattern: from the tips of the five cusps emerge strong wrinkles, which practically meet in one central point" (Von Koenigswald, 1950, p. 60).

With that classification, and with von Koenigswald's view of the stratigraphic situation of his Djetis layers, 3 different hominids were known from the Lower Pleistocene of Java: *Pithecanthropus* or *Homo erectus modjokertensis, Pithecanthropus* or *Homo erectus dubius*, and *Meganthropus palaeojavanicus*. In spite of several isolated teeth, however, no bony fragment of fossil *Pongo* is yet known from the Pleistocene of Java.

A SECOND SPECIMEN: MANDIBLE C, 1960

The next important discovery that contributed considerably to the knowledge of *Pithecanthropus dubius* was a mandible found on the surface by a boy near the village of Mandingan in the surroundings of Sangiran. It is much more complete than that of 1939 and has been described in much detail by Sartono (1961). Von Koenigswald (1968) assigned it to *Pithecanthropus dubius*. The right branch of the mandible is preserved with the last two molars, both premolars, the canine, and the alveolae of one lateral and two central incisors. The first molar was evidently lost during the life of that individual (Fig. 2). This mandible C, or Sangiran 9, confirmed the characteristics of the 1939 mandible, now known as Sangiran 5, in particular the robustness of the body, the mesioventral extension of the linea obliqua, and the existence of a relatively large and high-situated foramen mentale. X-rays demonstrated that both premolars had two roots, mesial and distal, and the new discovery displayed a surprisingly small canine (unknown in the first specimen).

As Sartono (1961, pp. 8, 47) assumed from the fact that mandible C was discovered in a topographic position very close to but slightly higher than the diatomite layers

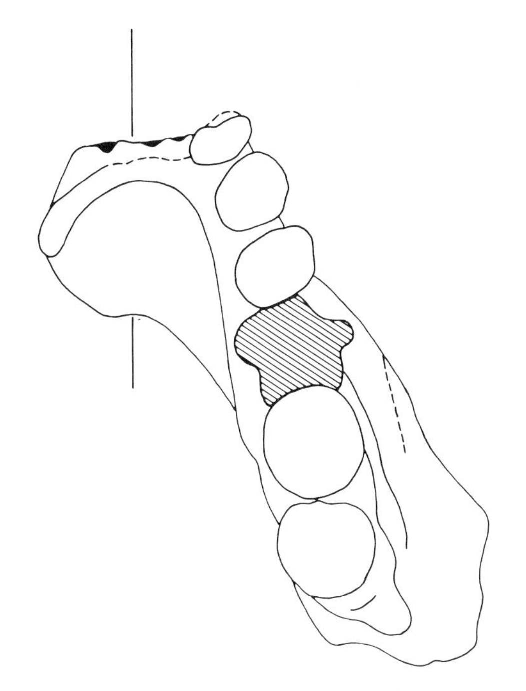

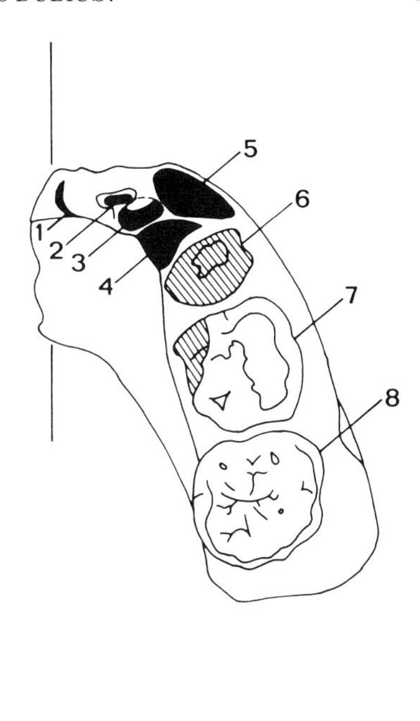

Fig. 2. Schematic comparison of mandibles Sangiran 5 (right) and Sangiran 9 (left) with an interpretation of the alveoli in the holotype of "*Pithecanthropus dubius* Koenigswald 1950." Natural size. Illustration: E. Pantak. 1, ? remnant of the alveolus of I_1 d; 2, remnant of the alveolus of I_2 d; 3, alveolus of C; 4, alveolus of the distolingual root of P_3 d; 5, alveolus of the mesiobuccal root of P_3 d; 6, damaged P_4 d; 7, M_1 d; and 8, M_2 d. Damaged areas are hatched.

within the middle of the Pucangan Formation, it should have come from the lower part of the Upper Black Clay. In 1970 he was able to determine the stratigraphic position of mandible C, based on adhering foraminifera, as Early Pleistocene. This dating has been reassessed by Siesser and Orchiston (1978), in the light of better known age limits for the foraminifera, as more than 1.6 m.y., which would mean that mandible C could well be of Pliocene age. Recent investigations on paleomagnetics carried out by Francois Semah (1982 a,b), however, point to a somewhat younger age of about 1.4 m.y. as a minimum. At any rate, the so-called *Pithecanthropus dubius* appears to be the oldest fossil hominoid from Java. The exact evaluation of its taxonomic and phylogenetic position is therefore very important for the understanding of the early evolution of man in Southeast Asia, and especially on Java.

MORPHOLOGICAL ANALYSIS

Does *Pithecanthropus dubius* really belong to the hominids? Or is it a fossil *Pongo* as Weidenreich supposed and Krantz (1975) suggested again? The two-rooted premolars and the wrinkled enamel of the occlusal surface of the cheek teeth could very well indicate some kind of fossil *Pongo*, known by isolated teeth from Java as well as from Southern China. If it is a hominid, to which subgroup does it belong?

In order to answer these questions, I have restudied the type of *Pithecanthropus dubius*, which is housed at the Senckenberg Museum at Frankfurt, as well as a cast of mandible C, the original of which belongs to the Geological Survey of Indonesia (Museum Geologi, Bandung). Weidenreich had only a cast of the type when he published the specimen in 1945. His very detailed description of the mandible and its dentition turns out to be essentially correct. But he was not sure about the interpretation of the alveolae of the anterior teeth and, in consequence, of the position of the symphysis.

Having the original and a cast of mandible C at hand, it is now easy to determine that Weidenreich's first interpretation, upon which he based his reconstruction of the mandible (1945, pl. 9, Figs. b 1–4), as well as the hypothesis that it might belong to an orang-like anthropoid, is wrong: the anterior root of P_3 is not situated mesially to the distal one but mesiobuccally. The shape of P_3 was therefore not sectorial and unicuspid as Weidenreich believed, and as seen in most non-human hominoids, but very much like that of a primitive human such as *Australopithecus afarensis*, as proven by mandible C. What Weidenreich regarded as the alveolus of the anterior root of P_3 is instead that of a small canine, and what he considered as "the broken and corroded (or otherwise damaged) root of the canine" is part of the spongiosa supporting the bottom of the alveolus of the lateral incisor. The anterior fracture of the type mandible of *Pithecanthropus dubius* is consequently not situated "lateral to the mid line" (Weidenreich, 1945, p. 56) but coincides more or less with the median section through the symphysis. Thus the interpretation of the morphology of the type mandible changes from a more apelike to a more humanlike appearance.

In order to review the taxonomic position of *Pithecanthropus dubius*, I have made comparisons of the two *dubius* mandibles with recent pongids as well as with various fossil hominoids, with australopithecines, *Meganthropus, Homo habilis, Homo erectus,* and *Homo sapiens*. These comparisons are based upon original material, casts, and the literature, and they are focused on 13 characters that seemed in the beginning to be potentially interesting for a reassessment of the taxonomic position of *Pithecanthropus dubius*.

The results of my investigation are expressed in Table 1. Looking at the distribution of characters, it becomes clear that some that have been emphasized in the literature as being of taxonomic value—like the relative size of M_3, enamel wrinkling, occurrence of a spina mentalis, and particularly an extended linea obliqua—turn out to be of no diagnostic value, at least in this context. But there are also typical characters of *Pithecanthropus dubius* that appear to indicate a certain hominid affinity, while others have a more pongid distribution.

CONCLUSIONS

Taken altogether, *Pithecanthropus dubius* is mainly characterized by morphological traits that are essentially human, like a small canine, absence of a simian shelf, lack of a sectorial unicuspid P_3, and a comparatively highly-situated foramen mentale. As for features of pongid affinity, like the two-rooted P_3 and P_4 and a shallow but distinct fossa genioglossi, these are shared with other early hominids, especially with the australopithecines. In this way we get a very clear-cut picture: *Pithecanthropus dubius* is definitely a hominid and not a pongid.

But is this the whole story? Is it satisfactory to assess the taxonomic position of fossils simply by comparing and counting characters? I don't think so. As Hennig (1950, 1966) has demonstrated, we have to differentiate between apomorphic and plesiomorphic characters, because only shared specialized characters may indicate a recent common ancestor and hence phylogenetic and taxonomic relationship. Specialization is a matter of adaptation, i.e., of function. Therefore, we have to think of function and of functional correlation of the characters involved.

In the case of *Pithecanthropus dubius*, the two-rooted premolars and the fossa genioglossi appear to be widespread among fossil and recent hominoids. Obviously these characters are symplesiomorphic and are therefore of no taxonomic value in this context. Concerning the other characters of hominid distribution, at least some of them seem to be functionally correlated. A small canine "needs" no simian shelf to strengthen mechanically the critical point of the symphysis. A small canine in the mandible should correspond to a small canine in the upper jaw, which in turn would not "need" a sectorial unicuspid lower third premolar as a honing antagonist.

Consequently, characters 1, 4, 5, and 9 (Table 1) seem to belong to one interrelated functional complex, which evidently has something to do with reduction of the anterior teeth, a typical trait of human evolution correlated with the achievement of upright posture. Therefore, these characters have to be regarded as synapomorphic hominid features. But as they are functionally correlated to each other, they do not count as several characters testing the hominid status of *Pithecanthropus dubius* independently. Such an independent test could be based on the relative position of the foramen mentale. The functional significance of this character is not clear to me, however, and has to be investigated further. Comparisons among recent pongids and humans show that

TABLE 1. Distribution of mandibular characters among fossil and recent hominoids compared to "Pithecanthropus dubius Koenigswald, 1950"

Character[1,2]	Taxon																"*P. dubius*", 1939	"*P. dubius*", 1960
	Pongo	*Gorilla*	*Pan*	*Ramapithecus*	*Sivapithecus*	*Dryopithecus*	*Proconsul*	*Ouranopithecus*	*Gigantopithecus*	*A. afarensis*	*A. africanus*	*A. robustus*	*Meganthropus*	*H. habilis*	*H. erectus*	*H. sapiens*		
1 Small canine	–	–	–	±	±	–	±	±	±	±	⊞	⊞	⊞	⊞	⊞	⊞	⊞	⊞
2 P_3 two-rooted	⊕	⊕	±	⊕	⊕	⊕	⊕	⊕	⊕	+(–)	⊕⃞	⊕⃞	–	–	–	–	⊕⃞	⊕⃞
3 P_4 two-rooted	⊕	⊕	⊕	⊕	⊕	⊕	⊕	⊕	⊕	⊕⃞	⊕⃞	⊕⃞	–	–	–	–	⊕⃞	⊕⃞
4 P_3 not sectorial	–	–	–	–	–	–	–	–	–	⊞	⊞	⊞	⊞	⊞	⊞	⊞	⊞	⊞
5 P_3 bicuspid	–	–	–	–	–	–	–	–	⊞	⊞	⊞	⊞	⊞	⊞	⊞	⊞	⊞	⊞
6 Enamel wrinkling[3]	++	–	+	–	M_3	–	–	M_3	–	M_3	M_3	M_3	+	M_3	M_3	–	+	?
7 M_3 shorter than M_2	±	±	+	–	–	–	–	–	–	–	–	–	?	–	–(+)	–(+)	?	+
8 Thick enamel	+	–	–	+	+	–	+	+	+	+	+	+	+	+	+	+	+	+
9 Simian shelf	+	+	+	+	+	+	⊟	+	+	±	⊟	⊟	⊟	⊟	⊟	⊟	⊟	⊟
10 Fossa genioglossi	⊕	⊕	⊕	⊕	⊕	⊕	⊕	⊕	⊕	⊕⃞	⊕⃞	⊕⃞	⊕⃞	–	–	–	?	⊕⃞
11 Spina mentalis	±	+	+	–	–	?	+	±	–	±	+	+	+	+	+	+	+	?
12 F. mentale low	+	+(–)	+(–)	+	±	+	+	+	+	⊟	⊟	⊟	⊟	⊟	⊟	–(+)	⊟	⊟
13 Linea obliqua ext.	+(–)	±	–	–	+	+	+	–	±	?	?	–	?	+	±	±	+	+

[1]Occurrence of characters: +, present; –, absent; ±, mixed or uncertain; –(+) or +(–), usually (sometimes); ++, extreme; ?, not known.
[2]Distribution: ○, character of mainly pongid distribution; □, character of mainly hominid distribution; ○⃞, character of mainly pongid distribution occurring among early hominids.
[3]M_3, enamel wrinkling mainly on M_3.

a low situation of the foramen mentale has nothing to do with the size of the canine or its root.

Whether the so-called *Pithecanthropus dubius* has to be regarded as an early australopithecine offshoot that migrated to Southeast Asia and/or as an early *Homo erectus* (Wolpoff, 1975, p. 24), and whether this taxon could serve as an ancestor of later *Homo erectus* from Java and/or of *Meganthropus* are matters for further investigation, hopefully to be aided by new fossil discoveries in Java.

LITERATURE CITED

Hennig, W (1950) Grundzüge einer Theorie der phylogenetischen Systematik. Berlin: Deutscher Zentralverlag.

Hennig, W (1966) Phylogenetic systematics. Urbana: University of Illinois Press.

Krantz, GS (1975) An explanation for the diastema of Javan erectus skull IV. In RH Tuttle (ed.): Paleoanthropology. Morphology and Paleoecology. The Hague: Mouton, pp. 361–372.

Sartono, S (1961) Notes on a new find of a *Pithecanthropus* mandible. Publikasi Teknik Seri Paleontologi, Bandung, *2:*1–51.

Sartono, S (1970) On the stratigraphic position of *Pithecanthropus* mandible C. Proc. Inst. Teknologi Bandung *4(4):*91–102.

Semah F (1982a) Chronostratigraphie et Paleomagnetisme du Plio-Pleistocene de Java. Application à l'age des sites à Pithecanthropes. 1er Congr. Internat. Paléont. Humaine, Nice, Prétirage *2:*542–558.

Semah, F (1982b) Pliocene and Pleistocene geomagnetic reversals recorded in the Gemolong and Sangiran Domes (Central Java). Mod. Quatern. Res. SE Asia *7:*151–164.

Siesser, WG, and Orchiston, DW (1978) Micropaleontological re-assessment of the age of *Pithecanthropus* mandible C from Sangiran, Indonesia. Mod. Quatern. Res. SE Asia *4:*25–30.

Von Koenigswald, GHR (1950) Fossil Hominids from the Lower Pleistocene of Java. Rep. 18th Sess. Int. Geol. Congr., Great Britain, 1948, *Part IX:*59–61.

Von Koenigswald, GHR (1968) Observations upon two *Pithecanthropus* mandibles from Sangiran, Central Java. Proc. Koninkl. Nederl. Akad. Wetensch., Ser. B *71:*99–107.

Weidenreich, F (1945) Giant Early Man from Java and South China. Anthropol. Pap. Am. Mus. Nat. Hist. *40:*1–134.

Wolpoff, MH (1975) Some aspects of human mandibular evolution. In JA McNamara, Jr. (ed): Determinants of Mandibular Form and Growth. Ann Arbor: University of Michigan Press, pp. 1–64.

PLATES

View of Gallery 1 at the opening of "Ancestors: Four Million Years of Humanity."

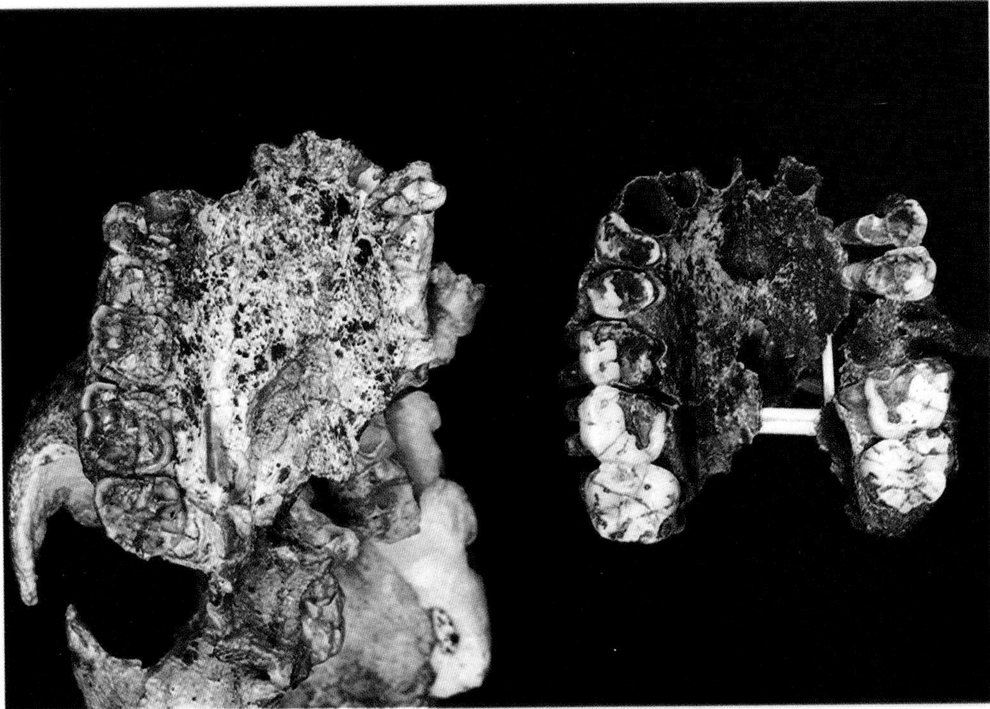

Plate 1. Sts. 71 (partial skull, *A. africanus*) and Stw. 53 (maxilla, *H. habilis*). Oblique right frontal view (above); occlusal view.

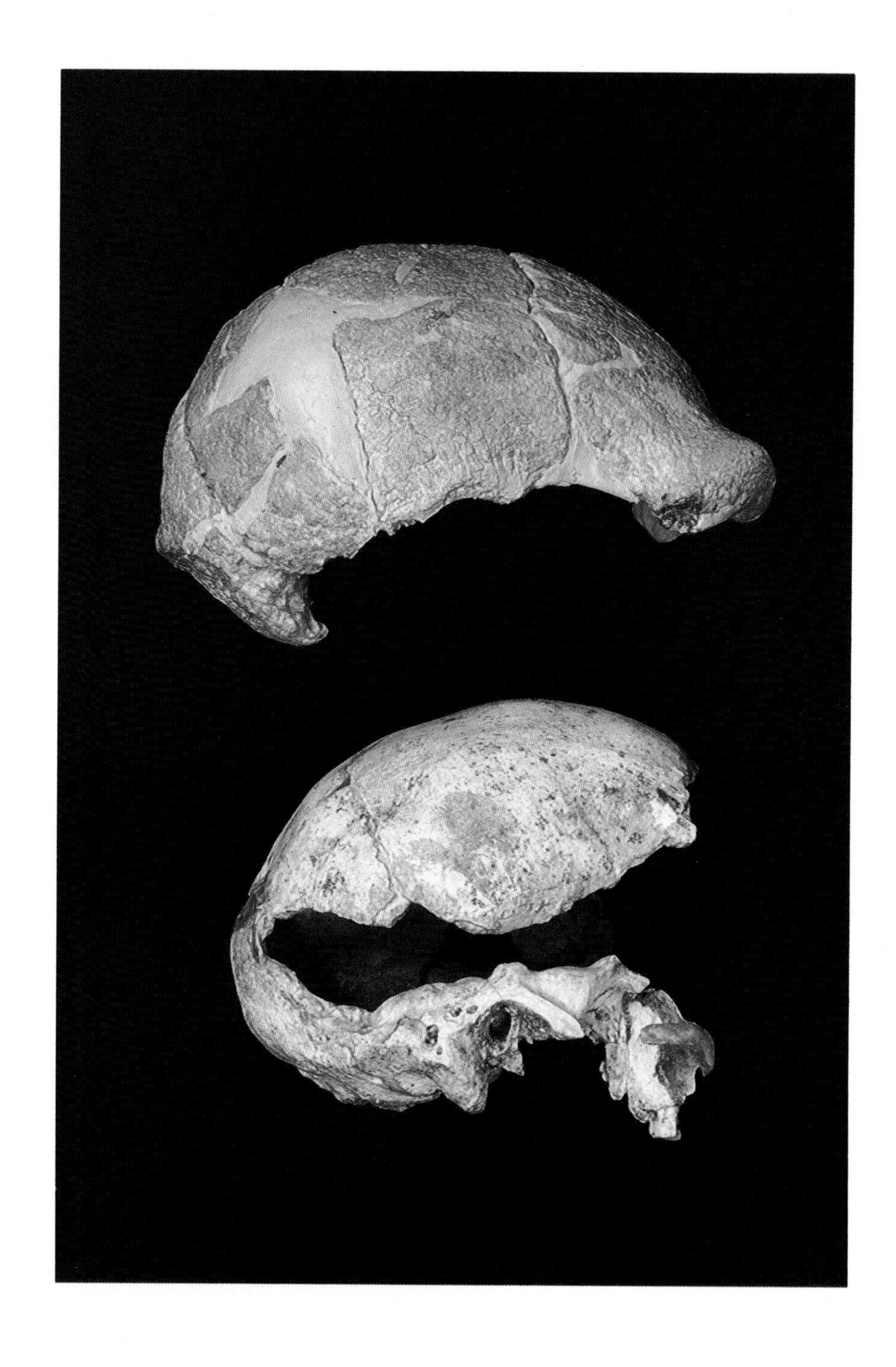

Plate 2. Saldanha 1 (archaic *Homo sapiens*; above) and Salé 1 (cf. *Homo erectus*), right lateral views.

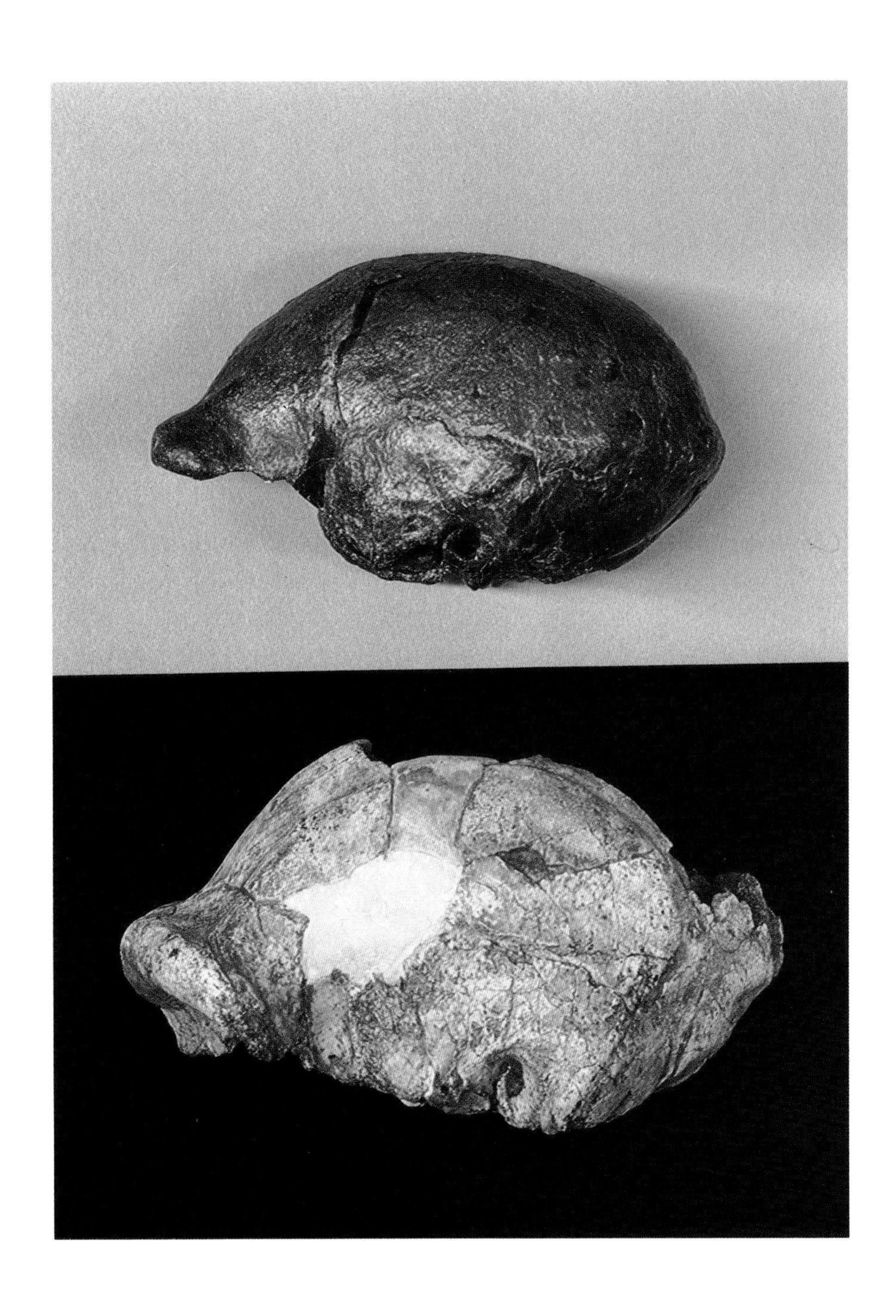

Plate 3. Sangiran 2 (above) and O.H. 9, *Homo erectus*, left lateral view.

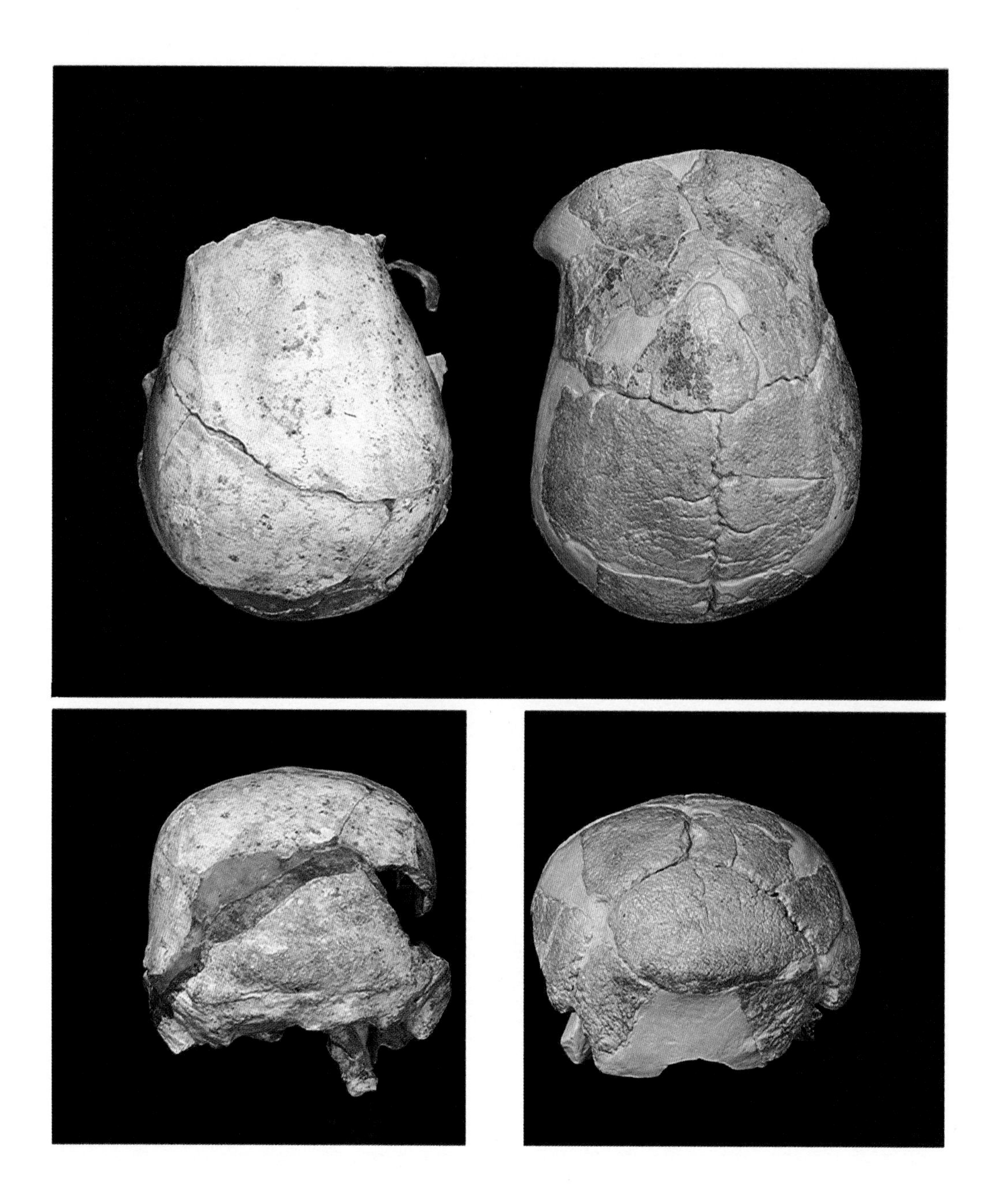

Plate 4. Saldanha 1 (archaic *Homo sapiens*; right) and Salé 1 (cf. *Homo erectus*); dorsal view (above); occipital view.

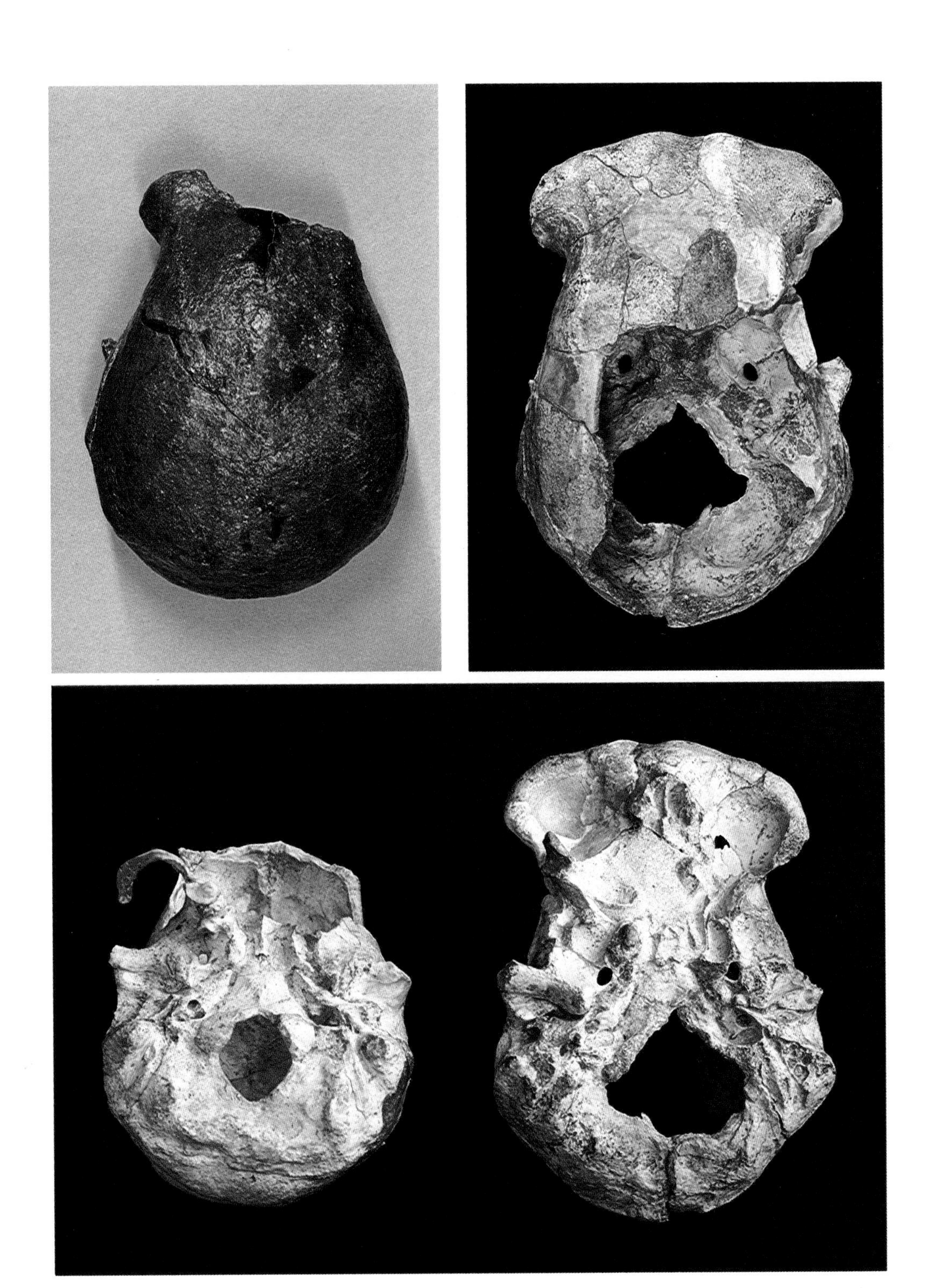

Plate 5. Above, Sangiran 2 (left) and O.H. 9, *Homo erectus*, dorsal view. Salé 1 (cf. *Homo erectus*; left) and O.H. 9, basal view.

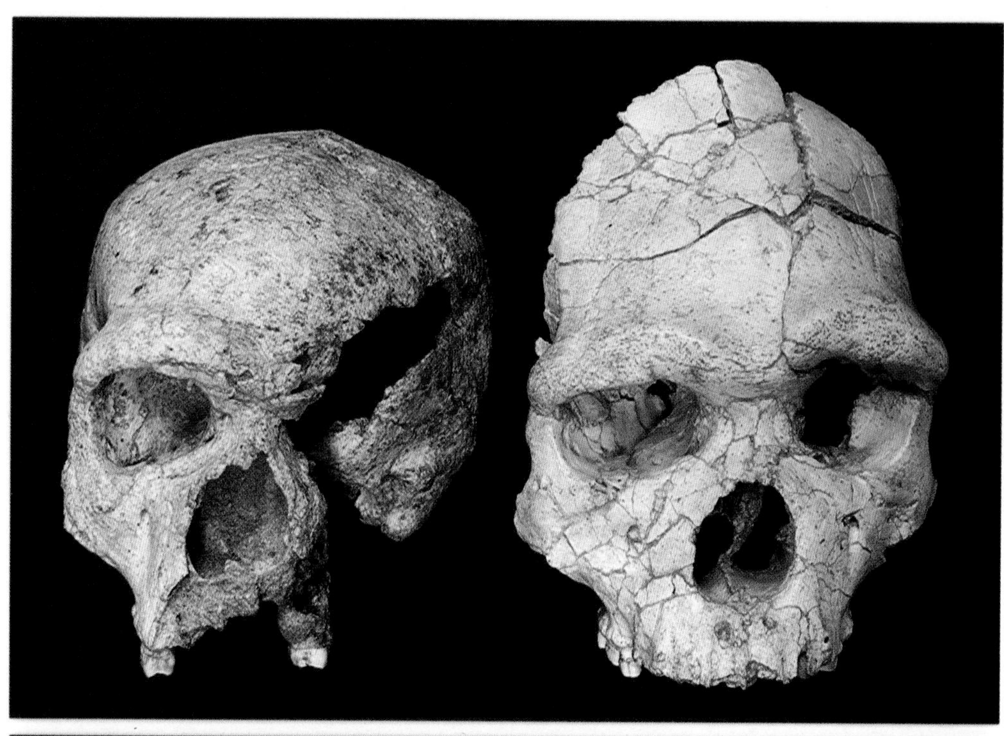

Plate 6. Steinheim 1 (left), Arago 21 (archaic *Homo sapiens*); (above) frontal view; occlusal view.

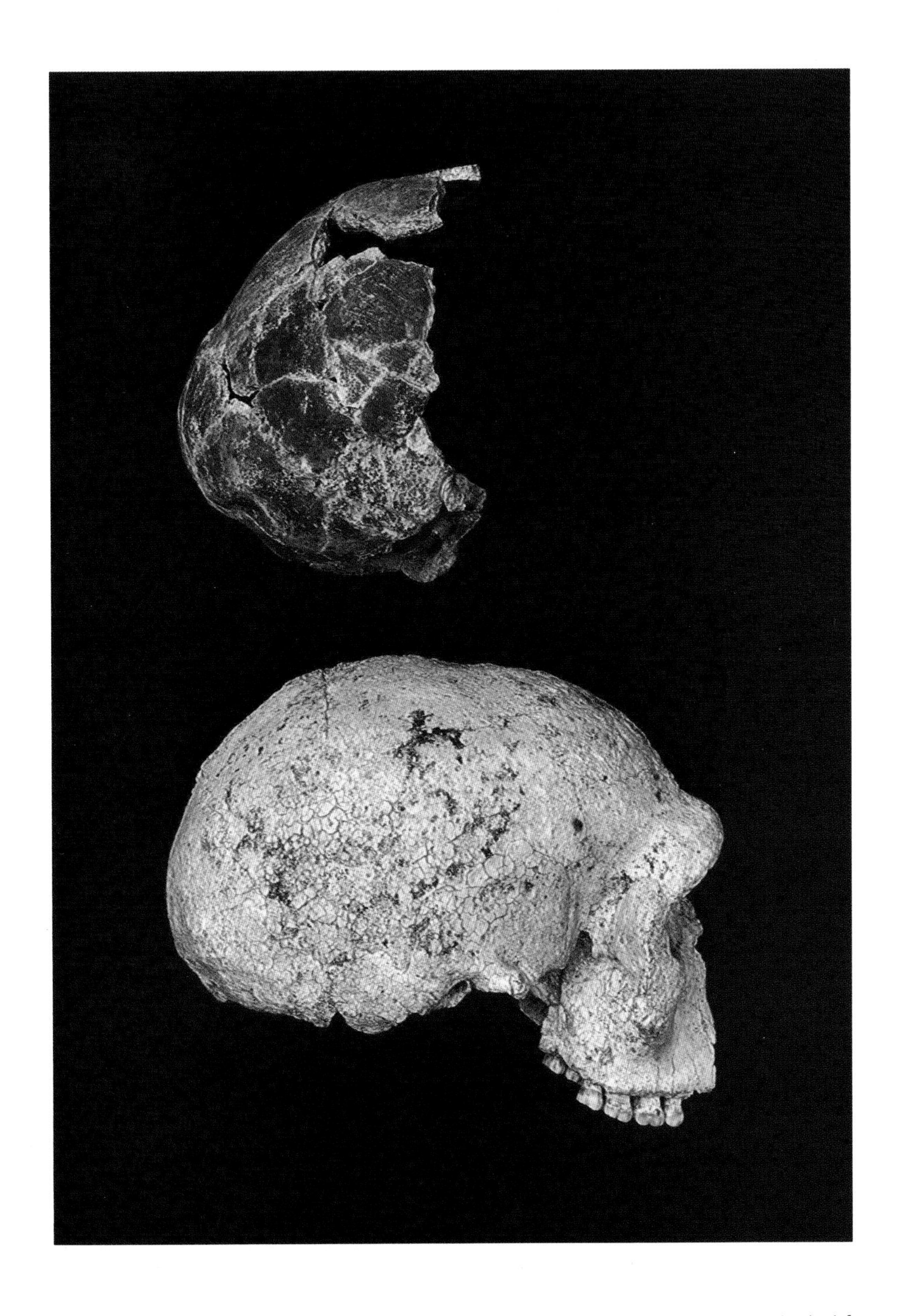

Plate 7. Biache 1 (above, *Homo sapiens cf. neanderthalensis*), Steinheim 1 (archaic *Homo sapiens*), right lateral view.

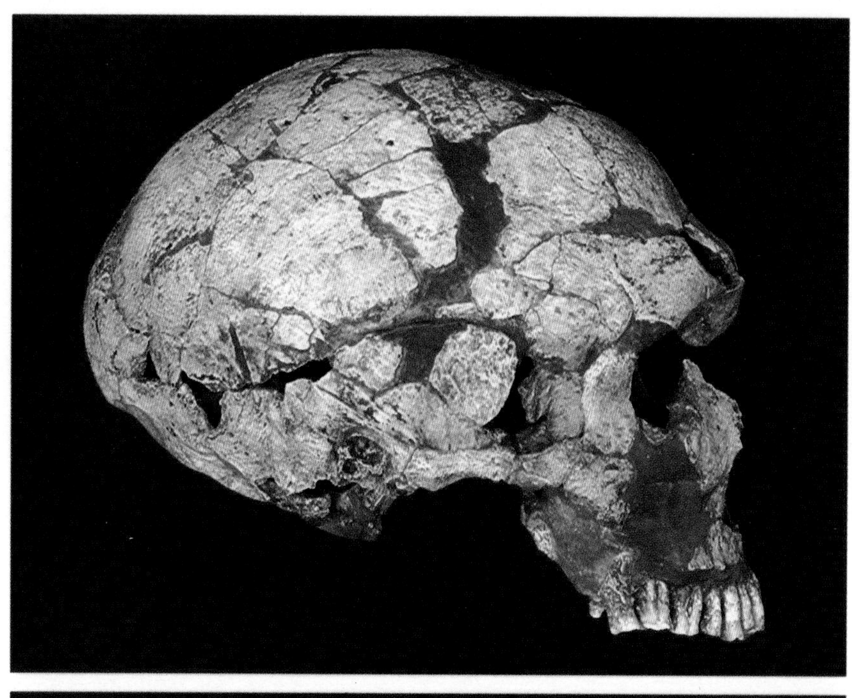

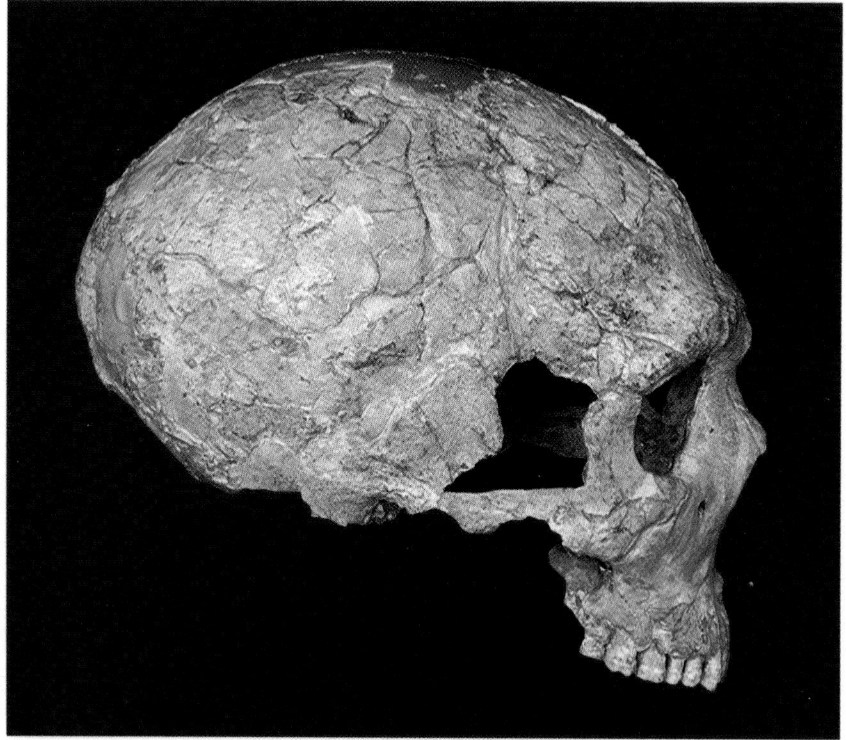

Plate 8. La Ferrassie 1 (above) and Amud 1 *(Homo sapiens neanderthalensis)*, right lateral view.

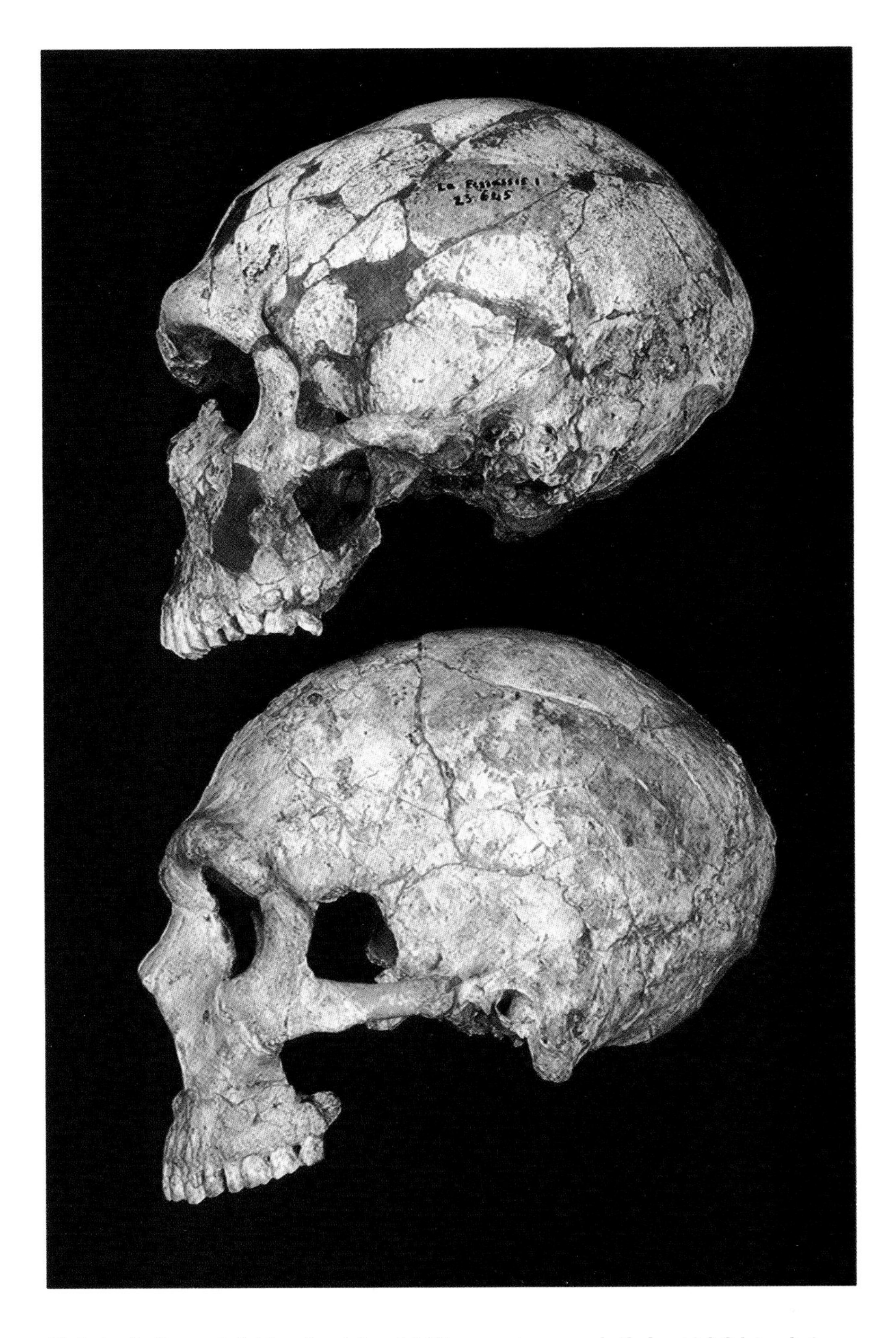

Plate 9. La Ferrassie 1 (above) and Amud 1 *(Homo sapiens neanderthalensis)*, left lateral view.

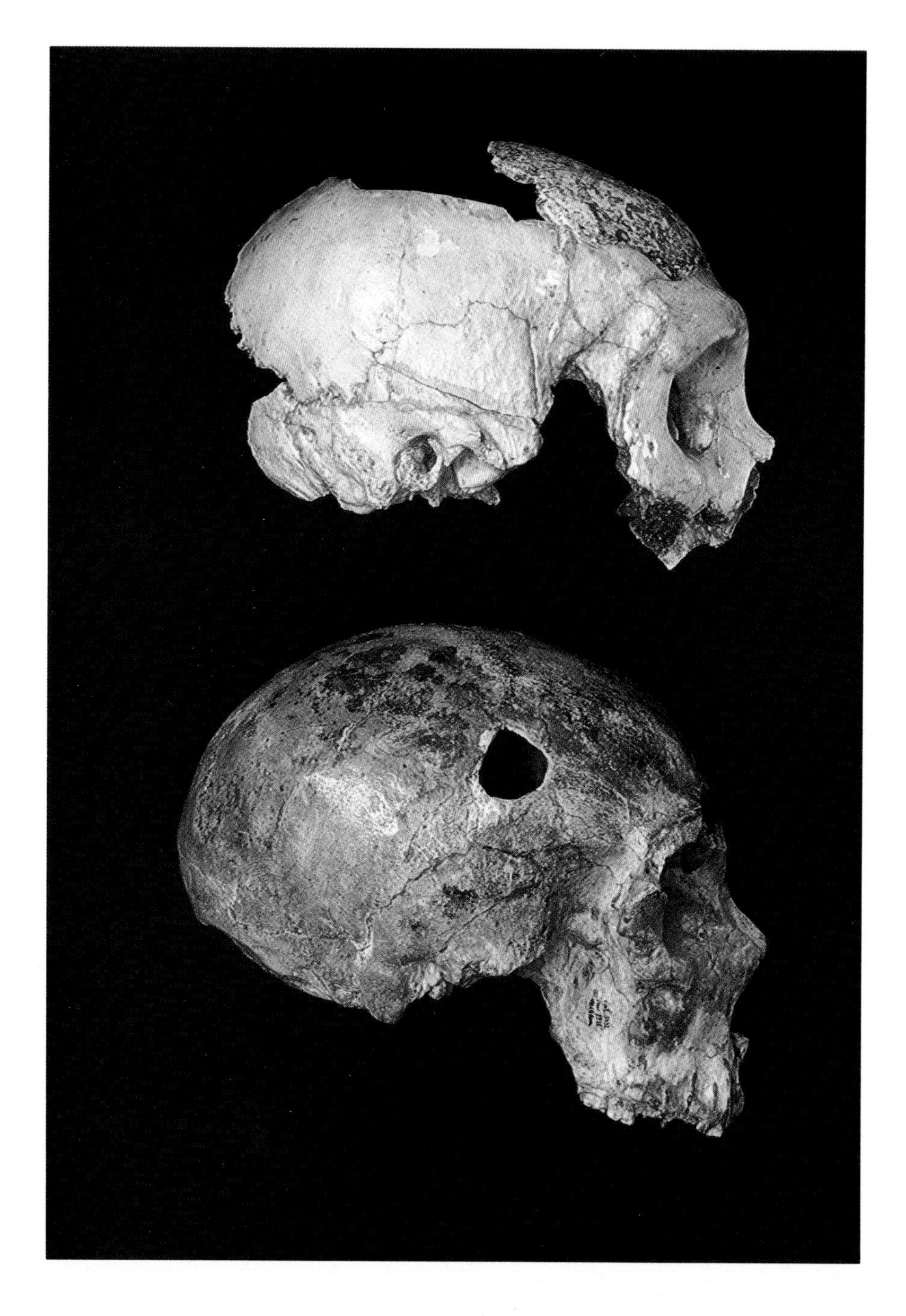

Plate 10. Krapina C/3 (above) and Saccopastore 1 *(Homo sapiens neanderthalensis)*, right lateral view.

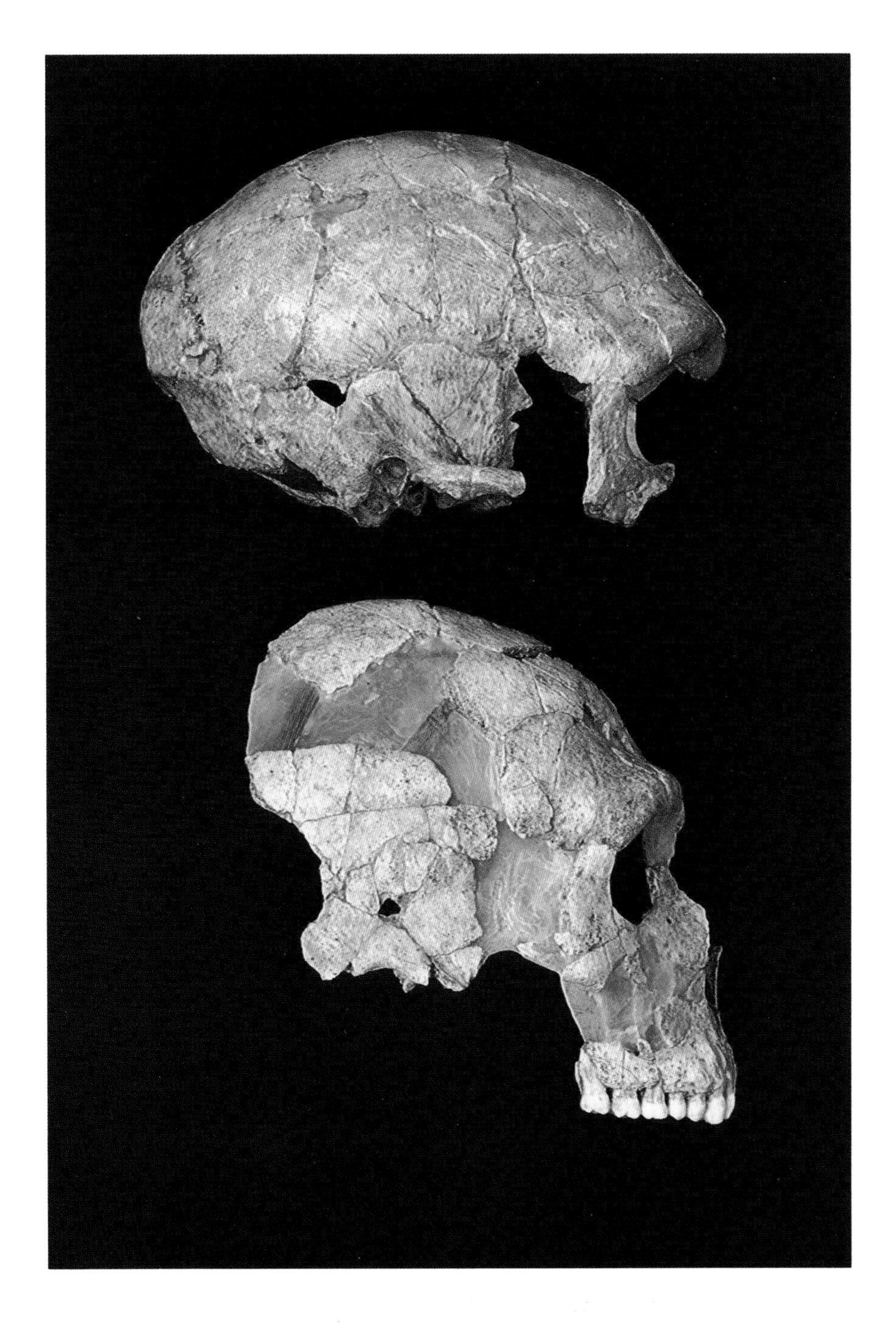

Plate 11. La Quina H5 (above) and Saint-Césaire 1 *(Homo sapiens neanderthalensis)*, right lateral view.

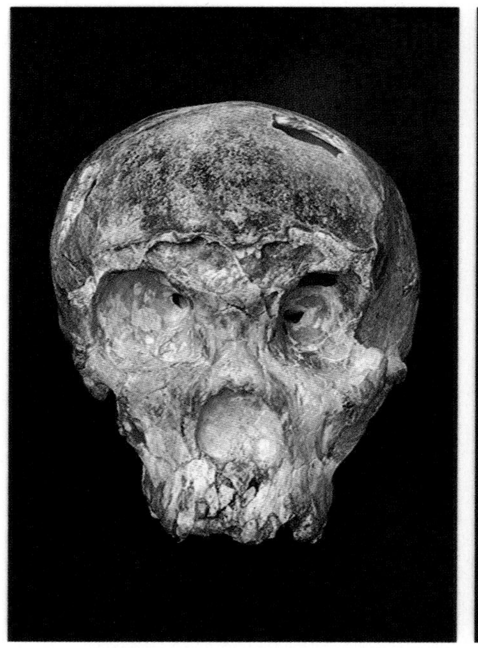

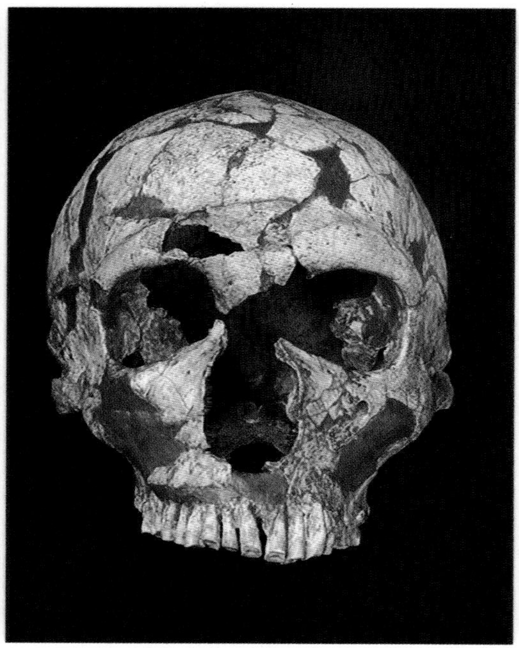

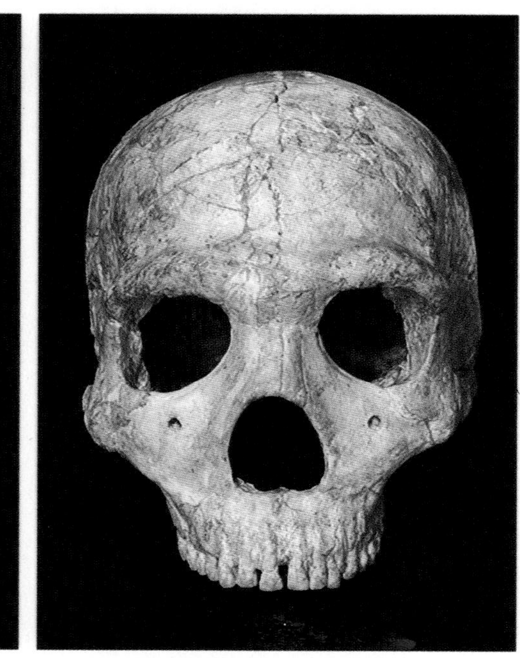

Plate 12. *Homo sapiens neanderthalensis*, frontal view, left to right: Saccopastore 1, La Ferrassie 1, Amud 1.

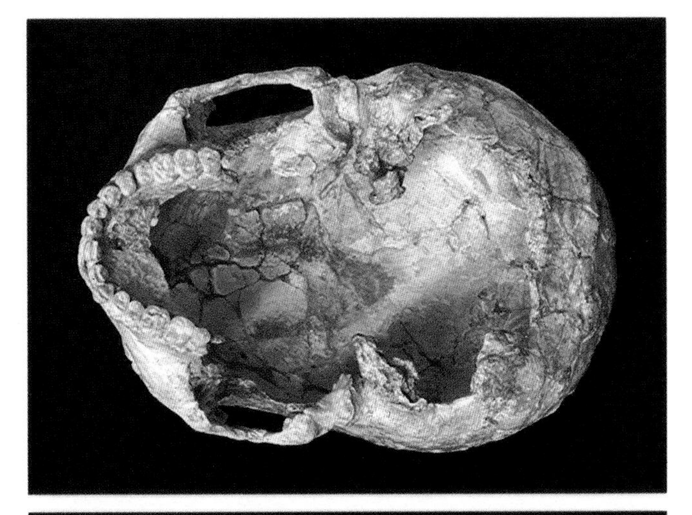

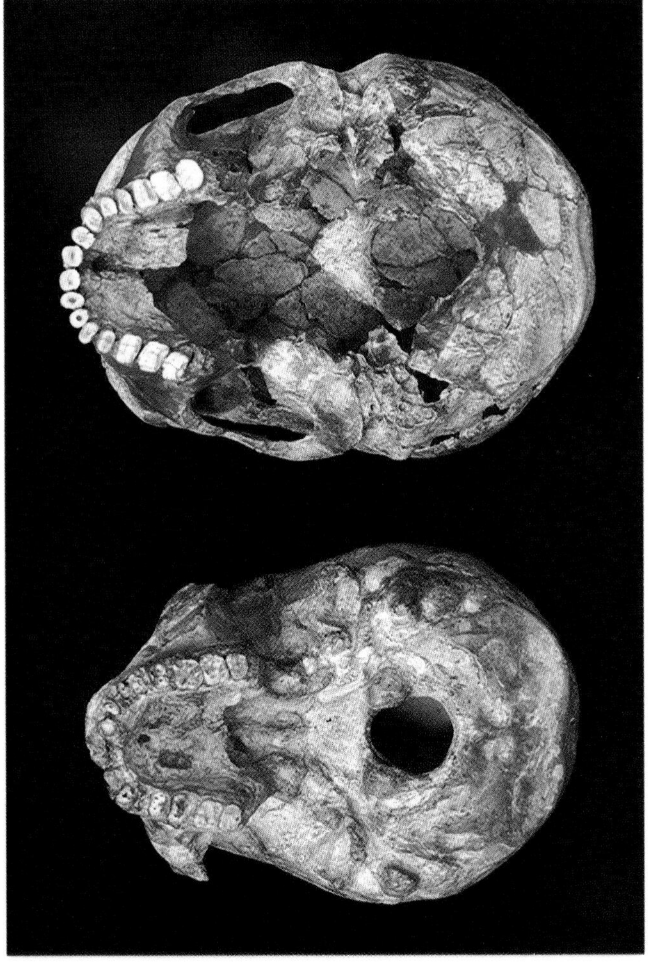

Plate 13. *Homo sapiens neanderthalensis*, basal view, left to right: Saccopastore 1, La Ferrassie 1, Amud 1.

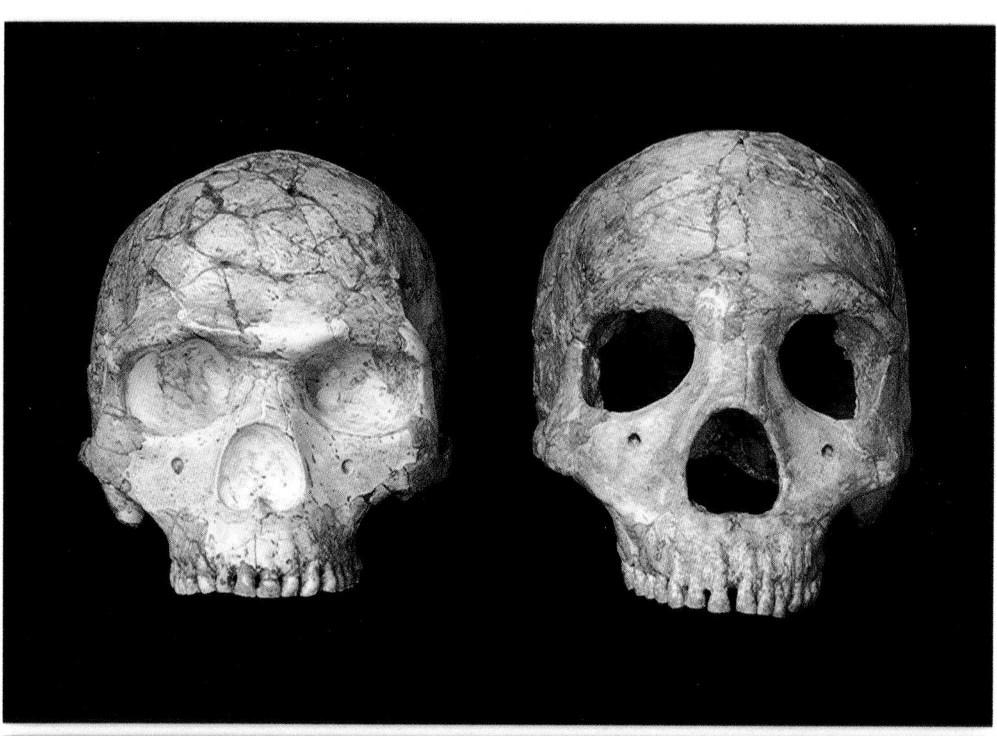

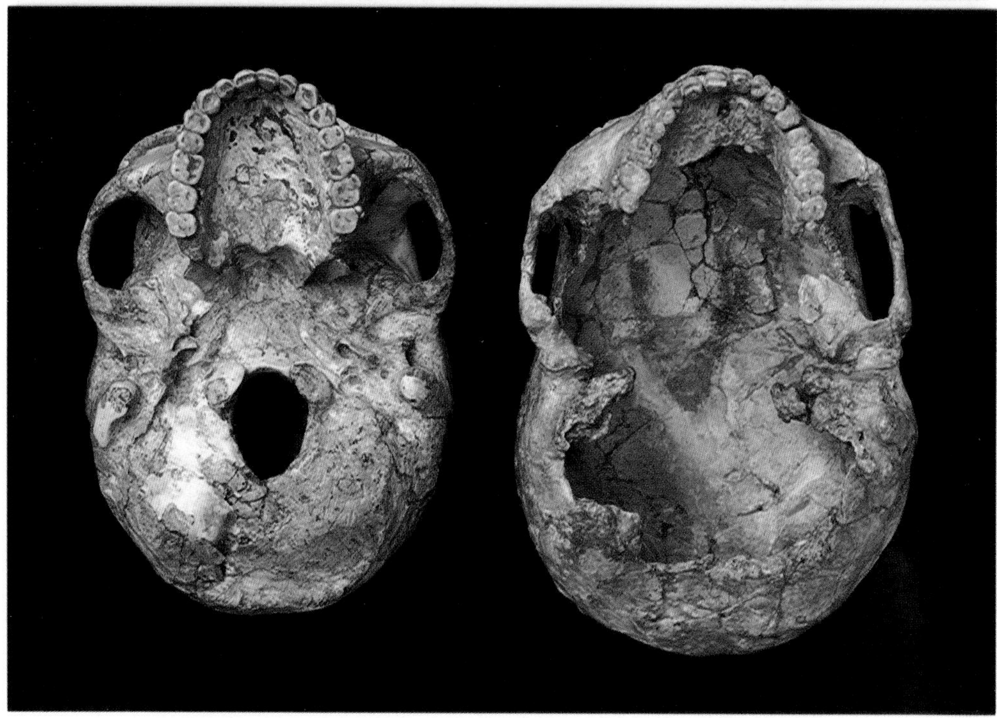

Plate 14. Skhūl 5 (left, *Homo sapiens sapiens*) and Amud 1 *(Homo sapiens neanderthalensis)*; (above) frontal view; basal view.

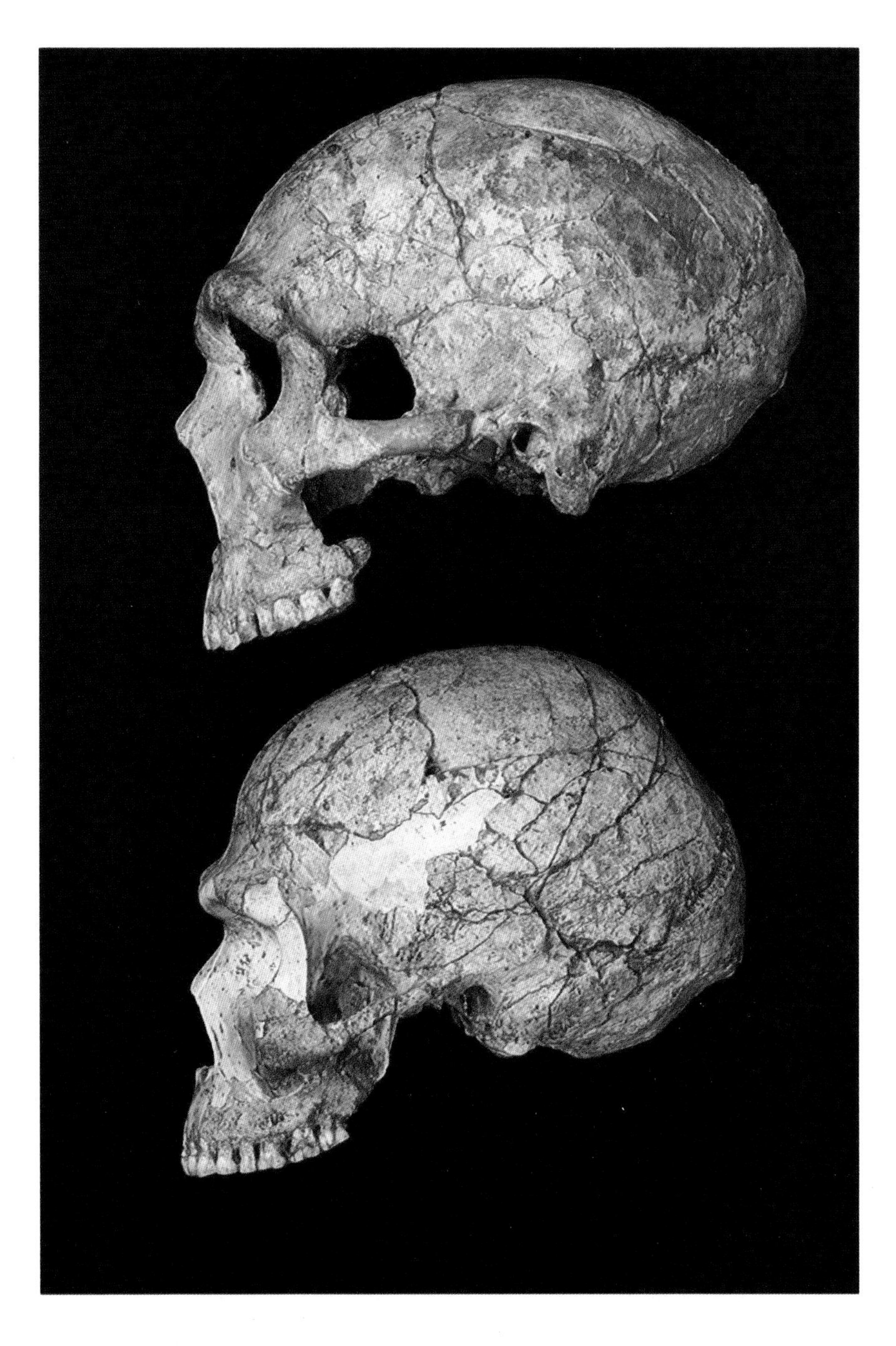

Plate 15. Skhūl 5 (below, *Homo sapiens sapiens*) and Amud 1 *(Homo sapiens neanderthalensis)*, left lateral view.

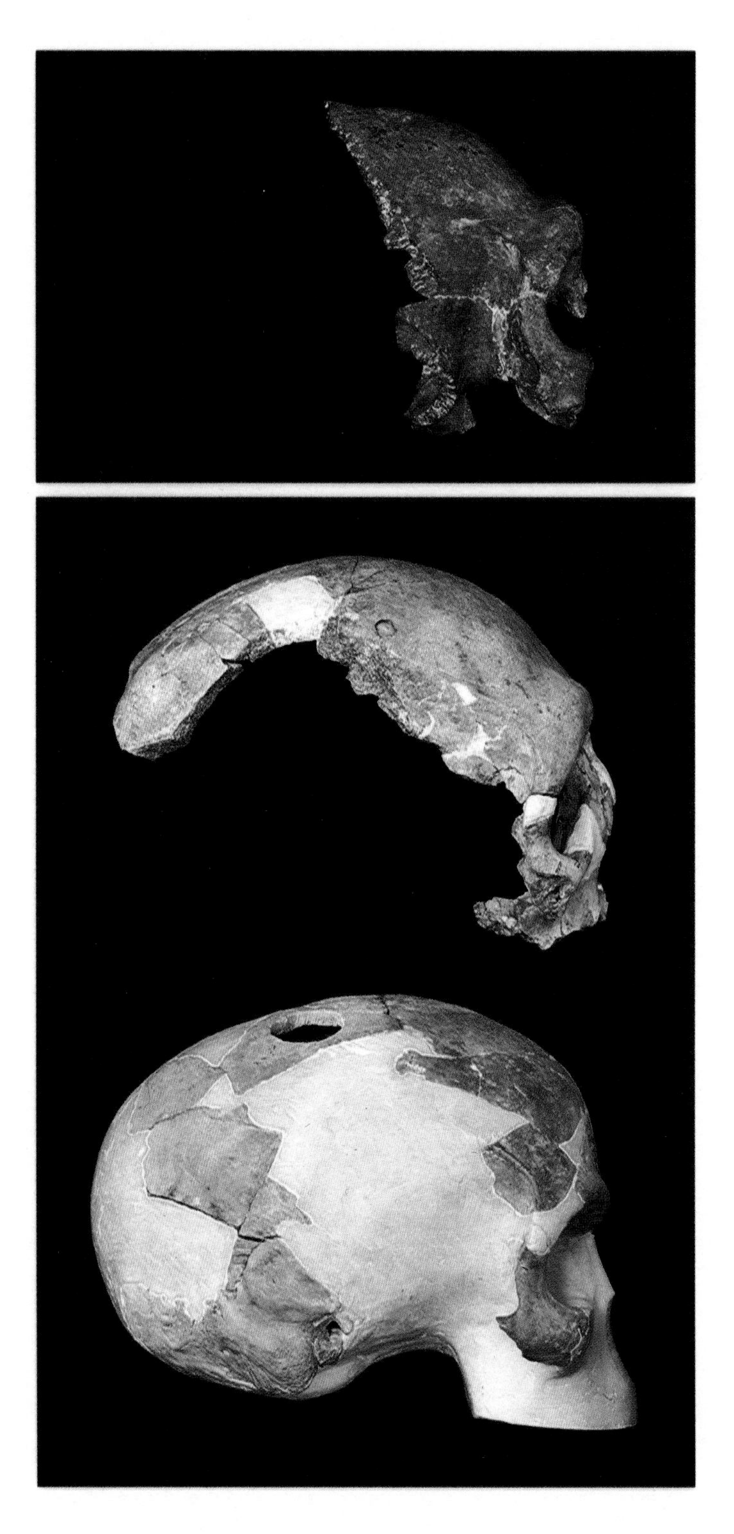

Plate 16. Zuttiyeh 1 (top, *Homo sapiens* subsp. indet.), Florisbad 1 (*Homo sapiens* subsp. indet.), Border Cave 1 (bottom, *Homo sapiens sapiens*), right lateral view.

Ancestors: The Hard Evidence, pages 245–248

New Chinese *Homo erectus* and Recent Work at Zhoukoudian

Wu Rukang (Woo Ju-Kang)
Institute of Vertebrate Paleontology and Paleoanthropology, Academia Sinica, Beijing 28, People's Republic of China

ABSTRACT The partial cranium, additional skull and mandible fragments, and isolated teeth from Hexian are the most important remains of *Homo erectus* recovered in China in recent years. The cranium resembles those from Zhoukoudian in most morphological and metrical features, but it has less postorbital constriction and a high temporal squama. These progressive features suggest comparison with the youngest Peking Man skull, and faunal evidence also indicates correlation with the uppermost layers of Zhoukoudian. Ongoing excavations in the latter site have not yet yielded any new human fossils, but they have produced a wealth of data about the cultural assemblage, fauna, paleogeography, and paleoclimatology. A variety of dating methods have been applied to these sites and to others producing Chinese *H. erectus*. The Yuanmou teeth may be only 600,000 years old, close in age to the Chenjiawo (Lantian) mandible, while the Gongwangling (Lantian) cranium could be 750,000–1 million years old. The Zhoukoudian deposits with human fossils (Layers 3–11) appear to span most of the interval between 230,000 and 450–500,000 years ago. Correlations to the oxygen isotope stages permit more detailed hypotheses as to the ages of these sites and layers.

INTRODUCTION

In recent years, China has yielded a number of new *Homo erectus* fossils from several localities. Small groups of isolated teeth were found in Yunxi (two teeth) and Yunxian (four teeth) Counties in Hubei Province (Wu and Dong, 1980, 1983) and in Xichuan (13 teeth; Wu and Wu, 1982) and Nanzhao (four teeth; Qiu et al., 1982) Counties in Henan Province. The most significant new discovery, however, was a partial skull found in Lontandong Cave (Hexian County, Anhui Province) in November, 1980 (Huang et al., 1981). The recovery of the Hexian skull marked the first time that a *Homo erectus* cranium was found in East or Southeast China. This find affords us fresh evidence in the study of the evolution of *H. erectus* in China and in Asia generally, and it may help us to understand the question of local variation in this species as well. At the same time, work has recommenced at the world-renowned site of Zhoukoudian (Chou-k'ou-tien) Locality 1, which has yielded the largest known sample of *Homo erectus*. Moreover, a variety of dating methods has been applied to these sites and to others with *H. erectus* in China.

HEXIAN MAN

The Hexian fossil site (118° 20′ E, 31° 45′ N) is located on the north slope of Wangjishan Hill in the Taodian Commune of Hexian County, which lies north of the Yangzi River. The locality was first found in 1973, when a cave containing abundant animal fossils was exposed by local peasants during the construction of a canal on the side of the hill slope. The contents of the cave were soon brought to the attention of local geologists. Later, due to their recognition of the fossils' importance, paleontologist Huang Wanbo and others from IVPP, together with local archeologists, went to the site for excavation. During the two excavations in 1980 and 1981, the following *Homo erectus* fossils were unearthed: a nearly complete calvaria, two skull fragments (part of a right supraorbital and a piece of parietal) that were not from the same individual as the calvaria, a fragmentary mandibular body with two molar teeth in situ, and nine isolated teeth. The material thus represents several individuals, all derived from yellowish brown sandy clay. The site was attributed to the Middle Pleistocene on the basis of biostratigraphic evidence.

Wu and Dong (1982) published a brief description of the find. The Hexian skull is that of a young male individual. Many morphological details that resemble those of Zhoukoudian *Homo erectus* skulls can be seen in the Hexian specimen, such as the low cranial vault, inferiorly-located maximum cranial breadth, flattened frontal bone, sagittal keel, developed supraorbital tori and occipital torus, thick cranial bones, and well-defined angle at the border of the occipital and nuchal planes. The cranial measurements of the Hexian skull also document its similarity to those of Peking Man in many respects. The cranial capacity of the Hexian skull is about 1,000 ml. Nevertheless, certain progressive features have been noted in this skull, such as the postorbital constriction being less than that in Peking Man and the temporal squama high with an arched parietal margin similar to that of skull V from Zhoukoudian. Thus, it is suggested that the Hexian skull may be best compared to the later forms of *Homo erectus* from Zhoukoudian. Detailed description and comparative study of Hexian remains are in progress.

Besides *Homo erectus*, 47 species of associated mammalian fossils have been identified. The Hexian assemblage contains several typical members of the *Ailuropoda–Stegodon* fauna widespread in South China. In addition, there are many northern members of the Zhoukoudian fauna, such as *Trogontherium cuvieri*, *Megaloceros pachyosteus*, and *Cervus grayi*, and even some eastern mountainous forms, such as *Anourosorex squamipes* and *Blarinella quadraticauda*. Such a fauna seems to represent a cold climate. Xu and You (1984) believe that the Hexian Man fauna could be correlated with Layers 3–4 of Locality 1 of Zhoukoudian, which yielded Peking Man skull V.

ZHOUKOUDIAN SITE RESEARCH

At the famous Zhoukoudian Locality 1, site of Peking Man, we have undertaken new excavations every year since 1979. So far, about 5,000 m^3 of hard deposits have been excavated. Animal fossils numbering some 2,000 specimens of more than 40 species and more than 5,000 stone artifacts have been unearthed, but as yet no human fossils have been found. Our excavation has now reached the 8th and 9th layers. As the hominid fossils were concentrated in the 10th and 11th layers, we expect to find more Peking Man fossils in future excavations.

We have made more detailed study of the strata of the cave deposits and have a clearer understanding of the history of development and formation of the cave and the changes in the entrance when Peking Man lived in it. From 1979 to 1981 we made a comprehensive study from different aspects of the Peking Man site and the Zhoukoudian area, including paleogeographical environments, animals and plants, paleosoils, developmental history of the cave, and different methods of dating of the layers of the Peking Man cave. A monograph consisting of all the results of these studies has been compiled and will soon be published by the Science Press in Beijing. Summaries have been provided by Liu (1983) and by Wu and Lin (1983).

NEW DATING RESULTS ON CHINESE *HOMO ERECTUS*

The two Yuanmou incisors have been believed to be evidence of the oldest *Homo erectus* found in China. In 1977, it was reported that the Yuanmou fossils were 1.7 million years (m.y.) old (Li et al., 1977) or 1.63–1.64 m.y. (Cheng et al., 1977) on the basis of paleomagnetic dating. However, Liu and Ding (1983) restudied the biostratigraphy, lithostratigraphy, and magnetostratigraphy of the Yuanmou formation and reached the conclusion that "Yuanmou Man" might not be older than 730,000 years BP, possibly only 500–600,000.

The two sites of Lantian Man may be somewhat older. From paleomagnetic studies, the Gongwangling locality has been suggested by Ma et al. (1978; see also Cheng et al., 1978) to lie in the latest Matuyama (750–800,000 years BP), while Chenjiawo would be younger, in the early Brunhes (650,000 years BP). On the basis of correlation with global climatic fluctuation, Xu and You (1982) attempted to date four North Chinese faunal assemblages. They suggested that, given the paleomagnetic results and the probable occurrence of an intense cold phase between Gongwangling and Chenjiawo, the former might be older than deep sea O^{18} stage 22, thus close to 1 m.y. At Chenjiawo, a triplet of red paleosols above the fauna was correlated to O^{18} stage 15, and as the Brunhes–Matuyama boundary (stage 20) occurs lower in the section, the hominid-bearing levels were correlated to stages 16–17, dating ca. 590–650,000 years BP. Extending this analysis to include studies of Chinese loess sections, Liu and Ding (1984) gave the Gongwangling cranium an age of 730–800,000 years and suggested that it is the earliest *Homo erectus* so far found in China. They estimated the age of the Chenjiawo mandible as 500–590,000 years BP.

From 1979 to 1981, numerous methods have been applied to dating the different layers of the Peking Man Cave. The deposits in this cave (see, e.g., Liu, 1983) are generally divided into 13 fossiliferous layers, with the old "basal gravel" now divided into layers 14–17. The main dating results are presented in Table 1.

Xu and You (1982; see also Xu and Ouyang, 1982) used these dates as a starting point to correlate the several layers at Locality 1 with O^{18} stages. They suggested that layer 10 (cold) equated to stage 12 (beginning about 425,000 years BP), layers 8–9 to stage 11, layers 6–7 to stage 10, layer 5 to stage 9, and layers 3–4 to stage 8 (ending about 240,000 years BP). However, a more direct but still tentative chronological correlation of early human fossil horizons in China with the loessic and deep-sea records was presented recently by Liu and Ding (1984). They dated the same layers at Zhoukoudian between 128–590,000 BP. Their lower limit is consistent with the chronometric dates, but the upper limit is much too young by comparison with dates from other sources.

The youngest well-preserved Chinese *Homo erectus* is Hexian Man. Liu and Ding (1984) suggested its age to lie between 150–400,000 years BP. Xu and You (1984) analyzed the faunal assemblage, finding the closest comparison with layers 3–4 (but not 5) at Zhoukoudian. On this basis, they correlated Hexian also with O^{18} stage 8 (240–280,000 years BP), which seems to be more likely.

Further work is under way by colleagues in different Chinese laboratories.

LITERATURE CITED

Cheng, G, et al. (1977) Discussion on the age of *Homo erectus yuanmouensis* and the early Matuyama event. Dizhi Xuebao *1:*34–43.

Cheng, G, et al. (1978) Dating of Lantian Man. Symposium on Paleoanthropology. Beijing: Science Press, pp. 151–157.

Guo, S, Zhou, S, Meng, W, Zhang, P, Sun, S, Hao, X, Liu, S, Zhang, F, Hu, R, and Liu, J (1980) Fission track dating of Peking Man. Kexue Tongbao *25:*770–772 (see also pp. 535–536).

Huang, W, Fang, D, and Ye, Y (1981) Preliminary study on the fossil hominid skull and fauna of Hexian, Anhui. Vertebrata PalAsiatica *20:*248–256.

Li, P, Chien, F, Ma, X, Pu, C, Hsing, L, and Chu, S (1977) Preliminary study on the age of Yuanmou Man by paleomagnetic technique. Scientia Sinica *20:*645–664.

TABLE 1. Some absolute dates obtained on the Peking Man deposits

Layer	Years	Dating Method and Reference
1–3	230,000 +30,000 −23,000; 256,000 +62,000 −40,000	Uranium-series[1,2]
4	290,000	Thermoluminescence[3]
6–7	350,000	Uranium-series[2]
7	370–400,000	Paleomagnetism[4]
8–9	420,000 + >180,000 − 100,000 ; >400,000	Uranium-series[1,2]
	462,000 ± 45,000	Fission-track[5]
10	520–620,000	Thermoluminescence[3]
12	>500,000	Uranium-series[2]
13–17	>730,000	Paleomagnetism[4]

References: [1]Zhao et al., 1979; [2]Xia, 1982; [3]Pei et al., 1979; [4]Qian et al., 1980; [5]Guo et al., 1980; see also Liu, 1983.

Liu, T, and Ding, M (1983) Discussion on the age of "Yuanmou Man." Acta Anthropol. Sin. *2:*40–48.

Liu, T, and Ding, M (1984) A tentative chronological correlation of early human fossil horizons in China with the loess-deep sea records. Acta Anthropol. Sin. *3:*93–101.

Liu, Z (1983) Le remplissage de la grotte de l'homme de Pékin, Choukoutien—Locality 1. L'Anthropol. *87:*163–176.

Ma, X, Qian, F, Li, P, and Ju, S (1978) Paleomagnetic dating of Lantian Man. Vertebrata PalAsiatica *16:*238–244.

Pei, J, et al. (1979) Thermoluminescence ages of quartz in ash materials from *Homo erectus pekinensis* site and its geological implications. Kexue Tongbao *24:*849.

Qian, F, et al. (1980) Magnetostratigraphic study on the cave deposits containing fossil Peking Man at Zhoukoudian. Kexue Tongbao *25:*359.

Qiu, Z, Xu, C, Zhang, W, Wang, R, Wang, J, and Zhao, C (1982) A human fossil tooth and fossil mammals from Nanzhao, Henan. Acta Anthropol. Sin. *1:*109–117.

Wu, M (1983) *Homo erectus* from Hexian, Anhui found in 1981. Acta Anthropol. Sin. *2:*109–115.

Wu, R, and Dong, X (1980) The fossil human teeth from Yunxian, Hubei. Vertebrata PalAsiatica *18*:142–149.

Wu, R, and Dong, X (1982) Preliminary study of *Homo erectus* remains from Hexian, Anhui. Acta Anthropol. Sin. *1:*2–13.

Wu, R, and Dong, X (1983) Des fossiles d'*Homo erectus* découverts en Chine. L'Anthropol. *87:*177–183.

Wu, R, and Lin, S (1983) Peking Man. Sci. Amer. *248:*86–94.

Wu, R, and Wu, X (1982) Human fossil teeth from Xichuan, Henan. Vertebrata PalAsiatica *20:*1–9.

Xia, M (1982) Uranium-series dating of fossil bones from Peking Man Cave—Mixing Model. Acta Anthropol. Sin. *1:*191–196.

Xu, Q, and Ouyang, L (1982) Climatic changes during Peking Man's time. Acta Anthropol. Sin. *1:*79–90.

Xu, Q, and You, Y (1982) Four post-Nihowanian Pleistocene mammalian faunas of North China: Correlation with deep-sea sediments. Acta Anthropol. Sin. *1:*180–190.

Xu, Q, and You, Y (1984) Hexian fauna: Correlation with deep-sea sediments. Acta Anthropol. Sin. *3:*62–67.

Zhao, S, Xia, M, Wang, S, et al. (1979) Uranium-series dating of Peking Man. Kexue Tongbao *25:*447.

Ancestors: The Hard Evidence, pages 249–254

The Phylogenetic Position of Olduvai Hominid 9, Especially as Determined From Basicranial Evidence

Wolfgang O. Maier and Abel T. Nkini

Zentrum der Morphologie (Dr. Senckenbergische Anatomie), Klinkum der Universität, D-6000 Frankfurt am Main 70, Federal Republic of Germany

ABSTRACT The skull proportions of Olduvai Hominid 9 (*Homo erectus*) have been investigated with the help of computer tomography. O.H. 9 differs from both australopithecines and *Homo sapiens* in having a straight tribasilar bone, which appears to be pushed into the cranial cavity. These features seem to correspond with the flat and elongated shape of the whole braincase of *Homo erectus*. Phylogenetic consequences of these findings are briefly discussed.

INTRODUCTION

The fossil hominid skull of O.H. 9 was discovered in 1960 by L.S.B. Leakey in Upper Bed II of the Olduvai Gorge in Tanzania (Leakey, 1961; Heberer, 1963). It was the first find of a *Homo erectus*-like creature in Africa, and it proved to be much older than the previously known Southeast Asian representatives of this taxon; the most likely absolute date seems to be about 1.1 million years (m.y.) (Rightmire, 1979). Apart from some comments by Tobias (1968), it was only Rightmire (1979) who provided a fairly detailed anatomical description of this specimen, which is so often depicted in textbooks of paleoanthropology. With the help of computer tomography, we attempted to further elucidate the proportions of this important hominid fossil.

HISTORICAL BACKGROUND

It is our impression that the systematic and phylogenetic evaluation of *Homo erectus* has changed quite remarkably within the last quarter century. However, this has not primarily been the result of increased knowledge about this taxon, but rather an indirect consequence of the drastic changes in paleoanthropology altogether. Up to the fifties, *Homo erectus* (formerly also called *Pithecanthropus* and *Sinanthropus*) was a synonym for archaic man ("archanthropines" of Heberer, 1956). This view was certainly much influenced by the classic studies of Weidenreich, who explicitly stated that "*Sinanthropus* and *Pithecanthropus* are the most primitive hominids so far known" (1943, p. 273).

Australopithecines were still considered with caution if not suspicion at that time. Weidenreich (1943) refused to accept australopithecines as forerunners of the "pithecanthropines," and at best would accept them as an independent offshoot of the common anthropoid stock of the late Tertiary. Although accepting some hominid-like features in the mandibular joint and in tooth morphology, he clearly expressed his view that "the skulls of *Australopithecus africanus* and *Paranthropus robustus* are skulls of apes and not of hominids" (1943, p. 268). This led to the result that Simpson (1945) and even Vallois (1955) classified the australopithecines with the Pongidae and not with the Hominidae. Apart from Broom and Dart, it was mainly Heberer (1956) and LeGros Clark (1967) who strongly spoke in favour of the australopithecines ("praehominines") being in the direct lineage of human evolution; for Heberer it was rather the pithecanthropines ("archanthropines") that were a side-branch. But in general, *Pithecanthropus* and *Sinanthropus* were widely accepted as "Early Man," directly leading to modern *Homo sapiens* with the "neanderthaloids" (or only the "preneanderthaloids") as an intermediate stage. As far as the morphological status of *Homo erectus* was concerned, its primitiveness was pointed out many times in various papers of Weidenreich; it was believed to be right in between some supposed anthropoid ancestors and modern man (Boule and Vallois, 1954).

However, elaboration of morphological status constitutes only one, although relevant, part of phylogenetic reconstruction; the other part is fitting these findings into a chronological frame, and providing good biological arguments for a supposed model of the evolutionary process. Biostratigraphy did not yield a very reliable geological framework for understanding human evolution at that time. During the 1960s and 1970s, paleoanthropology has undergone a spectacular, nearly explosive development in several respects. First, various methods of absolute dating have helped to establish a fairly reliable chronology for Plio-Pleistocene time; second, hundreds of new fossils have been recovered from many old and new localities, often stemming from well dated horizons; third, the technique of excavating has been much refined in order to provide as much taphonomic information as possible. Therefore, the main interest of paleoanthropology was directed toward understanding the complicated picture of Plio-Pleistocene hominid evolution in Africa, roughly covering the time span between 4 and 1 m.y. (Isaac and McCown, 1976; Coppens et al., 1976; Leakey and Leakey, 1978).

The gracile australopithecines and *Homo habilis* (Leakey et al., 1964) have become the central figures of scientific discussion of the last fifteen years. *Homo erectus*, whose presence in Africa had also been confirmed by important finds, was decidedly shifted toward the more recent and modern end of human history as a side effect of the more detailed understanding of Plio-Pleistocene events. Considering *H. erectus* as a "missing link" between modern man and anthropoids appears to us now as a very strange idea. As there seems to be hardly any temporal overlap between *Homo* paleospecies, our present attitude has also found its expression in some proposed phylogenetic trees, viewing *Homo erectus* simply as an intermediate evolutionary grade between *H. habilis* and *H. sapiens* (Tobias, 1973, 1976; Maier, 1975). One of the main problems for Weidenreich (1943)—namely, to recognize not only the descendants but also the forerunners of *H. erectus*—would seem to be resolved by this hypothesis.

However, it is quite evident that this evolutionary assumption is not yet based on precise morphological analysis. It has been lamented by a number of authors that the rapid increase of fossil discoveries was not accompanied by morphological studies according to the standards set forth by Weidenreich (1936, 1943, 1951) and Tobias (1967). Most finds were only briefly characterized and illustrated by photographs. It was Weidenreich (1943) who warned that "neglect of morphological features in studying the relationship of hominid types and the attempt to replace this by geological and chronological considerations does not secure a usable solution but only leads back to morphology again" (p. 243). On the other hand, it must be admitted that morphology has not yet been able to provide a satisfactory theoretical framework in which to study the human skull, comparable to the way modern biomechanics provides a sound basis for the study of postcranial skeletal parts (Pauwels, 1965; Kummer, 1972; Preuschoft, 1971, 1973, 1978). The reason for this deplorable situation may be that the head skeleton is more complicated and is influenced by many heterogeneous factors. It appears to be necessary, however, not only to regard the skull as an aggregate of features and characters, but also to understand it as a functional unit, or at least as a system of functional complexes (van der Klaauw, 1952).

THE CRANIUM OF O.H. 9

This skull is presently housed at the Department of Anatomy of the University of Frankfurt to be restudied by A.T. Nkini. We took the opportunity to further clean the specimen and to analyze its structures with the assistance of computer tomography. Computer tomography (CT) provides accurate optical serial sections, which help to study the relative positions of different skull structures. It is not the purpose of the present study to provide new descriptive data, but rather to go into the as yet untouched problem of the overall proportions of this skull, mainly of the shape and relative position of its cranial axis. This approach seems to be all the more important, since few data on the basicranium of the whole taxon *Homo erectus* are available. With the exception of the not yet fully published skull of Sangiran 17 (Jacob, 1975; Sartono, 1975) and of the Solo skulls VI and XI (Weidenreich, 1951), no Asian representatives of *H. erectus* have preserved intact skull bases. The East African specimens KNM-ER 3733 and 3883

(Leakey and Walker, 1976; Clarke, 1977) have not yet been studied in detail either.

The basicranial axis constitutes the linkage between the facial skull and the braincase, and many morphologists have directed their interest to this area (Huxley, 1867; Kummer, 1952; Starck, 1953; Biegert, 1957; Hofer, 1965). It is generally thought to be a character complex that sensitively reflects the specific modes of skull proportion, although biomechanical understanding of it is only beginning (Demes, 1982). Here we are mainly puzzled by the question of how the peculiar overall skull shape of *Homo erectus* (O.H. 9) is reflected by the basicranium.

CRANIAL PROPORTIONS OF O.H.9 AND MODERN HOMINOIDS

The superposition of three sagittal CT sections (paramedian, midorbital, and articular plane) is shown in Figure 1. Details of the method and its problems are discussed in Maier and Nkini (1984). The sections have been oriented on an approximate Frankfurt Horizontal (FH) plane. To point out only the most peculiar features at this time, we note the straight and horizontal tribasilar bone, entirely situated above the FH plane; the high position of the foramen magnum and of the shallow posterior cranial fossa is directly correlated with this situation. The remains of the basisphenoid seem to be highly elevated above the medial cranial fossae, and the chiasmatic region of the presphenoid projects into the anterior cranial fossa only to descend steeply to the well preserved cribriform plate; although it is not fully preserved, we assume that the planum sphenoideum must have strongly leaned anteriorly. The endocranial cast reflects this peculiar shape of the anterior cranial fossa by its beak-like frontal lobe and especially its medioventral orbital rostrum (Fig. 2). Thus, one sees a very well pronounced "*bec encephalique*" (in the sense of French authors), which is generally considered to be a primitive hominid trait (Vlček, 1969).

An even better understanding of the peculiar shape of the basicranium of O.H. 9 can be gained by comparisons with paramedian sections of a pongid (*Gorilla*) and of a modern human skull (Fig. 3). O.H. 9 shares with the pongids the straight tribasilar bone, although it is in a distinctly lower position in the latter; the nuchal plane and the foramen magnum are differently oriented in the apes. Modern man shows typical basal kyphosis at the sellar region, the clivus being strongly deflected posteriorly; this results in a low position of the foramen magnum and of the voluminous posterior cranial fossa. Both in modern humans and in the gorilla, the basioccipital is situated distinctly below the level of the outer acoustical porus (but obviously for different reasons), whereas in O.H. 9 they are at about the same level. The presphenoid of modern man is relatively high, but not as high as in O.H. 9. In both the pongids and modern man the sphenoidal plane runs nearly horizontally, but the cribriform plate is on a different level, and the pongids form a distinct olfactory fossa in front of the receding frontal lobes; in *Homo erectus* the small olfactory fossa is situated below the fairly well molded frontal lobes. The projecting supraorbital and interorbital structures form a massive interorbital pillar above the low roof of the nasal cavity in both pongids and O.H. 9. Closer inspection reveals that there exist some differences in the exact orientation of this structure between pongids and O.H. 9, while it is nearly completely reduced in modern man. As revealed by new computer tomograms done at the University Hospital at Utrecht (thanks to Drs. Wind and Zonnefeld), moderately sized frontal sinuses are developed within the glabellar torus.

COMPARISONS TO OTHER FOSSIL HOMINIDS

After having pointed out a number of peculiar features and proportions of the basicranium of O.H. 9 by comparison with pongids and modern man, we must think about their systematic and phylogenetic meaning. Unfortunately, we have very scarce evidence about the basicranial complex in fossil hominids altogether, so it is difficult to properly appreciate natural variation and to illustrate evolutionary transformations. Weidenreich (1943) repeatedly commented on the flatness and primitive form in his "*Sinanthropus*" material, although in all of them the central parts of the basicranium were missing. Two of the later Pleistocene Solo skulls (VI and XI) have relatively complete cranial bases; again the occipital parts of the basicranium appear relatively flat, and the kyphosis of the tribasilar bone is only feebly developed. Some mistakes in Weidenreich's interpretation of the basicranium of Solo man have been pointed out by Maier and Nkini (1984). As judged from a cast of the fairly complete skull from Java (Sangiran 17), there exists also a kind of pushing inward of the basicranial complex. Clarke (personal communication) has noticed similar, although less pronounced, characteristics in the East African find

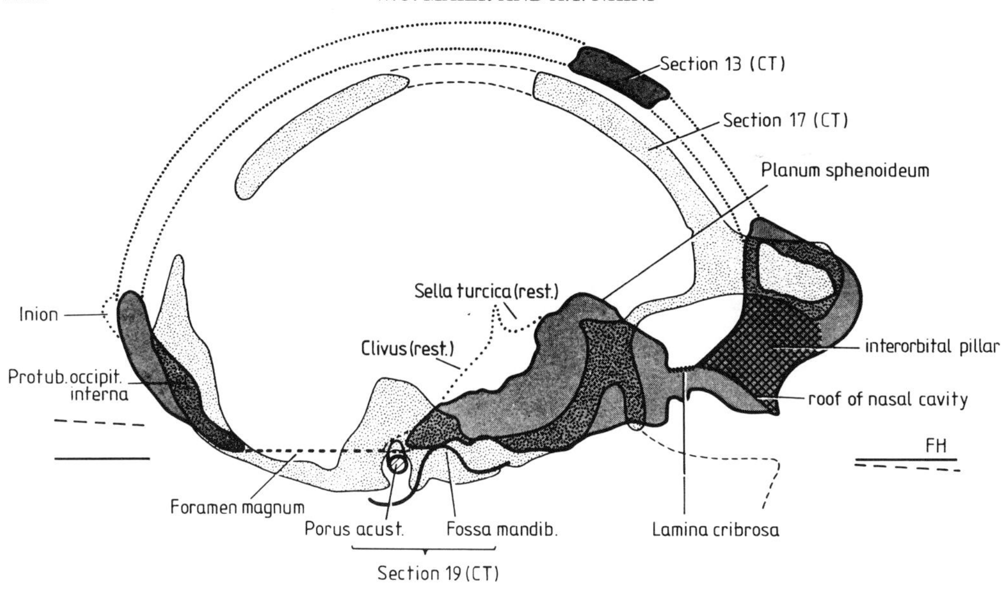

Fig. 1. Sagittal sections of Olduvai Hominid 9 (O.H. 9) based on computer tomograms. Three CT sections are superimposed to demonstrate the special relationships of various skull structures. Section 13 (darkly stippled) runs through the cribriform plate; Section 17 (lightly stippled) runs through the middle of the orbits; Section 19 (thick lines) cuts the outer meatus acousticus and the mandibular fossa. The diagram is oriented on an approximate Frankfurt Horizontal (FH) plane.

KNM-ER 3733. It would therefore seem to be a common, although somewhat variable, feature of all known "pithecanthropines" to have a flattened and stretched basicranium. It is most likely that this character is correlated with the specific elongation of the whole braincase and with a peculiar poise of the head. However, it seems unlikely to us that the structure of the pharynx exerts such a strong mechanical influence as to dominate the shape of the basicranium, as suggested by Laitman et al. (1979). It also appears unlikely that distortion and plastic deformation (for which we see hardly any evidence in O.H. 9) should have worked in the same direction in all known specimens of *Homo erectus* s.l.

Various Neanderthal skulls clearly show evidence for some flattening of the cranial basis as well (Keith, 1915; Sergi, 1948), but Howell (1951) demonstrated that this is only true for the latest and most specialized representatives, and not for the older forms from Skhūl. The largely complete basicranium of the Rhodesian skull presents an essentially modern appearance, although the mediosagittal section published by Pycraft (1928, Fig. 3) cannot be quite correct in its anterior parts.

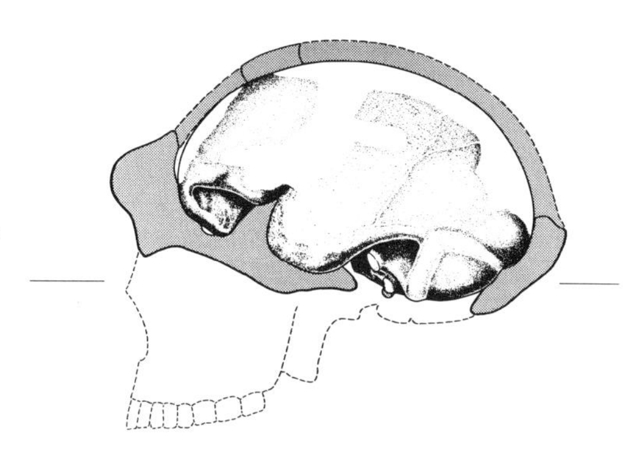

Fig. 2. The drawing of an endocast of O.H. 9 is projected onto a mediosagittal CT section of the same skull. Note the low and elongated overall shape of the endocast, and the protruding orbital rostrum of the frontal lobe. The stippled outlines of the steep and high facial skull are based on a reconstruction.

A very surprising result is found when O.H. 9 is compared with *Australopithecus* and *Homo sapiens* (Fig. 4). The mediosagittal sections of the robust *A. boisei* (Tobias, 1967) and the gracile *A. africanus* (LeGros Clark, 1967) present about the same proportions. It is evident that *Australopithecus* possessed a relatively well deflected clivus, being about intermediate between O.H. 9

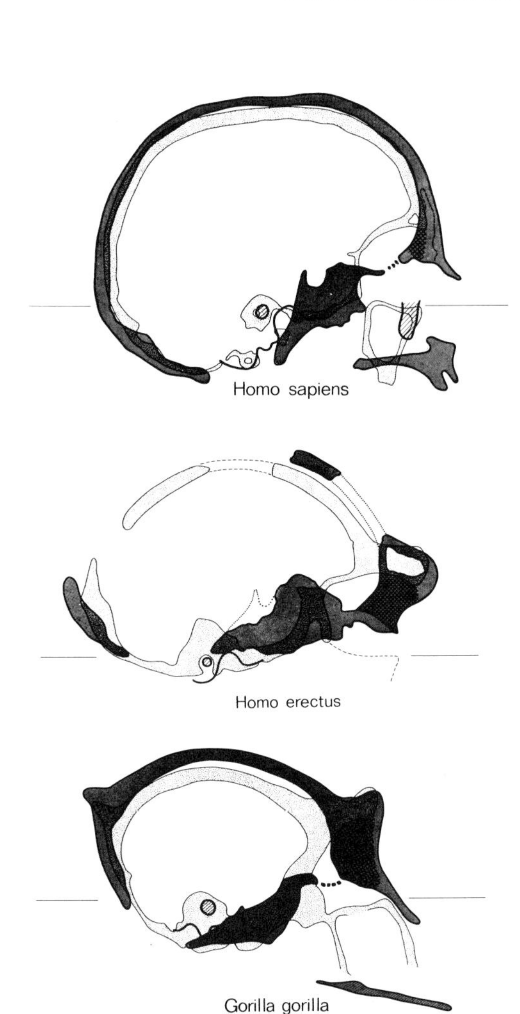

Fig. 3. Sagittal skull sections of a gorilla (bottom), *Homo erectus* (O.H. 9), and a modern human (top), all prepared by computer tomography as in Figure 1 and all calibrated to about the same size. The different skull proportions and relative positions of a number of skull structures are illustrated by this comparison. Note especially the stretched contour and elevated position of the tribasilar bone in O.H. 9.

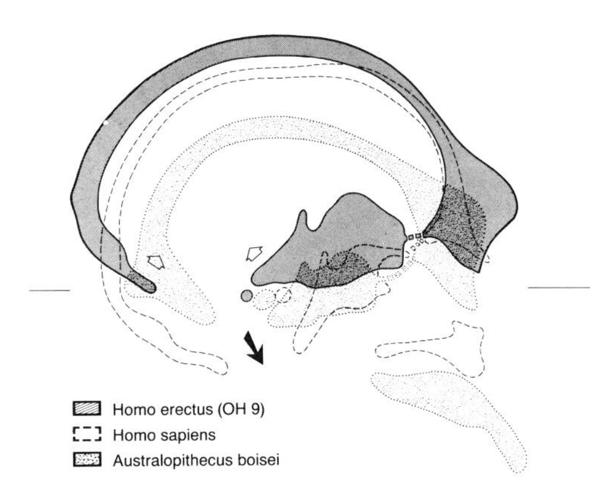

Fig. 4. Comparison of mediosagittal sections of *Homo erectus* (O.H. 9), *Australopithecus boisei*, and *Homo sapiens*. It is surprising to see that the basicranial axis of O.H. 9 is less deflected than that of *Australopithecus*, *Homo erectus* being quite aberrant in some respects.

and modern man in this respect; in consequence, its posterior cranial fossa and its foramen magnum are distinctly below the FH-plane. These results are partly at variance with those of Dean and Wood (1982).

SYSTEMATIC CONCLUSIONS

Given the poor present state of our knowledge about the functional principles controlling skull shape, it is certainly difficult to draw any reliable systematic conclusions from the data presented. Moreover, phylogenetic hypotheses can never be based on a single character or character complex alone. But single characters must not be in contradiction with such hypotheses, otherwise phylogenetic models are falsified by the hard evidence of individual fossil form. At this point, Weidenreich (1943) himself has to be criticised when he characterized his methodological approach as follows: "Studies along such a line as I have followed in this paper make it possible to recognize the single phases of the evolution of the hominid skull in its general form as in its detail, independent of the momentary state of a structure as it may appear in a given specimen. In so doing, one arrives at the conclusion that evolution, on the whole, follows a certain strictly-kept line without any indications of particular chance variations outside the adopted pattern" (p. 242). Real and properly understood specimens have to constitute the basis of phylogenetic theories, otherwise the criterion of falsifiability is suspended. In the case of O.H. 9 and other *Homo erectus* specimens, the peculiar shape of the basicranium, which also reflects a specific ontogenetic program, leads to two alternative conclusions: either this whole group constitutes a specialized side branch and cul-de-sac, or a hypothesis of morphological transformation must be formulated to explain a possible adaptive way back into the evolutionary main stream of human ascent.

LITERATURE CITED

Biegert, J (1957) Der Formwandel des Primatenschädels. Morph. Jahrb. *98*:77–199.

Boule, M, and Vallois, HV (1954) Fossile Menschen. Baden-Baden: Verlag Kunst und Wissenschaft.

Clarke, RJ (1977) The cranium of the Swartkrans hominid SK 847 and its relevance to human origins. Unpublished Ph.D. Thesis, University of the Witwatersrand, Johannesburg.

Coppens, Y, Howell, FC, Isaac, GL, and Leakey, REF (eds) (1976) Earliest Man and Environments in the Lake Rudolf Basin. Chicago: University of Chicago Press.

Dean, MC, and Wood, BA (1982) Basicranial anatomy of Plio-Pleistocene hominids from East and South Africa. Am. J. Phys. Anthropol. *59:*157–174.

Demes, B (1982) Über die mechanische Beanspruchung der Schädel-basis von Primaten. Unpublished Ph.D. Thesis, University of Bochum.

Heberer, G (1956) Die Fossilgeschichte der Hominoidea. In H Hofer, AH Schultz, and D Starck (eds): Primatologia I. Basel: Karger, pp. 379–560.

Heberer, G (1963) Über einen neuen archanthropinen Typus aus der Oldoway-Schlucht. Z. Morph. Anthropol. *53:*171–177.

Hofer, H (1965) Die morphologische Analyse des Schädels des Menschen. In G Heberer (ed): Menschliche Abstammungslehre. Stuttgart: Fischer, pp. 145–226.

Howell, FC (1951) The place of Neanderthal man in human evolution. Am. J. Phys. Anthropol. *9:*379–415.

Huxley, TH (1867) On two widely contrasted forms of the human cranium. J. Anat. Physiol. *1:*60–77.

Isaac, GL, and McCown, ER (eds) (1976) Human Origins. Menlo Park: Benjamin Inc.

Jacob, T (1975) Morphology and paleoecology of early man in Java. In RH Tuttle (ed): Paleoanthropology. The Hague: Mouton, pp. 311–325.

Keith, A (1915) The Antiquity of Man. London: Williams and Norgate.

Kummer, B (1952) Untersuchungen über die ontogenetische Entwicklung des menschlichen Schädelbasiswinkels. Z. Morph. Anthropol. *43:*331–360.

Kummer, B (1972) Biomechanics of bone: Mechanical properties, functional structure, functional adaptation. In YC Fung, N Perrone, and M Anliker (eds): Biomechanics: Its Foundations and Objectives. Englewood Cliffs: Prentice Hall, pp. 237–271.

Laitman, JT, Heimbuch, RC, and Crelin, ES (1979) The basicranium of fossil hominids as an indicator of their upper respiratory systems. Am. J. Phys. Anthropol. *51:*15–34.

Leakey, LSB (1961) New finds at Olduvai Gorge. Nature *189:*649–650.

Leakey, MG, and Leakey, RE (eds) (1978) Koobi Fora research project I. Oxford: Clarendon Press.

Leakey, RE, and Walker, AC (1976) *Australopithecus*, *Homo erectus* and the single species hypothesis. Nature *261:*572–574.

LeGros Clark, WE (1967) Man-apes or ape-men. New York: Holt, Rinehart and Winston.

Maier, W (1975) Neue Aspekte frühmenschlicher Evolution. Verhandl. Ges. Anthrop. Humangenet. 1973. Stuttgart: Fischer, pp. 13–29.

Maier, W., and Nkini, AT (1984) Olduvai hominid 9: New results of investigation. Cour. Forsch. Inst. Senckenberg *69:*123–30.

Pauwels, F (1965) Gesammelte Abhandlungen zur funktionellen Anatomie des Bewegungsapparates. Berlin: Springer.

Preuschoft, H (1971) Body posture and mode of locomotion in early Pleistocene hominids. Folia Primatol. *14:*209–240.

Preuschoft, H (1973) Body posture and locomotion in some East African Miocene Dryopithecinae. In MH Day (ed): Human Evolution. London: Taylor and Francis, pp. 13–46.

Preuschoft, H (1978) Motor behavior and shape of the locomotor apparatus. In ME Morbeck, H Preuschoft, and N Gomberg (eds): Environment, Behavior, and Morphology: Dynamic Interactions in Primates. Stuttgart: Fischer, pp. 263–275.

Pycraft, WP (1928) Rhodesian man. Description of the skull and other human remains from Broken Hill. In WP Pycraft (ed): Rhodesian Man and Associated Remains. London: British Museum Publications, pp. 1–51.

Rightmire, GP (1979): Cranial remains of *Homo erectus* from Beds II and IV, Olduvai Gorge, Tanzania. Am. J. Phys. Anthropol. *51:*99–116.

Sartono, S (1975) Implications arising from Pithecanthropus VIII. In RH Tuttle (ed): Paleoanthropology. The Hague: Mouton, pp. 327–360.

Sergi, S (1948) The palaeanthropi in Italy: The fossil men of Saccopastore and Circeo. Parts I and II.Man *48:*61–64, 76–79.

Simpson, GG (1945) The principles of classification and a classification of mammals. Bull. Am. Mus. Nat. Hist. *85:*1–350.

Starck, D (1953) Morphologische Untersuchungen am Kopf der Säugetiere, besonders der Prosimier, ein Beitrag zum Problem des Formwandels des Säugerschädels. Z. Wiss. Zool. *157:*169–219.

Tobias, PV (1967) Olduvai Gorge, Vol. 2. The Cranium and Maxillary Dentition of *Australopithecus (Zinjanthropus) boisei.* Cambridge: Cambridge University Press.

Tobias, PV (1968) Middle and early Upper Pleistocene members of the genus *Homo* in Africa. In G Kurth (ed): Evolution and Hominisation. Stuttgart: Fischer, pp. 176–194.

Tobias, PV (1973) Implications of the new age estimates of the early South African hominids. Nature *246:*79–83.

Tobias, PV (1976) African hominids: Dating and phylogeny. In GL Isaac and ER McCown (eds): Human Origins. Menlo Park: Benjamin Inc., pp. 377–422.

Vallois, H (1955) Ordre des Primates. In PP Grassé (ed): Traité de Zoologie, Vol. 17. Paris: Masson, pp. 1854–2206.

van der Klaauw, CJ (1952) Size and position of the functional components of the skull. Arch. Néerl. Zool. Leiden.

Vlček, E (1969) Neanderthaler der Tschechoslowakei. Prague: Academia Verlag.

Weidenreich, F (1936) Observations on the form and proportions of the endocranial casts of *Sinanthropus pekinensis*, other hominids and the great apes: A comparative study of brain size. Palaeontologia Sinica, Ser. D, *7:*1–50.

Weidenreich, F (1943) The skull of *Sinanthropus pekinensis*: A comparative study on a primitive hominid skull. Palaeontologia Sinica, Ser. D, *10:*1–484.

Weidenreich, F (1951) Morphology of Solo Man. Anthrop. Papers Am. Mus. Nat. Hist. *43:*205–290.

Ancestors: The Hard Evidence, pages 255–264

The Tempo of Change in the Evolution of Mid-Pleistocene *Homo*

G.P. Rightmire
Department of Anthropology, State University of New York, Binghamton, New York 13901

ABSTRACT The extinct species *Homo erectus* can be defined morphologically, without reference to chronology or gaps in the fossil record. Frontal proportions, parietal measurements, occipital curvature and scale lengths, and anatomy of the glenoid cavity and tympanic plate all distinguish *Homo erectus* from later humans. These characters are present in mid-Pleistocene assemblages from Africa and Asia. The impression that *Homo erectus* is a morphologically stable species is strengthened by quantitative analysis of brain size. When cranial capacities for individual fossils are plotted against geological time, there is a slight trend, but this change is not significant. Other individuals identified as early *Homo sapiens* have cranial capacities that are larger than expected for (late) *Homo erectus*. Fossils such as Petralona, Broken Hill, and Omo 2 lie well above the regression line constructed for archaic hominids, and there is evidence that brain volume began to increase more rapidly toward the close of the Middle Pleistocene.

INTRODUCTION

In this presentation, it is my purpose to comment rather generally on patterns of human evolution in the mid-Pleistocene. Quite a lot of evidence bearing on this topic has come to light in Asia since work at the famous localities in China and Indonesia was disrupted by the Second World War. Spectacular discoveries relating to Pleistocene prehistory have also been made in Africa, and there have been numerous important additions to the European record. Many of these fossils have been viewed either as *Homo erectus* or as archaic representatives of *Homo sapiens*. In addition to the paleontological discoveries, there has been increasingly sophisticated attention paid to the details of geological context. Stratigraphic proveniences are often reasonably well established, and many (not all) of the fossils can be tied with some accuracy to an absolute time scale. This means that it is possible to comment not only about the morphology of ancient populations but also about the rates at which evolution has occurred.

Questions concerning the tempo of evolution have of course been much discussed in the recent literature of paleobiology. Claims for gradual, progressive change within lineages have been countered by the argument that species are relatively stable entities and that speciation is a rapid event, "punctuating" the pattern of stasis prevailing during much of macroevolutionary history (Gould and Eldredge, 1977; Stanley, 1979; Levinton, 1983). In the case of humans, both views are defended by reference to the (same) fossil evidence. A gradualist interpretation suggests that earlier *Homo* species have been transformed slowly into later ones, with no branching or extinctions. Such gradual change with continuity is said especially to characterize the emergence of *Homo sapiens* from *Homo erectus*. Other workers who are inclined to accept a punctuationalist stance see signs of character stasis in mid-Pleistocene populations. Whether *Homo sapiens* evolved from more archaic ancestors gradually, or during a relatively rapid period of transition late in the Middle Pleistocene, has not been settled, although a good deal of attention has been focused on the problem (Cronin et al., 1981; Rightmire, 1981).

Several issues are raised here. One is whether extinct *Homo* species can be defined as discrete biological units rather than as arbitrary divisions of a (single) lineage. To what extent may *Homo erectus* be regarded as a real species, distinct from earlier or later humans? If there is a set of characters that serves to diagnose *Homo erectus* in this sense, these traits should be enumerated. Another question concerns rates and the evidence needed to check for the presence of significant trends in the evolution of *Homo* species. Here I wish to comment particularly on changes in endocranial volume in *Homo erectus*. Finally, there is the issue of how the first representatives of our own species may be recognized. It has been suggested that several apparently early fossils such as Broken Hill, Ndutu, and the Omo individuals from Africa, Petralona and the Arago assemblage from Europe, and the Dali cranium from China may exhibit derived features shared with more modern humans (Stringer et al., 1979; Wu, 1981; Rightmire, 1983). These claims are reasonable. Continuing study of these important specimens should shed more light on the origin of early *Homo sapiens*.

GRADUAL EVOLUTION OF SPECIES

A traditional view is that new species may arise either by branching of a lineage to produce two (or more) descendant groups or, in the absence of splits, by the accumulation of changes through time within a single lineage. For example, Gingerich (1979) frames the history of the hominids as one of gradual character divergence in two lineages derived from a Pliocene ancestor. One lineage, containing species of *Australopithecus*, is presumed to have become extinct, while the second is thought to have produced successive species of the genus *Homo*, including modern humans. Within this second lineage, three species are recognized, but the boundaries between *Homo habilis* and *Homo erectus* or between *Homo erectus* and *Homo sapiens* are not clearly demarcated. Gingerich defines species as "arbitrarily divided segments of an evolving lineage that differ morphologically from other species in the same or different lineages." The difficulties associated with carving species out of what is assumed—given a reasonably complete fossil record—would be a continuous sequence of slowly changing forms have been further enumerated by Tobias (1978, 1980). Tobias notes that both dating and morphology should be employed in the effort to name and describe successive species, although he cautions against accepting Campbell's (1974) proposal that *Homo* taxa be delimited on strictly chronological grounds.

Other workers, who generally agree with partitioning of the *Homo* lineage into three species, have emphasized the gradual nature of evolutionary tempos within this lineage. Cronin et al. (1981) have recently reviewed much of the Plio-Pleistocene hominid evidence and have concluded that fossils displaying "intermediate" morphology are fairly numerous. Such transitional individuals are said to be common in the European Middle Pleistocene, and the material from Petralona and other localities is used to support a claim for steady change in populations linking (late) *Homo erectus* with (early) *Homo sapiens*. Wolpoff (1980, 1982) argues that the division between these species is arbitrary, and for Europe at least he prefers a chronological criterion (the "end of the Mindel glaciation") to mark the species boundary. A few of the European fossils may be *Homo erectus* by this reasoning, but the entire mid-Pleistocene assemblage is thought to be part of a single sexually dimorphic lineage, ancestral to later Neanderthals and more modern humans. "Clade features" linking early and late European fossils are said to include sagittal torus reduction, a square palate, orbit form, and midfacial prognathism. However, why such resemblances are important is not spelled out, either by reference to function or through wider comparative studies designed to show that these characters are not shared with early *Homo sapiens* in other regions.

A few anthropologists have carried this thinking a step further by arguing that there is no need to recognize separate species in the Middle Pleistocene. In Jelinek's (1980a) opinion, all of the Middle and Late Pleistocene hominids from Europe may be viewed as *Homo sapiens*, although this European lineage displays marked sex dimorphism. In another paper addressing the remains from Ternifine and other sites in northwest Africa, Jelinek (1980b) suggests that in Africa as well as in Europe there are general "evolutionary trends leading to *Homo sapiens*

sapiens." Only one species is represented, but at the same time local changes (trends) documented in the Maghreb are not the same as those occurring in Europe or the Middle East. While he does not discuss subspecies, Jelinek does point to the importance of the environment in shaping population differences. He sees evolution within *Homo sapiens* as a complex process, proceeding at different rates in different geographic regions.

More explicit hypotheses of gradual evolution with regional continuity have been advanced by Thorne and Wolpoff (1981) for southeast Asia. The latter workers describe *Homo erectus* from Indonesian localities as part of a "morphological clade," which also includes later *Homo sapiens* populations such as that from Kow Swamp, Australia. Two taxa are recognized, but these chronospecies are no more than segments of a morphological continuum. Evidence for continuity is found in trends toward facial prognathism, eversion of the malar bone, rounding of the lower orbital margin, and overall reduction of facial and dental dimensions. This idea of an ancestral relationship between southeast Asian *Homo erectus* and Australian *Homo sapiens* is not new, of course, but Thorne and Wolpoff attempt to place some distance between themselves and Weidenreich or Coon by rejecting the use of subspecific designations. Instead, they argue that "local genetic continuity" is not a necessary assumption of their model, which supposes that differences distinguishing "regional clades" will be maintained over long periods in peripheral areas of the range occupied by mid-Pleistocene hominids.

SPECIES AS DISCRETE ENTITIES

The concept of paleontological species as segments of a single lineage, arbitrarily defined by changing morphology or stratigraphic breaks (gaps in the record), has not been accepted in all quarters. Some students of human evolution have questioned the prevailing view that (all) *Homo* species are sequentially related within a framework of gradual, progressive change. In particular, the assumption that *Homo erectus* populations all across the Old World merged imperceptibly with early *Homo sapiens* has been sharply criticized by Eldredge and Tattersall (1975, 1982) and Delson et al. (1977). These authors point to several characters found in *Homo erectus* crania (an undivided supraorbital torus, sagittal keeling, a small mastoid process) that do not seem to be present in the skulls of early *Homo sapiens*. This choice of traits may be questioned (Rightmire, 1980), but if such non-shared morphological specializations can be identified, then at least some groups of *Homo erectus* are not likely to be the direct ancestors of *Homo sapiens*. Delson et al. (1977) seem to doubt that there can be any continuity between the two species, although Delson (1981) has recently suggested that early evidence for speciation may be found in Europe. He hypothesizes that populations of *Homo sapiens* may have evolved there first, as a result of isolation due to glacial conditions. What should be emphasized here is not the timing or geographic location of speciation, but rather the view that the transition from archaic to more modern forms may have taken place just once, in a restricted geographic province, rather than gradually in many areas.

Bonde (1981) has reviewed much of the current thinking about species in paleontology, and his conclusions as applied to the genus *Homo* also merit close consideration. Bonde supports the idea that species exist as coherent entities, not only in the present but also as historical units. Paleontological species, in accord with Wiley's (1978) revision of Simpson's (1961) definition, are single lineages, which cannot be subdivided arbitrarily. That such a species may survive one or more branching events is considered unlikely. Here Bonde sticks to the cladist principle that splitting of a lineage must always give rise to forms that, for purposes of formal classification, are distinct from the common stem or ancestor.

Application of this thinking to the hominids results in the recognition of at least four species of *Homo*, existing after about 1.5 million years (m.y.) BP. *Homo erectus* as known from the earlier African localities and from the Far East is viewed as one product of a split also yielding *Homo heidelbergensis*. This latter species is represented by the mid-Pleistocene remains from Europe. Branching of this European lineage is hypothesized to have given rise to two new stems, classified as *Homo neanderthalensis* and *Homo sapiens*. One problem here is that the later Middle Pleistocene hominids from Africa do not clearly belong in any of the taxa named, and their phylogenetic significance is left unresolved. Similar concerns may be raised with respect to remains from China. Also, many paleoanthropologists will object to placing the Neanderthals in a species apart from *Homo sapiens*. Linking these Late Pleistocene Europeans with our own species has been accepted practice for some time, although it is clear that the Neanderthals display some morphological characters which are unlike those of more modern humans. A solution discussed by Bonde is to reclassify all of these groups as allopatric or successive subspecies of *Homo sapiens*. By this reasoning, *Homo sapiens* has great antiquity and can be recognized as a lineage spanning the entire Pleistocene.

HOMO ERECTUS AS A REAL TAXON

Debate over the definition of paleontological species will surely continue, although many of our questions about the phylogeny of *Homo* cannot be answered with the information now available. I feel that at present there is a substantial body of evidence favoring retention of the taxon *Homo erectus*. That this species can be distinguished from most Late Pleistocene populations of *Homo sapiens* and from modern humans is not in doubt. Descriptions based on the principal (especially Asian) fossil assemblages and provided decades ago by Weidenreich, Le-Gros Clark, and others are quite adequate for this purpose, although of course these earlier studies take no account of new discoveries. Comparisons undertaken by Arambourg (1963) and by Tobias and von Koenigswald (1964) have helped to demonstrate similarities among the Asian and African representatives of *Homo erectus*, while comprehensive reviews documenting the distinctions of this entire body of material have been published by Howell (1978) and Howells (1980).

Recent work of my own is intended to build on these earlier findings by comparing crania from East Africa and Indonesia (Rightmire, 1984). While there are obvious differences of size within both assemblages, it is apparent that the better preserved skulls from these regions are broadly similar. Resemblances extend not only to general form and proportions of the vault but also to many small anatomical details. Frontal profiles are low, and there is frequently a keel in the midline. Brows are thick, even in smaller individuals that may be females. Postorbital constriction is marked. Parietal chord lengths are short compared to those of modern humans, and the temporal line follows a flat arc that may produce an angular torus at the parietal mastoid angle. Supramastoid crests tend to be strongly developed, especially in the larger Indonesian hominids. The cranium is relatively broad across the base, and the occipital bone is wide and sharply flexed. Its upper scale slopes forward and is usually shorter than the nuchal plane below. The transverse torus of the occiput is most projecting near the midline. Its lower border is defined by the superior nuchal lines, which meet at a central linear tubercle. There is no true external occipital protuberance. The glenoid cavity is narrowed to produce a deep medial recess. The entoglenoid process may be partly of sphenoid origin, but a sphenoid spine such as occurs in later *Homo sapiens* is not present. There is no raised articular tubercle. The tympanic plate is thick inferiorly. A petrosal spine is prominent, and the plate terminates medially in a blunt projection.

These and other traits serve to describe *Homo erectus* and to differentiate this taxon from *Homo sapiens*. In this sense, the description stands as a diagnosis. However, some of the same characters do appear in earlier *Homo* from Africa, and even in species of *Australopithecus*. Such traits are primitive retentions from a common ancestor and generally do not help to diagnose *Homo erectus* relative to other hominids or hominoids. Traits that appear to be derived for *Homo erectus* (e.g., cranial vault thickness, brain size) are fewer in number but are more useful in assessing the relationships of this species to other groups (Wood, 1984). Where such traits are unique (autapomorphic) and not shared with *Homo sapiens*, they suggest no direct ties with later humans. Proponents of phylogenetic analysis would argue that only to the extent that derived features of *Homo erectus* are shared with *Homo sapiens* is there evidence for a close relationship between the two species.

In any case, the African and Asian crania share many characters, and it is perhaps surprising that greater distinctions among *Homo erectus* assemblages are not apparent. Although this Pleistocene species was widespread geographically, one has to look closely for signs of consistent regional variation. Some differences—in the expression of midline keeling and parasagittal flattening, for example, or in the orientation of the supramastoid crest—are present, and the two crania from Turkana have vault bones that are thinner than those of most Asian specimens. However, the evidence for overall similarity is to my mind more compelling than that suggesting regional or chronological variation (but see Wood, 1984). Skull form of *Homo erectus* seems to have remained relatively unchanged over a long span of time. This species can be characterized morphologically, and there is no reason to resort to definitions based on dating or gaps in the fossil record.

TRENDS IN *HOMO ERECTUS*

If *Homo erectus* is accepted as a real taxon, distinct morphologically from later *Homo sapiens*, then it is possible to check for patterns of change occurring during the lifetime of this species. This can be done quantitatively, by measuring aspects of skeletal or dental form for individuals drawn from assemblages of different geological age. Of course, the success of such an approach to evolutionary trends will depend on how securely the fossils can be dated. While there are still problems, many of the more important specimens can now be placed tentatively in chronological frameworks. Figure 1 compares the stratigraphy for Olduvai Gorge in East Africa with that for Java. The positions for several of the Chinese localities are also indicated.

Material from Africa appears to be older than that from other regions, and the Turkana (Kenya) fossils are perhaps the earliest *Homo erectus* on record. Remains from Olduvai Gorge (Tanzania) are somewhat younger. The braincase of Hominid 9 from upper Bed II is 1.1 to 1.2 m.y. old and therefore postdates the Turkana crania by several hundred thousand years. Dates for the Asian hominids are less certain. Recent studies of fauna collected from Dubois' early excavations at Trinil suggest that this Indonesian locality may be older than neighboring sites such as Kedung Brubus. De Vos et al. (1982) also argue that the Kedung Brubus material resembles the archaic Javanese fauna known collectively as Jetis. If this is correct, then *Homo erectus* at Trinil should be older than hominids associated elsewhere with Jetis assemblages. However, this interpretation has been challenged by Bartstra (1983), and in any case biostratigraphic evidence does not presently allow firm comparison of Trinil with Sangiran, where most of the hominid fossils have been found. At Sangiran, *Homo* is present mainly in the Kabuh Formation, while a few individuals have apparently been recovered from Pucangan levels. Some questions raised by magnetic polarity determinations (Sémah, 1982) and radiometric dates (Nishimura et al., 1980; Suzuki et al., cited in Matsu'ura, 1982) remain to be resolved, but it now looks as though the Sangiran deposits are younger than reported by Ninkovich and Burckle (1978). New work has also been done at several important Chinese localities, and most or all of the Asian *Homo erectus* assemblages may be less than 1.0 m.y. in age (Pope, 1983; Wu, 1985).

One character of *Homo* that is of special interest in any discussion of evolutionary trends is brain size. It is widely recognized that the brain of *Homo habilis* is larger than that of australopithecine species, and there is a clear tendency for endocranial volume to increase in the *Homo* lineage (Tobias, 1971). This increase seems to be relative as well as absolute (Pilbeam and Gould, 1974; Holloway and Post, 1982; Martin, 1983). However, the situation with respect to indi-

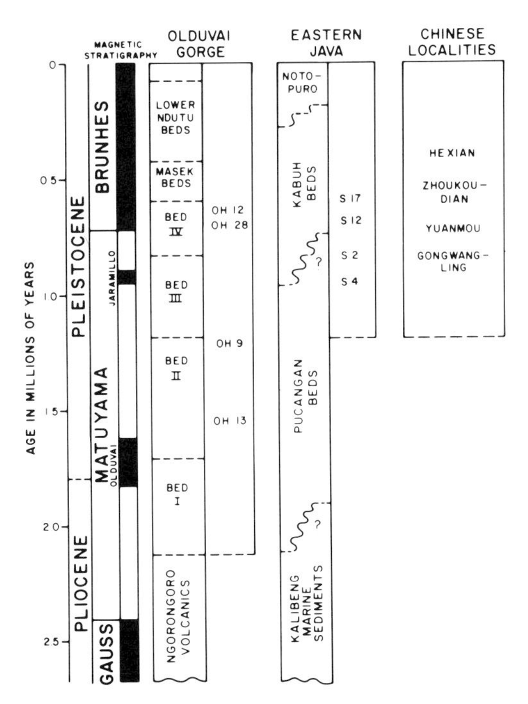

Fig. 1. Stratigraphic sequences and dating for Olduvai Gorge in East Africa and for localities in eastern Asia. Approximate stratigraphic proveniences for several of the more important hominid fossils are also indicated.

vidual species is not so straightforward. Whether *Homo erectus* exhibits a steady progression in cerebral development, or whether there may be a period of slow (even insignificant) change in the mid-Pleistocene, is still debated.

Information relevant to this question is provided in Table I. Here most of the more complete *Homo erectus* crania from Africa and Asia are listed, along with measurements of brain volume. These cranial capacities, along with estimates for geological age of the 20 specimens, can be analyzed using least squares regression. Fitting a linear regression for brain volume (y) on time (x) gives the relationship

$$y = 1033.38 - 134.83x,$$

which is plotted in Figure 2. Here there is a detectable trend, and the more recent hominids have larger brains. The regression coefficient indicates that brain size is increasing in *Homo erectus* at a rate of about 135 ml per 1.0 m.y. However, this rate seems low, and in fact the .95 confidence limits (-134.83 ± 150.26) calculated for the slope include zero. There is no evidence that the trend is statistically significant.

Of course, other characters should be investigated before one can generalize concerning patterns in evolution. Elsewhere, I have tried to examine selected cranial, mandibular, and dental measurements of *Homo erectus*, using a similar approach (Rightmire, 1981). This preliminary work can be questioned (Levinton, 1982, and reply by Rightmire, 1982a), but there are indications that mid-Pleistocene *Homo* populations display stasis in several of these traits. *Homo erectus* may have undergone little morphological change throughout most of its long history.

TABLE 1. Endocranial volume estimates for the more complete Homo erectus *crania from Asia and Africa*

Locality and specimen no.[1]		Cranial capacity (ml)	Reference
Salé		880	Holloway (1981a)
Hexian		1,025	Wu and Dong (1982)
Zhoukoudian	V	1,140	Weidenreich (1943)
Zhoukoudian	II	1,030	Weidenreich (1943)
Zhoukoudian	III	915	Weidenreich (1943)
Zhoukoudian	VI	850[2]	Weidenreich (1943)
Zhoukoudian	X	1,225	Weidenreich (1943)
Zhoukoudian	XI	1,015	Weidenreich (1943)
Zhoukoudian	XII	1,030	Weidenreich (1943)
Sangiran	10	855	Holloway (1981b)
Sangiran	12	1,059	Holloway (1981b)
Sangiran	17	1,004	Holloway (1981b)
Trinil	2	940	Holloway (1981b)
Olduvai	12	727	
Sangiran	2	813	Holloway (1981b)
Gongwangling		780	Woo (1966)
Sangiran	4	908	Holloway (1981b)
Olduvai	9	1,067	
East Turkana	3883	804	Holloway (personal communication)
East Turkana	3733	848	Holloway (personal communication)

[1]Specimens are listed in approximate chronological order, with younger localities first.
[2]Estimate only (see Weidenreich, 1943, p. 114).

EARLY *HOMO SAPIENS*

Throughout the preceeding discussion, I have tried to build a case for viewing *Homo erectus* as a real species, stable during a long time period. This taxon may exhibit trends in the evolution of some characters, but there is little evidence for rapid, progressive morphological change. Brain size for example, seems to increase in later members of the species, but at a relatively low rate. In statistical terms, this trend is not significant. If this picture of *Homo erectus* is generally correct, then an important question is whether there was some change in evolutionary tempos associated with the appearance of our own species late in the Middle Pleistocene. One response, which disregards or denies the case for stasis, holds that trends established in *Homo erectus* were simply continued in later populations. *Homo sapiens* then emerged gradually, in several different geographic regions. According to this scenario, already mentioned, fossils such as Petralona in Europe are intermediate in their morphology and represent populations that are transitional in the progression from archaic to more modern humans. An alternate view is that *Homo sapiens* arose as a result of rapid transformation of only one or a few groups, in a more restricted area.

This is of course an issue which cannot be resolved easily. However, there is now tentative agreement that not only Petralona but also other fossils are best regarded as early *Homo sapiens*. The fine Petralona cranium has been described by Stringer et al. (1979), who note the presence of both archaic and more sapient features. Observations of my own confirm that in several respects this specimen resembles *Homo sapiens*. The frontal profile is flattened, and the brows are heavy. This gives the skull an archaic appearance, although the frontal bone itself is relatively broad and exhibits

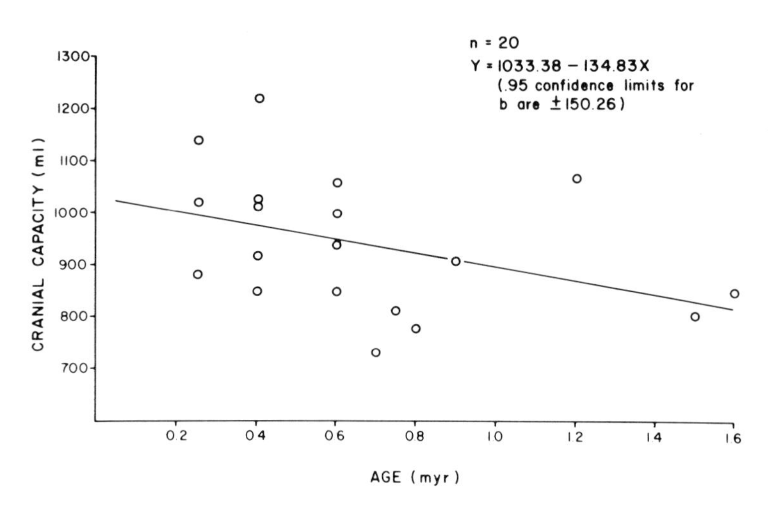

Fig. 2. Cranial capacity and geological age plotted for the *Homo erectus* specimens listed in Table 1. The regression line suggests that brain size increases in more recent crania, although the slope is not significantly different from zero.

less postorbital constriction than is usual for *Homo erectus*. Parietal chord length is greater than that recorded for nearly all African or Indonesian *Homo erectus* individuals. The occipital upper scale is vertical and exceeds the length of the nuchal plane by some 10 mm when both scale lengths are taken from the linear tubercle. Stringer et al. (1979) suggest that Petralona does not display "the posteriorly rotated nuchal region typical of *Homo erectus* fossils." I would concur, with the additional comment that some details of cranial base anatomy are modern. There is little development of a juxtamastoid process. The glenoid cavity is bounded anteriorly by an articular tubercle, but unfortunately the medial part of this cavity is obscured by stalagmitic coating. The relative contributions of the temporal and sphenoid to construction of the entoglenoid process cannot be observed. The tympanic plate appears to be thin, with a locally thickened styloid sheath.

Other European hominids are less well preserved. The important Arago assemblage consists of a face and parietal bone, two mandibles, postcranial remains, and isolated teeth. Opinion is divided as to which taxon is represented by the fossils, but there are clear indications that the Arago skull is unlike African or Asian *Homo erectus* (Stringer, 1981; Rightmire, 1982b). Material from Africa also supports a claim for the presence of *Homo sapiens* at an early date. Cranial capacity of the Broken Hill specimen is expanded well beyond the *Homo erectus* average, and other aspects (especially of occipital and cranial base morphology) suggest these regions to be essentially modern. Details of the sphenotemporal articulation and mandibular fossa are like those of recent humans. The Ndutu cranium has parietal walls that are more rounded than those of Broken Hill, while in other features of the occiput and temporal bone these two individuals are similar (Rightmire, 1983). Two incomplete skulls from the Omo Kibish Fm., discussed recently by Day and Stringer (1982), differ in some ways from each other but also point generally in the direction of more modern humans.

All of these hominids display some characters traditionally associated with *Homo erectus*. Such traits include a low frontal profile, a strong supraorbital torus, an angled occiput, and a broad cranial base. However, these features may have little diagnostic value if they are interpreted as retentions from an ancestor shared broadly by archaic *Homo* populations. Other details of parietal and occipital structure, the occipitomastoid region, the tympanic plate, and the glenoid cavity may reflect a derived condition shared only with *Homo sapiens*. If this is a fair reading of the evidence, then the European and African fossils, perhaps together with individuals such as Dali from China, should be lumped with later humans.

Questions concerning the tempo of change in these first populations of *Homo sapiens* can be addressed only with difficulty. The fossil record of the later Middle Pleistocene is still very sparse, even given the encouraging discoveries of the last decade. Another problem is dating. The caves at Petralona and Arago have been investigated extensively, and faunal materials from both sites have been studied in detail. However, neither biostratigraphic information nor other physical approaches to the cave contents have provided unequivocal results. Ages given for the hominids are subject to considerable uncertainty (Cook et al., 1982). Much the same is true for Africa and the Far East.

If it is accepted that these materials are of late or latest Middle Pleistocene date, and if all of the better crania are treated together without concern for sampling problems introduced by geography, then there is a strong indication that brain volume began to change more rapidly between 400,000 and 200,000 years BP. Even the smaller individuals from Ndutu and Dali have cranial capacities that place them well above the regression line drawn for *Homo erectus* in Figure 2. Larger skulls such as Petralona, Broken Hill, and Omo 2 depart even more strikingly from values expected for late *Homo erectus*. Indeed, these hominids have brains close to the average size for modern humans. Using reasonable estimates for geological age and cranial capacity (Table 2), it is possible to fit a second regression line for these representatives of early *Homo sapiens*. Given the small size of the sample and uncertainties of dating, this exercise surely can be questioned. Nevertheless, results are interesting. The slope of the least squares line calculated for six crania is close to 900 ml/m.y. This suggests a rate of brain increase far higher than that characteristic of *Homo erectus*. Presumably such a rate could be sustained for only a short period, and there is in fact good evidence to show that hominid crania had reached fully modern size by the onset of the Late Pleistocene.

TABLE 2. Endocranial volume and geological age for crania of early Homo sapiens[1]

Locality and specimen no.	Cranial capacity (ml)	Approximate date (yr)
Omo 2	1,430	130,000
Broken Hill	1,285	150,000
Petralona	1,200	200,000
Dali	1,120	200,000–300,000
Arago 21	1,166	200,000–400,000
Ndutu	1,100	200,000–400,000

[1]Information in the table can be used to relate brain size to time, and one least squares regression of age (calibrated in millions of years) on cranial capacity takes the form $y = 1423.74 - 899.61x$. Even with uncertainties of dating, this suggests rapid brain expansion in early *Homo sapiens* populations.

CONCLUSIONS

This study of earlier *Homo* crania casts doubt on any view of human evolution as occurring gradually, at a steady tempo throughout the Pleistocene. *Homo erectus* fossils from sites dispersed widely across the Old World are in fact quite similar morphologically. These hominids from Africa and Asia can be described as heavy browed, with little supratoral hollowing. Frontal narrowing is marked. The vault is low in outline, with an endocranial capacity that is appreciably smaller than that characteristic of *Homo sapiens*. The occipital bone is strongly flexed, and even in smaller individuals a transverse torus is developed. The occipital upper scale is inclined forward and is usually shorter than the nuchal plane below. Anatomy of the glenoid cavity and of the inferiorly thickened tympanic plate is similar in all *Homo erectus*, and there are consistent differences from the condition found in later humans.

This suite of features serving to differentiate *Homo erectus* from *Homo sapiens* occurs both in East African remains of Early Pleistocene antiquity and in assemblages from Indonesia that are probably of much more recent date. Here there is an indication that the species remained relatively stable for a long time. This suggestion is strengthened by the observation that at least one key character shows little quantitative change in a number of *Homo erectus* crania that can be measured. When endocranial volume is plotted against geological age, the least squares regression line computed for individual specimens has a slope indistinguishable from zero. There is some increase in brain size during the long history of the species, but this change is not statistically significant.

Although many workers have claimed to see evidence for gradual evolution, with continuity between successive populations in areas such as northwestern Africa, western Europe, China, and Southeast Asia, these claims usually are based on comparisons involving features of uncertain significance and assessments of overall similarity. There may be little attempt to isolate characters that are diagnostic of the local "clade." Instead, general resemblances of face form or vault shape are cited, without evidence to demonstrate that such traits link local assemblages but not others. Trends in facial or dental reduction that are said to be characteristic of sequences in Southeast Asia, for example, may be equally apparent in other regions, where presumably they need not document any special relationship between ancestor–descendant populations. I do not mean to deny here that continuity between archaic and more modern people can be demonstrated for some area(s) of the Old World. It is quite likely that *Homo erectus* did evolve into *Homo sapiens* somewhere, but careful analysis of characters is needed to show this convincingly.

Crania that can be identified as early *Homo sapiens* are now known from Africa, Europe, and the Far East. These fossils exhibit features particularly of the occipital bone and skull base that are derived in comparison to the *Homo erectus* condition. Most also have brains that are larger than expected, even for late *Homo erectus*, and there is some evidence to show that endocranial volume expanded rapidly toward the close of the Middle Pleistocene. There is too little information to allow speculation about where evolutionary tempos may have quickened first. It is not clear that any split giving rise to distinct, contemporary species actually occurred. Probably some populations advanced rapidly while others became extinct. In any case, it now seems less likely that trends established in archaic Middle Pleistocene *Homo* were simply continued in the same gradual fashion as *Homo sapiens* appeared. The story of human evolution in this time period may turn out to be more complex than anticipated.

ACKNOWLEDGMENTS

Many individuals and institutions assisted me with the research on which these findings are based. I am most grateful for access to the fossils, for funding provided by the National Science Foundation (BNS80-04852, BNS82-17396), and for the opportunity to participate in the opening of the "Ancestors" exhibition.

LITERATURE CITED

Arambourg, C (1963) Le gisement de Ternifine, II. *L'Atlanthropus mauritanicus*. Arch. Inst. Paléontol. Hum., Mém. *32:*37–190.

Bartstra, G-J (1983) The vertebrate-bearing deposits of Kedungbrubus and Trinil, Java, Indonesia. Geol. Mijnbouw *62:*329–336.

Bonde, N (1981) Problems of species concepts in palaeontology. In J Martinelli (ed): Concept and Method in Paleontology. Barcelona: University of Barcelona, pp. 19–34.

Campbell, B (1974) A new taxonomy of fossil man. Yrb. Phys. Anthropol. *17:*194–201.

Cook, J, Stringer, CB, Currant, AP, Schwarcz, HP, and Wintle, AG (1982) A review of the chronology of the European Middle Pleistocene hominid record. Yrb. Phys. Anthropol. *25:*19–65.

Cronin, JE, Boaz, NT, Stringer, CB, and Rak, Y (1981) Tempo and mode in hominid evolution. Nature *292:*113–122.

Day, MH, and Stringer CB (1982) A reconsideration of the Omo Kibish remains and the *erectus–sapiens* transition. 1er Cong. Internat. Paléont. Hum. (Prétirage). Nice: C.N.R.S., pp. 814–846.

Delson, E (1981) Paleoanthropology: Pliocene and Pleistocene human evolution. Paleobiol. *7:*298–305

Delson, E, Eldredge, N, and Tattersall, I (1977) Reconstruction of hominid phylogeny: A testable framework based on cladistic analysis. J. Hum. Evol. *6:*263–278.

de Vos, J, Sartono, S, Hardja-Sasmita, S, and Sondaar, PY (1982) The fauna from Trinil, type locality of *Homo erectus*: A reinterpretation. Geol. Mijnbouw *61:*207–211.

Eldredge, N, and Tattersall, I (1975) Evolutionary models, phylogenetic reconstruction and another look at hominid phylogeny. In FS Szalay (ed): Approaches to Primate Paleobiology. Basel: Karger, pp. 218–242.

Eldredge, N, and Tattersall, I (1982) The Myths of Human Evolution. New York: Columbia University Press.

Gingerich, PD (1979) The stratophenetic approach to phylogeny reconstruction in vertebrate paleontology. In J Cracraft and N Eldredge (eds): Phylogenetic Analysis and Paleontology. New York: Columbia University Press, pp. 41–77.

Gould, SJ, and Eldredge, N (1977) Punctuated equilibria: The tempo and mode of evolution reconsidered. Paleobiol. *3:*115–151.

Holloway, RL (1981a) Volumetric and asymmetry determinations on recent hominid endocasts: Spy 1 and 2, Djebel Ihroud 1, and the Salé *Homo erectus* specimens, with some notes on Neandertal brain size. Am. J. Phys. Anthropol. *55:*385–393.

Holloway, RL (1981b) The Indonesian *Homo erectus* brain endocasts revisited. Am. J. Phys. Anthropol. *55:*503–521.

Holloway, RL, and Post, DG (1982) The relativity of relative brain measures and hominid mosaic evolution. In E Armstrong and D Falk (eds): Primate Brain Evolution: Methods and Concepts. New York: Plenum, pp. 57–76.

Howell, FC (1978) Hominidae. In VJ Maglio and HBS Cooke (eds): Evolution of African Mammals. Cambridge, Massachusetts: Harvard University Press, pp. 154–248.

Howells, WW (1980) *Homo erectus*—who, when and where: A survey. Yrb. Phys. Anthropol. *23:*1–23.

Jelinek, J (1980a) European *Homo erectus* and the origin of *Homo sapiens.* In L-K Königsson (ed): Current Argument on Early Man. Oxford: Pergamon, pp. 137–144.

Jelinek, J (1980b) Variability and geography. Contribution to our knowledge of European and North African Middle Pleistocene hominids. Anthropologie (Brno), pp. 109–114.

Levinton, JS (1982) Estimating stasis: Can a null hypothesis be too null? Paleobiol. *8:*307.

Levinton, JS (1983) Stasis in progress: The empirical basis of macroevolution. Ann. Rev. Ecol. Syst. *14:*103–137.

Martin, RD (1983) Human Brain Evolution in an Ecological Context. 52nd James Arthur Lecture on the Evolution of the Human Brain. New York: American Museum of Natural History.

Matsu'ura, S (1982) A chronological framing for the Sangiran hominids. Bull. Nat. Sci. Mus. (Tokyo) *8:*1–53.

Ninkovich, D, and Burckle, LH (1978) Absolute age of the hominid-bearing beds in eastern Java. Nature *275:*306–308.

Nishimura, S, Thio, KH, and Hehuwat, F (1980) Fission-track ages of the tuffs of the Pucangan and Kabuh formations, and the tektite at Sangiran, central Java. In S Nishimura (ed): Physical Geology of Indonesian Island Arcs. Kyoto: Koto University, pp. 72–80.

Pilbeam, DR, and Gould, SJ (1974) Size and scaling in human evolution. Science *186:*892–901.

Pope, GG (1983) Evidence on the age of the Asian hominidae. Proc. Nat. Acad. Sci. *80:*4988–4992.

Rightmire, GP (1980) *Homo erectus* and human evolution in the African Middle Pleistocene. In L-K Königsson (ed): Current Argument on Early Man. Oxford: Pergamon, pp. 70–85.

Rightmire, GP (1981) Patterns in the evolution of *Homo erectus.* Paleobiol. *7:*241–246.

Rightmire, GP (1982a) Reply to Levinton. Paleobiol. *8:*307–308.

Rightmire, GP (1982b) The Tautavel hominids and *Homo erectus* from Olduvai Gorge. 1er Cong. Internat. Paléont. Hum. (Prétirage). Nice: C.N.R.S., pp. 798–813.

Rightmire, GP (1983) The Lake Ndutu cranium and early *Homo sapiens* in Africa. Am. J. Phys. Anthropol. *61:*245–254.

Rightmire, GP (1984) Comparisons of *Homo erectus* from Africa and southeast Asia. Cour. Forsch. Inst. Senckenberg *66* (in press).

Sémah, F (1982) Chronostratigraphie et paleomagnetisme du Plio-Pleistocene de Java. Application a l'age des sites a Pithecanthropes. 1er Cong. Internat. Paléont. Hum. (Prétiroge). Nice: C.N.R.S., pp. 542–558.

Simpson, GG (1961) Principles of Animal Taxonomy. New York: Columbia University Press.

Stanley, SM (1979) Macroevolution: Patterns and Process. San Francisco: Freeman.

Stringer, CB (1981) The dating of European Middle Pleistocene hominids and the existence of *Homo erectus* in Europe. Anthropologie (Brno) *19:*3–14.

Stringer, CB, Howell, FC, and Melentis, J (1979) The significance of the fossil hominid skull from Petralona, Greece. J. Archaeol. Sci. *6:*235–253.

Thorne, AG, and Wolpoff, MH (1981) Regional continuity in Australasian Pleistocene hominid evolution. Am. J. Phys. Anthropol. *55:*337–349.

Tobias, PV (1971) The Brain in Hominid Evolution. New York: Columbia University Press.

Tobias, PV (1978) The earliest Transvaal members of the genus *Homo* with another look at some problems of hominid taxonomy and systemtics. Z. Morphol. Anthropol. *69:*225–265.

Tobias, PV (1980) A survey and synthesis of the African hominids of the late Tertiary and early Quaternary periods. In L-K Königsson (ed): Current Argument on Early Man. Oxford: Pergamon, pp. 86–113.

Tobias, PV, and von Koenigswald, GHR (1964) A comparison between the Olduvai hominines and those of Java and some implications for hominid phylogeny. Nature *204:*515–518.

Weidenreich, F (1943) The skull of *Sinanthropus pekinensis.* Palaeontol. Sin. (new series D) *10:*1–291.

Wiley, EO (1978) The evolutionary species concept reconsidered. Syst. Zool. *27:*17–26.

Wolpoff, MH (1980) Cranial remains of Middle Pleistocene European hominids. J. Hum. Evol. *9:*339–358.

Wolpoff, MH (1982) The Arago dental sample in the context of hominid dental evolution. 1er Cong. Internat. Paléont. Hum. (Prétirage). Nice: C.N.R.S., pp. 389–410.

Woo, J (1966) The skull of Lantian man. Curr. Anthropol. *7:*83–86.

Wood, B (1984) The origin of *Homo erectus.* Cour. Forsch. Inst. Senckenberg *66* (in press).

Wu, R, and Dong, X (1982) Preliminary study of *Homo erectus* remains from Hexian, Anhui. Acta Anthropol. Sin. *1:*2–13.

Wu, R (1985) New Chinese *Homo erectus* and recent work at Zhoukoudian. In E Delson (ed): Ancestors: The Hard Evidence. New York: Alan R. Liss, Inc., pp. 245–248.

Wu, X (1981) A well-preserved cranium of an archaic type of early *Homo sapiens* from Dali, China. Scienta Sin. *24:*530–541.

Ancestors: The Hard Evidence, pages 265–267

Later Middle Pleistocene Hominids

Jeffrey T. Laitman
Department of Anatomy, Mount Sinai School of Medicine, New York, New York 10029

SUMMARY OF THE DISCUSSION

The study of hominid evolution is largely a search to understand the changes and transitions that have occurred in the fossil record. Questions on when and where one group of hominids evolved into another, or how we classify and position individual specimens themselves, are at the very center of our investigations. As our discussions approach the fossil material from the later part of the Middle Pleistocene, and the emergence of *Homo sapiens* itself, these questions and the debates surrounding them take on added impetus.

The discussion part of this session did not focus upon any single issue. Rather, a number of topics were addressed, including those on the age, condition, and reconstruction of various specimens, as well as those on the taxonomic and phylogenetic position of controversial hominids such as Arago or Petralona. Although no dramatic new positions were advanced, the participants did reemphasize and clarify their arguments. A summary of the major points raised during the discussions, grouped under somewhat general headings, is given in the following paragraphs.

Arago

A number of questions were addressed to Henry and Marie-Antoinette de Lumley and Chris Stringer concerning the age, systematic position, and reconstruction of the Arago cranium. In response to questions on the age of Arago, Henry de Lumley detailed the stratigraphy of the site. He emphasized that there are 15 meters of deposit at Arago between two stalagmitic horizons. The lower stalagmitic floor has been dated by electron spin resonance (ESR) at about 700,000 years. The upper horizon, dated by both ESR and uranium disequilibrium, is about 350,000 years old. The level at which the partial cranium (Arago 21 face) was found is dated at 450,000 years. This level is in a zone devoid of any external ground water, thus making it an essentially closed system. As a result, the dating on the specimen is believed to be very reliable. The dating on the parietal bone of the same individual is slightly younger. This younger date, however, is believed by de Lumley to be due to the location of the parietal in an open system area affected by external ground water. In response to a related question, de Lumley noted that the age of the specimens is indeed an important element in his division of European Middle Pleistocene hominids into three subgroups.

The de Lumleys were quite clear in asserting their belief that specimens such as Mauer and Arago represent European *Homo erectus*. Marie-Antoinette de Lumley stressed what she saw as strong morphological similarities between Arago 21 and Sangiran 17, the only Indonesian *Homo erectus* with considerable portions of the face preserved. She emphasized the similarities of these specimens, stressing the flattening of the vault and the low and projecting area of the midface as being regions of particular importance.

In response to questions, Chris Stringer discussed the possibilities of distortion and alternative reconstructions of the Arago 21 face. He noted that the large degree of facial prognathism shown by Arago could be due to distortion that resulted in the face having been pulled too far anteriorly. Some of the differences between Arago and Petralona, as he also noted in his presentation, could be reduced or even eliminated in an alternative reconstruction. Not surprisingly, Marie-Antoinette de Lumley disagreed strongly with the suggestion that any distortion is present in Arago and stood by her reconstruction. It is regrettable that her manuscript did not arrive in time for inclusion in this volume.

Dating of Salé and Other North African Hominids

Jean-Jacques Hublin discussed the difficulties in dating much of the North African sample. He noted that although the Ternifine specimens are the oldest, the material being from the beginning of the Middle Pleistocene, there is no definite dating of the sites. Correlation with some East African faunas places it at around 500,000 to 600,000 years or possibly even older. Specimens such as those from Salé, Sidi Abderrahman, and the Thomas Quarries are dated at about 400,000 years. The Rabat specimen is particularly difficult to date, due to uncertainties over its geological provenance; Hublin's best estimate is around 200,000 years. The more recent material from Jebel Irhoud is associated with a Mousterian industry, and on the basis of micro-faunal evidence may be between 100,000 and 40,000 years old. Specimens such as Dar-es-Soltan are almost modern, found in Aterian levels. Hublin noted that these North African specimens do not show any areas of morphologic discontinuity as can be found in the European material.

Steinheim

In response to a question regarding to what extent the appearance of Steinheim is due to distortion, Karl Adam stressed that the degree of distortion in Steinheim is not as pronounced as some workers have made it appear. Adam noted that his recent examination of the cranium with the use of computerized tomography (CT) scanning has shown that deformation—save in the obvious twisting of the face and neurocranium or the damage on the left side—is not overly extensive.

The Skull Base

A number of questions were raised concerning the appearance of the basicranium in later Middle Pleistocene hominids. Stringer stressed that Petralona and Broken Hill appear to have basicranial flexion that is largely modern in appearance. He further noted that the reduction of facial prognathism seen in Petralona or Broken Hill, as well as changes in the angulation of the tympanic and petrous bones, are also related to the acquisition of basicranial flexion.

Number of Hominid Lineages in Europe

Both Hublin and Stringer responded to questions on the number of lineages in Europe before the last glaciation. Both were in basic agreement that there were *not* two contemporary lineages in Europe during this period. Hublin stressed the existence of one lineage in Europe ranging from an older group encompassing Petralona, Vértesszöllös, Bilzingsleben, and Arago that eventually led to the Neandertals. Hublin also emphasized that he did not see any pre-sapiens fossils in Europe.

While Stringer was in overall agreement with Hublin, he stressed that he is somewhat more cautious in recognizing Neandertal characters in specimens like Petralona. Stringer noted, however, that there is nothing in the morphology of Petralona or Arago that would deny them the position of being ancestral to Neandertals. He also noted that, were it not for their geographic location, there is nothing in the morphology of Bodo (from Ethiopia) or Broken Hill (from Zambia) that would deny them a similar position of being ancestral to Neandertals.

ASSESSING RELATIONSHIPS AMONG LATER FOSSIL HOMINIDS: THE VALUE OF BASICRANIAL EVIDENCE

Assessing relationships among hominids of the latter portion of the Middle Pleistocene is—as evident from our session—no simple task. Whether hominids such as Arago, Petralona, or Salé should be placed in *Homo erectus* or *Homo sapiens* remains a debatable issue. Scientists involved in these ongoing discussions have advanced their views—as well as their compendia of traits to support those views—with equal fervor. Neither the taxonomic affinities nor the phylogenetic relationships of these hominids will be settled by our exchanges here. These questions will likely continue unabated for some time.

The major benefit of this session for me lies not in the acceptance or rejection of any specific phylogenetic scheme, but rather in furthering my interest in understanding more about which areas of the cranium could be particularly useful in assessing relationships. For example, as I listened to the list of the 80 traits that aligned a specimen with *Homo sapiens* or the 40 that placed it with *Homo erectus*, I could not help but wonder about the value of many of the traits being analyzed. What is the variability among living populations with regard to certain of the characters chosen for study? Are entire regions of the cranium itself more plastic than others during development or in response to disease? In short, are all regions of the cranium of equal value in determining relationships, or are there features that may be of greater diagnostic importance than others in assessing the affinity of fossil hominids?

There are, of course, no simple or direct answers to these questions. Nevertheless, some regions of the skull have come under increasingly closer examination in recent years. One area that has emerged as being of particular importance is that of the cranial base. This region has received considerable attention throughout our symposium, being discussed in detail in presentations, for example, by Olson on the Plio-Pleistocene hominids; by Maier on Early Pleistocene specimens such as O.H. 9; and, within our own session, particularly by Hublin in his analysis of Salé.

The cranial base exhibits a number of features that make it of particular importance in discussing relationships among fossil hominids. Perhaps most important among these is that, overall, the basicranium shows considerably less variability than do many regions of the face or neurocranial vault. Indeed, many clinical studies have shown that while extensive deformations of the face or vault are usually compatible with life, abnormalities of certain areas of the base frequently are not (see Bosma, 1976). The special nature of the base is due to many factors, including the distinctive embryology of large portions of the base (derived from cartilage rather than membranous bone), and the fact that many inviolable nervous and vascular structures pass through the basicranial foramina. The basicranium may thus be a relatively stable area as compared with other portions of the skull. Basicranial alteration may indicate change of a more substantial nature than the more frequent, and easily achieved, changes in the more plastic parts of the skull.

My own studies have examined the aspect of the base known as basicranial flexion. This feature has been the focus of considerable attention due to its value in understanding developmental, comparative, and evolutionary relationships among both living and fossil primates (see reviews by Schulter, 1976; Sirianni and Swindler, 1979). For example, the highly flexed skull of modern adult *Homo sapiens* has been shown to be a characteristic trait of our species. Our analyses have shown that the pattern found among modern *H. sapiens* is significantly different from that shown either by Plio-Pleistocene australopithecines or by certain specimens usually attributed to *H. erectus*, such as KNM-ER 3733. Examination of the more intact later Middle Pleistocene hominids, such as Petralona, Broken Hill, or Steinheim, indicate that these specimens exhibit basicranial flexion largely identical to that of modern *H. sapiens*. Less complete hominids, such as Salé, also show a number of basicranial features characteristic of the highly flexed condition of modern *H. sapiens* (Laitman, 1981; Laitman et al., 1979; Laitman and Crelin, 1980; Laitman and Heimbuch, 1982, 1984).

Our studies thus indicate strong similarities in the basicranial morphology of a number of later Middle Pleistocene hominids and modern *H. sapiens*. As emphasized, basicranial flexion is not merely another "trait," but an important character which may prove of great value in interpreting relationships. Nevertheless, we must always be cautious about relying upon any single character in order to classify a specimen. As Tobias emphasized in his symposium contribution, dependence upon any single trait or region would limit one's ability to comprehend the "total morphological pattern" of a specimen. With this *caveat* in mind, it must be also stressed that it is equally in error to believe that the mere number of traits—often of dubious or unspecified biological meaning—is sufficient to define a specimen's taxonomic or phylogenetic position. What is needed before we can reach a better understanding of relationships among fossil hominids is a greater appreciation of those characteristics—like basicranial flexion—that may be of important diagnostic value. When we attain that level we will understand considerably more about the functional anatomy of fossil hominids, and perhaps be able to answer some of the key questions that still elude us.

LITERATURE CITED

Bosma, JF (ed) (1976) Symposium on Development of the Basicranium. Washington, D.C.: U.S. Government Printing Office.

Laitman, JT (1981) The value of the chondrocranial base in assessing the taxonomic affinity of fossil hominids. Am. J. Phys. Anthropol. *54*:242.

Laitman, JT, and Crelin, ES (1980) An analysis of the Salé cranium: A possible early indicator of a modern upper respiratory tract. Am. J. Phys. Anthropol. *52*:245–246.

Laitman, JT, and Heimbuch, RC (1982) The basicranium of Plio-Pleistocene hominids as an indicator of their upper respiratory systems. Am. J. Phys. Anthropol. *59*:323–344.

Laitman, JT, and Heimbuch, RC (1984) The basicranium and upper respiratory system of European and African *Homo erectus*. Am. J. Phys. Anthropol. *63*:180.

Laitman, JT, Heimbuch, RC, and Crelin, ES (1979) The basicranium of fossil hominids as an indicator of their upper respiratory systems. Am. J. Phys. Anthropol. *51*:15–34.

Schulter, FP (1976) Studies of the basicranial axis. A review. Am. J. Phys. Anthropol. *44*:453–468.

Sirianni, JE, and Swindler, DR (1979) A review of postnatal craniofacial growth in Old World monkeys and apes. Yrbk. Phys. Anthropol. *22*:80–104.

Ancestors: The Hard Evidence, pages 268–271

A Review of Recent Research on Heidelberg Man, *Homo erectus heidelbergensis*

Reinhart Kraatz
Geologisch-Paläontologisches Institut, Universität Heidelberg, D-6900 Heidelberg 1, Federal Republic of Germany

ABSTRACT Recent research on the Heidelberg jaw and the faunal fossil record collected from the fluvial sandbeds of the former meander of the Neckar River (where the mandible was found) provides evidence not only about the local paleoecology in the early Middle Pleistocene (Cromerian) but also on the behavior of this oldest member of the European *Homo erectus* group. New metrical measuring studies of the dental arch of the Heidelberg jaw and its comparison with other fossil hominid jawbones confirm the inclusion of "*Homo heidelbergensis*" into the *Homo erectus* group and indicate at the same time a gradual evolution of the hominid lineage.

ÜBERSICHT Neuere Untersuchungen des Unterkiefers von Mauer und des faunistischen Fossilmaterials aus den fluviatilen Sedimentschichten der ehemaligen Neckarschlinge (worin der Unterkiefer von Mauer gefunden wurde) liefern Hinweise auf die Palökologie seines Lebensraumes im Altpleistozän (Cromer II) und das Verhalten dieses ältesten Angehörigen der europäischen *Homo erectus*-Gruppe. Neue Meßwerte des Zahnbogens und der Vergleich mit anderen fossilen Unterkiefern der Hominidae festigen die Zuordnung des "*Homo heidelbergensis*" innerhalb der *Homo erectus*-Gruppe und lassen gleichzeitig eine gradualistische Evolution der hominiden Stammesreihe erkennen.

"Fossils demonstrate that evolution is a fact."
George Gaylord Simpson, 1983.

HISTORICAL REVIEW

In 1907 the jawbone of the "Heidelberg Man" was discovered in the former "Grafenrain" sandpit. This is near the village of Mauer, which is situated south of Heidelberg at the eastern side of the Upper Rhine Graben Valley, located in a region called "Kraichgau" in southwestern Germany.

Otto Schoetensack, Professor of Paleontology and Paleoanthropology at the University of Heidelberg, had predicted for more than 20 years that a fossil human bone would be found there (Kraatz and Querner, 1967). He had planned this discovery strategically and supervised the collecting of bone fragments in several sandpits in the surroundings of Mauer. Schoetensack wrote in his monograph (1908) that the sediments in which the mandible was embedded could have been formed in the Tertiary period. He put this fossil in the genus *Homo*. By doing so, Schoetensack was the first scientist worldwide to rank such an old fossil among real hominids that walked on two legs; he thus made himself many enemies in his lifetime.

Almost sixty years later, the late paleanthropologist G. Heberer (1961, 1969) and then B.G. Campbell (1964) placed all "pithecanthropines" in the genus *Homo* and classified the members of this group as subspecies of the species *H. erectus*, so that the now valid and almost universally recognized name of Heidelberg Man is *Homo erectus heidelbergensis* (see also Querner, 1968). This human fossil was documented absolutely clearly by Schoetensack; he drew an exact location plan, took good photographs and X-rays of the specimen, and provided a notarized document. Schoetensack supervised the collecting of fossils for 25 years until he died in 1912. The Geological Institute of the University of Heidelberg continued his work during the next 50 years until after 90 years of exploitation the sandpit was closed in 1962. Almost the total fossil record of the "Mauer Fauna" belongs to the collections of this institute.

RECENT RESEARCH

Having studied all our fossils collected in the former Grafenrain Sandpit, I recently finished the manuscript of a catalogue of the complete fossil record. There are some 5,000 bones, teeth, horns, and antlers of a tropical fauna dated to the early Middle Pleistocene, including one human fossil and some 20 bone fragments and pebbles used or altered by hominids.

For almost 20 years the existing material has been studied anew with a variety of methods and viewpoints. The most important papers discuss:

1) sedimentation of the former meander of the Neckar River, and evolutionary history of the Neckar (Rücklin and Schweizer, 1971; Meier-Hilbert, 1974; Schweizer and Kraatz, 1982);
2) the Mauer Fauna (Schütt, 1969a,b, 1970; W. von Koenigswald, 1973, 1982);
3) dating of the locality by its fauna and matrix (Oakley, 1958, 1964; Fleischer et al., 1965; Oakley et al., 1971; W. von Koenigswald, 1982);
4) possible tools of the Heidelberg Man (Pelosse, 1966; Pelosse and Kraatz, 1976);
5) and, last but not least, the human mandible (Howell, 1960; G.H.R. von Koenigswald, 1968; Day, 1977; Runge, 1977; Puech et al., 1980; Puech, 1982; Czarnetzky, 1983; Roth, 1983).

The Possible Artifacts

In the 1950s and 1960s it was suggested that chunks of Lower Triassic sandstone (the so-called Nasenschaber) were used as tools (Querner, 1968). This is in part a misinterpretation, for most of the specimens are not tools but have a natural form. Their edges show the tectogenetic pattern caused by Late Pliocene movements.

In 1970 the late prehistorian H. Lindner presented a "pebble tool" of basalt which he had collected in the area of the former sandpit in 1968, although the pit had already been closed at that time. I consider such "surface evidence" (in which some of the Nasenschaber has to be included) as absolutely invalid. (The basaltic pebble tool belongs to the collection of the late GHR von Koenigswald.)

Jean Léon Pelosse (1966; Pelosse and Kraatz, 1976) discovered dents on the fossil bones of larger animals that were caused neither by routine carnivore biting nor by pebbles when transported in the Neckar River. Pelosse holds that these bones could have been worked by *Homo erectus heidelbergensis.*

At the end of the 1970s, Henry de Lumley (Paris) studied all cherts collected in the sand layers where the Heidelberg mandible had been discovered. He identified two specimens that had been intentionally altered (not yet published).

Comments and New Results on Morphology of the Mauer Mandible

As pointed out by, among others, the late Professor G.H.R. von Koenigswald, the second molar of the Mauer jaw is larger than either the first or the third molar. This may be additional evidence that *Homo heidelbergensis* is a member of the *Homo erectus* group (see G.H.R. von Koenigswald, 1968). Moreover, the Heidelberg jaw does not have a gap between the third molar and the ascending ramus. Trinkaus and Howells (1979) are of the opinion that such a gap is a characteristic feature of the Neanderthals, and its absence removes this fossil from consideration as a member of that group.

Puech (1980, 1982; Puech et al., 1980) studied dental wear at high magnification; his findings showed that the buccal surfaces of the molars evince particular signs of abrasion (Fig. 1). This indicates that Heidelberg Man was a vegetarian rather than a carnivore. The heavily worn incisors must also have been used for nonfeeding activities similar to those of the living eskimos. Puech (1980) came to the conclusion that *H. e. heidelbergensis* hunted only small animals, presumably living in small groups along the Neckar River and in the wooded hinterland.

Helga Roth (1983) has developed a new method of studying dental arch evolution by plotting the perpendicular length of the jaw

Fig. 1. Microphotographs of the buccal surface of the second molar right. Peels and photos by P.F. Puech.

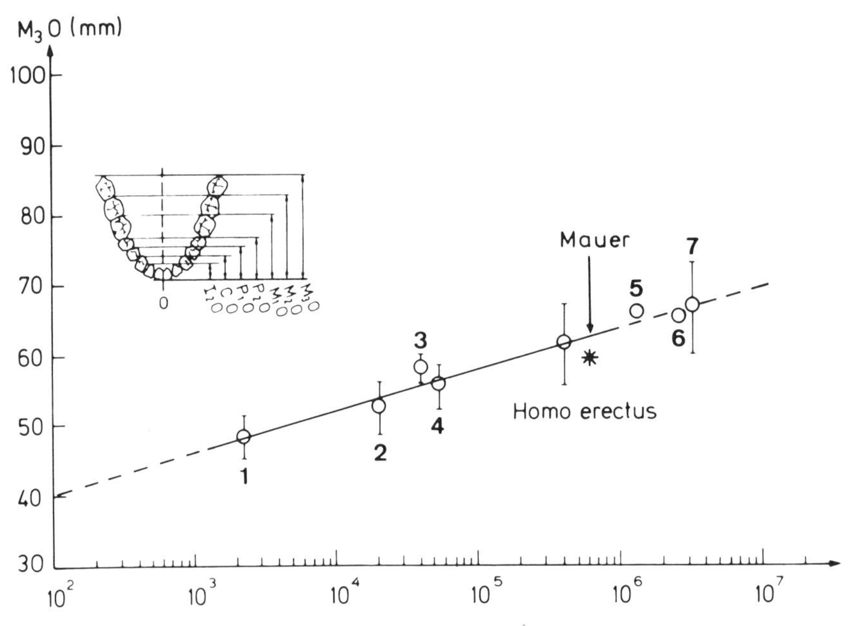

Fig. 2. Regression of dental arch length to M_3 on geological age in hominids. Key: 1 = 12 modern European mandibles; 2 = 14 Upper Paleolithics; 3 = three Israeli "Protocromagnons"; 4 = 13 Neanderthals; *Homo erectus* = 12 specimens; 5 = one *Homo habilis*; 6 = one *Australopithecus africanus*; 7 = two Afar australopithecines. x-axis = geological age in years (logarithmic scale); y - axis = third molar perpendicular arch length (in mm), corresponding to the inset diagram. (After Roth, 1983.)

from infradentale to the rear of each tooth against time (logarithmically). She compared the dental arches of all European fossil hominid mandibles with a selection of earlier specimens and found that *Homo heidelbergensis* fits well into the European *Homo erectus* group. Figure 2 reproduces her graph of jaw length to M_3, with a regression of this variate on time ($y = a\ln x + b$, where y is jaw length and x is geological age; (see

Fig. 3. The Heidelberg mandible linked with the Olduvai hominid skull O.H. 9. Photo by Schacherl, GPIH.

Roth, 1983). To a certain degree the curve shows the gradual evolution of dental reduction in the human family (Fig. 2).

Finally, it is interesting to note that the Early Pleistocene cranium O.H. 9 from Olduvai Gorge, Tanzania preserves both glenoid fossae, into which the condyles of the Heidelberg jawbone fit precisely (Fig. 3). The implications of this fit will be discussed by Kraatz et al. (in preparation).

LITERATURE CITED

Campbell, BG (1964) Quantitative taxonomy and human evolution. In SL Washburn (ed): Classification and Human Evolution. London: Methuen and Co Ltd., pp. 50–74.

Czarnetzki, A .(1983) Belege zur Entwicklungsgeschichte des Menschen in Südwestdeutschland. In H Müller-Beck (ed): Urgeschichte in Baden-Württemberg. Stuttgart: Konrad Theiss, pp. 217–240.

Day, MH (1977) Guide to Fossil Man. London: Cassel.

Fleischer, RL, Price, PB, and Walker, RN (1965) Applications of fission tracks and fissiontrack-dating to anthropology. General Electric Res. Lab., Report No. 65-RL-3878 M, Schenectady, New York.

Heberer, G (1961) Die Abstammung des Menschen. In F Gessner (ed): Handbuch der Biologie, Band IX (2). Konstanz: Athenaion, pp. 245–328.

Heberer, G (1969) Der Ursprung des Menschen. Unser gegenwärtiger Wissensstand. Stuttgart: Fischer.

Howell, FC (1960) European and northwest African Middle Pleistocene hominids. Curr. Anthropol. *1:*195–232.

Koenigswald, GHR von (1968) Die Geschichte des Menschen, 2nd ed. Berlin: Springer.

Koenigswald, W von (1973) Veränderungen in der Kleinsäugerfauna von Mitteleuropa zwischen Cromer und Eem (Pleistozän). Eiszeitalter und Gegenwart *23/24:*159–167.

Koenigswald, W von (1982) Zur Gliederung des Quartärs. In J Niethammer and F Krapp (eds): Handbuch der Säugetiere Europas—Band 2/1 Rodentia II. Wiesbaden: Akad. Verl. Ges., pp. 15–17.

Kraatz, R, and Querner, H (1967) Die Entdeckung des *Homo heidelbergensis* durch Otto Schoetensack vor 60 Jahren. Ruperto Carola, Heidelberg *42:*178–183.

Kraatz, R, Maier, W, and Nkini, AT: Zur Stellung des Frühmenschen *Homo erectus.* Natur und Museum, Frankfurt: Kramer. (In preparation).

Meier-Hilbert, G (1974) Sedimentologische Untersuchungen fluviatiler Ablagerungen in der Mauerer Neckarschleife. Heidelberger geogr. Arb. *40:*201–218.

Oakley, KP (1958) Application of fluorine, uranium and nitrogen analysis to the relative dating of the Rhünda Skull. Neues Jahrb. Geol. Pal., Mh. *1958:*130–136.

Oakley, KP (1964) Frameworks for Dating Fossil Man. London: Weidenfeld and Nicholson.

Oakley, KP, Campbell, BG, and Molleson, TI (1971) Catalogue of Fossil Hominids. Part II: Europe. London: British Museum (Natural History).

Pelosse, JL (1966) Traces possible d'action humaines sur des os fossiles du gisement d'*Homo heidelbergensis.* Bull. Assoc. Franç. Et. Quat., Paris *4:*247–250.

Pelosse, JL, and Kraatz, R (1976) Une Pièce osseuse façonnée du gisement de Grafenrain (*Homo heidelbergensis*). *Quaternaria 19:*93–105.

Puech, PF (1980) *Homo heidelbergensis* lointain ancêtre était droitier et vivait dans les bois. Montpellier: Midi Libre.

Puech, PF (1982) L'usure dentaire de l'Homme de Tautavel. Prétirage, 1er Cong. Internat. Paléontol. Hum. Nice: C.N.R.S., pp. 249–275.

Puech, PF, Prone, A, and Kraatz, R (1980) Microscopie de l'usure dentaire chez l'Homme fossile: Bol alimentaire et environnement. CR Acad. Sci. Paris, *290*(D):1413–1416.

Querner, H (1968) Stammesgeschichte des Menschen. Stuttgart: Kohlhammer.

Roth, H (1983) Comparaison statistique de la forme des arcades alvéolaire et dentaire des mandibules des hominidés fossiles. Mus. Nat. d'Hist. Natur., Musée de l'Homme, Lab. de Paléontologie humaine et de Préhistoire, Mémoire 17, 2 vol., Paris.

Rücklin, H, and Schweizer, V (1971) The Geology and Geomorphology of Heidelberg and Its Surroundings. In G Müller (ed): Sedimentology of Parts of Central Europe. Guidebook 7th Intern. Sediment. Cong., 1971. Frankfurt: Kramer, pp. 337–344.

Runge, B (1977) Morphologische Untersuchungen der Zahnkronen der Funde von Mauer, Oberkassel, Bad Nauheim und Trebur. In U Schäfer (ed): Beiträge zur Odontologie paläolithischer, meso- und neolithischer Menschenfunde. Giessen: Anthropol. Inst. Univ. Giessen.

Schoetensack, O (1908) Der Unterkiefer des *Homo heidelbergensis* aus den Sanden von Mauer bei Heidelberg. Ein Beitrag zur Paläontologie des Menschen. Leipzig: Engelmann.

Schütt, G (1969a) *Panthera pardus sickenbergi* n. subsp. aus den Mauerer Sanden. N. Jb. Geol. Paläont., Mh. *1969:*299–310.

Schütt, G (1969b) Untersuchungen am Gebiß von *Panthera leo fossilis* (v. Reichenau 1906) und *Panthera leo spelaea* (Goldfuss 1810). N. Jb. Geol. Paläont., Abh. *134:*192–220.

Schütt, G (1970) Nachweis der Säbelzahnkatze *Homotherium* in den altpleistozänen Mosbacher Sanden (Wiesbaden/Hessen). N. Jb. Geol. Paläont., Mh. *1970:*187–192.

Schweizer, V, and Kraatz, R (1982) Kraichgau und südlicher Odenwald. In MP Gwinner (ed): Sammlung Geologischer Führer, Band 72. Berlin-Stuttgart: Gebr. Borntraeger.

Trinkaus, E., and Howells, WW (1979) The Neanderthals. Sc. Amer. *24:*118–133.

Ancestors: The Hard Evidence, pages 272–276

The Chronological and Systematic Position of the Steinheim Skull

Karl Dietrich Adam
Staatliches Museum für Naturkunde in Stuttgart, Geologisch-Paläontologische Abteilung, D-7000 Stuttgart, Federal Republic of Germany

ABSTRACT The Steinheim skull was discovered in 1933 by Karl Sigrist in his father's gravel pit, where the specimen was embedded in Pleistocene sands of the river Murr. On the basis of the rich occurrence of *Elephas antiquus* and the fossil record of other mammals, these older sands belong to the Mindel-Riss- or Holstein interglacial. The so-called "Waldelefanten-Sande" were subsequently overlain by gravels, yielding especially remains of *Elephas primigenius fraasi.* This is a transitional or intermediate form between *Elephas trogontherii* and *Elephas primigenius,* which allows one to put these younger gravels, known as "Steppenelefanten-Kiese," into the Riss or Saale glacial.

The Steinheim skull shows some archaic features, but also characteristics which are obviously advanced. The specimen named *Homo steinheimensis* by Fritz Berckhemer (1936) can be considered as an early Middle Pleistocene form among the ancestors of modern man, to whom it seems to be more closely related than to the Neandertals of the Late Pleistocene. The Steinheim man of Holstein interglacial age can be regarded as a transitional form between *Homo erectus* of the Early Pleistocene and *Homo sapiens* of the late Würm glacial at the end of the Ice Age.

ABRISS Der 1933 von Karl Sigrist entdeckte Steinheimer Urmenschen-Schädel stammt aus den durch zahlreiche Funde von *Elephas antiquus* gekennzeichneten älteren Sanden der Murr, die nach ihrer Säugetier-Fauna dem Mindel-Riss oder Holstein-Interglazial zugehören. Diese unteren Waldelefanten-Sande werden von jüngeren Kiesen überlagert, in deren Fossilbestand *Elephas primigenius fraasi* vorherrscht; es ist eine Übergangsform von *Elephas trogontherii* zu *Elephas primigenius,* welche die als Steppenelefanten-Kiese bezeichnete obere Steinheimer Hauptfundschicht ins Riss- oder Saale-Glazial zu stellen erlaubt.

Der Schädel des Steinheimer Urmenschen zeigt neben manchen primitiven Zügen auch derart progressive Merkmale, dass man den von Fritz Berckhemer (1936) *Homo steinheimensis* benannten Fund als Beleg einer frühen, mittelpleistozänen Form unter den Vorfahren des heutigen Menschen betrachten kann, dem er näher zu stehen scheint als dem jungpleistozänen Neandertaler. Der Steinheimer des Holstein-Interglazials erweist sich demnach als eine Übergangsform, als ein Bindeglied zwischen *Homo erectus* des älteren Pleistozäns und *Homo sapiens* des jüngeren Würm-Glazials.

The Steinheim skull was found on July 24, 1933, in the Sigrist gravel pit (Fig. 1) at Steinheim on the Murr by Karl Sigrist, son of the owner. It was removed carefully under the direction of the curator of the Württembergische Naturaliensammlung, Fritz Berckhemer (Fig. 2; Berckhemer, 1933a,b, 1934b; Adam, 1975b). The specimen is housed in the Staatliches Museum für Naturkunde in Stuttgart, the former Württembergische Naturaliensammlung.

The skull was discovered about 7 m below the surface, embedded in Pleistocene sands of the river Murr. According to the mammalian fauna from the same site, these deposits must be regarded as interglacial. On the basis of the rich occurrence of *Elephas antiquus*, these sands are known as "Waldelefanten-Sande" and have been put into the Holstein interglacial, Albrecht Penck's well-known Mindel-Riss interglacial (Adam, 1954a,b). The interglacial deposits in the Steinheim gravel pits were subsequently overlain by younger gravels, the so-called "Steppenelefanten-Kiese," yielding *Elephas primigenius fraasi*, a form intermediate between *trogontherii* and *primigenius*, which is characteristic for the early Riss or Saale glacial in Steinheim on the Murr as well as for the gravels of the Gösel and Pleisse rivers near Markkleeberg, south of Leipzig, one of the most famous Paleolithic sites in Middle Europe (Baumann and Mania, 1983).

The sands and gravels at Steinheim on the Murr, with their rich mammalian finds, represent river deposits resulting from Pleistocene tectonics. They are restricted to the neighborhood of this small Swabian town and must be regarded as the reaction of the river Murr to the sinking of the land surface, which continued over the period from the Holstein interglacial until the Riss glacial, and this in spite of climatic change (Wagner, 1929, 1934). Such climatic change is perceivable in a zone between the "Waldelefanten-Sande" and the "Steppenelefanten-Kiese" by the gradual change of flora and fauna, especially by the occurrence of forest and steppe elephants representing migrations in summer as well as in winter (Adam, 1961).

In addition to the abundant *Elephas* remains there is also a great number of mammalian forms that give insight into the climate as well as the chronological position of the deposits (Adam, 1977). This applies to the whole fluviatile sequence of deposits, but especially to the strata where the major finds were made, the "Waldelefanten-Sande" and the "Steppenelefanten-Kiese." Their rich fossil record also allows a comparison with other mammalian faunas of southwest Germany; that from the Holstein interglacial bone breccia of the Heppenloch, at the entrance to a cave above Gutenberg in the Swabian Alb, is of special importance (Adam, 1975a).

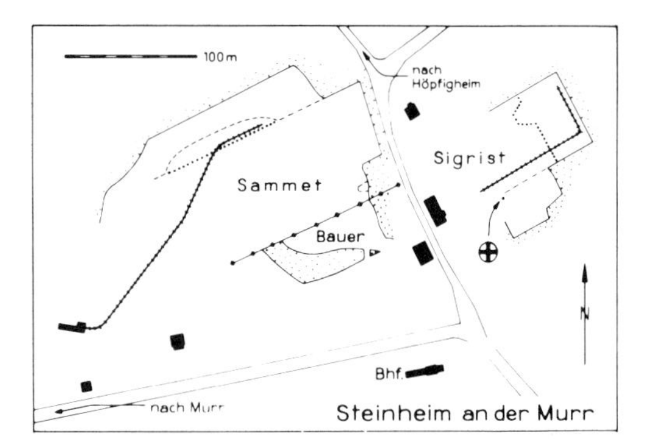

Fig. 1. Plan of the Sigrist, Bauer, and Sammet gravel pits at Steinheim on the Murr in 1930–1936, indicating findspot of the Steinheim skull (circled cross). The position of the several excavation faces is shown for three years: 1930 (heavy dots), 1932 (dashed line), and 1936 (thin solid line). Stippling indicates areas remaining to be excavated in 1936.

Fig. 2. The Steinheim skull as discovered in the Sigrist pit on July 24, 1933, lying on its right side and partially covered by sand.

To determine the chronological position of the Steinheim sands and gravels within the Pleistocene time scale, it is important to have proof for another, younger occurrence of *Elephas antiquus*, and that occurs in the Sammet gravel pit in a layer (above the glacial "Steppenelefanten-Kiese") that can be considered as belonging to the Eem interglacial, as documented in the Untertürkheim travertines near Stuttgart (Adam and Berckhemer, 1983). In addition to this, there were for a short period of time even older gravels visible in the Sigrist gravel pit, which yielded a very few remains of steppe elephants, probably indicating a climatic change within the Holstein interglacial.

The age of the Steinheim sands and gravels, especially of the two main strata, may be determined not only in regard to their fossil record, but also through the sequence of all the Quaternary deposits at this locality. On the basis of the stratigraphic sequence, Fritz Berckhemer (1934a) was already able to derive a minimum age for the "Waldelefanten-Sande" and therefore also for the Steinheim skull. From the so-called "Vereisungskurve"—a sequence of cold and warm periods during the Pleistocene—which was established by Wolfgang Soergel and enjoyed almost worldwide acceptance at that time, it was apparent that there must have been a warm period within the Riss glacial. However, on the basis of the well documented fauna, it was clear that the Steinheim warm phase could not have been just an interstadial, but must have been an interglacial period, namely, the Mindel-Riss interglacial.

This dating was recently confirmed by the extensive studies of Gert Bloos on the origin of the fluviatile deposits in the lower Murr valley as well as of the "Fliesserden" and "Lösse" in the Steinheim area. Moreover, he was able to interpret the whole sequence of geological events in the environs of Steinheim, thus confirming on geological grounds (Bloos, 1977) the paleontologically-based dating of the main strata presented here. The incontestably great scientific value of the Steinheim skull is unquestionably enhanced by the fact that it can be so securely correlated to the late Holstein or Mindel-Riss interglacial. It is in the context of this sound paleontological and geological dating, and not on the basis of the often doubtful chemical and physical dating results, that the morphology of the skull should be interpreted (Szabo and Collins, 1975).

Fritz Berckhemer, who named the specimen *Homo steinheimensis,* recognized early that in spite of its Middle Pleistocene age it must be related more closely to modern man than to the Neandertals of the later Pleistocene (Berckhemer, 1936; Oakley et al., 1971; but compare Stringer, 1985). Some of the morphological features are archaic, for example the relatively small cranial capacity of about 1,100 cm^3, weak mastoid, and the considerable development of the torus supraorbitalis, well separated from the frontal squama by a broad sulcus supratoralis. Others are obviously advanced, such as the large frontal sinus (Fig. 3), vertical parietals (Fig. 4), deep nasal root, nearly straight course of the superior orbital margin, upward bend of the upper jaw margin from the alveolar plane to the zygoma, pronounced canine fossa, and relatively small M^3 (see color Plates 6 and 7). Some of these were mentioned by Fritz Berckhemer a short time after discovery. It is these advanced or derived features that should be used to determine the phylogenetic position of the Steinheim skull.

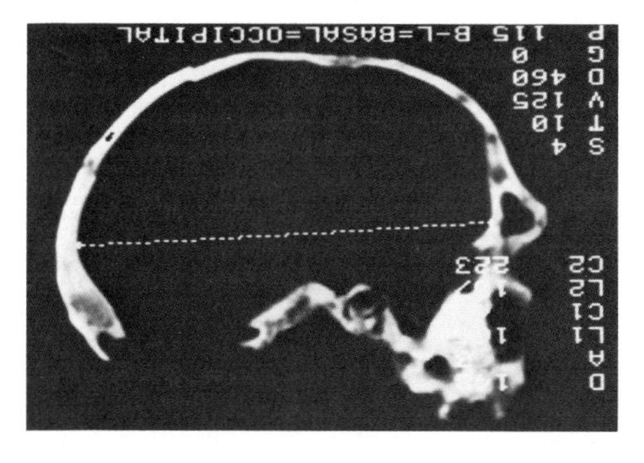

Fig. 3. Computerized tomogram of the Steinheim skull, showing cranial vault thickness and large frontal sinus development. The skull was tomographed with the assistance of the Ludwigsburg Hospital.

Fig. 4. Occipital view of the Steinheim skull after recent preparation (1983); note vertical parietal walls.

For paleoanthropology as well as paleontology in general it is the derived characteristics and not the primitive retentions that are important in classification. Therefore, *Homo steinheimensis* should not be included in the *Homo erectus* group: this specimen is rather a transitional form or connecting link between *Homo erectus* of the Early Pleistocene and *Homo sapiens* of the Würm glacial at the end of the Ice Age (Wolpoff, 1980).

Neither other parts of the skeleton nor tools (as in the gravels of Swanscombe on the Thames, with a stratigraphically and morphologically comparable skull) were found in the sands of Steinheim on the Murr (Brothwell et al., 1964; Freising, 1972; Adam, 1973). The small skull is the only human document. It represents a woman of small size who, according to her teeth, did not survive her third decade. Her face was reconstructed by Gerassimov (1964). Gieseler (1974) has argued that the skull exhibits serious injuries, especially to the strongly damaged left side (Fig. 5), which were inflicted during life by her fellow men, obviously with a "blunt instrument"(shades of Agatha Christie!). This intentional and mortal violence led to the breaking of the bony wall and sinuses. The skull was then removed from the body, and (as already recognized by Berckhemer, 1937), the foramen magnum was forcibly enlarged to open up the skull base (Fig. 6). Finally, one can assume the brain was removed.

The doubts recently expressed by Czarnetzki (1983) against this well-considered hypothesis of post-mortem enlargement of the skull base are without merit. If this damage had occurred through taphonomic or diagenetic causes after embedment, the missing pieces of the left side would have been found nearby during the painstakingly detailed collection. A post-depositional rearrangement of the skull seems impossible given the nature of its excavation and preservation.

The damage to this cranium, as with that seen in such fossils as the Monte Circeo Neandertal and as practiced by modern headhunters, was not a result of hunger or because the brain was regarded as a delicacy. Such a procedure on the dead body may have involved magical or religious activity, and it represents evidence of the degree of intellectual evolution 250,000 years ago (Kipp, 1980; Adam, 1984). Except for the Mauer lower jaw of Heidelberg man (Kraatz, 1985) discovered in older Pleistocene sands of the river Neckar in the same region of Baden-Württemberg, the well-preserved Steinheim specimen is the most important German find of fossil man (Adam, 1982).

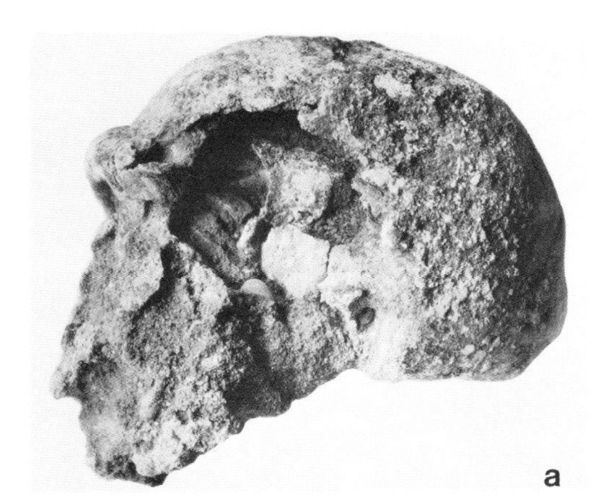

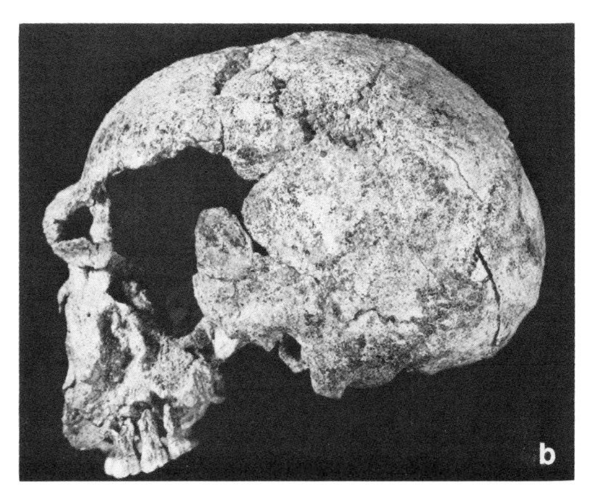

Fig. 5. Left lateral views of the Steinheim skull: a) as found; b) after 1983 preparation; brought to same scale. For right lateral view, see Plate 7 in color section.

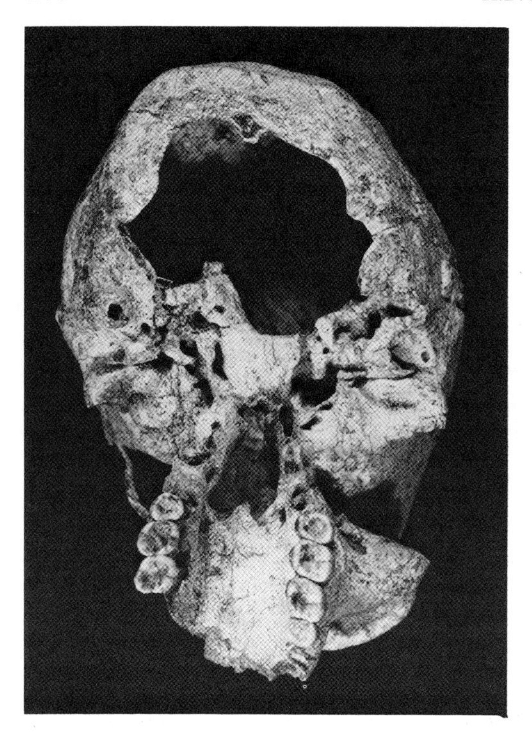

Fig. 6. Basal view of Steinheim skull after 1983 preparation, showing damage to foramen magnum region. For occlusal and facial views, see color Plate 6.

LITERATURE CITED

Adam, KD (1954a) Die zeitliche Stellung der Urmenschen-Fundschicht von Steinheim an der Murr innerhalb des Pleistozäns. Eiszeitalter u. Gegenwart *4/5:*18–21.

Adam, KD (1954b) Die mittelpleistozänen Faunen von Steinheim an der Murr (Württemberg). Quaternaria *1:*131–144.

Adam, KD (1961) Die Bedeutung der pleistozänen Säugetier-Faunen Mitteleuropas für die Geschichte des Eiszeitalters. Stuttgarter Beitr. Naturkde. *78:*1–34.

Adam, KD (1973) Die "Artefakte des *Homo steinheimensis*" als Belege urgeschichtlichen Irrens. Stuttgarter Beitr. Naturkde. Serie B *6:*1–99.

Adam, KD (1975a) Die mittelpleistozäne Säugetier-Fauna aus dem Heppenloch bei Gutenberg (Württemberg). Abh. Karst- u. Höhlenkde. Reihe D *1:*i–iv, 1–247.

Adam, KD (1975b) Das Urmensch-Museum Steinheim an der Murr. Steinheim an der Murr: Stadtverwaltung.

Adam, KD (1977) Die mittelpleistozänen Schotter der unteren Murr (Baden-Württemberg) und ihre Säugetier-Faunen. Jber. Mitt. oberrhein. geol. Ver. Neue Folge *59:*83–89.

Adam, KD (1982) Der Mensch im Eiszeitalter. Funde aus dem Pleistozän des Neckarlandes. Stuttgarter Beitr. Naturkde. Serie C *15:*3–17, 26–53, 70.

Adam, KD (1984) Der Mensch der Vorzeit. Führer durch das Urmensch-Museum Steinheim an der Murr. Stuttgart: Konrad Theiss Verlag.

Adam, KD, and Berckhemer, F (1983) Der Urmensch und seine Umwelt im Eiszeitalter auf Untertürkheimer Markung. Ein Beitrag zur Urgeschichte des Neckarlandes. Stuttgart: Bürgerverein Untertürkheim.

Baumann, W, and Mania, D (1983) Die Paläolithischen Neufunde von Markkleeberg bei Leipzig. Berlin: VEB Deutscher Verlag der Wissenschaften.

Berckhemer, F (1933a) Notiz über den Fund eines Urmenschenschädels in den Schottern von Steinheim a. d. Murr. Palaeontol. Z. *15:*224.

Berckhemer, F (1933b) Ein Menschen-Schädel aus den diluvialen Schottern von Steinheim a. d. Murr. Anthropol. Anz. *10:*318–321.

Berckhemer, F (1934a) Der Steinheimer Urmensch und die Tierwelt seines Lebensgebietes. Aus d. Heimat *47:*101–115.

Berckhemer, F (1934b) Wie der Urmenschenschädel von Steinheim a. d. Murr gefunden wurde. Kosmos *31:*242–246.

Berckhemer, F (1936) Der Urmenschenschädel aus den zwischeneiszeitlichen Fluss-Schottern von Steinheim an der Murr. Forsch. u. Fortschr. *12:*349–350.

Berckhemer, F (1937) Bemerkungen zu H. Weinert's Abhandlung "Der Urmenschen-Schädel von Steinheim." Verh. Ges. phys. Anthropol. *8:*49–58.

Bloos, G (1977) Zur Geologie des Quartärs bei Steinheim an der Murr (Baden-Württemberg). Jber. Mitt. oberrhein. geol. Ver. Neue Folge *59:*215–246.

Brothwell, DR, Campbell, BG, and Castell, CP (1964) The Swanscombe Skull. A Survey of Research on a Pleistocene Site. London: Royal Anthropological Institute of Great Britain and Ireland.

Czarnetzki, A (1983) Zur Entwicklung des Menschen in Südwestdeutschland. In H Müller-Beck (ed): Urgeschichte in Baden-Württemberg. Stuttgart: Konrad Theiss Verlag.

Freising, H (1972) Die "Steingeräte" des Steinheimer Urmenschen. Jh. Ges. Naturkde. Württemb. *127:*50–51.

Gerassimov, MM (1964) Menschen des Steinzeitalters. Moskau: Verlag "Nauka."

Gieseler, W (1974) Die Fossilgeschichte des Menschen. Stuttgart: Gustav Fischer Verlag.

Kipp, FA (1980) Die Evolution des Menschen im Hinblick auf seine lange Jugendzeit. Stuttgart: Verlag Freies Geistesleben.

Kraatz, R (1985) A review of recent research on Heidelberg Man, *Homo erectus heidelbergensis.* In E Delson (ed): Ancestors: The Hard Evidence. New York: Alan R. Liss, Inc., pp. 268–271.

Oakley, KP, Campbell, BG, and Molleson, TI (1971) Catalogue of Fossil Hominids. Part II: Europe. London: Trustees of the British Museum (Natural History).

Stringer, C.B. (1985) Middle Pleistocene Hominid Variability and the Origin of Late Pleistocene Humans. In E. Delson (ed): Ancestors: The Hard Evidence. New York: Alan R. Liss, Inc., pp. 289–295.

Szabo, BJ, and Collins, D (1975) Ages of fossil bones from British interglacial sites. Nature *254:*680–682.

Wagner, G (1929) Junge Krustenbewegungen im Landschaftsbilde Süddeutschlands. Beiträge zur Flussgeschichte Süddeutschlands I. Öhringen: Verlag der Hohenloheschen Buchhandlung F. Rau.

Wagner, G (1934) Der Fundort des Steinheimer Urmenschen. Aus d. Heimat *47:*97–101.

Wolpoff, MH (1980) Cranial Remains of Middle Pleistocene European Hominids. J. Hum. Evol. *9:*339–358.

Ancestors: The Hard Evidence, pages 277–282

A Late Rissian Deposit in Rome: Rebibbia–Casal de'Pazzi

Amilcare Bietti
Departimento di Biologia Animale e dell'Uomo, Sezione di Antropologia, Università di Roma "La Sapienza," Città Universitaria, 00100 Rome, Italy

ABSTRACT The locality is situated in the lower valley of the Aniene River, in an area of growing urbanization. The large archaeological deposit, covering over 2,000 m^2, has been excavated now for several years under the direction of a multidisciplinary team. The fauna recovered includes elephant, rhino, hippo, cervids, carnivores, and birds. A fragment of a rather thin (10 mm) human parietal was discovered in 1983. As other deposits nearby (Monte Circeo, Torre in Pietra) appear to span a long time interval, it is important to have a better idea both of the age of the deposit (possibly mid to late "Riss") and of its span: that is, how much time is represented? To that end, the large sample of Mousterian–"Pontinian" artifacts (524) recovered from 19 5 × 5 m squares was subjected to χ^2 analysis to determine if the great variation in patination was correlated with typology or horizon. No significant difference from independence was determined in three separate tests. The apparent groupings in Bordes indices according to patina appear, therefore, to be due to sampling bias effects. The assemblage appears closer in several features to the lower level (m) of Torre in Pietra than to level d or to Monte Circeo, but only one handaxe was recovered, while level m of Torre in Pietra is usually considered Acheulian due to its numerous bifaces. Whether these differences are due to "cultural" or "functional" distinctions is as yet uncertain.

INTRODUCTION

The site of Rebibbia–Casal de'Pazzi was discovered in 1981 after a series of surveys done by the Soprintendenza Archeologica di Roma in this zone, where a general building plan was outlined by the City Council of Rome. The lower valley of the Aniene River—i.e., the last few km before its confluence with the Tiber—has yielded numerous anthropological and archaeological remains at the sites of Ponte Mammolo, Sedia del Diavolo, Monte delle Gioie, and, most famous of all, Saccopastore. The Rebibbia–Casal de'Pazzi site represents the latest discovery of archaeological materials deposited by the various fluvial episodes of the Aniene, the bed of which has repeatedly changed through time, due to volcanic activity in the Rome area. Figure 1 shows the position of the new site in relation to the others in this region. It is situated on the right bank of the river, in an area of increasing urbanization. Other sites, such as Saccopastore, Sedia del Diavolo, and Monte delle Gioie, have been completely covered by buildings in the past several years.

The first finds consisted of some tusks and molars of *Palaeoloxodon antiquus*, teeth of *Bos*, and some flint tools, essentially sidescrapers and unretouched flakes. A rescue excavation was begun under the direction of A.P. Anzidei of the Soprintendenza Archeologica di Roma, with the help of Mr. M. Ruffo. It was immediately clear that the size of the site (more than 2,000 m^2) required a much larger effort, and a regular excavation was started and is still in progress, with the help of A.G. Segre for geology and stratigraphy, P. Cassoli for the fauna, and A. Bietti for archaeology. A preliminary report was presented at the colloqium on the Paleolithic of

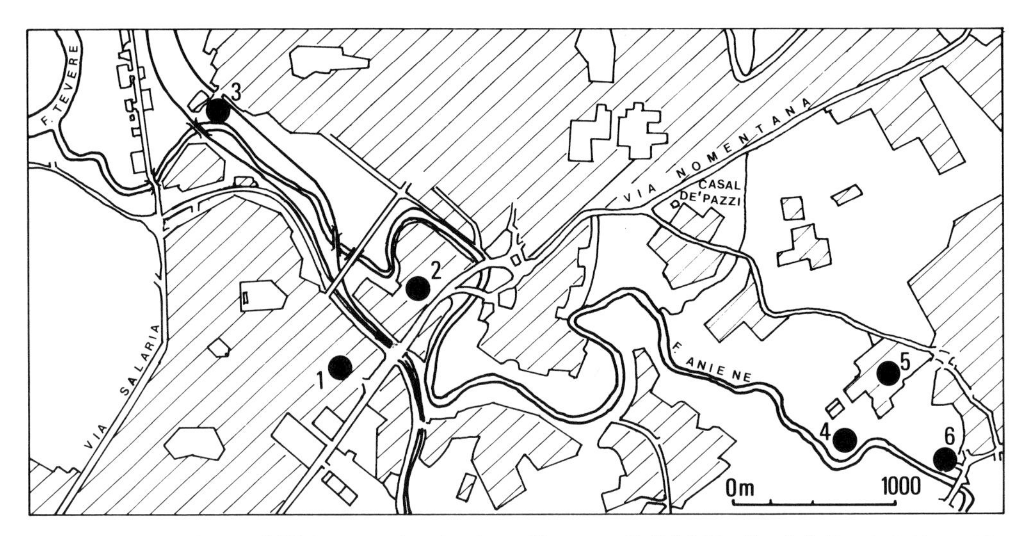

Fig. 1. Location of several Pleistocene sites in the lower valley of the Aniene, Rome, Italy. 1, Sedia del Diavolo; 2, Saccopastore; 3, Monte delle Gioie; 4, Ripa Mammea; 5, Rebibbia–Casal de'Pazzi; 6, Ponte Mammolo. The hatched areas represent urban expansion (from Segre, 1983).

the Latium held in Rome during October, 1982 (Anzidei et al., 1984).

STRATIGRAPHY, FAUNA, AND HUMAN PALEONTOLOGY

The Rebibbia–Casal de'Pazzi excavation lies on the so called "Middle Terrace" of the Aniene river valley (Saccopastore lies on a terrace about 10 m below). The stratigraphic sequence is presented in Figure 2, based on the 1982 work by Segre (Segre, 1983). The excavation area has been divided into 5 × 5 m squares. The faunal remains consist essentially of *Palaeoloxodon antiquus, Hippopotamus amphibius, Dicerorhinus* (probably *D. hemitoechus*), *Cervus elaphus, Dama* sp., *Capreolus capreolus, Hyaena crocuta, Canis lupus*, and some birds, such as *Anser albifrons, Anas penelope, A. strepera*, and *A. crecca*.

One of the most recent discoveries, during 1983, was a fragment of a human parietal (Fig. 3) near the eastern edge of the excavated area. Stratigraphically, this is from the lower part of the deposit, layer 4 in Figure 2B. This fragment is presently under study at the University of Rome by P. Passarello, G. Manzi, and L. Salvadei: according to their preliminary analyses, the parietal prominence on the exocranic face and the terminal branches of the meningeal artery are clearly evident. The fragment (75 × 76 mm) is rather thin (about 10 mm). Is this a suggestion of a "*sapiens*" character?

It is clear that such a suggestion is closely related to the problem of the chronological classification of the site or, better, being a fluvial deposit, of a reasonable time-span estimation of the archaeological layers. An indication of a possibly large time interval could be given by the fact that the faunal remains and artifacts showed different degrees of patina, as discussed below. However, the faunal remains, at least at the present stage of analysis, do not suggest substantial differences throughout the deposit. A first general tentative date comes from geological considerations: the archaeological deposit overlies the "lithoid" tuff formation, which is estimated to be about 350,000 years old. On the other hand, the position of the Rebibbia–Casal de'Pazzi site on the middle terrace of the Aniene valley is very similar to those of Sedia del Diavolo and Monte delle Gioie, and all three of these sites could therefore be attributed to a middle–final stage of the Rissian glacial (Anzidei et al., 1984).

We are just starting, in collaboration with the Geochemical Institute of the University of Rome, a program of absolute dating of the animal bones of the deposit with different patinae, both with the racemization method and the differences in fluorine content. In my opinion, at present, it is the only sensible method for determining the time span of the archaeological deposit, as well as also providing a reasonable age estimate for the parietal fragment.

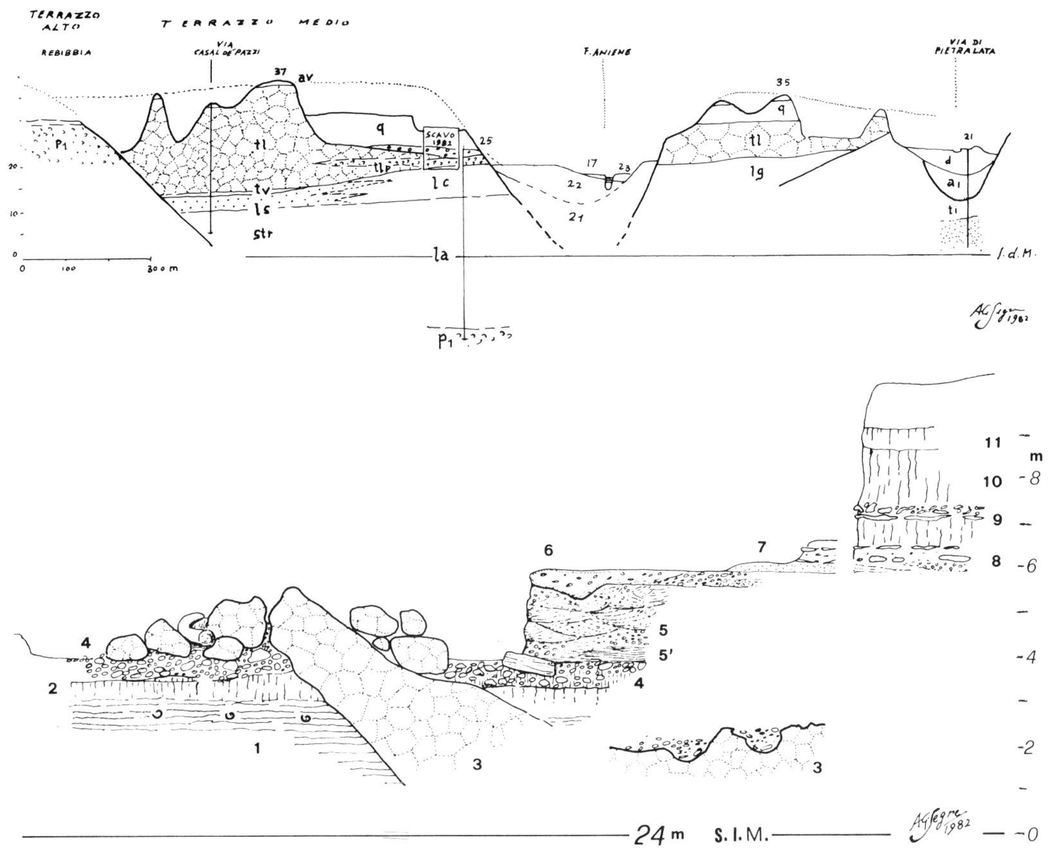

Fig. 2. Stratigraphic sections through the valley of the Aniene from Rebibbia to Pietralata (from Anzidei et al., 1984). A) General section across the valley, with the location of the 1982 excavation ("Scavo 1982") at Casal de'Pazzi, left of center. From below, the layers include: a lacustrine horizon (lc), the "lithoid" tuff (tl), the sands and gravels yielding the site itself, and a tuffitic paleosol (q) with traces of plants and travertine layers. B) Detailed section at the site, with a base at 24 m above sea level. Layer 1, silty lacustrine horizon; 2, paleosol; 3, "lithoid" tuff, partly eroded by the river; 4, base of the archaeological deposit, containing redeposited fragments of layers 1 and 3; 5, continuation of archaeological deposit, with cross-bedding of river deposits; 6–7, silts and sands with rare remains of *Cervus elaphus*; 8–10, tuffitic paleosol with travertines; 11, capping brown soil layer.

ARCHAEOLOGY AND ITS IMPLICATIONS FOR INTRA-SITE CHRONOLOGY

The lithic industry (Fig. 4) is on flakes and is essentially Mousterian of "Pontinian" facies, with side-scrapers of various types, some with Quina retouch (No. 2) denticulated tools (No. 9), end-scrapers (No. 10), and borers (No. 5). Only one hand axe (No. 11) has been found after about a year of excavations, as well as one heavy-duty flint chopping tool. There is also a fragment of elephant bone intentionally worked at the extremities. Generally speaking, the whole industry is very similar to the ones of Sedia del Diavolo and Monte delle Gioie, which have been classified as Rissian "Protopontinian" (Taschini, 1967).

We have tried to approach more precise chronological distinctions in the archaeological deposit by analyzing the distribution of the artifact patinae according to the stratigraphic layers. At least five types of patina have been recognized on the artifacts from Rebibbia: from a sharply fresh one (graded 0) up to a very worn one (4). Only the large number of tools (several hundred) and faunal remains coming from a careful, modern excavation raised such problems: no mention of comparable variation has been reported, for instance, from the sites of Sedia del Diavolo and Monte delle Gioie, where the arti-

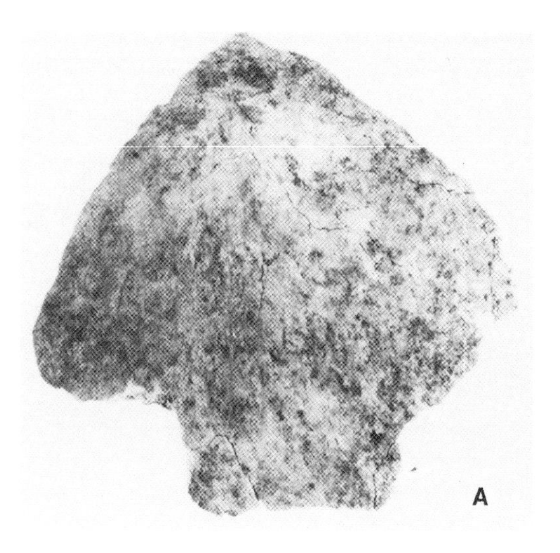

Fig. 3. Fragment of hominid parietal recovered in 1983 in layer 4. A, external view; B, internal view.

facts collected were rather scarce and, moreover, resulted from old excavations.

We therefore needed a more detailed structure of the archaeological layer (horizons 4 and 5 of Fig. 2B). This has been done by M. Ruffo and there is clear evidence of different phases of deposition by the river: gravel of medium–small size and sand; as is customary for a river deposit, these components are spread out in extended layers or occur in more restricted lenses.

In the present communication, we have considered the artifacts (524) coming from the squares which show a rather high density of artifacts. We used the non-parametric χ^2 test in order to examine the degree of correlation (or of independence) among the patinae of the artifacts (five classes) and the different layers (four classes). The resulting total χ^2 value is 8.3; with 12 degrees of freedom, this gives a probability of independence of about 50%. There is, therefore, essentially no correlation between stratigraphic units and the degree of patina of the artifacts, as one could have expected from the type of deposit. The next problem, then, is to investigate if there are significant typological distinctions among the tools with different patinae.

The analysis has been performed using the standard type list (Bordes, 1961), and the detailed results will be published elsewhere. In Table 1 are given the Bordes indices for the various patinae: the most worn (3 and 4) have been grouped together for reasons of statistical consistency.

At a glance, one can see that the typological pattern of the tools is rather "Pontinian," in the sense of Taschini (1979). In particular, the results for the tools with fresh patina (0 and 1) are very close to the ones obtained for the lower (Würmian) layers of Grotta Guattari at Monte Circeo (Taschini, 1979, p. 223): the side-scrapers dominate, the "facetting" is moderate or low, and the Levallois technique is almost absent. On the other hand, the tools with patina 2 show a consistent decrease of scrapers together with an increase of notched tools, while there is an increase of the "Upper Paleolithic" group (III) for the tools with patinae 3 and 4. The question now is if these results can be interpreted as a significant indication of chronological (and/or cultural) differences among the industrial assemblages characterized by different patina. I am personally inclined to think that these differences are due more to sampling bias effects: in fact, only a "stripe" of the deposit has been analyzed.

A χ^2 test has again been performed for five groups of tools and four patinae, and has given about 6% probability of independence.

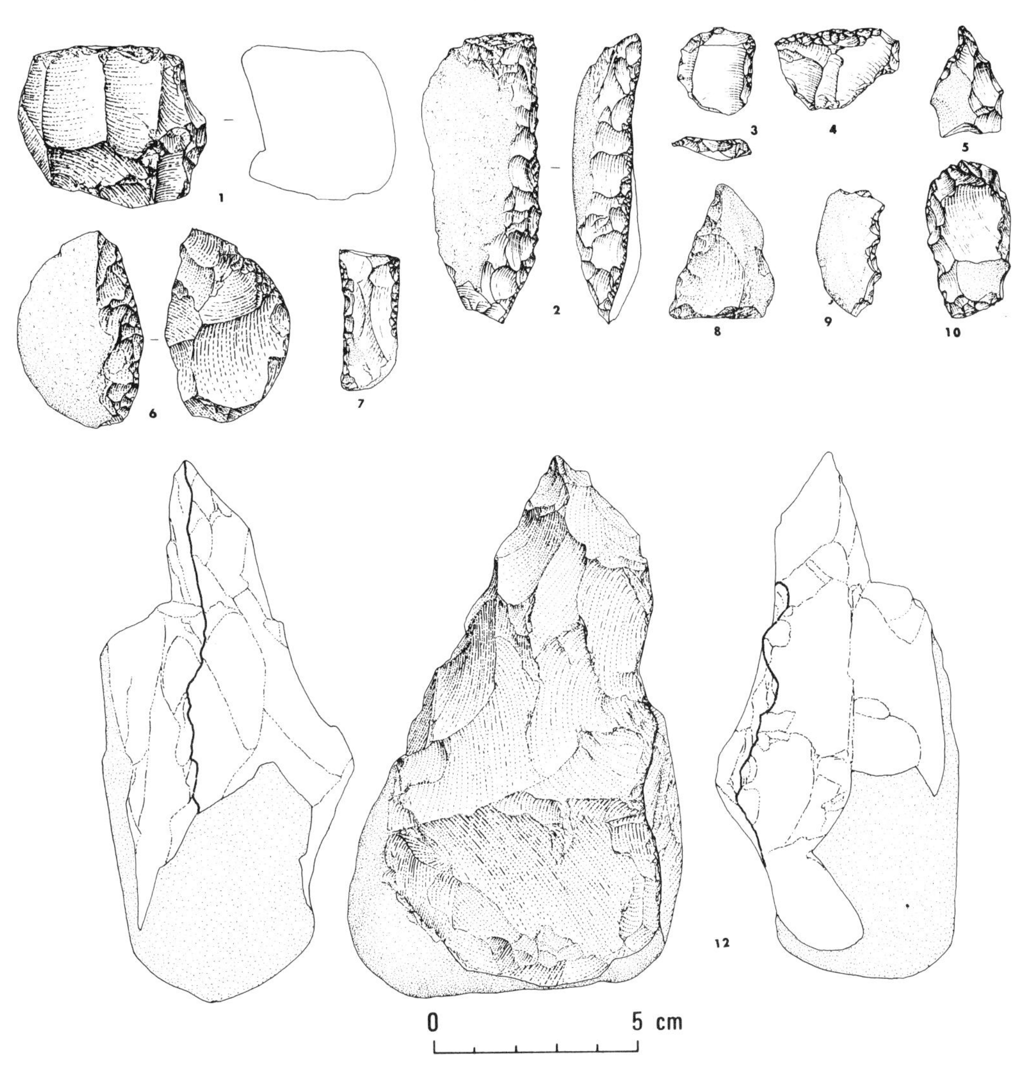

Fig. 4. A selection of artifacts from Rebibbia–Casal de'Pazzi. 1, prismatic core; 2, single convex side-scraper with Quina retouch; 3, straight side-scraper on Levallois flake; 4, transverse side-scraper; 5, borer; 6, side-scraper with Quina retouch and *dos aminci*; 7, double side-scraper; 8, simple straight side-scraper; 9, denticulate; 10, end-scraper; 12, limestone hand axe (the sole example from the site). From Anzidei et al. (1984).

This is certainly a low value, but it is still not statistically significant, and, moreover, the Cramer association coefficient is only 0.16. Another χ^2 test for the distribution, according to the different patinae, of the various side-scrapers of the Bordes list (simple, double, convergent, etc.) is even less significant: the probability of independence comes out to be of the order of 60%. We thus see that there is no significant evidence for a difference between the patina of the artifacts and their typological classification.

CONCLUSIONS

The "Mousterian" character of the Rebibbia industry is certainly interesting in connection with a possible "*sapiens*" attribution of the parietal fragment. In fact, only the handaxe, which is essentially unique in all the lower Aniene valley, and the chopping tool look definitely "foreign" in the whole assemblage. This circumstance, and the typological indices shown in Table 1, should therefore reinforce the "Protopontinian" identification proposed by Taschini (1967),

TABLE 1. Bordes (1961; Sireix and Bordes, 1972) indexes for four patination stages of the Rebibbia–Casal de'Pazzi industry

	Patination stage			
Indexes	0	1	2	3–4
IF[1]				
Without cortex	3.85	25.5	33.3	16.7
With cortex	1.7	13.1	17.9	8.9
IL[3]				
"Real"[2]	1.2	2.3	—	—
"Essential"[4]	—	—	—	—
IR[5]				
"Real"	54.0	40.3	26.9	37.8
"Essential"	68.4	58.9	39.3	54.1
IC[6]				
"Real"	14.4	12.9	5.9	7.2
"Essential"	18.7	18.9	9.3	10.4
II[7]				
"Real"	54.0	40.3	28.4	37.8
"Essential"	68.4	58.9	41.6	54.1
III[8]				
"Real"	3.6	6.8	5.4	10.5
"Essential"	4.6	10.0	7.9	14.9
IV[9]				
"Real"	4.8	2.3	3.9	4.8
"Essential"	6.1	3.3	5.6	6.9
IV + notches[10]				
"Real"	10.8	11.4	23.1	14.5
"Essential"	13.7	16.4	33.7	20.7

[1]Index of facetted butts (vs. all butts).
[2]Including all 63 types of tools defined by Bordes (1961).
[3]Index of Levallois débitage.
[4]Excluding Levallois flakes and points, and types 45–50.
[5]Index of "racloirs" (side-scrapers).
[6]"Charentian" Index: relative percentage of "limaces," convex and transverse side-scrapers.
[7]Index of Mousterian types.
[8]Index of Upper Paleolithic types (burins, backed knives, endscrapers, borers, and truncated blades).
[9]Index of denticulates.
[10]Index of denticulates + notches.

essentially parallel to the supposed existence of a "Protomousterian" or a "Protocharentian" of Riss age in France.

Actually, even if the typological distribution of the tools looks very similar, there are marked technical differences between the industry of Rebibbia–Casal de'Pazzi and the classical Pontinian of Grotta Guattari (Taschini, 1979) or even the Riss–Würm industry of layer d of Torre in Pietra (Piperno and Biddittu, 1978). The Levallois technique is practically unknown; the indices shown in Table 1 refer only to a few "atypical" Levallois flakes, with no points at all. The "facetting" index seems rather high, but, in contrast with Grotta Guattari and Torre in Pietra layer d, one has to consider that a large quantity of the heels of these flakes are broken, removed, or not recognizable, and these categories of heels are traditionally excluded from the computation of the indices. In addition, the cores are essentially shapeless, and no "classical" discoidal core has been found, again in contrast to Grotta Guattari and the d level of Torre in Pietra. Moreover, the indices given in Table 1 (especially for the artifacts with patina 2), as well as some of their technical characters, are consistent with the ones from the lower level (m) of Torre in Pietra, but there is a significant difference: the latter site yielded several handaxes, so that it is traditionally attributed to the Acheulian.

Should we think that Rebibbia–Casal de'Pazzi represents a different "culture" from that of Torre in Pietra, as is the traditional opinion, or should we explain the distinctions in terms of a different activity or "adaptation" of the same human group, which only 30 km away consistently produced handaxes? The question is completely open. Chronology is of paramount importance, and we hope that the absolute dating that we have in progress, as well as the complete study of the industries, the fauna, and the parietal fragment will shed some light on this interesting problem.

LITERATURE CITED

Anzidei, AP, Bietti, A, Cassoli, P, Ruffo, M, and Segre, AG (1984) Risultati preliminari dello scavo in un deposito Pleistocenico in località Rebibbia–Casal de'Pazzi (Roma). Atti. XXIV Riun. Sci. Ist. Ital. Preist. Protostoria Lazio, Rome, 1982. Florence: Parenti, pp. 131–139.

Bordes, F (1961) Typologie du paléolithique ancien et moyen. Bordeaux: Delmas.

Piperno, M, and Biddittu, I (1978) Studio tipologico ed interpretazione dell'industria acheuleana e premusteriana dei livelli m e d di Torre in Pietra (Roma). Quaternaria *20:*441–536.

Segre, AG (1983) Geologia quaternaria nella bassa valle dell'Aniene (Roma). Riv. Antropol., Suppl. *62:*87–98.

Sireix, M, and Bordes, F (1972) Le Mousterien de Chinchon (Gironde). Bull. Soc. Préhist. France *69:*324–336.

Taschini, M (1967) Il "Protopontiniano" Rissiano di Sedia del Diavolo e Monte delle Gioie (Roma). Quaternaria *9:*301–319.

Taschini, M (1979) L'industrie lithique de Grotta Guattari au Mont Circé (Latium): Definition culturelle, typologique et chronologique du Pontinien. Quaternaria *21:*179–247.

Ancestors: The Hard Evidence, pages 283–288

Human Fossils From the North African Middle Pleistocene and the Origin of *Homo sapiens*

J.J. Hublin
LA 49 du C.N.R.S., Laboratoire de Paléontologie des Vertébrés et de Paléontologie Humaine, Université Paris VI, 75230 Paris Cedex 05, France

ABSTRACT A chronological framework of the Middle Pleistocene fossil hominids from North Africa remains difficult to establish. The Ternifine (now Tighenif) site in Algeria was likely formed at the very beginning of the Middle Pleistocene. In Morocco, a set of human fossils (Thomas quarries I and III, Salé, Sidi Abderrahman) has been recovered from continental deposits broadly contemporary with the maximum (G2) of the Anfatian transgression. The most recent data lead us to give this transgression an age around 400,000 years BP. The Rabat individual is somewhat younger. In the Salé specimen that is the most complete, posterior cranial morphology displays abnormal features in the nuchal area, and thus its rear skull aspect should not be taken into account in taxonomic discussions. On the other hand this fossil displays a mosaic of "*erectus*" and "*sapiens*" features, which does not support the hypothesis of two diverging clades, at least in Africa.

INTRODUCTION

North Africa has provided an important series of Middle and Late Pleistocene hominids. Besides the human fossils of Ternifine (now Tighenif) in Algeria, most of these discoveries occurred in Morocco, and this area has become one of the most interesting for the study of *Homo erectus* and early *Homo sapiens* evolution. Unfortunately, an important proportion of these fossils result from accidental discoveries, which can explain two characteristics of these specimens: they are often fragmentary and difficult to date. Moreover, despite several attempts (Biberson, 1970; Brébion, 1980; Debenath et al., 1982) the Pleistocene stratigraphy of the Moroccan Atlantic coast is not really correlated with the European chronology, and "absolute" (chronometric) dates are very scarce.

The various Pleistocene continental deposits of the Moroccan coast are interstratified between several marine transgressions. For a long time, the only chronometric dates for the Middle Pleistocene part of this sequence have been provided by the U^{230}/Th^{234} method on marine shells. These dates (established by Stearns and Thurber, 1965) were used by Jaeger (1981) to attempt a correlation of the various transgressions with the isotopic temperature graph established by Shackleton and Opdyke (1973). Nevertheless, the Stearns and Thurber ages were already heavily criticized by Kaufman et al. (including Thurber himself) in 1971, and during the last ten years, new data have been provided confirming that this chronological framework is probably wrong.

Jaeger (1981) considered that the Middle Pleistocene hominids from North Africa should be separated into two main chronological groups: an older, consisting of the Tighenif material; and a younger, including the Sidi Abderrahman, Salé, and Rabat material. The Thomas quarries specimens would fall between the two major groups. According to him the Salé cranium, which is the most complete of these fossils, should be considered as an evolved *Homo erectus*,and a possible phyletic relationship between the latter group and the first *Homo sapiens* in this area was unlikely because of the very short time separating them. The purpose of this paper is to briefly discuss the age of these fossil hominids and some of the features of the Salé skull.

CHRONOLOGY

The microfauna from Tighenif indicates an age more recent than Ubeidiya, and Jaeger (1981) has proposed a date around 600,000 years BP for this site. But according to Geraads (1981) the fauna is possibly early Middle Pleistocene or late Early Pleistocene, that is to say around 0.6–1.0 million years (m.y.) BP.

The Thomas quarries near Casablanca (Morocco) have been numbered I, II, and III. Only the first and the third have yielded fossil hominids. The two sites appear to be nearly contemporary (Geraads et al., 1980) and should be placed either in the late Amirian (Jaeger, 1981) or at the very beginning of the Tensiftian (Geraads et al., 1980), two successive continental stages in the local geochronology. Therefore, the Thomas quarries specimens could be chronologically near the Salé skull, which is considered to be of the same age as the Sidi Abderrahman mandible, i.e., early Tensiftian (Biberson, 1961; Jaeger, 1981). Virtually all these fossil hominids seem to be contemporary with the maximum of a marine transgression that comes between Amirian and Tensiftian near the coast. The dating of this Anfatian (G2) transgression is the central problem in giving these specimens an accurate age.

About Salé, Jaeger (1981) used the Stearns and Thurber dating already discussed, which was "more than 200,000 BP" for the Anfatian; this became simply "200,000 BP" in some papers. In this hypothesis, the Anfatian should be correlated to the 11th, 9th, and 7th Shackleton and Opdyke (1973) oxygen isotope stages, the latest of these stages corresponding to the maximum of the transgression. This level is represented by a marine terrace around +30 m above present sea level. It is the warmest local episode of the Middle Pleistocene, and the major part of the marine deposits of the Rabat area were formed during this period. An equally important terrace is also known in the Canary Islands, +28 m above the present sea level, with volcanic horizons above and below the marine deposits. On the north coast of Gran Canaria, the underlying tephrite has been dated by K–Ar at 0.548 and 0.529 ± 0.02 m.y. BP and the overlying hauyne phonolite at 0.297 and 0.326 ± 0.03 m.y. BP (Lietz and Schmincke, 1975). According to Jaeger (personal communication), the age of the Anfatian transgression by correlation could thus be about 400,000 years BP. This increased age of the Anfatian is supported by the age of the succeeding Harounian and Ouljian transgressions. Recently Hoang et al. (1978) have provided ^{230}Th/^{234}U dates between 140,000 and 120,000 years BP for the Ouljian 1, considerably older than previously believed.

Therefore, the Salé group can probably now be dated at close to 400,000 years BP, rather than near 200,000 years BP. The age of the Rabat specimen has been much debated (Biberson, 1961), but in any case it is more recent than Salé: either middle Tensiftian or possibly falling within the Harounian regression.

Concerning dating we can conclude: 1) that the Middle Pleistocene hominids of the Maghreb thus do not divide into two discrete groups, one at the beginning and the other at the end of this period; and 2) that the Salé-Sidi Abderrahman-Thomas quarries group is rather old, compared with the age of the first modern humans known in this area.

THE SALÉ SKULL

The Salé cranium displays a mixture of primitive and "*sapiens*" features, which led Jaeger (1975a,b, 1981) to consider it as an "evolved *Homo erectus*." Besides features like the position of the spheno-occipital synchondrosis (Laitman and Crelin, 1980; Laitman, 1981), the *sapiens* characters of Salé are essentially the development of the parietal eminences and the morphology of the occipital bone (see for example Wolpoff, 1980, or Bräuer, 1984). In posterior view (Fig. 1A), the parietal bosses are already well differentiated while the lateral walls of the skull are still diverging downward. The occipital shows a rounded outline in lateral view. The *planum nuchae* is short and low. The *torus occipitalis transversus* is poorly developed, wide (max. 25 mm) but flat. All these features confer on the specimen its apparently modern appearance. When one examines the

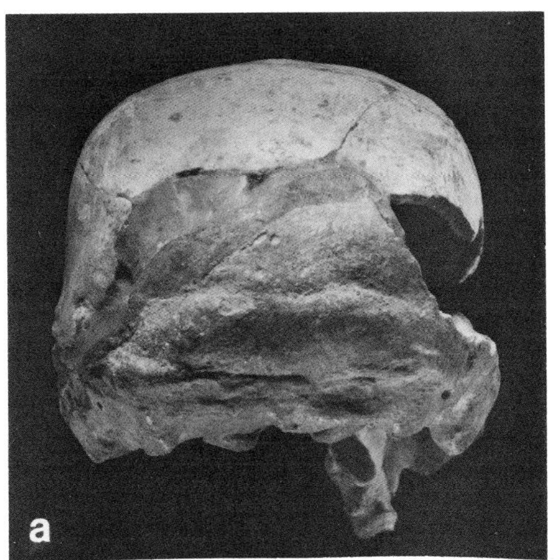

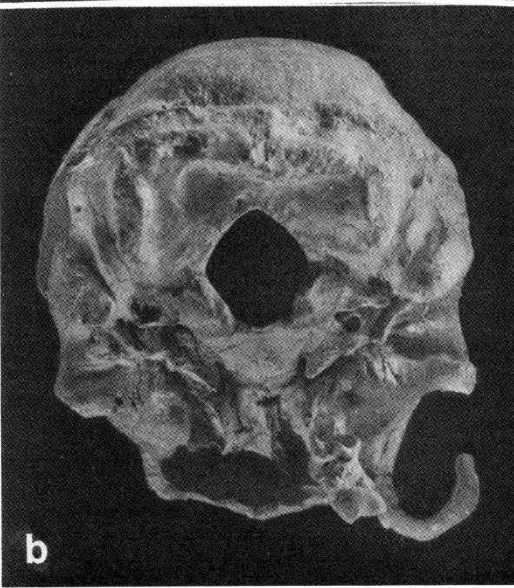

Fig. 1. a) Posterior view of the Salé skull. b) Nuchal area of the Salé skull. Photographs by C. Tarka, American Museum of Natural History.

measurements of the Salé cranium, most of the "*sapiens*" dimensions also involve the inion.

However, the *planum nuchae* of this individual is very peculiar and clearly shows abnormal or pathological features (Fig. 1B). While the *sulcus supratoralis* (i.e., *linea nuchae suprema*) is well defined, forming a regular arch inflexed in its medial portion, the *planum occipitale* is not well delimited from the *planum nuchae*. The latter displays a completely asymmetrical pattern, and it is not possible to recognize most of the muscular impressions at their usual position. On this area there is a wide and irregular thickening of bone at the level of the *linea nuchae inferior*. Where the bone is usually thin, it is here between 1 and 2 cm thick. Tomography shows that this structure is formed by a thickening of the diploic tissue and not by a thickening of the external table (which is the case for the *torus occipitalis transversus*). It does not seem to be a structure added onto a normal *squama occipitalis*.

Below this strong swelling, which passes only 8 mm from the edge of the *foramen magnum*, the insertions of the *m. rectus capitis posterior major* and *minor* are rather regular. But it is not possible to clearly delineate the usually extended attachment of the *m. semispinalis capitis* on this bony chaos. Perhaps on the left this muscle inserted on a raised rectangular area, about 25 mm long, just below the torus. In this hypothesis, it would be almost 20 mm distant from the medial plane! On the right side this muscle was probably attached in two irregular depressions nearer to the median plane. Laterally, the secondary antero-inferior branch of the *linea nuchae inferior* forms a strong relief perpendicular to the swelling and joining the juxtamastoid ridge.

Various pathological explanations have been proposed for such an extraordinary muscular and bony variation, but none of them is fully convincing. This sort of bony bridge has also been described on modern skulls as the development of a proatlas (Ferembach, 1959). During this abnormal embryological development, bone would have been formed into the somite which provides the nuchal muscles.

Considering the lack of clear *linea nuchae superior, tuberculum linearum*, and *crista occipitalis externa*, it is very difficult to decide where to locate inion on this skull. Nevertheless as inion represent the upper limit of the nuchal area in the medial plane[1], it should be at the upper limit of this distubed area, as it has been placed in the previous studies (Jaeger, 1981, Fig. 11-8). This determination gave an unusually low inion: 30

[1]See comments in Hublin (1978a).

mm for the inion–opisthion distance is indeed too short, not only for a *Homo erectus*, but also for an early *H. sapiens*. Compared with a sample of 11 *H. erectus*[2] (from Sangiran, Zhoukoudian, and Ngandong), Salé is significantly different for this measurement (in a t-test, $P < 0.02$). But compared with a sample of ten early *Homo sapiens*[2] (Broken Hill, Preneandertals, Neandertals, and "Protocromagnoïds"), it is equally significantly different ($P < 0.02$). Furthermore, the two samples yield almost the same means (49.27 mm and 49.75 mm, respectively) and cannot be separated by this metric feature. Salé is of course significantly different from the compound sample ($P < 0.01$). Considering the morphology of this area and these metrical data, the shortness of the *planum nuchae* should be considered as an abnormal characteristic of this individual rather than as a modern feature.

Therefore, in taxonomic discussions about the Salé cranium, it is necessary to ignore all the characters of this area as well as measurements involving inion, or at least to be very cautious in using them. The length of the nuchal plane, the position of the opisthocranion, and the occipital angle have no real meaning. The external projection of the endinion is situated only 2 mm below the previously determined inion, but, for the same reasons, this surprisingly modern feature should be put aside. The *torus occipitalis transversus* (which develops between the *linea nuchae superior* and *suprema*) is never very projecting when this area is wide (Hublin, 1978b). One can wonder how this structure would have looked if the *planum nuchae* had been normally extended.

Considering the rest of the dimensions (Jaeger, 1975b) and the morphology, the specimen falls within the range of variation of "*Homo erectus*" as it has been tentatively defined in recent publications (Day and Stringer, 1982; Howell, 1978; Howells, 1980; Wolpoff, 1980). Unfortunately, all these definitions involved many features of the occipital bone that cannot be used here. But the Salé specimen remains a very small brained individual: 930–960 cm^3 according to Jaeger (1975b) or 860 cm^3 according to Holloway (1981). The skull is long and low, with a ratio of the basion-bregma height to the estimated total length[3] of 0.61. In the parietal area, the bregma-asterion distance versus biasterionic breadth is 1.08. The frontal bone is very narrow (min. frontal breadth: 77 mm), receding and with a median keeling reaching a strong bregmatic prominence. The maximum breadth of the skull (estimated at 139 mm) is located at the level of the *crista supramastoidea*. The temporal squama is low with a relatively flat superior border. There is a marked angulation between the tympanic and petrous portions of the temporal. The skull bones are thick; despite the lack of a *torus angularis*, the thickness of the parietal reaches 15 mm near the asterion. The cerebellum is small, and the *eminentia cruciformis* is very low on the internal face of the occipital. The meningeal veins display a relatively simple pattern (Saban, 1982, 1984).

The small size of the cranium, its relatively weak muscular markings, and its only moderately marked temporal lines may be partly due to sexual dimorphism. However, the Salé skull does present some genuine advanced features, such as the proportions of the basisphenoid and basioccipital[4], the relative gracility of the tympanic bone, and above all the development of the parietal bosses.

CONCLUSIONS

In western Europe we are lucky enough to recognize a very derived group of *Homo sapiens*, and on the basis of even a few apomorphic features it is easy to include most of the material in this species because we have to put it into *Homo sapiens neanderthalensis*. In other words, the "Anteneandertals" are "Preneandertals" (see discussion in Hublin, 1982). Thus, we can at least define a minimal comprehension for the taxon *Homo sapiens* in this area. In Africa the problem is somewhat more difficult.

In putting together the Asian and African material we are not able to find any autapomorphy of *Homo erectus* (see Rightmire, 1985). Just to mention anatomical areas preserved on Salé, features considered typical of this group, such as the very strong occipi-

[2]Measurements in Hublin (1978b).

[3]The anterior part of the frontal bone being missing, the total length has been estimated after addition of the length of the natural endocast, thickness of the occipital, and thickness of the Thomas quarry III hominid supraorbitary area. This specimen, roughly contemporary with Salé, fits it rather well by its dimensions.

[4]With the restriction that the only available elements of comparison in *Homo erectus* are primitive specimens such as KNM-ER 3733 and 3883 or Olduvai H. 9 (Laitman, personal communication).

tal torus, the *torus angularis*, the great robustness of the tympanic, and the fissure between tympanic and mastoid, appear to be regional characteristics of the Asian fossils, rare or unknown in Africa. Other traits are plesiomorphic (for example, the lack of a true *protuberantia occipitalis externa*[5]), and still others are simply wrong (for example the "small mastoid"[5]). As to the thickening of the cranial bones or the development of strong superstructures on the rear of the skull, the fossil evidence shows that these features have gone through a reversal in *Homo sapiens* (Hublin, 1978b). And, finally, most of the features used to separate the two taxa are unfortunately metric ones, which are, by definition, continuous and show some overlap between the two groups.

No feature can exclude such a specimen as Salé from the ancestry of modern man. Taking into account only the shared derived features, Salé should be considered as a primitive *Homo sapiens*, although most of its characters are "*erectus*" features. But, by the same reasoning and assuming very few reversals[6], the Ngandong specimens (which have a more elevated vault and more modern proportions of the cranial bones than the earlier Indonesian *Homo erectus*) or even the Zhoukoudian specimens should be included in *Homo sapiens*, unless their advanced (*sapiens*-like) features are considered as the result of a simple parallelism. Thus, the only two alternatives are either that "*sapiens*-like" characters evolved in parallel in Africa and Eurasia or that some derived feature of Asian *H. erectus* underwent reversal, if that group gave rise to early *H. sapiens*. If what are called "early *Homo sapiens*" are only evolved *Homo erectus* with a more or less enlarged brain and a still great robusticity of the skull and the teeth, according to this methodology all *Homo erectus* should be included into *Homo sapiens*, as has been already proposed (Jelinek, 1981; Thoma, 1973). The two species should thus be only two grades in an anagenetic process regionally differentiated in the speed of acquisition of the *sapiens* characters (even if, in certain areas, some populations feebly participated in the evolution of their anatomically modern successors).

The documentation of an evolutionary stasis in this lineage would be the only way to maintain the two specific denominations (with the understanding that they are grades). This question is still open, partly because of an uncertain chronological framework (Allen, 1982; Bilsborough, 1976; Day, 1982; Kennedy, 1983; Rightmire, 1981, 1985; Wolpoff and Nkini, 1985). But, if this stasis hypothesis is rejected, then keeping the concept of *erectus* as a species is tantamount to leaving such a mosaic specimen as Salé (and many others) to wander forever in taxonomic limbo.

LITERATURE CITED

Allen, LL (1982) Stasis vs. evolutionary change in *Homo erectus*. Am. J. Phys. Anthropol. *57*:166 (Abstract).

Biberson, P (1961) Le cadre paléogéographique de la préhistoire du Maroc atlantique. Publ. Serv. Antiq. Maroc *16*:1–235.

Biberson, P (1970) Index-cards on the marine and continental cycles of the Moroccan Quaternary. Quaternaria *13*:1–76.

Bilsborough, A (1976) Patterns of evolution in Middle Pleistocene hominids. J. Hum. Evol. *5*:423–439.

Bräuer, G (1984) A craniological approach to the origin of anatomically modern *Homo sapiens* in Africa and implications for the appearance of modern Europeans. In FH Smith and F Spencer (eds): The Origins of Modern Humans: A World Survey of the Fossil Evidence. New York: Alan R. Liss, Inc., pp. 327–410.

Brébion, P (1980) Corrélations entre les terrasses marocaines atlantiques et le Pleistocene méditerranéen dans la chronologie glaciaire. Bull. Mus. Natn. Hist. Nat. Paris *C2*:17–24.

Day, MH (1982) The *Homo erectus* pelvis: Punctuation or gradualism? 1er Congr. Intern. Paleont. Hum. (Prétirage) Nice:C.N.R.S., pp. 411–421.

Day, MH, and Stringer, CB (1982) A reconsideration of the Omo Kibish remains and the *erectus–sapiens* transition. 1er Congr. Intern. Paleont. Hum. (Prétirage) Nice:C.N.R.S., pp. 814–846.

Debenath, A, Raynal, JP, and Texier, JP (1982) Position stratigraphique des restes humains paléolithiques marocains sur la base des travaux récents. C.R. Acad. Sci. Paris *294*:1247–1250.

Delson, E, Eldredge, N, and Tattersall, I (1977) Reconstruction of hominid phylogeny: A testable framework based on cladistic analysis. J. Hum. Evol. *6*:263–278.

Ferembach, D (1959) A propos d'un pont osseux anormal sur l'écaille occipitale d'un crâne fossile. Arch. Anat. Pathol. *7(2)*:173–175.

Geraads, D (1981) Bovidae et Giraffidae (Artiodactyla, Mammalia) du Pléistocène de Ternifine (Algérie). Bull. Mus. Natn. Hist. Nat. Paris *C3*:47–86.

Geraads, D, Beriro, P, and Roche, H (1980) La faune et l'industrie des sites à *Homo erectus* des carrières Thomas (Maroc). Précisions sur l'age de ces Hominidés. C.R. Acad. Sci. Paris D *291*:195–198.

Hoang, CT, Ortlieb, L, and Weisrock, A (1978) Nouvelles datations $^{230}Th^{234}U$ de terrasses marines "oul-

[5]See the list of autapomorphies in Delson et al. (1977).

[6]Features such as the thickening of the cortex of the long bones, the thickening of the inner and outer table of the cranial bones, the development of strong superstructures from this outer table with various morphological and metrical consequences, or the robustness of the tympanic are likely bound together and should not be computed as different characters.

jiennes" du sud-ouest du Maroc et leurs significations stratigraphique et tectonique. C.R. Acad. Sci. Paris D *286*:1759–1762.

Holloway, RH (1981) Volumetric and asymmetry determinations on recent hominid endocasts: Spy I and II, Djebel Irhoud 1 and the Salé *Homo erectus* specimens, with some notes on Neandertal brain size. Am. J. Phys. Anthropol. *55*:385–393.

Howell, FC (1978) Hominidae. In VJ Maglio and HBS Cooke (eds): Evolution of African Mammals. Cambridge, Massachusetts: Harvard University Press, pp. 154–248.

Howells, WW (1980) *Homo erectus*—Who, when and where: A survey. Yrbk. Phys. Anthropol. *23*:1–23.

Hublin, JJ (1978a) Anatomie du centre de l'écaille occipitale, le problème de l'inion. Cah. Anthropol. *1978*:65–83.

Hublin, JJ (1978b) Le torus occipital transverse et les structures associées: Évolution dans le genre *Homo*. Thèse, 3eme cycle, Université Paris VI.

Hublin, JJ (1982) Les Anténéandertaliens: Présapiens ou Prénéandertaliens? Geobios, Mem. Spéc. *6*:345–357.

Jaeger, JJ (1975a) The mammalian fauna and hominid fossils of the Middle Pleistocene of the Maghreb. In KW Butzer and GL Isaac (eds): After the Australopithecines. The Hague: Mouton, pp. 299–418.

Jaeger, JJ (1975b) Découverte d'un crâne d'hominidé dans le Pleistocène moyen du Maroc. Problèmes Actuels de Paléontologie-Evolution des Vertébrés. Paris: C.N.R.S., pp. 897–902.

Jaeger, JJ (1981) Les hommes fossiles du Pléistocène moyen du Maghreb dans leur cadre géologique, chronologique et paléoécologique. In BA Sigmon and JS Cybulski (eds): *Homo erectus* Papers in Honor of Davidson Black. Toronto: University of Toronto Press, pp. 159–264.

Jelinek, J (1981) Was *Homo erectus* already *Homo sapiens*? Les Processus de l'hominisation. Paris: C.N.R.S., pp. 85–89.

Kaufman, A, Broecker, WS, Ku, TL, and Thurber, DL (1971) The status of U-series methods of mollusk dating. Geochim. Cosmochim. Acta *35*:1155–1183.

Kennedy, GE (1983) A morphometric and taxonomic assessment of a hominine femur from the lower member, Koobi Fora, Lake Turkana. Am. J. Phys. Anthropol. *61*:429–436.

Laitman, JT (1981) The value of the chondrocranial base in assessing taxonomic and phylogenetic relationships among fossil hominids. Am. J. Phys. Anthropol. *54*:242.

Laitman, JT, and Crelin, ES (1980) An analysis of the Salé cranium: A possible early indicator of a modern upper respiratory tract. Am. J. Phys. Anthropol. *52*:245.

Lietz, J, and Schmincke, HU (1975) Miocene-Pliocene sea level changes and volcanic phases on Gran Canaria (Canary Islands) in the light of new K–Ar ages. Palaeogeography, Palaeoclimatology, Palaeoecology *18*:213–239.

Rightmire, GP (1981) Patterns in the evolution of *Homo erectus*. Paleobiol. *7*:241–246.

Rightmire, GP (1985) The tempo of change in the evolution of mid-Pleistocene *Homo*. In E Delson (ed): Ancestors: The Hard Evidence. New York: Alan R. Liss, Inc., pp. 255–264.

Saban, R (1982) Les empreintes endocrâniennes des veines ménigées moyennes et les étapes de l'évolution humaine. Ann. Paléontol. *68*:171–220.

Saban, R (1984) Anatomie et évolution des veines méningées moyennes chez les hommes fossiles. Paris: CTHS, La Documentation Française.

Shackleton, NJ, and Opdyke, ND (1973) Oxygen-isotope and paleomagnetic stratigraphy of equatorial Pacific core V28-238: Oxygen-isotope temperature and ice volumes on a 10^5 year and 10^6 year scale. Quat. Res. *3*:39–55.

Stearns, CE, and Thurber, DL (1965) Th^{230}/U^{234} dates of the Late Pleistocene marine fossils from the Mediterranean and Moroccan littorals. Quaternaria *7*:29–42.

Thoma, A (1973) New evidence for the polycentric evolution of *Homo sapiens*. J. Hum. Evol. *2*:529–539.

Wolpoff, MH (1980) Paleoanthropology. New York: Alfred A. Knopf.

Wolpoff, MH, and Nkini, AT (1985) Early and early Middle Pleistocene hominids from Asia and Africa. In E Delson (ed): Ancestors: The Hard Evidence. New York: Alan R. Liss, Inc., pp. 202–205.

Ancestors: The Hard Evidence, pages 289–295

Middle Pleistocene Hominid Variability and the Origin of Late Pleistocene Humans

C.B. Stringer
Department of Palaeontology, British Museum (Natural History), London SW7 5BD, England

ABSTRACT The European Middle Pleistocene hominid record has been supplemented by a number of important recent finds. Although the traditional dating framework now appears inadequate, some absolute dates are becoming available. However, chronological control is still not good enough to properly test available evolutionary models. These models centre on the possible existence of *H. erectus* in Europe and the extent of modern or Neanderthal characters in the fossils themselves. The author favours a model in which the European hominids are divided into two groups. The first group (Mauer, Vértesszöllös, Bilzingsleben, Arago, Petralona) may be relatively earlier in date and contains specimens with archaic characters and little sign of exclusive synapomorphies with either Neanderthals or modern humans. The second group (Swanscombe, Biache, Bourgeois-Delaunay, Suard, and perhaps Steinheim and Pontnewydd) is predominantly or wholly late Middle Pleistocene in age and can be more positively aligned with Neanderthals. Some evidence of continuity of characters between the two groups can be recognized, but the author prefers to classify the "archaic" group with similar hominids from Africa (e.g., Bodo, Broken Hill) and perhaps Asia (Dali). This group is morphologically close to an hypothesised morphotype for the common ancestor of Neanderthals and modern humans. Europe appears to record only the evolution of the Neanderthal lineage during the Middle and early Late Pleistocene, while Africa and perhaps Asia record the evolution of anatomically modern *H. sapiens*.

INTRODUCTION

The European Middle Pleistocene hominid record has received increasing attention in recent years as new fossils have been discovered and new dating techniques applied. The material has been reviewed in detail elsewhere (e.g., Oakley et al., 1971; Howells, 1980; Wolpoff, 1980a; Anonymous, 1981; Cook et al., 1982), so only a brief introduction will be provided here, with detailed references to only the most relevant or recent discussions of the material.

Using a commonly accepted chronostratigraphic division of the Pleistocene (Butzer and Isaac, 1975; Kukla, 1978), the Middle Pleistocene spans the period between about 730,000 years (730 ky) and 128,000 years (128 ky) ago. Although there is archaeological evidence suggesting early human occupation, at least in southern Europe, the only known European fossil that might represent an Early Pleistocene hominid is the dubious cranial fragment from Venta Micena, Spain. Of the main hominid finds, the Mauer mandible, probably associated with a late "Cromerian" fauna like that of Mosbach, is likely to be the oldest, with a probable age in excess of 450 ky. Somewhat younger are finds dated within the "Mindel" faunal complex, such as Arago (Tautavel) and Vértesszöllös, perhaps ca. 400 ky. Hominids assigned to the Holsteinian/Hoxnian/Mindel-Riss complexes, such as Bilzingsleben, Swanscombe, Steinheim, and perhaps Montmaurin, may in fact derive from distinct stages covering a period of ca. 200 ky in the later Middle Pleistocene (Kukla, 1978; Cook et al., 1982). Recent absolute dating of the Bilzingsleben site suggests the hominid fragments may be relatively early in this sequence. The well preserved Petralona cranium has been dated as Early Pleistocene by some workers (i.e., pre-Brunhes, >730 ky—Kurtén and Poulianos, 1981) or as late Middle Pleistocene by others (ca. 200 ky—Hennig et al., 1982), but the actual age may lie between these extremes, perhaps approximating that of the morphologically comparable Arago 21 hominid (Stringer, 1984). Similar dating problems now surround the Vértesszöllös hominids, since recent absolute age determinations of ca. 200 ky are considerably younger than had been expected from the associated fauna (Cook et al., 1982). A number of French hominid sites are assigned to the latest Middle Pleistocene ("Riss" complex), including material dated at less than 200 ky. These include Biache, La Chaise (Bourgeois-Delaunay and Suard), and Fontéchevade.

EVOLUTIONARY MODELS

Many evolutionary models have been proposed to explain variation in the European hominid record. Some of these can be summarised as follows.

a. Presapiens Model

This would require the existence of separate ancestors for Neanderthals and modern humans during the Middle Pleistocene (Vallois, 1954). In this model, the Swanscombe and Steinheim hominids would lie on separate lineages ancestral to modern *H. sapiens* and Neanderthals, respectively.

b. Preneanderthal or Early Neanderthal Model

This model suggests that the divergence of modern humans and Neanderthals occurred in Europe or an adjoining area at the end of the Middle Pleistocene or early in the Late Pleistocene. In this model, the European Middle Pleistocene hominids would represent an undifferentiated stock of "archaic" *H. sapiens* or generalised Neanderthals showing the gradual establishment of distinctive modern or late Neanderthal characters (Breitinger, 1957; Howell, 1960; Le Gros Clark, 1964).

c. *H. erectus-sapiens* Models

These suggest that the species *H. erectus* existed in the European Middle Pleistocene and evolved into, or was replaced by, *H. sapiens*. (i) In a gradual evolutionary model of this type, chronological boundaries between the two species may be proposed (e.g., "end of Mindel" (Wolpoff, 1980b) or 300 ky (Campbell, 1972). (ii) A further development of this model suggests that sexual dimorphism is perhaps the major component of variation in the Middle Pleistocene sample, such that specimens with a robust *erectus*-like cranial morphology are male individuals (e.g., Bilzingsleben, Petralona), while those with a more gracile morphology are females (e.g., Swanscombe, Steinheim) of the same lineage (Wolpoff, 1980b). (iii) Another variant of this model parallels the presapiens model except that two broadly contemporaneous groups are envisaged in the Middle Pleistocene, representing *H. erectus* in eastern Europe (Bilzingsleben, Vértesszöllös, Petralona) and *H. sapiens* in western Europe (e.g., Swanscombe, Steinheim). Further evolution in the latter group leads to premodern forms (e.g., Ehringsdorf) and preneanderthal forms (Vlček, 1978).

d. Primitive *sapiens*-Neanderthal Model

This model also suggests that there were two main groups in the European Middle Pleistocene. An earlier *erectus*-like group (e.g., Arago, Petralona, Bilzingsleben) gave way by replacement or, more probably, evolution to an early Neanderthal group (e.g., Swanscombe, Biache, and perhaps Steinheim and Pontnewydd) in the later Middle Pleistocene (Stringer, 1981; Cook et al., 1982; Hublin, 1982).

TESTING THE MODELS

All of the above models require an adequate relative dating framework for the Middle Pleistocene hominids in order that they can be properly tested. Models a, c(ii), and c(iii) require the demonstration of the coexistence of two distinct species or morphologies, whereas the other models require a relative ordering of the relevant hominids. The dating problems of the European Middle Pleistocene record are still formidable, and even a relative ordering for the hominids cannot yet be established, let alone an absolute placement for each find (Cook et al., 1982). For this reason, most of the models cannot be fully tested, and the chronological classification systems of c(i) are not practicable (Stringer, 1981). The specific fossils claimed as members of the presapiens lineage of model a (Swanscombe and Fontéchevade) have been interpreted more plausibly as early Neanderthals, although the Fontéchevade frontal fragment is still a problematic specimen (Hublin, 1982; Stringer et al., 1984). Regarding model c(ii), I believe that the range in size, robusticity, and occipital morphology between specimens such as Petralona and Steinheim far exceeds that indicated for dimorphic *H. erectus* and Neanderthal samples (Stringer, 1981, 1984), and therefore variation in the European Middle Pleistocene hominids must be predominantly attributed to population differences, rather than to sexual dimorphism (see Fig. 1). While this might appear to support model c(iii), there is no evidence as yet to demonstrate clearly that the two distinct morphologies ("*erectus*-like" and "*sapiens*-like") in fact coexisted in Europe. Another look at this problem is needed, since a similarly high degree of variation is also present in African hominids of this age (e.g., Bodo vs. Ndutu).

My own preference would still be for model d, because the specimen that is most certainly dated to the earlier Middle Pleistocene (Mauer) does not, in my opinion, show any clear derived characters shared with either Neanderthals or modern humans. I am prepared to argue the same point for most of the Arago material (the Arago 2 mandible can be interpreted as Neanderthal-like, however) and the Vértesszöllös, Bilzingsleben, and Petralona fossils. Equally, the specimens that are most certainly dated to the late Middle Pleistocene (Biache, Bourgeois-Delaunay, Suard) do show clear, derived Neanderthal characteristics, and I am prepared to argue the same point for the Swanscombe and Ehringsdorf fossils, with the possible addition of the Steinheim, Pontnewydd, and Atapuerca specimens.

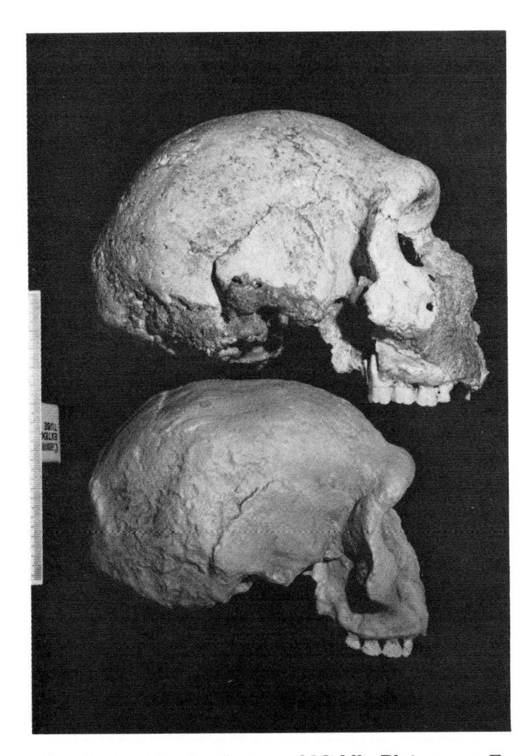

Fig. 1. Right lateral view of Middle Pleistocene European hominid crania showing range of variation: Petralona (incompletely prepared) above, Steinheim (cast) below.

CLASSIFICATION OF THE EUROPEAN HOMINIDS

Having stated my preference with regard to evolutionary models, it is necessary to discuss once again my reasons for questioning the existence of the taxon *H. erectus* in the known European record (I do not doubt its potential existence earlier in the European record). First, it is necessary to decide whether European specimens such as Mauer, Arago, Vértesszöllös, Bilzingsleben, and Petralona can be classified together, or whether they should be referred to more than one hominid group. There are obvious problems in comparing isolated mandibles or cranial fragments, which is why I have preferred to compare the fragmentary specimens to the more complete material from Arago and Petralona. Thus, an important preliminary question is to decide whether the Arago and Petralona fossils are similar enough to be classified together.

A new and skillfully produced composite reconstruction of the Arago 21/47 cranium has been studied and classified as *H. erectus*, while the Petralona specimen has been considered to be more Neanderthal-like (Vandermeersch, 1985). The main distinctions appear to be in the narrower, flatter, frontal bone (and associated smaller endocranial volume), greater facial prognathism, and angular torus development of the Arago specimen, while the Petralona cranium appears to show more advanced or Neanderthal-like features in pneumatisation, morphology of the maxilla, nose and supraorbital torus, endocranial expansion, and reduced prognathism. As already indicated (Hemmer, 1982; Stringer, 1984) some of the supposed "*erectus*" characters of the Arago cranium could be attributed to distortion remaining in the reconstruction, and certainly differences in prognathism and nasal form between Arago 21 and Petralona could be reduced or virtually eliminated in an alternative reconstruction. The differences in pneumatisation, supraorbital torus form, and presence of an angular torus would remain, however, but it could be argued that in other respects (palate size and shape, dental size, overall cranial thickness) the Petralona specimen is the more archaic of the two. Certainly, it is unlikely that the missing Arago occipital bone could have been more archaic and less Neanderthal-like than that of the Petralona cranium (except, perhaps, if it resembled that of Bilzingsleben). For me, at least, the similarities between these specimens far outweigh their differences, and "advanced" characters they share resemble Late Pleistocene hominids but differ from those of Asian *H. erectus*. These "advanced" characters include endocranial morphology, increased mid-facial prognathism, lack of ectocranial buttressing, increased cranial height, distinctive parietal arch shape, and, perhaps, reduced total prognathism (Hemmer, 1982; Holloway, 1982; Stringer, 1984).

If, therefore, the Arago and Petralona specimens can be regarded as representing the same hominid population, it is not difficult to extend this group by morphological similarities to include the Mauer mandible (Aguirre and de Lumley, 1977) and the cranial and dental material from Bilzingsleben and Vértesszöllös, although in several respects the latter specimen does appear rather "advanced" (Wolpoff, 1980b, 1982; Stringer, 1980, 1981, 1984). Overall, the group is characterised by a cranial robusticity (e.g., in supraorbital and occipital torus development, bone thickness, muscularity) greater than in early or late Neanderthals or modern humans, and this can be interpreted as retained from an ancestral population that was even more *erectus*-like in morphology. However, the "advanced" features of the Petralona and Arago (and Vértesszöllös?) material assume a greater significance in my view than retained archaic characters that are more marked in the Bilzingsleben specimens (for which we lack facial, basicranial, temporal, or parietal parts that appear "advanced" in Petralona or Arago).

NEANDERTHAL CHARACTERS IN THE EARLY EUROPEAN HOMINIDS

Having attempted to establish the "archaic" *sapiens* character of these assumed earlier Middle Pleistocene specimens, it is still necessary to discuss whether synapomorphies with Neanderthals (or modern humans) also exist in them, as these would not be expected under model d unless a transitional population leading to Neanderthals was being sampled.

A number of Neanderthal autapomorphies have been proposed, and the relevant ones can be summarised as a high degree of midfacial projection and associated anteriorly placed dentition, voluminous and projecting nasal opening, double arched and pneumatised supraorbital torus, cranial shape subpherical ("*en bombe*") in occipital view, large occipitomastoid crest relative to mastoid process, highly curved occipital plane, and double arched occipital torus with suprainiac fossa. Other features common in Neanderthals include the inflated and highly pneumatised maxillae, taurodont molars, H–O pattern of mandibular foramen, anterior mastoid tubercle, lambdoid flattening, and the high position of the auditory meatus relative to the zygomatic process root (for further discussion see Stringer et al., 1984). The distribution of these and other features in Late Pleistocene hominids is graphically displayed in Figure 2.

Of the above list of Neanderthal characters, it is possible to identify an anteriorly placed dentition only in the Arago 2 mandible of the "archaic" European group, while midfacial prognathism in Arago 21 and Petralona, although marked, does not reach typical Neanderthal levels and is less developed than in Broken Hill 1 (Stringer, 1984). Petralona certainly possesses a Neanderthal-like supraorbital torus and "inflated" maxilla, but with an even higher degree of pneumatisation. Given the morphology of Broken Hill 1 and Bodo 1, it may be that the level of pneumatisation and the maxillary form of Petralona were more widespread in the robust Middle Pleistocene hominids, and therefore they do not represent genuine synapomorphies with Neanderthals (but compare de Bonis and Melentis, 1982). However, the supraorbital torus form is much more specifically Neanderthal-like. The nasal bones of Petralona are projecting, and the nasal opening is certainly absolutely large, yet when scaled against the overall size of the massive face, relative nasal size is similar to average modern values, unlike that of Bodo and most Neanderthals, where it remains distinctly larger. In the parietal and occipital region of Petralona there are no obvious Neanderthal characters, and this is also more generally true for the Arago, Bilzingsleben, and Vértesszöllös specimens, although the curved occipital plane of the latter fossil is certainly marked (Hublin, 1982). Overall, then, it might be argued that an incipient Neanderthal morphology is present in the Petralona cranium, but most of the relevant characters can be matched in Middle Pleistocene specimens generally, and it is with these other fossils (probably including Broken Hill 1, Bodo, and perhaps Dali)

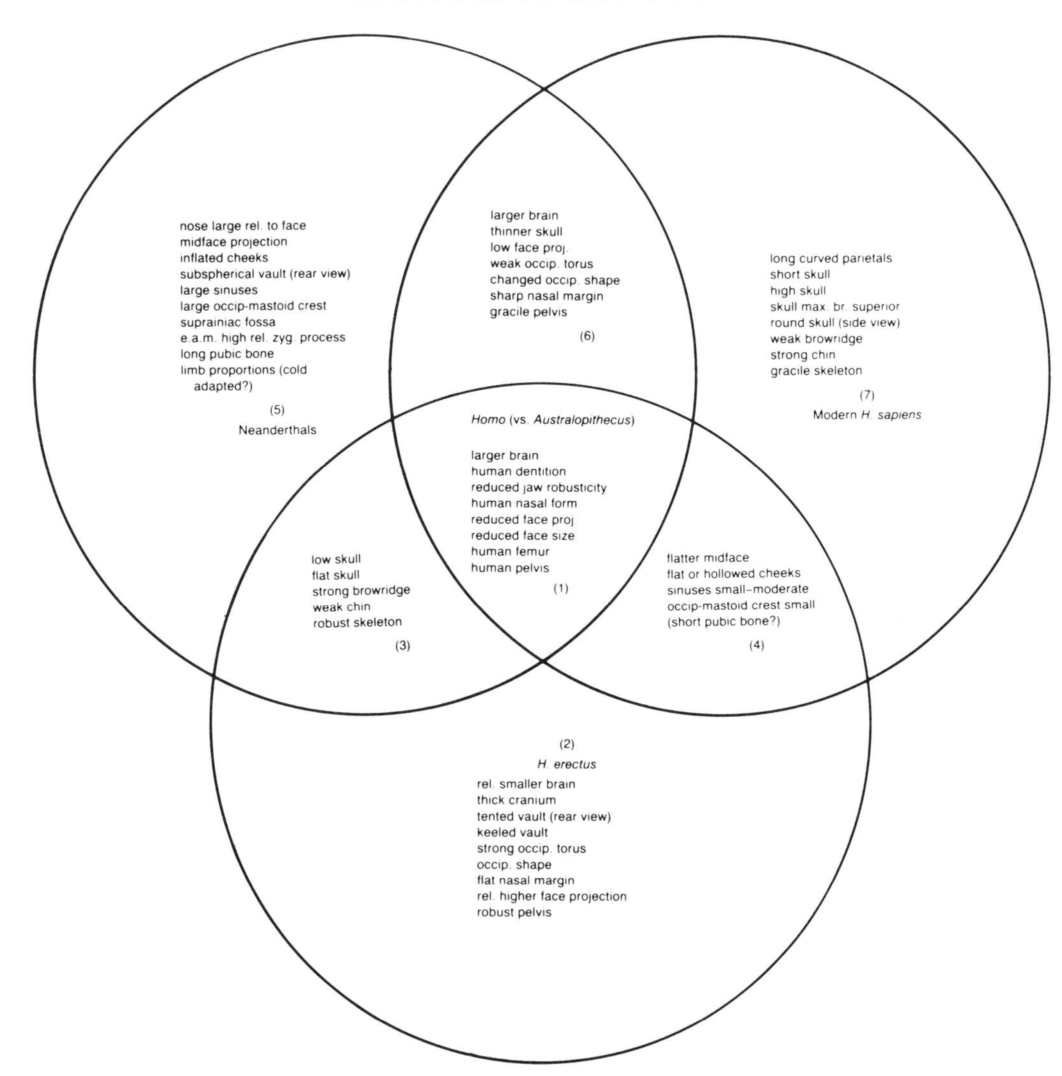

Fig. 2. Simplified Venn diagram showing characters typical of *H. erectus*, Neanderthals, and modern humans. Group 1 characters are synapomorphies for *Homo* generally. Group 3 can be interpreted as plesiomorphies retained in *H. erectus* and Neanderthals but lost in modern humans; group 4 as plesiomorphies retained in *H. erectus* and modern humans but lost in Neanderthals; group 6 as synapomorphies of Neanderthals and modern humans, or characters evolved in parallel. The "archaic" European hominids mainly show combinations of group 2, 3, or 4 characters. The late Middle Pleistocene hominids of Europe show more group 5 or 6 characters and may postdate the divergence of Neanderthals and modern humans.

that the principal allegiance of Petralona and the European "archaic" hominid group lies. This primitive group may be close to the common ancestor of Neanderthals and modern humans and could be classified as a primitive grade or subspecies of *H. sapiens* or as a distinct species (e.g., *H. heidelbergensis*).

Looking at the possible or probable late Middle Pleistocene European sample, it is evident that the Swanscombe fossil has a Neanderthal-like occipital torus and suprainiac fossa, and probably had a prominent occipitomastoid crest, while the Biache, Suard, and Bourgeois-Delaunay fossils show such features much more strongly, even to the extent of overall cranial shape in Biache. Mandibular material from the La Chaise sites also shows Neanderthal features, but it is only in the Pontnewydd upper molars that a significant level of taurodontism has yet been recognised. The Steinheim specimen is more problematic, since it is poorly preserved and distorted, yet a suprainiac fossa is present (Hublin, 1982). However, for me, at least, this specimen sits less easily in the early Neanderthal group, since it apparently departs in a number of plesiomorphous retentions from the typical Neanderthal condition (e.g., in supraorbital form, cranial shape, basicranial morphology and flexion, lower level of midfacial prognathism, and occipitomastoid morphology).

CONCLUSIONS

Any evolutionary scheme to explain European Middle Pleistocene hominid variation at our present state of knowledge is likely to be an oversimplification. However, the most appropriate scheme has two main hominid groups. The "archaic" group resembles *H. erectus* in a number of respects and lacks significant autapomorphies found in either Neanderthals or modern humans. The second group contains late Middle Pleistocene hominids and does show synapomorphies with Neanderthals. It may have been derived from members of the other group by local evolution. The placement of individual specimens such as Petralona and Steinheim is difficult, and relative dating cannot yet demonstrate whether all members of the "archaic" group in fact predated all members of the Neanderthal-like group. Sexual dimorphism does appear to have contributed significantly to variation in the samples, but population differences are probably the primary source of variation.

Regarding the evolution of anatomically modern *H. sapiens*, the fact that Europe appears to record only the evolution of Neanderthal derived characters during the Middle and early Late Pleistocene suggests that modern humans evolved elsewhere. In western Europe, the Saint-Césaire fossil at last provides clear evidence that some Neanderthals, at least, were simply too late and too specialised to have contributed to the evolution of modern hominids in the area. No genuinely transitional fossils displaying combinations of Neanderthal and modern derived characters have been found in Europe or southwest Asia, although it is true that in eastern Europe and southwest Asia less typical Neanderthal fossils are present. It is unclear whether their somewhat greater resemblance to modern humans is due to a greater retention of plesiomorphous characters, gene flow from contemporaneous anatomically modern populations, or to an *in situ* transition from a Neanderthal-like to a modern human morphology (Stringer et al., 1984).

It is only in Africa that fossils displaying a clear mosaic of non-modern and modern characteristics exist in the later Middle Pleistocene or early Late Pleistocene, and the presence of actual anatomically modern hominids in the early Late Pleistocene seems probable. Fossils such as Florisbad, Djebel Irhoud, and Omo-Kibish 2 do seem to represent the kind of morphological precursors of modern humans that are missing from the late Middle Pleistocene or early Late Pleistocene of Europe, and none could be described as Neanderthal-like in terms of synapomorphies. The evidence from the Klasies River Mouth, Border Cave, and Omo-Kibish 1 hominid fossils indicates an anatomically modern presence in southern and northeast Africa in the early Late Pleistocene. However, while the chronological position of the former fossils (KRM) is most secure, they are also the most fragmentary, and while the latter two sites provide unequivocal evidence of anatomically modern hominids, they are less certainly dated. Taken together, however, they may be used to place the origin of modern humans in Africa during the timespan of the Middle Stone Age.

Whether there was a single African centre of origin for modern humans, followed by a radiation to Asia and Europe later in the Late Pleistocene, or whether other non-European centres of origin existed is unclear. While the Zuttiyeh, Zhoukoudian, Hexian, and Ngandong fossils do not appear to this author to be transitional between archaic and modern humans, the relationships of the Dali specimen are less easy to determine, partly because of the lack of good data. The specimen may be part of the "archaic" Middle Pleistocene hominid group described earlier, lacking significant synapomorphies with either Neanderthals or modern humans. Alternatively, it might indeed represent a direct ancestor for modern humans in the area. Distinct local centres of origin for Australian populations may also be indicated by the high variability of Late Pleistocene samples. This may be a reflection of morphological changes acquired after the initial colonisation of Australia, of gene flow from more archaic populations in Southeast Asia, or a sign of derivation from ancestors already distinct from those of modern Africans and Europeans. Only a careful analysis of genuine local "clade" characters and their continuity through time can resolve this problem.

Finally, it is hoped that the wider, but careful, application of cladistic methods to the study of later Pleistocene hominid evolution will further clarify relationships. It should also lead to the replacement of the present wide use of subspecific categories based mainly on geographic or chronological criteria (Campbell, 1972) by a more meaningful taxonomy based on morphological characters. This in turn may lead to the restriction of the specific name *H. sapiens* to the groups sharing the derived characters of modern humans. It would probably then be necessary on morphological grounds to return the Neanderthals to the status of a distinct species. Given some recent interpretations of the European evidence as recording the separate evolution of the Neanderthal lineage for some 300,000 years, followed by a brief period of coexistence of late Neanderthals and modern humans, this suggestion may not be as extreme as it sounds.

ACKNOWLEDGMENTS

I am most grateful to the American Museum of Natural History, and especially to its Director, Dr. T.D. Nicholson, and to Drs. I. Tattersall and E. Delson for the opportunity to take part in the enjoyable and stimulating events associated with the "Ancestors" symposium.

LITERATURE CITED

Aguirre, E, and de Lumley, MA (1977) Fossil men from Atapuerca, Spain: Their bearing on human evolution in the Middle Pleistocene. J. Hum. Evol. *6:*681–688.

Anonymous (1981) Les Premiers Habitants de l'Europe (Exhibition Catalogue). Paris: Lab. Préhist. Musée de l'Homme.

Breitinger, E (1957) Zur phyletischen evolution von *Homo sapiens*. Anthropol. Anz. *21:*62–83.

Butzer, KW, and Isaac, GL (eds) (1975) After the Australopithecines. The Hague: Mouton.

Campbell, BG (1972) Conceptual progress in physical anthropology: Fossil man. Ann. Rev. Anthropol. *1:*27–54.

Cook, J, Stringer, CB, Currant, AP, Schwarcz, HP, and Wintle, AG (1982) A review of the chronology of the European Middle Pleistocene hominid record. Yrbk. Phys. Anthropol. *25:*19–65.

de Bonis, L, and Melentis, J (1982) L'homme de Petralona: Comparaisons avec l'homme de Tautavel. 1er Congr. Internat. Paleont. Hum. (Prétirage). Nice: C.N.R.S., pp. 847–874.

Hemmer, H (1982) Major factors in the evolution of hominid skull morphology. Biological correlates and the position of the Anteneandertals. 1er Congr. Internat. Paléont. Hum. (Prétirage). Nice: C.N.R.S., pp. 339–354.

Hennig, GJ, Herr, W, Weber, E, and Xirotiris, NI (1982) Petralona cave dating controversy. Nature *299:*281–282.

Holloway, RL (1982) *Homo erectus* brain endocasts: Volumetric and morphological observations with some comments on the cerebral asymmetries. 1er Congr. Internat. Paléont. Hum. (Prétirage). Nice: C.N.R.S., pp. 355–366.

Howell, FC (1960) European and northwest African Middle Pleistocene hominids. Curr. Anthrop. *1:*195–232.

Howells, WW (1980) *Homo erectus*—Who, when and where: A survey. Yrbk. Phys. Anthropol. *23:*1–23.

Hublin, JJ (1982) Les anténéandertaliens: Présapiens ou prénéandertaliens? Geobios Mém. Spéc. *6:*345–357.

Kukla, G (1978) The classical European glacial stages: Correlation with deep-sea sediments. Trans. Nebr. Acad. Sci. *6:*57–93.

Kurtén, B, and Poulianos, AN (1981) Fossil Carnivora of Petralona Cave (status 1980). Anthropos (Athens) *8:*9–56.

Le Gros Clark, WE (1964) The Fossil Evidence for Human Evolution, 2nd ed. Chicago: University of Chicago Press.

Oakley, KP, Campbell, BG, and Molleson, TI (1971) Catalogue of Fossil Hominids, Vol. 2. Europe. London: British Museum (Natural History).

Stringer, CB (1980) The phylogenetic position of the Petralona cranium. Anthropos (Athens) *7:*81–95.

Stringer, CB (1981) The dating of European Middle Pleistocene hominids and the existence of *Homo erectus* in Europe. Anthropologie (Brno) *19:*3–14.

Stringer, CB (1984) Some further notes on the morphology and dating of the Petralona hominid. J. Hum. Evol. *12:*731–742.

Stringer, CB, Hublin, JJ, and Vandermeersch, B (1984) The origin of anatomically modern humans in western Europe. In FH Smith and F Spencer (eds): The Origin of Modern Humans. New York: Alan R. Liss, Inc., pp. 51–136.

Vallois, HV (1954) Neandertals and praesapiens. J. R. Anthropol. Inst. *84:*111–130.

Vandermeersch, B (1985) The origin of the Neandertals. In E Delson (ed): Ancestors: The Hard Evidence. New York: Alan R. Liss, Inc., pp. 306–309.

Vlček, E (1978) A new discovery of *Homo erectus* in Central Europe. J. Hum. Evol. *7:*239–251.

Wolpoff, MH (1980a) Paleoanthropology. New York: Knopf.

Wolpoff, MH (1980b) Cranial remains of Middle Pleistocene European hominids. J. Hum. Evol. *9:*339–358.

Wolpoff, MH (1982) The Arago dental sample in the context of hominid dental evolution 1er Congr. Internat. Paléont. Hum. (Prétirage). Nice: C.N.R.S., pp. 389–410.

Ancestors: The Hard Evidence, pages 296–300

Late Pleistocene Human Fossils and Evolutionary Relationships

Eric Delson
Department of Anthropology, Herbert H. Lehman College, City University of New York, Bronx, New York 10468 and Department of Vertebrate Paleontology, American Museum of Natural History, New York, New York 10024

The final scientific session began with a series of papers on the best-known form of extinct Late Pleistocene humans, the Neanderthals. Bernard Vandermeersch continued the trend of interpretation of the European later Middle Pleistocene fossils which dominated the previous session: namely, that most of these present derived features linking them to the true Neanderthals. Vandermeersch termed the European Eemian and most later Saale ("Riss") fossils such as Biache "preneandertals," as contrasted with some even earlier specimens (Swanscombe, Steinheim, Arago, and Petralona), which may lie on the Neanderthal lineage. In discussion, he was asked not about these earliest Neanderthals, but about the significance of the youngest well-dated specimen, from St. Césaire. He noted that this fossil, with typical Neanderthal morphology (as now agreed by all workers, it seems), was recovered from the upper of two Châtelperronian layers at the site. According to Vandermeersch, this culture, or phase, is quite restricted, with only a few well-documented sites in a small range. It is contemporary with the earliest Aurignacian, from which no human fossils are known, although Vandermeersch expects to find moderns (Cro-Magnons) associated with such levels. In sum, he said that the importance of St. Césaire was to show that Neanderthals lived longer than had been thought, disappearing only at the beginning of the Upper Paleolithic. He also offered his opinion that the Châtelperronian is not only derived from the Mousterian (a long-mooted point), but that it could well be included as the last phase of the Middle Paleolithic. More detailed discussions of this question have recently been offered by Harrold (1983) and by Clark Howell (1984), the formal chairman of this session. Although Howell was unable to prepare the summary for this volume, his introduction to Smith and Spencer's recent book (Howell, 1984) on the origins of modern humans covers many of the points raised briefly here; other papers in that volume also offer different perspectives on these problems as well.

The second speaker was Jakov Radovčić, who reviewed Gorjanović-Kramberger's work at Krapina, especially in light of the search for modern humans contemporary with the Neanderthals there. He concluded that there is no evidence for such contemporaneity anywhere in Europe as yet, but suggested that the early occurrence of Aurignacian-like artifact assemblages in central and eastern Europe might indicate a "source area" for the Cro-Magnons. He was asked about the evidence for an Upper Paleolithic earlier in central than in western Europe and replied that although terminology is still variable, the early Aurignacian or Szeletian may be as old as 40,000 years BP, based on data from Istálloskö, Hungary. The results of the recent study of Bacho Kiro cave (Bulgaria) have been summarized by Ginter and Kozłowski (1982) and provide further information on this problem. The oldest layers in the cave (14–12) yield Middle Paleolithic assemblages of Mousterian type; a date of > 47,500 years BP has been obtained from layer 13. Layer 11 yielded artifacts of Upper Paleolithic facies that have been assigned to the Bachokirian, typologically similar to the assemblage from the dated level at Istálloskö. The warm layer 11 at Bacho Kiro produced a radiocarbon date on charcoal of > 43,000 years BP, but Mook (1982) suggests that because there was some activity in the sample, a 1-σ age-range may be provided: namely, $50{,}000\ ^{+9{,}000}_{-4{,}000}$ years BP. This places layer 11 between 45,000 and 60,000 years BP, contemporary with Neanderthals at Shanidar and in France. A single dP_3 ("dm_1") from this level is larger than moderns and is said by Ginter and Kozłowski (1982) to be possibly transitional between Neanderthals and moderns in some (undefined) ways—this point is not made by Glen and Kaczanowski (1982) in their description of the human remains.

Ralph Holloway summarized his studies of Late Pleistocene endocast size and morphology, concluding that Neanderthals continue to be maligned in terms of having less well-developed brains than anatomically modern humans. His findings support and extend the view that Neanderthals had brains larger, on average, than early moderns and with well-developed Broca's and Wernicke's areas. In response to a question about the Neanderthal occipital lobe, he said that although this area is not significantly different in morphology from that of moderns, it may have been relatively larger. He mentioned ongoing work with native Australians that suggests a relatively larger visual cortex than in Caucasians, combined with "tremendous capacity for visuo-spatial integration and solving visual problems at a very, very early age." Holloway concluded that "in a romantic sense, I like to think of Neandertals out there on the periglacial tundra facing all sorts of difficulties, not only with keen eyesight but also keen olfactory senses and so forth."

Jean-Louis Heim reviewed his work on problems of sex determination for Neanderthal crania. As much of this work has been published already (Heim, 1981–1982, 1983) no manuscript was submitted for this volume, but some of the major conclusions can be summarized briefly. The Neanderthals, for the first time in human paleontology, provide a series of specimens that can be accurately sexed on the basis of postcranial morphology. In turn, this allows a more precise evaluation of patterns of dimorphism in the skull of individuals of known sex. By comparison to modern humans, Neanderthals show a rather higher degree of sexual dimorphism in most cranial dimensions—length, breadth, and height. Overall size is the best feature for cranial determination of sex, but that requires a sample for comparison. Of individual measurements for the Neanderthals, cranial length (glabella–opisthocranion) shows very strong dimorphism: the female to male ratio is 92.7, vs. 94 for Upper Paleolithic Europeans and 98 for living humans. Among other features, bregma is farther forward in females than males, so the forehead is higher (but the supraorbital torus protrusion does not vary with sex); occipital bunning is higher in females and the mastoid process is weaker; and the parietal segment of the sagittal arc is greater than the frontal in females, with the reverse true for males.

Erik Trinkaus and Fred Smith completed the review of Neanderthals by examining their evolutionary fate. They postulated that a "transition" of some sort occurred from the typical Neanderthal morphology seen in Europe and Western Asia during the Eem and earlier Weichsel to modern human morphology, but were undecided as to whether this was an *in situ*, gradual evolution or the result of a migration/invasion. They evaluated the types of characters that changed during this transition, how much change occurred, and the possible functional explanations for such change. In response to a question, Trinkaus elaborated on his suggestion that Neanderthals might have had a longer gestation period than living humans (this work has now been published, see Trinkaus, 1984). Briefly, he indicated that Neanderthal pubic bones differ markedly from those of early moderns or living humans, while other aspects of the pelvis do not. The resulting larger pelvic aperture would permit passage of a skull some 20% larger than that permitted by modern pelves. In terms of brain size and neonatal growth patterns of modern humans, that would mean an additional 2–3 months of fetal development. This fits well with estimates of how long human gestation "should" be, based on brain/body scaling and mammalian development patterns.

Smith was then asked how, in light of the divergent nature of Neanderthal morphology, he could conceive of a transformation of this morphology into that of early *Homo sapiens sapiens* in only 5–10,000 yr. He indicated that in central Europe there is a clear picture of gradual change from earlier to later Neandertals through earlier and later moderns (see also Smith, 1984). In western Europe, this gradual modification is not clear, with St. Césaire being typically Neanderthal but late, although there then is a fair gap before the first well-dated moderns appear. Smith agreed that there was probably some gene flow from outside Europe, especially as modern-like populations occur in southern Africa by 65,000 years BP at least, and then in the Near East by about 40,000 years BP, but argued that continuity within Europe appears very strong. Trinkaus added that he felt the differences among most workers in this area were a result of varying emphasis on gene flow vs. regional continuity, and that there was little reason to expect that the pattern of replacement would have been identical from western Europe across to central Asia, much less farther east and south. The use of terms such as "transition" by Trinkaus and Smith may tend to polarize this discussion even farther, because they carry the implication of continuity. Howells (1976) discussed the alternative models of migration and local evolution explicitly, and various workers have paid lip-service to the complementary nature of these processes and their probable mixture in the Neanderthal case. Nonetheless, most authors have either sidestepped the issue (as do Trinkaus and Smith here) or taken a stand at one end of the philosophical "morphocline." Stringer (1982) has tried to suggest one path to a partial solution, involving the development of clear predictive hypotheses to be tested by fossils and archeology, and further work in this direction would be useful.

Ron Clarke moved the discussion out of Europe and back to Africa with a report on his recent research at the Florisbad locality. No new hominid remains have been recovered, but an early Middle Stone Age layer has been partly excavated *in situ* and should be chronometrically dated soon; it appears that the skull and other remains are of late Middle Pleistocene age. Clarke also was able to prepare a more accurate reconstruction of the Florisbad cranial fragments, resulting in a rather more archaic face. He was asked to clarify his views on the relationship and relative "modernness" of the Florisbad, Ngaloba, and Border Cave crania. He repeated that both Florisbad and Ngaloba (as judged from the cast) are less anatomically modern than Border Cave, which is very modern.

Alan Thorne discussed the "Origin of the Australians," bringing the audience much closer to the present but far afield geographically. Due to the press of other matters (see Lewin, 1984), he was unable to submit a manuscript, but some of the ideas he presented are to be found in Wolpoff et al. (1984) and the papers by Thorne cited therein; I shall summarize only his presentation here. Thorne argued that two regional morphs of long duration occurred in eastern Asia: a northern variant from China (Zhoukoudian through modern East Asia and the Americas) and a southern Javanese line ("Java Man" through Ngandong and the robust Australian fossils). Of the second group, Sangiran 17 shows a complex of features of the face (orbit shape, brow ridge form, and details of the floor of the nasal orifice, palate, and malar) that "are Australian in the morphological if not geographical sense." Moreover, Thorne noted, features of the asterion region and of the nuchal crest and torus are surprisingly modern. The Kow Swamp crania, between 10,000 and 16,000 yr old, show much similarity to the Javanese fossils. Kow Swamp 5 is "one of a group of Late Pleistocene Australian hominids which Peter Brown (Univ. New England) has suggested show artifical cranial deformation. He argues, in part from ethnographic analogy, that these people were deforming the shape of children's heads by manual or pedal pressure, over a period of about six months. I think he is probably right, but I'm not totally convinced yet of the arguments, particularly as there doesn't seem to be any long-term binding involved." From the new site of Coobool Creek, near Kow Swamp, Brown has recovered about 130 individuals, some with and others without deformation. Another large series comes from the most recently found site, Willandra Lakes in New South Wales. Using an experimental electron spin resonance approach, the oldest specimen, WLH 50, is far older than 30,000, perhaps something like 60,000 years old. It is rather similar to the Solo crania, with vault bone thickness between 14 and 17 mm (in part due to advanced individual age) and very low maximum breadth. For these people to have entered Australia from Indonesia, either via New Guinea and the north or through Timor and into the south, said Thorne, "involves a series of crossings, the maximum one being of the order of 80 to 100 km. It is most likely that we are dealing with purposive human behavior and the making of watercraft."

The second, northern lineage is documented, according to Thorne, not only at Zhoukoudian, but also at Dali and Maba of late Middle Pleistocene age, and in later Pleistocene fossils from the Upper Cave in north China and from Ziyang and Liukiang in the far south. In Australia, the Keilor, Lake Mungo, and several Willandra Lakes hominids in the 32,000–20,000 yr range represent the same lineage. Fossils of intermediate geographical position are known from Wajak, Okinawa, and Tabon (Philippines). In Australia, these more gracile populations show no evidence of cranial deformation; they are often buried with a sequence of cremation, bone smashing, and reburning and then interment; and they are associated with a much more complex cultural assemblage than is found with the robust forms. Thorne suggested that the use of bamboo as a major cultural raw material by Indonesian hominids (see also Pope, 1983; Pope and Cronin, 1984) might have led to its utilization in simple raft-building, as is still done today in China. He thus saw the two groups independently crossing large water barriers to reach different areas of Australia before contacting each other. In conclusion, Thorne said that "an interesting question is that if people are moving across ocean gaps of a substantial nature in East and Southeast Asia, is this also the mechanism which may involve the processes leading to the occupation of the Americas?"

In discussion, Thorne was asked to comment on Birdsell's trihybrid theory of Australian population formation. Originally, Birdsell (1949) proposed a succession of three human entries into Australia: first, the so-called Oceanic Negritos, rain-forest people seen in Tasmania and northeast Queensland; second, the "Murrians," best represented in the Murray River Valley of southeastern Australia; and third, the Carpentarians. This last group, said Thorne, can be separated out as very recent immigrants from New Guinea and Indonesia into northern Australia, where they mixed with already present populations in a broad band. Given the coastal movement of people, Thorne continued, the Tasmanians might fit with the coastal southern Australian "gracile" group, such as Keilor and Lake Mungo. The earliest immigrants were not these people, for Thorne, but the robust group whose descendants are the Murrians. Ron Clarke asked Thorne to comment on what he saw as a similarity between the Ngaloba hominid and the oldest specimen from Willandra Lakes, WLH 50. Thorne agreed that from a brief look at the Ngaloba cast, there were "very many detailed characteristics which I would see, from my part of the world."

Finally, Arun Sonakia gave a short presentation on a new find from India. Sonakia had not been invited to speak originally, but as he had new information of potentially great interest, the organizers decided to grant him some time. As it turned out, this talk was one of the highlights of the symposium, representing almost the first public announcement of what paleoanthropologists have long sought—a truly archaic hominid from the Indo-Pakistan subcontinent. Much of a calvarium was recovered by Sonakia from a conglomeratic layer in the Narmada Valley, which he dated to the late Middle Pleistocene. The skull vault is low, the bone relatively thick, and the supraorbital torus well developed. Sonakia has suggested alignment with late *Homo erectus* populations or archaic *H. sapiens* as known in Europe, but not Neanderthals. Gyani Badam of Pune, working in the U.S. on Pleistocene mammals, visited the A.M.N.H. in the summer of 1984 and commented on the age of the deposits. As discussed in his 1979 book, he argued that it is unlikely that this hominid is older than early Late Pleistocene or possibly latest Middle Pleistocene. Badam thought that the faunal assemblage listed by Sonakia is mixed, with some later Middle Pleistocene forms possibly associated with younger taxa. He noted that a radiocarbon date of about 32,000 years BP was obtained on molluscs from the cemented sandy gravels near Devakachar (Badam, 1979, p. 178). Moreover, several of those who observed a cast of the Narmada hominid in New York thought that it was rather more similar to late archaic *Homo sapiens* than to *Homo erectus*, although they agreed with Sonakia that it was not a Neanderthal. I would guess that a date between 100,000–150,000 years and association with a Middle Stone Age cultural tradition might be expected eventually. Further study of the morphology of Narmada Man and of the faunal and possible cultural association, as well as a better estimate of the age of the deposits, are awaited eagerly. This area will certainly be a focus of much attention over the coming years, and the description of its first paleoanthropological fruit was a fitting climax to the "Ancestors" symposium.

LITERATURE CITED

Badam, GL (1979) Pleistocene Fauna of India. Pune: Deccan College.

Birdsell, JB (1949) The racial origin of the extinct Tasmanians. Rec. Queen Victoria Mus., Lauceston *2:*105–122.

Ginter, B, and Kozłowski, JK (1982) Conclusions. In JK Kozłowski (ed): Excavation in the Bacho Kiro Cave (Bulgaria), Final Report. Warsaw: Panstwowe Wydawnictwo Naukowe, pp. 169–172.

Glen, E, and Kaczanowski, K (1982) Human remains. In JK Kozłowski (ed): Excavation in the Bacho Kiro Cave (Bulgaria), Final Report. Warsaw: Panstwowe Wydawnictwo Naukowe, pp. 75–79.

Harrold, FB (1983) The Châtelperronian and the Middle–Upper Paleolithic transition. In E Trinkaus (ed): The Mousterian Legacy. Oxford: British Archaeol. Reports S164, pp. 123–140.

Heim, J-L (1981–1982) Le dimorphisme sexuel du crâne des hommes de Néandertal. L'Anthropol. *85/86:*193–218, 451–469.

Heim, J-L (1983) Les variations du squelette post-crânien des hommes de Neandertal suivant le sexe. L'Anthropol. *87:*5–26.

Howell, FC (1984) Introduction. In FH Smith and F Spencer (eds): The Origins of Modern Humans. New York: Alan R. Liss, Inc., pp. xiii–xxii.

Howells, WW (1976) Explaining modern man: Evolutionists versus migrationists. J. Hum. Evol. *5:*577–596.

Lewin, R (1984) Extinction threatens Australian anthropology. Science *225:*393–394.

Mook, WG (1982) Radiocarbon Dating. In JK Kozłowski (ed): Excavation in the Bacho Kiro Cave (Bulgaria), Final Report. Warsaw: Panstwowe Wydawnictwo Naukowe, p. 168.

Pope, GG (1983) Evidence on the age of the Asian Hominidae. Proc Natl. Acad. Sci. U.S.A. *80:*4988–4992.

Pope, GG, and Cronin, JE (1984) The Asian Hominidae. J. Hum. Evol. *13:*377–396.

Smith, FH (1984) Fossil hominids from the Upper Pleistocene of Central Europe and the origin of modern Europeans. In FH Smith and F. Spencer (eds): The Origins of Modern Humans. New York: Alan R. Liss, Inc., pp. 137–209.

Stringer, CB (1982) Towards a solution to the Neanderthal Problem. J. Hum. Evol. *11:*431–438.

Trinkaus, E (1984) Neandertal pubic morphology and gestation length. Curr. Anthropol. *25:*509–514.

Wolpoff, MH, Wu, X, and Thorne, AL (1984) Modern *Homo sapiens* origins: A general theory of hominid evolution involving the fossil evidence from East Asia. In FH Smith and F Spencer (eds): The Origins of Modern Humans. New York: Alan R. Liss, Inc., pp. 411–483.

Ancestors: The Hard Evidence, pages 301–305

A New Reconstruction of the Florisbad Cranium, With Notes on the Site

Ronald J. Clarke
Palaeoanthropology Research Group, Department of Anatomy, University of the Witwatersrand Medical School, Parktown 2193 Johannesburg, South Africa

ABSTRACT The hominid cranium from the Middle Stone Age site of Florisbad was faultily reconstructed by Dreyer just after its discovery in 1932. Some published interpretations and measurements on this cranium are therefore incorrect. A more accurate reconstruction has now been made by the author. Excavations and research conducted by the author at Florisbad since 1981 have uncovered a Middle Stone Age occupation floor and clarified the stratigraphy. The cranium shows obvious signs of gnawing by a hyaena. It was discovered with other bones in the eye of a spring at the level of the lowest peat, Peat I. It compares well with the ca. 120,000 year old Ngaloba cranium and the ca. 130,000 year old Omo II calvaria. There is a strong similarity to the Broken Hill cranium, compared to which Florisbad is more advanced. Thus the Florisbad cranium could bc in the age range of 100–200,000 years old.

INTRODUCTION

Forty-five kilometres northwest of Bloemfontein in the Orange Free State of South Africa is situated the Hagenstad Saltpan, on the southeast margin of which is a large sand dune. Thermal springs emerge through the dune from the dolerite and Ecca Shale beneath. The springs were developed early this century as therapeutic baths by the owner, Floris Venter, and hence came to be known as Florisbad. In 1912, while enlarging one of the baths, Floris Venter uncovered many fossils and stone tools that were preserved as curios by his wife Martha. These antiquities subsequently came to the notice of Dr. Robert Broom, who studied them and described several extinct species (Broom, 1913). In 1917, 1926, 1928, and 1932, Professor T.F. Dreyer conducted excavations at the site. In February 1932, while they were excavating a large spring eye, Mr. G. Venter discovered a hominid right upper third molar, and Dreyer shortly recovered from the same spring eye a partial hominid calotte. Sieving of the excavated material resulted in the retrieval of parts of the face (Dreyer, 1938). Unfortunately, this cranium was never properly described, and I recently observed that the face was incorrectly reconstructed by Dreyer (1935). Certain of the subsequent interpretations and measurements on this cranium (Dreyer, 1935; Drennan, 1935; Galloway, 1937; Rightmire, 1978) are thus unreliable.

THE FLORISBAD CRANIUM

Reconstruction

In 1981, I was appointed by the National Museum in Bloemfontein to conduct new excavations at Florisbad, and in 1983 I was asked to restore the cranium, which had been seriously damaged when it was improperly moulded in solid silicone rubber by a museum artist in the 1970s. In trying to remove the one-piece solid rubber mould from the calotte, he not only broke the fossil but smashed parts of it irrevocably into powder. I was able to replace many paper-thin fragments of inner table in their correct position, but some small pieces of the vault are missing.

While cleaning the Florisbad cranium, I found that parts of the facial bones had been obscured with plaster of Paris and shellac and that a large frontal sinus had been completely filled with plaster of Paris. All of this plaster I have now removed. I have used the minimum of plaster reconstruction to fill the gaps between the facial fragments and have left the face separate from the calotte to facilitate study.

The facial fragments had not been damaged, but while cleaning and examining them I found that Dreyer had made three great errors in his reconstruction. First, he had joined together the left and right palatal fragments, apparently not observing that although the incisive fossa and canal are present on the left they are entirely absent on the right. In other words, several millimetres of right palate adjacent to the sagittal midline are missing. Dreyer's reconstruction had thus made both nasal cavity and palate much too narrow. Second, Dreyer had positioned the zygomatic fragment unnaturally, such that the lateral surface of the frontal process was rotated anteriorly, and thus the medial edge of the facial surface was rotated posteromedially. This gave the face a peculiarly pronounced infraorbital hollow. Third, although not one of the six facial fragments makes an indisputable contact with any other nor with the frontal, Dreyer had joined each fragment to its neighbour. The total effect made the face extremely small and modern in appearance. As there is no trace of zygomaticofrontal suture, it is clear that the right zygomatic fragment cannot be in contact with the frontal, although the morphology indicates that they approach each other closely. Similarly, the surface contours of the zygomatic fragment and the adjacent palatal portion of the right maxilla indicate that they probably do contact in the region of the inframalar notch, although there is no definite join. The nasal bones also probably contact the adjacent left and right frontal processes of the maxilla, although there are no clear joins, but none of these can contact the frontal in the glabella region. There has to be a gap of 2 or 3 mm. The right frontal process of the maxilla could not have contacted the right palatal portion along the lateral nasal margin. By joining these, Dreyer caused the nasal opening to be unnaturally short in height.

In my reconstruction (Figs. 1, 2), I have left a gap of about 7 mm between the two portions. The fact that there are no positive joins between any of these facial fragments means that reconstruction is difficult and can for the most part be no more than educated guesswork based on surface contours and anatomical comparisons. It is clear also that most measurements on the face are meaningless, as we cannot know the height and breadth of the orbits and nasal cavity, the length of the nasal bones, position of nasion or nasospinale, or facial angles. I cannot claim that my reconstruction is faultless, as there are no guidelines except for the sagittal midline and anatomical comparison; it is, however, clearly more realistic than Dreyer's, which, because of the anatomical errors just mentioned, was fallacious.

Morphological Description

Now that the palate is broader, the nasal cavity larger, and the zygomatic bone more correctly positioned, the face has a much larger and more archaic appearance than that of Dreyer's reconstruction. The Florisbad face now approximates more to that of the Broken Hill face, although it is not as massive.

The Florisbad calotte is thick (greatest thickness 12 mm at bregma, 9.5 mm on the frontal, and 9 mm on the parietal) and displays well-fused sutures. The frontal squame slopes gently superoposteriorly from a supraorbital margin that is about 18 mm thick lateral to the prominent glabella and thins to circa 9 mm laterally. Although the supraorbital margin is thick, it is not formed into a supraorbital torus, nor is there an ophryonic groove, but posterior to glabella and extending laterally to a position behind the supraorbital notch is a supraglabellar depression. Laterally the frontal squame slopes almost imperceptibly backward from the supraorbital margin. Galloway (1937) was not correct in his assertion that there is a "well defined sulcus supraorbitalis, which runs upward and laterally from the supraorbital notch." In fact, there is only an extremely faint sulcus. Internally the supraorbital region behind and to the right of glabella has an extensive frontal sinus measuring 31.5 mm laterally and 21 mm posteriorly. This sinus extends superiorly to just beneath the arcus superciliaris and across the sagittal midline slightly into the left supraorbital region. To the left of glabella the

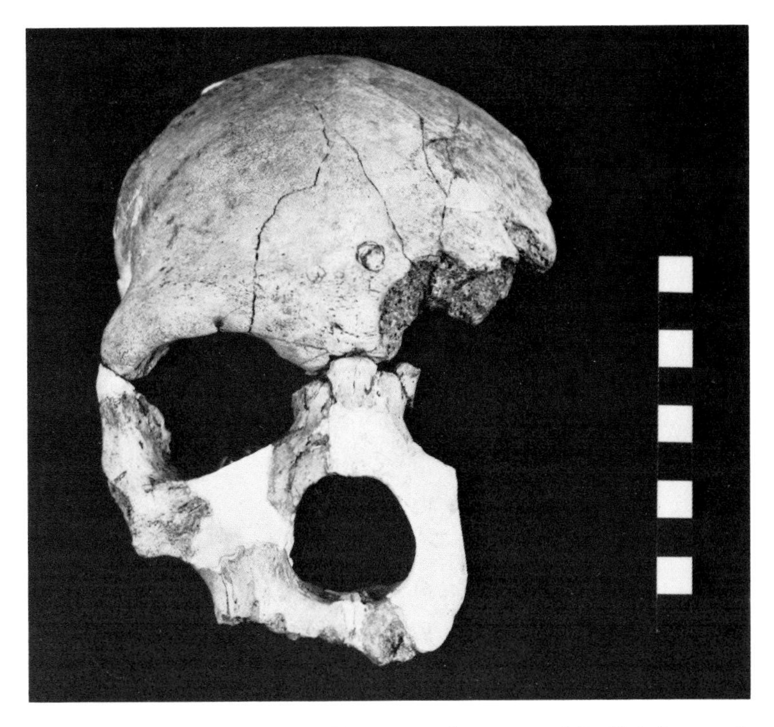

Fig. 1. Florisbad cranium, facial view. Reconstructed areas are white. Note large and small punctate depressions above glabella. Scale bar is in cm.

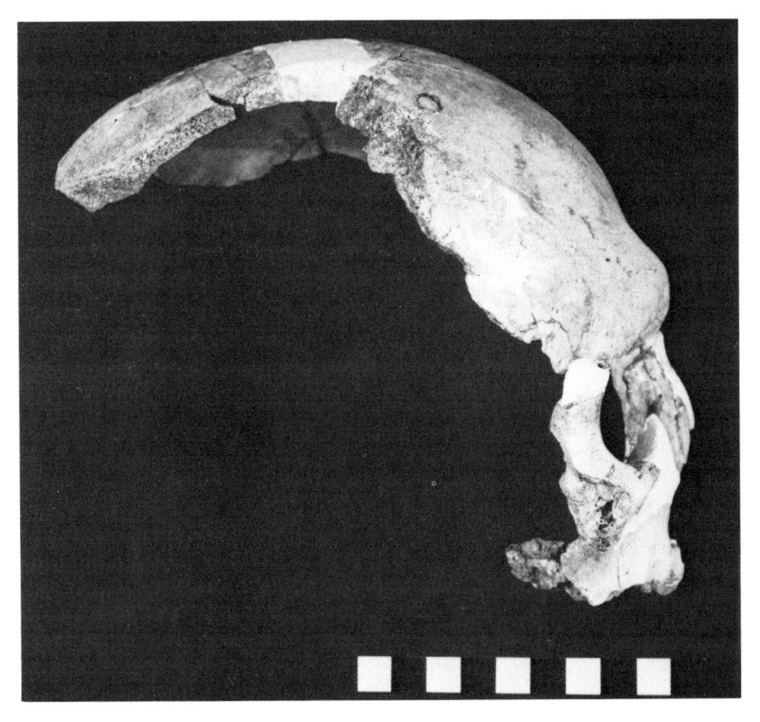

Fig. 2. Florisbad cranium, right lateral view. Note circular punctate depression at back of frontal. Scale bar is in cm.

roof of only a small sinus is preserved, and above it is solid bone.

Taphonomy and Other Damage

The Florisbad calotte, in common with many other bones from the same deposit, shows clear signs of having been gnawed by a hyaena. The frontal has three punctate impressions made by carnivore teeth, and the left side of the calotte displays several graze marks made by carnivore gnawing that lead laterally toward the chewed left lateral margin of the calotte. The supraorbital margin on the left has been chewed away. Twenty millimetres posterior to glabella and just left of the sagittal midline is a punctate circular depression made by a hyaena tooth. It is about 9 mm in diameter. In this depression the outer table has been pushed into the diploë such that it is hinged on its anterolateral margin and sunken to a depth of 4 mm posteromedially. A long graze mark runs into this deep margin from a position 34 mm posteromedially. Also just behind glabella but to the right of the sagittal midline is a 5-mm-wide slight depression of the outer table made by a carnivore tooth. The third carnivore tooth depression is 8 mm in diameter, situated 9 mm anterior to the coronal suture and about 17 mm medial to the right temporal line. Sampson (1974, p. 181) discussed the two larger depressions but concluded that they were not made by carnivore canines as they are too large and far apart. He was assuming that if they were carnivore marks they should have been made by opposite canines in one bite and therefore would have been close together. He did not refer to the very distinct marks of chewing along the left side of the calotte, and it is clear that the two large punctate depressions were made by separate bites during chewing of each side of the calotte after the death of the hominid.

Posterior to the coronal suture on the right lateral margin of the preserved portion of parietal, a large (27 mm^2) portion of parietal has been pointlessly cut away and destroyed for purposes of amino acid racemization dating of the cranium by Protsch (1975) and Bada. I say "pointlessly" because a letter written by Jeffrey Bada on December 2, 1974 to the National Museum director stated that a large portion of parietal was destroyed before the potential results were seen to be minimal. Incidentally, Rightmire (1978) was mistaken in his statement that the upper portions of both nasal bones had been sent to the British Museum for dating and substituted with casts. I found that both upper nasal portions are still complete and original.

GEOLOGICAL SETTING, DATING, AND ARCHAEOLOGY OF THE FLORISBAD SITE

The deposits consist of 6 to 8 metres of sands interleaved with four organic layers referred to as "peats." It was from the level of the lowest organic layer, Peat I, that Dreyer obtained the cranium along with quantities of faunal remains in a similar state of preservation (see Dreyer, 1938). My recent excavations uncovered a Middle Stone Age living floor adjacent to the springs, between Peats II and III. It was considerably higher and geologically more recent than Peat I. A date obtained by J.C. Vogel on charcoal from this living floor revealed that it was greater than 43,700 years, i.e., beyond the limits of C-14 dating (Pta-3465). Additional dates on an organic clay between this sand layer and Peat II were also in the infinite range: greater than 47,200 years (insoluble fraction) and greater than 44,600 years (soluble fraction). We are currently having the peats dated by Uranium Series method, and a preliminary analysis of Peat I has indicated a date well in excess of 100,000 years.

In an unpublished section of the spring deposits, Dreyer indicated the Florisbad skull in what he called a debris cone of the Western Eye at the level of Peat I. Our current excavations have shown that this particular eye erupted through the overlying levels into Peat III. It is therefore not impossible that the skull was originally in one of these higher levels and through spring disturbance was redeposited. However, in our extensive excavations of the undisturbed higher levels, we found not one bone or tooth that was of similar preservation to those recovered by Dreyer from the spring eyes. In addition, the faunal remains from Dreyer's excavation tend to be of large size and show obvious indications of having weathered on the surface and of being gnawed by carnivores. These contrast markedly with the bones that we excavated from the Middle Stone Age occupation level above Peat II. Here the bones tend to be small, fragmented, and show obvious signs of having been accumulated and broken up by man. We did not find at this level—even in the vicinity of the spring eyes—any bones taphonomically similar to those recovered by Dreyer. This suggests that the spring eye bones were deposited under conditions different to those from the undisturbed Middle Stone Age occupation of higher levels, and we feel there is little reason to doubt that Dreyer's bones and the hominid did originate from the general level of Peat I.

The Middle Stone Age occupation floor has now become one of the most significant aspects of the Florisbad site. It represents a butchery site, with the remains of broken and smashed animal bones, principally of bovids but with hippo also common. Also present are areas of tool chipping debris and a hearth. Kathleen Kuman, who is now analysing this occurrence as well as the old collections of Florisbad artifacts, has found at least 15 sets of artifacts that can be re-fit together and represent the level encompassed by a short-term occupation. In the coming year, these finds will be published in detail, along with an evaluation of the earlier, poorly documented work at Florisbad. The scientists involved with me in this multidisciplinary project include: K. Kuman, (University of Pennsylvania); Professor E. van Zinderen Bakker (Institute for Environmental Sciences, Bloemfontein), who is analysing the pollens obtained from 8 m of deposits; and Dr. Moshe Inbar (University of Haifa), now responsible for sedimentology. In addition, a new dating programme has been initiated with the cooperation of Drs. J. Vogel, J. Kronfeld, J. Bischoff, J. Gowlett, and Prof. Wölfli.

DISCUSSION

Morphologically, the Florisbad cranium appears similar to but somewhat more advanced than the Broken Hill cranium. It is extremely similar to the Ngaloba cranium from Tanzania, dated to 120,000 ± 30,000 years (Day et al., 1980), that was also associated with Middle Stone Age artifacts. It has similarities also with the Omo II calvaria (Day, 1969), dated to about 130,000 years (Butzer et al., 1969). A fragment of hominid frontal bone excavated from Layer 16 of the Middle Stone Age II level at Klasies River Mouth appears more modern in its morphology than that of Florisbad (Singer and Wymer, 1982). According to Shackleton (1982), the deposits in which the 16425 frontal was found can be dated to either 80,000 or 100,000 years ago. In the same publication, however, Butzer suggests that the Middle Stone Age II Layer 16 is ca. 115,000 years BP. The cranium from Border Cave, which has been claimed as dating to 90–110,000 years ago (Beaumont, 1980), is completely modern in morphology and bears no comparison with the Florisbad cranium, which has more archaic morphology.

The results of our research on the Florisbad cranium and site are currently being prepared for publication as a monograph. Meanwhile, it can be said that all indications suggest that the Florisbad cranium seems to belong to an archaic form of *Homo sapiens* that lived in Africa between 100,000 and 200,000 years ago and that Ngaloba and Omo II are other representatives of this form of man. They possibly represent the next stage of human development after the *Homo erectus/sapiens* transition seen in the hominids from Broken Hill and Saldanha.

LITERATURE CITED

Beaumont, PB (1980) On the age of Border Cave hominids 1–5. Palaeontol. Africana *23:*21–33.

Broom, R (1913) Man contemporaneous with extinct animals in South Africa. Ann. So. Afr. Mus. *12:*13–16.

Butzer, KW, Brown, FH, and Thurber, DL (1969) Horizontal sediments of the lower Omo Valley: The Kibish Formation. Quaternaria *11:*15–29.

Day, MH (1969) Omo human skeletal remains. Nature *222:*1135–1138.

Day, MH, Leakey, MD, and Magori, C (1980) A new hominid fossil skull (L.H. 18) from the Ngaloba Beds, Laetoli, northern Tanzania. Nature *284:*55–56.

Drennan, MR (1935) The Florisbad skull. S. Afr. J. Sci. *32:*601–602.

Dreyer, TF (1935) A human skull from Florisbad, Orange Free State, with a note on the endocranial cast, by C.U. Ariëns Kappers. Koninklijke Akademie van Wetenschappen te Amsterdam, Proceedings *35*(1):3–12.

Dreyer, TF (1938) The archaeology of the Florisbad deposits. Argeologiese Navorsing van die Nasionale Museum, Bloemfontein *1:*65–77.

Galloway, A (1937) The nature and status of the Florisbad skull as revealed by its non-metrical features. Am. J Phys. Anthropol. *23:*1–16.

Protsch, RRR (1975) The absolute dating of Upper Pleistocene subsaharan fossil hominids and their place in human evolution. J. Hum. Evol. *4:*297–322.

Rightmire, GP (1978) Florisbad and human population succession in southern Africa. Am. J. Phys. Anthropol. *48:*475–486.

Sampson, CG (1974) The Stone Age Archaeology of Southern Africa. New York: Academic Press.

Shackleton, NJ (1982) Stratigraphy and chronology of the KRM deposits: Oxygen isotope evidence. In R Singer and J Wymer (eds): The Middle Stone Age at Klasies River Mouth in South Africa. Chicago: University of Chicago Press.

Singer, R, and Wymer, J (eds) (1982) The Middle Stone Age at Klasies River Mouth in South Africa. Chicago: University of Chicago Press.

Ancestors: The Hard Evidence, pages 306–309

The Origin of the Neandertals

Bernard Vandermeersch
Laboratoire d'Anthropologie (U.A. 376 du C.N.R.S.), Université de Bordeaux I, 33405 Talence, France

ABSTRACT The inclusion of the Neandertals within *H. sapiens* reduced the question of their origins to one of subspeciation, or local geographic isolation, rather than independent evolution from a different species. New fossil discoveries and the application of phylogenetic systematic methods have permitted the clarification of many points about the early differentiation of this lineage. Most important was the separation of distinctive derived traits from those (such as "large brow ridges") that merely *described* not only Neandertals but most archaic hominids. Diagnostic Neandertal traits include features of the vault and face, especially of the occipital, temporal, and sub-orbital regions. The presence of at least one such derived character in a fossil that pre-dates the Würmian "classic" Neandertals qualifies that fossil to be included in the lineage. On that basis, no African or Asian fossils (other than some from the early Würm in the Near East and central Asia) can be considered as directly linked to the Neandertal group. On the other hand, all European Riss-Würm fossils (and, to a lesser degree, those of the Riss, such as Biache) present a typically Neandertal morphology; they may be termed preneandertals. Of the still earlier fossils, Swanscombe and Steinheim are clearly on the lineage, as are probably Arago and Petralona, which present a Neandertal maxillary morphology. On the other hand, Bilzingsleben appears from occipital structure to be closest to the Asian *Homo erectus*, while Vértesszöllös might represent the "archaic *Homo sapiens*" morphotype expected at the base of the Neandertal lineage. Study of fossils along this lineage reveals a mosaic pattern of development of Neandertal synapomorphies but clearly documents the differentiation of the clade well before the end of the Middle Pleistocene.

INTRODUCTION

The notion of Neandertal has undergone numerous variations since the discovery of the first fossil at Neandertal in 1856. This fossil was first considered by some anatomists (Virchow and Meyer, for example) to be a recent and pathological type. It has been recognized as a representative of an extinct hominid variety only since about 1865, thanks to Huxley (1863). In 1864, King established a particular species, *Homo neanderthalensis.*.

At the beginning of this century, discoveries multiplied in Europe. Since 1920 in the Near East (Galilee), in East Africa (Broken Hill), and in Asia (Solo), fossils with some resemblances to the European Neandertals were brought to light, reinforcing the idea that the general characteristics of this man existed in a way all over the ancient world. The term "Paleanthropian" was coined in 1918 by Elliot Smith and permitted the regrouping of all the fossils. This is how the notion of evolutionary phase was born, as developed by Hrdlička (1927). From this perspective, humanity as a whole went through a succession of phases during its evolution: Archanthropian, Paleanthropian, and Neanthropian. Each phase was characterized by a general architecture upon which could be superimposed secondary morphological traits, differing regionally. The Neandertals of Europe became only a variety (or population) inside a vast, basically homogeneous group. In this context the problem of the origin of the group was simple: the "Paleanthropians" derived from the "Archanthropians," most of which are classified today as *Homo erectus*.

The discovery of fossil man in Palestine disturbed this evolutionary scheme. Whether interpreted as hybrids or as humans of modern type, their probable age implied the contemporaneity of two lines: the Neandertals and another leading to modern humans of the Upper Paleolithic. The possibility of interbreeding imposed a revision of the taxonomic status of the Neandertals, which were reclassified as the subspecies *Homo sapiens neanderthalensis*; in turn, this had consequences for phylogeny. In fact, subspeciation is the result of the (generally geographical) isolation of populations within a species. The diverse subspecies of one species cannot be directly descended from another species. Consequently, the Neandertal lineage must have originated from a population of archaic *Homo sapiens* not from *Homo erectus*.

Since 1950, the efforts that have been made in the analysis of fossils and in bringing precision to the interpretation of characters reestablished order. Little by little, thanks to the diffusion of phylogenetic (cladistic) systematics, the distinction between derived characters of the Neandertals and those plesiomorphic ones shared either with *Homo erectus* or with other *Homo sapiens* became more precise. In addition, new discoveries—e.g., Salzgitter-Lebenstedt, La Chaise, Biache-Saint-Vaast, Tautavel—completed the chronological series of European fossils that preceded the classical Neandertals of the early Würm. Consequently, it is easier to understand the origin and evolution of the Neandertal lineage.

DIAGNOSTIC NEANDERTAL MORPHOLOGY

It is evident that the recognition of the derived and unique characters of any lineage is a necessary first step in the reconstitution of that lineage. In the Neandertal case there was long a confusion between discriptive characteristics and distinctive or (diagnostic) characteristics. Boule and Vallois (1952) wrote: "The great development of the brow ridges characterizes all the known skulls of *Homo neanderthalensis*"[1]. This is not accurate because the *torus supraorbitalis* also characterizes all the known skulls of *Homo erectus*. This torus is not considered as a Neandertal characteristic, but only as an archaic feature of the genus *Homo*, which can be found in the Neandertal population. There are no Neandertals without a torus, but the supraorbital torus does not make a Neandertal.

What are then the distinctive characteristics that until now were found only among typical Neandertals of the Würm? Without going into detail, we can cite the following (see Plates 8–13):

1) A high cranial capacity in a long, wide, and low skull of oval transverse contour, with a large face, both long and wide.

2) A diagnostic occipital architecture, particularly the *torus occipitalis*, which is always subdivided in the sagittal region into two lips limiting a supra-iniac fossa with a transverse elongation. The characteristics of this region were well described by Hublin (1978).

3) A distinctive morphology of the temporal region: in particular, the external auditory meatus lies at the level of the zygomatic process; the petro-tympanic crest (*crista petrosa*) begins at the most inferior point of the tympanic and has no contact with the mastoid process; the styloid is far from the edge of the auditory meatus; the digastric groove (*incisura mastoidea*) is closed anteriorly (see Vandermeersch, 1981).

4) The morphology of the sub-orbital region comprises an extended maxilla and a flat malar, oblique backwards and outwards. The cheek bones are "reduced and retreating" (Boule, 1911–1913); this disposition was particularly well described by Sergi (1947).

In addition to the general architecture of the skull, the occipital, temporal, and maxillary regions are certainly the ones possessing the most pronounced diagnostic characteristics. These appear also on the post-cranial bones, particularly on the shoulder blade, the arm and forearm, the ilium,

[1]In this edition the Neandertals were still considered as a separate species.

the thigh bone, and the ankle bone. But these bones are rarely found in pre-Würm periods, and they will not be taken into further consideration. Nonetheless, we can consider that when we find one or more diagnostic characteristics on a fossil prior to the Würm, this fossil belongs to the Neandertal lineage.

Before examining the pre-Würm fossils, it is necessary to establish a precise geographical distribution of the typical forms from the Würm. Recent work has confirmed the presence of Neandertals in the Middle East (Plates 12–15; Suzuki and Takai, 1970; Vandermeersch, 1981; Trinkaus, 1983). The Neandertal population extended east to Uzhbekistan, at Teshik-Tash. On the other hand, African and East Asian fossils belonged to another population, certainly more ancient. This is the case of the Galilee (Zuttiyeh) skull (Plate 16), which is certainly pre-Würmian (Gisis and Bar Yosef, 1974), and which possesses none of the particular characteristics of this group (Piveteau, 1957; Vandermeersch, 1981). This is also true for the Rhodesian skull, which is even more ancient and generally considered as a transitional form between *Homo erectus* and *Homo sapiens* in Africa (Wolpoff, 1980). In Asia the Solo skulls, which have sometimes been grouped with the Neandertals in the Paleanthropic group, are in fact *Homo erectus* (Santa Luca, 1980); further, the partial vault of Maba in China, despite certain resemblances, also cannot be classified as Neandertal. It is true that the cranial regions most diagnostic of the Neandertals are not present in the Maba skull, but it is more similar to the Dali cranium and probably corresponds to a still archaic *Homo sapiens* variety (Wolpoff, 1980). Consequently, there are no Neandertals in Asia and Africa, and there also are no older fossils (even in the Near East) that can be considered preneandertal.

PRE-WÜRM EUROPEAN FOSSILS AS POSSIBLE NEANDERTAL ANCESTORS

Only Europe remains. Let us review the discoveries there of fossils from before the Würm glaciation, which can be regarded as the period of the classic Neandertals.

All the European fossils of the Riss-Würm interglacial show a morphology very close to the classic Neandertal pattern; they are all "Preneandertals": La Chaise (Bourgeois-Delaunay), Saccopastore, Salzgitter-Lebenstedt, Gibraltar, Ehringsdorf, etc. The Riss fossils are also of this type, but with less pronounced Neandertal characteristics, as is the case with La Chaise (Suard) and Biache-Sainte-Vaast (Plate 7; Vandermeersch, 1978).

During the Riss glaciation and the following interglacial period, Europe was populated by a form very close to the classical Neandertals, and there were no other hominids in this region. Even the partial vault from Fontéchevade posseses no characteristics that would allow it to be included in another lineage (Vandermeersch et al., 1976).

The problem is with the most ancient fossils. But most of them are unfortunately badly dated (compare also Stringer, 1985).

Swanscombe: The rear part of the skull found there could be of Mindel-Riss age (Oakley, 1957). It has a *torus occipitalis* and a juxtamastoid crest of Neandertal type.

Steinheim: The skull dates from the Mindel-Riss (Adam, 1954). The *torus occipitalis* extends transversely and is somewhat extended superiorly. Just as for Neandertals, the maximum protrusion of the torus is located on both parts of a slight depression corresponding to the supra-iniac fossa (Plates 6 and 7).

Vértesszöllös: This occipital comes from a site dated from the Mindel (Kretzoi and Vertes, 1965). Thoma (1966) has described this piece as having a *torus occipitalis* highly extended superiorly. This torus is different from those of Würm Neandertals but not far from that of Steinheim.

Petralona: We cannot discuss the age of this fossil. It is probably from the Middle Pleistocene, and a pre-Riss date seems likely, but it is impossible to be sure. The *torus occipitalis* is pronounced and extended. The face, especially the maxillary region and the morphology of the cheek bones, resembles the Neandertals.

Arago: This is another Middle Pleistocene fossil, which could be of Mindel age. Without going into details, attention should be paid to the Neandertal characteristics of the maxillary region. "The morphology of this region resembles the classic Neandertal face" (Spitery, 1982) (Plate 6).

Bilzingsleben: The cranial fragments from this site date to the Mindel-Riss (Vlcek and Mania, 1977). The occipital fragment has a torus that is very pronounced but hardly extended vertically. There is no supra-iniac fossa, and the point of maximum protrusion lies along the sagittal plane. This morphology is invariably characteristic of *Homo erectus*.

We thus have for the pre-Riss period a series of fossils belonging either to the Mindel or to the Mindel-Riss. They originate in Britain, France, Germany, Hungary, and Greece. The most common cranial region is the occipital, which is preserved in all samples except Arago; consequently, it draws most of our attention. Hublin (1978) discussed the transformation of a large and superiorly extended *torus occipitalis*, like that from Vértesszöllös, toward a typical Neandertal morphology via a vertical contraction, together with a progressive accentuation of the supra-iniac fossa. This morphological trend includes all the occipital fragments except Bilzingsleben. The morphology that would characterize the start of this evolution, and which is possibly represented by Vértesszöllös, does not yet show Neandertal characteristics. But these are already present in Steinheim. The latter is therefore already a pre-Neandertal type, whereas Vértesszöllös belongs to the ancient *Homo sapiens* population from which the Neandertals are derived. When we examine the face, the Arago and Petralona specimens already show Neandertal characteristics in the maxillary area, which is not so for Steinheim.

CONCLUSIONS

Europe shows a highly diversified morphology before the Riss period: Bilzingsleben, at least according to the available fragments, seems rather more connected with *Homo erectus* as known from Asia. Vértesszöllös could represent an archaic *Homo sapiens* population, but the other fossils all show Neandertal synapomorphies. These characters are not as developed as in the Riss-Würm or Würm hominids, but they are already clearly identifiable. They vary in their development from sample to sample. Furthermore, the fossils show a mosaic of different Neandertal characteristics: the maxillary area of Steinheim is not extended, whereas that of Arago is.

Consequently, there was a considerable degree of polymorphism at the beginning of the Neandertal lineage. This diversity lessens over time, and the general morphology became progressively more similar to the classic Neandertal type. However, these early fossils have to be linked to the Neandertal lineage and show that its characteristics were clearly established before the end of the Middle Pleistocene.

LITERATURE CITED

Adam, KD (1954) Die mittelpleistozänen Faunen von Steinheim an der Murr (Württemberg). *Quaternaria* *1*:131–144.

Boule, M (1911–1913) L'homme fossile de La Chapelle-aux-Saints. *Ann Paléontol.* *6*:111–172; 7:3–192; *8*:1–67.

Boule, M, and Vallois, HV (1952) Les hommes fossiles. Paris: Masson.

de Bonis, L, and Melentis, J (1982) L'Homme de Pétralona, comparaison avec l'Homme de Tautavel. 1er Congr. Internat. Paléont. Hum. Prétirage. Nice: CNRS, pp. 847–874.

Gisis, I, and Bar Yosef, O (1974) New excavations in Zuttiyeh cave, Wadi Amud, Israel. *Paléorient* *2*:175–180.

Hrdlička, A (1927) The Neanderthal phase of Man. J. Roy. Anthropol. Inst. 57:249–269.

Hublin, JJ (1978) Le torus occipital transverse et les structures associées. Evolution dans le genre *Homo*. Thèse 3me cycle, Univ. Paris VI.

Huxley, TH (1863) Man's Place in Nature.

King, W (1864) The reputed fossil man of the Neanderthal. Quart. J. Sci. (London) *1*:88–97.

Kretzoi, M, and Vertes, L (1965) Upper Biharian (Intermindel) Pebble-industry occupation site in Western Hungary. Curr. Anthropol. *6*:74–87.

Oakley, KP (1957) Stratigraphical age of the Swanscombe skull. Am. J. Phys. Anthropol. *15*:253–260.

Piveteau, J (1957) Traité de Paléontologie. VII. Primates, Paléontologie Humaine. Paris:Masson.

Santa Luca, AP (1980) The Ngandong Fossil Hominids. Yale Univ. Publ. Anthropol. *78*:1–175.

Sergi, S (1947) Sulla morfologia della "facies anterior corporis maxillae" nei paleantropi di Saccopastore e del Monte Circeo. Riv. Antropol. *35*:401–408.

Smith, G.E. (1918) Essays on the Evolution of Man. London: Oxford University Press.

Spitery, J (1982) La face de l'Homme de Tautavel. 1er Congr. Internat. Paléont. Hum. (Prétirage). Nice: C.N.R.S., pp. 847–874.

Stringer, CB (1985) Middle Pleistocene hominid variability and the origin of Late Pleistocene humans. In E Delson (ed): Ancestors: The Hard Evidence. New York: Alan R. Liss, Inc., pp. 289–295.

Suzuki, H, and Takai, F (1970) The Amud Man and his Cave Site. Tokyo: Univ. of Tokyo Press.

Thoma, A (1966) L'occipital de l'homme mindélien de Vértesszöllös. L'Anthropol. *70*:495–534.

Trinkaus, E (1983) The Shanidar Neandertals. New York:Academic Press.

Vandermeersch, B, Tillier, AM, and Krukoff, S (1976) Position chronologique des restes humains de Fontéchevade. IX Cong. Un. Int. Sci. Pre. Protohist. (Prétirage) Coll. IX. Le peuplement anténeandertaliens de l'Europe. Nice: C.N.R.S., pp. 19–26.

Vandermeersch, B (1978) Le crâne pré-wurmien de Biache-Saint-Vaast (Pas de Calais). In Les origines humaines et les époques de l'intelligence. Paris:Masson (Fondation Singer-Polignac), pp. 153–159.

Vandermeersch, B (1981) Les hommes fossiles de Qafzeh (Israël). Paris:C.N.R.S.

Vlček, E, and Mania, D (1977) Ein neue Fund von *Homo erectus* in Europa: Bilzingsleben (D.D.R.). Anthropologie (Brno) *15*:154–169.

Wolpoff, M (1980) Paleoanthropology. New York: Knopf.

Ancestors: The Hard Evidence, pages 310–318

Neanderthals and Their Contemporaries

Jakov Radovčić
Geološko-Paleontološki Muzej, Zagreb, Yugoslavia

ABSTRACT Early in the history of paleoanthropology, the idea became entrenched that there were anatomically modern humans (or their clear ancestors) living contemporaneously with Neanderthals in Europe or nearby. This idea was given prominence with regard to the finds from Krapina, discovered at the turn of the century. Gorjanović-Kramberger, the excavator and describer of these fossils, had a clear conception of the process of evolution and argued strongly that Neanderthals, at Krapina and elsewhere, were an ancient variety of humans, rather than a "primitive" side-branch. Some more recent workers have suggested that certain Krapina specimens—particularly the A cranium—were evidence for the contemporaneity of moderns with Neanderthals at this site. In fact, the A cranium is from a higher stratum than most of the other remains, but it does show some more "modern" morphological features. This could be due to its juvenile nature or perhaps to its being a "transitional" specimen between the Neanderthals and later Europeans, as our new reconstruction suggests. In either case, it may not be fully contemporary with the bulk of the Krapina Neanderthals. Later central European fossils also show something of a mosaic of Neanderthal and modern features, and they predate both the last Neanderthals and the first occurrence of moderns farther west. It may be that anatomically modern Europeans originated in central Europe from a Neanderthal stock, while true contemporaries of the Neanderthals in other regions gave rise to different modern populations.

The characteristics and taxonomy of the Neanderthal man have been written about most extensively, but often with but little originality. New finds belonging to his family have become more numerous than legitimate new thoughts. Today it is no more the question of a single or a couple of Neanderthal skulls, as in the time of Darwin, but of a large and important section of man's antiquity, documented ever more geologically, paleontologically, and anthropologically. But the distressing part is that the more there is, the less prehistory seems to know what to do with it. Of speculations there have been indeed enough, but most of them so far have led not into the sunlight but rather into a dark, blind alley from which there appears to be no exit.

Hrdlička, 1930

The Neanderthals are known from a relatively good and well-dated fossil record, and they may be the closest to us in time of all "archaic" hominids. If we cannot solve the "Neanderthal problem" and thereby arrive at an understanding of the relationship of Neanderthals to "modern" *Homo sapiens* (Trinkaus and Howells, 1979; Stringer and Burleigh, 1981), there would seem little hope of resolving any of the more complex issues concerning hominid evolution. The "Neanderthal problem" is also an important one for general evolutionary studies since its interpretation has a bearing on the wider issue of the rate and mode of hominid (and non-hominid) evolution (Gould and Eldredge, 1977; Cronin et al., 1981).

Stringer, 1982

INTRODUCTION

Ziman (1968) has argued that progress in a given field of science is marked by a change in consensus within the scientific community. In other words, a majority of what Ziman calls the "invisible college" is converted from an old to a new set of beliefs about a particular phenomenon. If this is a valid description of progress, one can question whether this century has seen progress toward understanding the Neanderthal phenomenon.

A recent popular account of the Neanderthal question could be seen in the *Newsweek* article (April 23, 1984) concerning the scientific gathering and symposium that marked the opening of a display of fossils in the American Museum of Natural History. That exhibit was envisaged to demonstrate the hard (fossil) evidence for the human evolutionary journey from the remote past. The *Newsweek* article focuses in part on the Neanderthals and expresses a common belief, often repeated throughout the history of paleoanthropology, that Neanderthals and early modern humans were contemporaries. This view, accepted by many scientists for more than a hundred years, effectively implies that modern humans must have come from outside Europe and replaced the Neanderthals "perhaps by bludgeoning them out of existence."

Conversely, the same article stressed that some scientists have believed that Neanderthal genes still flow through modern humans." Thus, the Neanderthals' "fate" remains one of the many puzzles confronting this field of science (see Trinkaus and Smith, 1985). It has been suggested that ideas regarding the phyletic status of Neanderthals follow a geographic pattern. For example, Trinkaus and Howells (1979) assert that "there are [those] in the U.S.S.R. and elsewhere in eastern Europe, and many in the U.S." who favor "a restatement of the old 'Neanderthal phase' hypothesis" and others, mainly from the West, who "ascribe the disappearance of the Neanderthals not to local evolution but to invasion by new peoples of modern form." The fact is that both of these theoretically and geographically divergent views can be traced to the historic development of paleoanthropology within *central* Europe.

The idea of a Neanderthal phase in human evolution is usually ascribed to Aleš Hrdlička, who presented this hypothesis in several papers (Hrdlička, 1914, 1927, 1930). Although he deserves credit for the most complete formulation of the unilinear idea concerning the place of the Neanderthals in human ancestry (Spencer and Smith, 1981; Spencer, 1984), the original concept could be traced back much earlier. The concept originated at the very beginning of this century (Brace, 1964, 1981) and is found in the papers of Gorjanović-Kramberger (1904a) and Schwalbe (1906). Both thought that human evolution had essentially proceeded in a stepwise fashion through distinct morphological stages of development, with the Neanderthals being the ancestral stage to modern humans. The appraisal suggested by Gorjanović-Kramberger and Schwalbe was opposed mainly by Boule (1913), who concluded that Neanderthals were not antecedents of Europeans, that more modern hominids lived contemporaneously with Neanderthals in Europe, and that this "*sapiens*" lineage could be traced, independently of the Neanderthals, to an earlier period. For various reasons, the "side-branch" interpretation of the Neanderthals and the claims of more modern-appearing contemporaries can also be traced to earlier central European workers, such as Klaatsch (1923; Klaatsch and Hauser, 1910) and Adloff (1907a,b,c, 1908a,b, 1910), among others.

However, by the 1920s, Hrdlička was virtually alone, standing against the roster of scientists opposed to his interpretation of the Neanderthals (Spencer and Smith, 1981; Spencer, 1984). He presented the Huxley Lecture of 1927, refuting the claims for Neanderthal contemporaries thus:

> .... If he lived in Europe, coexisting with the Neanderthaler, where are his remains, and why did he not prevail sooner over his inferior cousin? His traces, it will be recalled, never, in Europe or elsewhere, precede or coexist with, but always follow the Mousterian (Hrdlička, 1927).

Contrary to Hrdlička's expectations, the Huxley Lecture was not successful in turning the tide of general scientific opinion. The next several decades are represented by efforts to change the reconstruction of European prehistory by attempting to locate the "cousin," and according to Vallois's Huxley Lecture of 1954:

> .... One fact at all events seems now to be established: the European *Homo sapiens* is not derived from the Neanderthal men who preceded him. His stock was for long dis-

tinct from and had evolved, under the name *Praesapiens*, in a parallel to theirs. Long-debated, the Presapiens forms are thus not a myth. They did exist. The few remains of them we possess are the tangible evidence of the great antiquity of the phylum that culminates in modern man (Vallois, 1954).

THE KRAPINA REMAINS AND THE SEARCH FOR NEANDERTHAL CONTEMPORARIES

The recovery of the remains of a fossil human population in the rock shelter in the town of Krapina, between 1899 and 1905, provided the first picture of a pre-Upper Paleolithic group numbering more than a dozen individuals (and, according to some recent estimates, as many as 80 individuals (Trinkaus, 1977; Wolpoff, 1978, 1979). In order to demonstrate the antiquity of this sample, Gorjanović was the first scholar to apply the fluorine method of relative dating to show that the hominids and the "Diluvial" faunal remains at Krapina were contemporary. This was some fifty years before the same technique was used in England to discredit the antiquity of Piltdown. Although the material is exceedingly fragmentary, original studies of the Krapina Neanderthals by Gorjanović-Kramberger (1899–1918) and more recently by many scholars (Brace, 1962, 1979; Smith, 1976b, 1982, 1984; Trinkaus, 1977, 1978; Wolpoff, 1979, 1980a; and others) attempt to view this material as a single, normally varying biological population. Some workers both in the past and today are willing to accept this Neanderthal sample as one that approaches more recent forms of man in some of its features and shows resemblance to other Neanderthal fossils in others. When Gorjanović-Kramberger discovered the Krapina fossils in 1899 he was probably the first scholar who recognized that despite their "crude and brutish" morphology Neanderthals were best considered as representations of a fossil variety of *Homo sapiens* (Gorjanović-Kramberger, unpublished notebooks; Malez, 1970).

For some thirty years after his first discovery at Krapina, Gorjanović-Kramberger tried to perceive the Neanderthals as direct ancestors of modern man. This attitude was contrary to the general climate of opinion of his time. Many European anthropologists reasoned with an idealized Aristotelian morphotypic approach. From their primarily medical backgrounds, they viewed the mixture of supposedly *sapiens* features with supposedly Neanderthal features as the result of a battle fought between an Aurignacian and a Mousterian group for the possession of the Krapina rock shelter (Brace, 1964; Brace and Montagu, 1977). It is interesting to point out that in 1918 Gorjanović-Kramberger was already well aware of the predominant scientific climate of paleoanthropology to that time, noting:

> There have been distinguished scholars, such as Cuvier for example, who considered the Diluvium as a period of general catastrophe and therefore believed that no creature could survive this event. Thus, according to his opinion, no human could survive either. Regarding man, Cuvier said accordingly: "L'homme fossile n'existe pas." We mention this here because the concepts of such distinguished scientists could obscure and cover, to the detriment of science, real knowledge. This has happened, frequently, in the matter of human existence throughout the Diluvium (1918, p. 162; all translations by Radovčić).

In such a scientific climate it is not surprising that the bones of the first Neanderthal, and thus of his eventual contemporaries, were ascribed to distant Cossacks who were wandering through Europe at the beginning of the last century. Immediately after this interpretation, a more serious case for a Neanderthal contemporary was made when the clearly Pleistocene Cro-Magnon remains were found. Thus, the 1892 Anthropological Congress in Ulm, Germany reached a consensus that there were no fossil humans in Europe (Neanderthals were not considered human), and therefore the German anthropologist Ranke (1896) could have used Kollman's words, "there are no European fossil races" (Gorjanović-Kramberger, 1918).

Gorjanović-Kramberger was absolutely certain of the importance of his numerous Neanderthal skeletal remains from Krapina. For example, in 1918 he noted, the Krapina fossils should "satisfy some experts, namely Virchow, that there was no case for the claim that fossil remains were pathological people." In the same paper he quotes Klaatsch: "I do not know whether the Neanderthal problem could be solved and recognized without the Krapina cranial fragments." Yet Klaatsch (1902, 1923; Klaatsch and Hauser, 1910), Adloff (1907c, 1910) and others insisted on promoting the Neanderthals as a side-branch of human ancestry, or on identifying a "more advanced type" in the Krapina collection.

Contrasting with these opinions at the very beginning of this century, Gorjanović-Kramberger (1904b) attempted to understand both variation within the Krapina sample and the mechanisms for evolutionary and morphological changes underlying the differences between Neanderthals and modern humans. For instance, consider his views regarding the morphology of the supraorbital torus. In 1904, and later in 1918, he wrote:

> I would point out that the skulls of apes (orangutans, gorillas, etc.) exhibit those thick supraorbital tori as well. This feature cannot be considered as a phylogenetic resemblance between man and apes, but is rather a phenomenon which is caused by head-shape, particularly by the receding forehead and by the influence of some strong masticatory muscles. We see, for example, that the juvenile orang does not exhibit the heavy brow-ridges which are so characteristic of adult individuals. A similar conclusion is also valid for man and this could be determined from the juvenile skull from Krapina.

As a second example, in 1904 he wrote:

> Later on, during the Upper Diluvium when man acquired great intelligence he learned how to produce more advanced tools for his survival and therefore there was no need for those strong muscles. Thus the forehead became steeper and jaws and temporal bones became more gracile.

These are only two examples of Gorjanović-Kramberger's perception of the evolutionary process, demonstrating his reasoning regarding Neanderthal ancestry. It was several years later, in 1913, that Hrdlička developed his suggestion that changes in "masticatory function" were an integral cause of later patterns of human cranial evolution. Therefore, it was really Gorjanović's idea that was subsequently developed by Brace (1962, 1979) and later expressed in various forms by several works (e.g., Brose and Wolpoff, 1971; Frayer, 1978; Smith, 1976b, 1982; Endo, 1966; Russell, 1983; etc.).

Thus, for Gorjanović-Kramberger, morphology was changeable, and for morphological changes he attempted to understand the geochronological framework as an important factor in timing the origin of specific adaptations. In the first decade of this century, there were only a few hominid fossils that he could put in a sequence, on the basis of their morphological traits and the chronology suggested by the associated faunal remains. Gorjanović-Kramberger was well aware that an accurate understanding of the stratigraphic and chronologic relationship between the Neanderthals and later fossils was *sine qua non* of any serious discussion and interpretation of the fossil material. Using the available data, he proposed a morphological continuum ranging from Krapina and Taubach to fossils from Neanderthal, Spy, and Šipka, and finally the Moravian sites such as Brno and Předmostí, followed by some Neolithic crania which were found in the cave of Oborzykow in 1902.

The Krapina Neanderthal sample has traditionally been considered to show evidence for two "human types" represented within the collection, one being Neanderthal and the other being "advanced." For instance, in the crania, interest has generally centered on Krapina A because of its supposed modern characteristics. This was expressed in Škerlj's (1958) suggestion that the A cranium is proof of a more advanced hominid living contemporaneously with the Krapina Neanderthals. Coon (1962) also stressed the cranium's modern appearance. However, as pointed out by Smith (1976b), the fact that the A cranium appears more modern than the adults from Krapina is not especially surprising. Many workers (e.g., Weidenreich, 1943; Vlček, 1964, 1967, 1970; Heim 1974, 1982; Hublin, 1980b) have demonstrated that young Neanderthals simply do not exhibit the fully developed features of adult Neanderthals. Therefore, Smith (1976b) explains the so-called modern characteristics of the Krapina A cranium on the basis of its young ontogenetic age rather than on its taxonomic relationships.

In the years when the stratigraphic record was only vaguely determined and man's antiquity was basically measured in "ante-Diluvium or Diluvium" operative units, Gorjanović-Kramberger was unusual as an experienced paleontologist with good perception of the meaning of geological time and with the understanding that morphology could be changed through time. This is indicated by the fact that he systematically recorded the stratigraphic position of all the Neanderthals and much of the fauna. Thus, prior to the establishment of Penck and Brückner's (1909) Quaternary chronology, in the rather short time sequence present in the Krapina shelter, he identified eight cultural layers and kept stratigraphic records by layer for almost all specimens that are derived from the site (Gorjanović-Kramberger, unpublished notebooks; Malez, 1970; Smith, 1976b; Wolpoff, 1979). Today, this more accurate stratigraphy could once again suggest the presence of a "more advanced" Neanderthal contemporary at Krapina.

We know that the A cranium comes stratigraphically high in the Krapina sequence, while the unquestioned Neanderthal specimens are much older than the A cranium, as most of them come from Gorjanović-Kramberger's "*Homo sapiens* zone" (i.e., the fourth level). Therefore, as Smith (1976b) pointed out, even if Krapina A is accepted as a representative of a more advanced hominid population than the stratigraphically lower Krapina Neanderthals, it does not mean that the two populations were contemporary.

Recently, Wolpoff (1980a) suggested that the Krapina A cranium represents a transitional specimen between Neanderthals and modern Europeans. The newly reconstructed Krapina A cranium (Minugh and Radovčić in preparation) could support Wolpoff's assessment on specific morphological grounds. Thus, the possible presence of post-Neanderthal humans in the Mousterian sequence at Krapina would seem to be better evidence for the *in situ* evolution of early Upper Paleolithic humans from European Neanderthals than for the presence of two contemporaneous groups. Whether the A cranium reflects juvenile Neanderthal morphology (Smith, 1976b) or is a transitional specimen (Wolpoff, 1980a), the presence of that morphology within the Krapina sequence tends to support the views of Brace (1962, 1964), Brose and Wolpoff (1971), Smith (1976b, 1982, 1984; Smith and Ranyard, 1980; Spencer and Smith, 1981), and others that Neanderthals and early Upper Paleolithic humans exhibited continuous morphological change into modern Europeans. These views would support much earlier claims made by Gorjanović and Hrdlička.

Another specimen from the large Krapina collection has been identified as a piece of evidence indicating the presence of advanced hominids, and thus Neanderthal contemporaries at Krapina. In their work on Combe Capelle, Klaatsch and Hauser (1910) established the taxon "*Homo aurignacensis Hauseri*" as an early form of modern man contemporaneous with Neanderthals. They also suggested that mandible K from Krapina was evidence for this type of advanced hominid at Krapina, basically because of its gracility. Gorjanović-Kramberger (1910a,b, 1913) responded rather strongly, showing that this specimen was no different from the other Krapina Neanderthals. He viewed existing morphological differences at the site as normal biological variation within a single species, a consequence of dealing with a large fossil sample.

The examples cited indicate that, instead of the presence of more modern Neanderthal contemporaries at Krapina, as suggested by scholars in the past (Klaatsch and Hauser, 1910; Klaatsch, 1928; Škerlj, 1958; Coon, 1962), recent investigations support the original remarks made by Gorjanović-Kramberger in many of his published reports.

NEANDERTHAL CONTEMPORARIES ELSEWHERE IN EUROPE

It is rather difficult for me to talk in a short paper about the Neanderthals and their contemporaries without a more serious attempt to clearly delineate some of the critical concepts involved. The term "Neanderthal" today seems rather well defined and morphologically outlined, while the term "modern" could be more puzzling. If "modern *Homo sapiens*" includes all existing variants of human populations, which is the scientific consensus, it is clear that "modern" is not restricted to Europeans alone. Since living Europeans have their recent and present contemporaries on the other continents, we can assume that early Europeans with different morphologies had contemporaries elsewhere as well. Ancient Africans, Asians, or Australians would have exhibited different morphological traits than do living inhabitants of those regions. If some of those ancient human populations were evolving into modern ones, why were Europeans the only ones who did not evolve? Mechanisms of human evolution are far from fully understood, but recent advances in paleoanthropology suggest some of the possible ways that morphological changes took place. If we limit our consideration to the geographically known area of Neanderthals, Europe and the Near East, we can see that Neanderthals *were* in the process of evolution.

If "*historia est magistra vitae*," almost eighty years after the first discovery at Krapina we are still facing similar if not identical problems. One hundred twenty years after the Neander valley discovery, and despite intensive efforts, paleoanthropology has not yet discovered Neanderthal contemporaries in Europe. During their long time-span—except at the very end of the Mousterian and Early Upper Paleolithic period, when cultural diversification and acceleration took place—there is no evidence that would clearly indicate coexistence of Neanderthal and non-Neanderthal morphology in Europe, the circum-Mediterranean, and Near East (i.e., the greater Mousterian cultural area).

Any evaluation of the relationships of the Neanderthals to their neighboring contemporaries involves a rather long consideration of their morphological, spatial, and temporal affinities to their predecessors (see Vandermeersch, 1985; Stringer, 1985) as well as to their African and East Asian contemporaries, and to their anatomically modern human successors. The Neanderthal sample is usually defined by a total morphological pattern that is established only in Europe, the Near East, and western Asia from the last interglacial to the middle of the last glacial (Coon, 1962; Howell, 1951; Hublin, 1978, 1982; Hrdlička, 1927; Santa Luca, 1978; Smith, 1976b, 1977; Trinkaus, 1983; Wolpoff, 1980a). As pointed out by many workers, estimates of the exact ages for the oldest and youngest Neanderthal samples vary, depending upon both the accuracy of available dates and the different opinions about the morphological, temporal, and geographical affinities of specimens within Europe and western Asia.

In the last few decades, there has been an increase in the Neanderthal sample from western Europe (de Lumley, 1973; Heim, 1976; Vandermeersch, 1976; Lévêque and Vandermeersch, 1980). Recent studies have also reevaluated many of the previously known specimens (Hublin, 1978, 1980a; Stringer, 1974, 1981, 1982; Stringer et al., 1984; Smith, 1976b; Trinkaus, 1983; Wolpoff, 1980a).

This period also yielded certain earlier specimens that were once considered "presapiens" forms. Changing interpretations of specimens such as Fontéchevade, Swanscombe, and Steinheim now show them to be within the expected range of variation of the Middle or early Late Pleistocene Europeans (Hublin, 1978; Wolpoff, 1980b). Although the problem of relative dating for many of these older specimens still remains (Cook et al., 1982; Stringer and Burleigh, 1981; Vandermeersch et al., 1976), the critical issue is the recognition of a single lineage that goes from the earliest inhabitants of Europe to the latest Neanderthals (Vandermeersch, 1978, 1985). The recent literature on the Neanderthals (e.g., Howells, 1976; Spencer and Smith, 1981; Smith, 1982, 1983a; Stringer, 1974, 1978, 1981, 1982, 1985; Stringer et al., 1979; Trinkaus, 1983; Trinkaus and Howells, 1979; Trinkaus and Smith, 1985; Vandermeersch, 1978, 1981, 1985; Wolpoff, 1980b, 1985; and many others) makes it evident that human paleontologists are far from reaching a consensus on the issues of their origin and relationships.

The most recently discovered Neanderthal remains in France are those from Saint-Césaire. Since they are found in association with tools of the Châtelperronian, the earliest western European Upper Paleolithic industry (Lévêque and Vandermeersch, 1980; Vandermeersch, 1981; Stringer et al., 1984), for many scholars this provides evidence that at least the latest western European Neanderthals had contemporaries in Europe; i.e., modern Europeans came into this region from elsewhere, introducing a new technology and different behaviors. However, western Europe has not yielded anatomically modern humans of this age, which would demonstrate the implausibility of a transition from Neanderthals to modern Europeans. Previously discovered "early" specimens (Combe-Capelle, Cro-Magnon) are too late in time to provide evidence of the *earliest* modern people in western Europe.

The morphological sequence of the Neanderthals and their successors in central Europe is a somewhat different situation (Jelínek, 1976, 1983). The most recent discoveries of late Neanderthals are those from Vindija (northwestern Yugoslavia), which are at least 40,000 years old (Malez et al., 1980; Wolpoff et al., 1981; Malez and Ulrich, 1982). Although these Vindija remains clearly exhibit Neanderthal morphology (by comparison with earlier Neanderthals from Krapina and sites in Czechoslovakia), they show a reduction in facial massiveness and other features more closely approaching the modern European condition than usually seen in Neanderthals. This sample is chronologically followed by the Aurignacian-associated specimens from Brno, Mladeč, Velika Pećina, and several other sites. In many features, these approach earlier Neanderthal specimens (Gorjanović-Kramberger, 1904; Hrdlička, 1914, 1927, 1930; Weidenreich, 1947; Jelínek, 1976, 1983; Smith, 1976a, 1982, 1983b, 1984; Wolpoff, 1980a, 1982). The age of the transition from the Neanderthals to later Europeans in central Europe may be older than the earliest modern specimens farther west (Smith, 1982). Rather than seeing the earlier passage of "invading Aurignacians," central Europe might have been the place of origin of the earliest modern Europeans. In the context of history, it is interesting that there has been no attempt to call these people "Mladečoids."

ACKNOWLEDGMENTS

I thank Drs. Eric Delson and Ian Tattersall for inviting me to participate in this stimulating and rewarding conference. With much gratitude, I acknowledge the constant interest, support, and encouragement of Professors Milford H. Wolpoff (University of Michigan) and Fred H. Smith (University of Tennessee). Ms. Karen Rosenberg read the manuscript and made numerous useful suggestions.

LITERATURE CITED

Adloff, P (1907a) Die Zähne des *Homo primigenius* von Krapina. Anat. Anz. *31*:273–282.

Adloff, P (1907b) Die Zähne des *Homo primigenius* von Krapina und ihre Bedeutung für die systematische Stellung desselben. Zeit. Morphol. Anthropol. *10*:197–202.

Adloff, P (1907c) Einige Besonderheiten des menschlichen Gebisses und ihre stammesgeschichtliche Bedeutung. Zeits. Morphol. Anthropol. *10*:106–121.

Adloff, P (1908a) Das Gebiss des Menschen und der Anthropomorphen Vergleichend-anatomische Untersuchungen. Zugleich ein Beitrag zur menschlichen Stammesgeschichte. Berlin: Julius Springer.

Adloff, P (1908b) Schlussbemerkung zu: "Die Zähne des *Homo primigenius* von Krapina." Anat. Anz. *32*:301–302.

Adloff, P (1910) Neue Studien über das Gebiss der diluvialen und rezenten Menschenrassen. Dtsch. Mtschr. Zhd. *28*:134–159.

Boule, M (1913) L'homme Fossile de La Chapelle-aux-Saints. Paris: Masson.

Brace, CL (1962) Refocusing on the Neanderthal problem. Am. Anthropol. *64*:729–741.

Brace, CL (1964) The fate of the "classic" Neanderthals: A consideration of hominid catastrophism. Curr. Anthropol. *5*:3–43.

Brace, CL (1979) Krapina, "classic" Neanderthals and the evolution of the European face. J. Hum. Evol. *8*:527–550.

Brace, CL (1981) Tales of the phylogenetic woods: The evolution and significance of evolutionary trees. Am. J. Phys. Anthropol. *56*:411–429.

Brace, CL, and Montagu, A (1977) Human Evolution, 2nd Edition. New York: MacMillan.

Brose, DS, and Wolpoff, MH (1971) Early Upper Paleolithic man and late Middle Paleolithic tools. Am. Anthropol. *73*:1156–1194.

Cook, J, Stringer, CB, Currant, AP, Schwarcz, HP, and Wintle, AG (1982) A review of the chronology of the European Middle Pleistocene hominid record. Yrbk. Phys. Anthropol. *25*:19–65.

Coon, CS (1962) The Origin of Races. New York: Knopf.

Cronin, JE, Boaz, NT, Stringer, CB, and Rak, Y (1981) Tempo and mode in hominid evolution. Nature *292*:113–122.

de Lumley, MA (1973) Anténéandertaliens et Néandertaliens du bassin Mediterranéen Occidental Européen. Études Quaternaires (Univ. Provence), Mém. 2.

Endo, B (1966) Experimental studies on the mechanical significance of the form of human facial skeleton. Jap. Fac. Sci. Univ. Tokyo, Sec. V (Anthropol.) *3*:1–106.

Frayer, DW (1978) Evolution of the Dentition in Upper Paleolithic and Mesolithic Europe. Univ. Kansas Publ. Anthropol. *10*:1–201.

Gorjanović-Kramberger, D (1899) Der paläolitische Mensch und sein Zeitgenossen aus dem Diluvium von Krapina in Croatien. Mitt. Anthropol. Gesell. Wien *29*:65–68.

Gorjanović-Kramberger, D (1901) Der Paläolitische Mensch und sein Zeitgenossen aus dem Diluvium von Krapina in Kroatien I. Mitt. Anthropol. Gesell. Wien *31*:164–197.

Gorjanović-Kramberger, D (1902) Der Paläolitische Mensch und sein Zeitgenossen aus dem Diluvium von Krapina in Kroatien II. Mitt Anthropol. Gesell Wien *32*:189–216.

Gorjanović-Kramberger, D (1904a) Der Paläolitische Mensch und sein Zeitgenossen aus dem Diluvium von Krapina in Kroatien III. Mitt. Anthropol. Gesell. Wien *34*:187–199.

Gorjanović-Kramberger, D (1904b) Potijče li moderni čovjek ravnó od dilúvijalonga Homo primigeniusa? I kongr. srp. lekara i prirodnjacka, pp. 1–8.

Gorjanović-Kramberger, D (1910a) *Homo aurignacensis* Hauseri in Krapina? Verh. Geolog. Reichsanst. *14*:312–317.

Gorjanović-Kramberger, D (1910b) Zur Frage der Existenz des *Homo aurignaciensis* in Krapina. Ber. geol. Kommis. Kroat. Slavon. (Zagreb), pp. 5–8.

Gorjanović-Kramberger, D (1913) Život i kultura diluvijalonga čovjeka iz Krapine u Hrvatskoj (Hominis diluvialis e Krapina in Croatia vita et cultura. Djela Jug. Akad. Znan. Umjet. *23*:1–54.

Gorjanović-Kramberger, D (1918) Pračovjek iz Krapine. Priroda *8*:162–165.

Gould, SJ, and Eldredge, N (1977) Punctuated equilibria: The tempo, and mode of evolution reconsidered. Paleobiol. *3*:115–151.

Heim, JL, (1974) Les hommes fossiles de La Ferrassie (Dordogne) et le problème de la définition des Néandertaliens classiques. L'Anthropol. *78*:81–112, 321–378.

Heim, JL (1976) Les Néandertaliens en Périgord. In H de Lumley (ed): La Préhistoire Française Vol. 1. Paris: Centre National de la Recherche Scientifique, pp. 578–583.

Heim, JL (1982) Les Enfants Néandertaliens de La Ferrassie. Paris: Masson.

Howell, FC (1951): The place of Neanderthal man in human evolution. Am. J. Phys. Anthropol. *9*:379–416.

Howells, WW (1976) Explaining modern man: Evolutionists versus migrationists. J. Hum. Evol. *5*:577–596.

Hrdlička, A (1914) The most ancient skeletal remains of man. Ann. Rep. Smithsonian Inst. (1913), pp. 491–552.

Hrdlička, A (1927) The Neanderthal phase of man. J. Roy. Anthropol. Inst. *57*:249–274.

Hrdlička, A (1930) The skeletal remains of early man. Smithsonian Misc. Coll. *83*:1–379.

Hublin, J-J (1978) Quelques caractères apomorphes du crane néandertalien et leur interprétations phylogénique. C. R. Acad. Sci., Paris, D *287*:923–926.

Hublin, J-J (1980a) A propos de restes inédits du gisement de la Quina (Charente): Un trait méconnu des néandertaliens et des anténeandertaliens. L'Anthropol. *84*:81–88.

Hublin, J-J (1980b) La Chaise Suard, Engis 2 et La Quina H. 18: Developpement de la morphologie occipitale externe chez l'enfant prénéandertalien et néandertalien. C. R. Acad. Sci., Paris, D *291*:669–672.

Hublin, J-J (1982) Les anténéandertaliens: Présapiens ou prénéandertaliens? Géobios, Mém. Spec. *3*:345–357.

Jelínek, J (1976) The *Homo sapiens neanderthalensis* and *Homo sapiens sapiens* relationship in central Europe. Anthropol. (Brno) *14*:79–81.

Jelínek, J (1983) The Mladeč finds and their evolutionary importance. Anthropol. (Brno) *21*:57–64.

Klaatsch, H (1902) Occipitalia und Temporalia der Schädel von Spy vergleichen mit denen von Krapina. Zeits. Ethnol. *34*:392–409.

Klaatsch, H (1923) The Evolution and Progress of Mankind. New York: Stokes.

Klaatsch, H, and Hauser, O (1910) *Homo aurignacensis Hauseri*, ein paläolithischer Skelett Fund aus dem unteren Aurignacien der Station Combe-Capelle bei Montferrand (Perigord). Praehist. Zeits. *1*:273–338.

Lévêque, F, and Vandermeersch, B (1980) Découverte de restes humains dans un niveau castelperronien à Saint-Césaire (Charente Maritime). C. R. Acad. Sci., Paris, D *291*:187–189.

Malez, M (1970) Novi pogledi na stratigrafiju krapinskog nalazišta. In M Malez (ed): Krapina 1899–1969. Zagreb: Jugoslavenska Akademija Znanosti i Umjetnosti, pp. 13–44.

Malez, M, Smith, FH, Rukavina, D, and Radovčić, J (1980) Upper Pleistocene fossil hominids from Vindija, Croatia, Yugoslavia. Curr. Anthropol. *21*:365–367.

Malez, M, and Ullrich, H (1982) Neuere paläeoanthropologische Untersuchen am Material aus der Höhle Vindija (Kroatien, Jugoslawien). Palaeont. Jugosl. *29*:1–44.

Penck, A, and Brückner, E (1909) Die Alpen im Eiszeitalter. Leipzig: Tauchnitz.

Ranke, J (1896) Der fossile Mensch und die Menschenrassen. Correspondenzblatt Deutsch. Gesell. Anthropol. Ethnol. Urgesch. *27*:151–156.

Russell, MD (1983) The Functional and Adaptive Significance of the Supraorbital Torus. Unpublished Ph.D. Thesis, University of Michigan, Ann Arbor.

Santa Luca, A (1978) A re-examination of presumed Neandertal-like fossils. J. Hum. Evol. *7*:619–636.

Schwalbe, G (1906) Studien zur Vorgeschichte des Menschen. Stuttgart: E. Schweizerbart.

Škerlj, B (1958) Were Neanderthalers the only inhabitants of Krapina? Bull. Sci. (Yougoslavie) *4*:44.

Smith, FH (1976a) A fossil hominid frontal from Velika Pecina (Croatia) and a consideration of Upper Pleistocene hominids from Yugoslavia. Am. J. Phys. Anthropol. *44*:127–134.

Smith, FH (1976b) The Neandertal remains from Krapina: A descriptive and comparative study. Univ. Tennessee Dep. Anthropol. Rep. Invest. *15*:1–359.

Smith, FH (1977) On the application of morphological "dating" to the hominid fossil record. J. Anthropol. Res. *33*:302–316.

Smith, FH (1982) Upper Pleistocene hominid evolution in South-Central Europe: A review of the evidence and analysis of trends. Curr. Anthropol. *23*:667–703.

Smith, FH (1983a) Behavioral interpretations of changes in craniofacial morphology across the archaic/modern *Homo sapiens* transition. In E Trinkaus (ed): The Mousterian Legacy: Human Biocultural Change in the Upper Pleistocene. Oxford: Brit. Archaeol. Repts. Intl. Ser. *S164*:141–164.

Smith, FH (1983b) On hominid evolution in south-central Europe. Curr. Anthropol. *24*:236–237.

Smith, FH (1984) Fossil Hominids from the Upper Pleistocene of central Europe and the origin of modern Europeans. In FH Smith and F Spencer (eds): The Origins of Modern Humans: A World Survey of the Fossil Evidence. New York: Alan R. Liss, Inc., pp. 137–209.

Smith, FH, and Ranyard, GC (1980) Evolution of the supraorbital region in Upper Pleistocene fossil hominids from South-Central Europe. Am. J. Phys. Anthropol. *53*:589–609.

Spencer, F (1984) The Neandertals and their evolutionary significance: A brief historical survey. In FH Smith and F Spencer (eds): The Origins of Modern Humans: A World Survey of the Fossil Evidence. New York: Alan R. Liss, Inc., pp. 1–49.

Spencer, F, and Smith, FH (1981) The significance of Aleš Hrdlička's "Neanderthal phase of man:" A historical and current assessment. Am. J. Phys. Anthropol. *56*:435–459.

Stringer, CB (1974) Population relationships of later Pleistocene hominids: A multivariate study of available crania. J. Archaeol. Sci. *1*:317–342.

Stringer, CB (1978) Some problems in Middle and Upper Plesitocene hominid relationships. In D Chivers and K Joysey (eds): Recent Advances in Primatology, Vol. 3. London: Academic Press, pp. 395–418.

Stringer, CB (1981) The dating of European Middle Pleistocene hominids and the existence of *Homo erectus* in Europe. Anthropol. (Brno) *19*:3–14.

Stringer, CB (1982) Towards a solution to the Neanderthal problem. J. Hum. Evol. *11*:431–438.

Stringer, CB (1985) Middle Pleistocene hominid variability and the origin of Late Pleistocene hominids. In E Delson (ed): Ancestors: The Hard Evidence. New York: Alan R. Liss, Inc., pp. 289–295.

Stringer, CB, and Burleigh, R (1981) The Neanderthal problem and the prospects for direct dating of Neanderthal remains. Bull. Brit. Mus. (Nat. Hist.), Geol. *35*:225–241.

Stringer, CB, Howell, FC, and Melentis, J (1979) The significance of the fossil hominid skull from Petralona, Greece. J. Archaeol. Sci. *6*:235–253.

Stringer, CB, Hublin, J-J, and Vandermeersch, B (1984) The origin of anatomically modern humans in western Europe. In FH Smith and F Spencer (eds): The Origins of Modern Humans. New York: Alan R. Liss, Inc., pp. 51–135.

Trinkaus, E (1977) The Neandertals from Krapina: An inventory of the lower limb remains. Zeits. Morph. Anthropol. *67*:44–59.

Trinkaus, E (1978) Functional implications of the Krapina Neandertal lower limb remains. In M Malez (ed): Krapinski Pračovjek i Evolucija Hominida. Zagreb: Jugoslavenska Akademija Znanosti i Umjetnosti.

Trinkaus, E (1983) The Shanidar Neandertals. New York: Academic Press.

Trinkaus, E, and Howells, WW (1979) The Neanderthals. Sci. Amer. *241*:118–133.

Trinkaus, E, and Smith, FH (1985) The fate of the Neanderthals. In E Delson (ed): Ancestors: The Hard Evidence. New York: Alan R. Liss, Inc., pp. 325–333.

Vallois, HV (1954) Neanderthals and presapiens. J. Roy. Anthropol. Inst. *84*:111–130.

Vandermeersch, B (1976) Les Néandertaliens en Charente. In H de Lumley (ed): La Préhistoire Française. Vol. 1. Paris: Centre National de la Recherche Scientifique, pp. 584–587.

Vandermeersch, B (1978) Les premiers Néandertaliens. La Recherche *9*:694–696.

Vandermeersch, B (1981) Les premiers *Homo sapiens* au Proche-Orient. In D Ferembach (ed): Les Processus de l'Hominisation. Coll. Internat. C.N.R.S. *599*:97–100.

Vandermeersch, B (1985) The origin of the Neandertals. In E Delson (ed): Ancestors: The Hard Evidence. New York: Alan R. Liss, Inc., pp. 306–309.

Vandermeersch, B, Tiller, AM, and Krukoff, S (1976) Position chronologique de restes humains de Fontéchevade. In A Thoma (ed): Le Peuplement Anténéandertalien de l'Europe. IXth Cong. UISPP (Prétirage). Nice: C.N.R.S., pp. 1–26.

Vlček, E (1964) Einige in der Ontogenese des modernen Menschen untersuchte Neandertalmerkmale. Zeit. Morph. Anthropol. *56*:63–83.

Vlček, E (1967) Die Sinus frontalis bei europäischen Neandertalern. Anthropol. Anz. *30*:166–189.

Vlček, E (1970) Étude comparative ontophylogénétique de l'enfant de Pech de l'Azé. Arch. Inst. Paleontol. Hum. Mém. *33*:149–178.

Weidenreich, F (1943) The skull of *Sinanthropus pekinensis*: A comparative study of a primitive hominid skull. Palaeontol. Sin. n.s. D, No. 10 (whole series No. 127).

Weidenreich, F (1947) Facts and speculations concerning the origin of *Homo sapiens*. Am. Anthropol. *49*:187–203.

Wolpoff, MH (1978) The dental remains from Krapina. In M Malez (ed): Krapinski Pračovjek i Evolucija Hominida. Zagreb: Jugoslavenska Akademija Znanosti i Umjetnosti, pp. 119–144.

Wolpoff, MH (1979) The Krapina dental remains. Am. J. Phys. Anthropol. *50*:67–114.

Wolpoff, MH (1980a) Paleoanthropology. New York: Knopf.

Wolpoff, MH (1980b) Cranial remains of Pleistocene European hominids. J. Hum. Evol. *9*:339–358.

Wolpoff, MH (1982) Comment on Upper Pleistocene evolution in South-Central Europe: A review of the evidence and analysis of trends. Curr. Anthropol. *23*:693.

Wolpoff, MH, Smith, FH, Malez, M, Radovčić, J, and Rukavina, D (1981) Upper Pleistocene hominid remains from Vindija Cave, Croatia, Yugoslavia. Am. J. Phys. Anthropol. *54*:499–545.

Ziman, JM (1968) Public knowledge: An essay concerning the social dimension of science. Cambridge: Cambridge University Press.

Ancestors: The Hard Evidence, pages 319–324

The Poor Brain of *Homo sapiens neanderthalensis*: See What You Please . . .

Ralph L. Holloway
Department of Anthropology, Columbia University, New York, New York 10027

ABSTRACT Neandertals have always been viewed as primitive, and they are currently viewed by some as large-brained but without language (or having a highly restricted one), and retaining primitive features in their brain endocasts. Their large size most probably relates to allometric and metabolic adaptations related to muscularity and living within a periglacial and/or tundra-like habitat, and not behavioral complexity. Their endocasts do not show "primitive" features if size, convolutional patterns, and asymmetries are considered together. Were modern living human hunters and gatherers to be judged on the basis of stone tool technology alone, they would probably be considered less advanced, "brain-wise," than Neandertals.

INTRODUCTION

Poor *Homo sapiens neanderthalensis.*[1] Surely no other ethnic group has had so many nasty slurs and insults thrown at itself than our distant cousins of some 40,000 to 50,000 years ago. All manner of pathological (*à la* Virchow), teratological (rickets, vitamin-D deficiency, e.g., Ivanhoe, 1970), imbecilia (whether from Italy, Ireland, Russia, etc., etc.), and evolutionary insults have been leveled at this early representative of a lineage whose brain was somewhat larger than our own. The final blow is the somewhat prevalent attitude that, based on computer decisions and a lack of art work, poor Neandertals were also mute, or at least babbling away with a highly restricted set of phonemes (Lieberman, 1975, 1976; Lieberman et al., 1972; Lieberman and Crelin, 1971; Marshack, 1976; Jaynes, 1976; cf. Falk, 1975; Wind, 1976; DuBrul, 1976). The movie "Quest for Fire," unless it was all in my imagination, even had them taking their women from behind, until finally some copulatory finesse was given them by the gracile folks, a gift surely good as fire.

The prominent brow ridges, the largish and broad nasal aperture, the large teeth, the lack of a mental process on the chin, the occipital bun, etc., etc., have produced many a bestial portrait, and indeed the very museum in which this symposium was held (through the efforts of McGregor) produced some beautiful reconstructions. Several generations of introductory anthropology students have been forced to imagine them as wearing three-piece suits, carrying attaché cases, and strap-hanging from our local IRT subways, with the vexing problem of trying to assess whether or not such a beast would fall in the range of modern human normal variation. Usually, I take the position, at least for my students, that yes, *I* would notice them, and regard them (the Neandertal "business men") as unusual in appearance. But for this gathering, and the setting, I will confine my remarks to the Neandertal brain, and what we know about it. This is an easy

[1] I am using *neanderthalensis* as a subspecific designation, meaning that I regard these *people* as fully *Homo sapiens*, differing from modern populations in a way analogous to the differences between, for example, Australian aborigines and Eskimos. In other words, I regard them as yet one more ethnic variation of *Homo sapiens*, being unable to digest or find palatable the hubris of a designation such as *H. s. sapiens*; a designation too wise for the realities . . . .

TABLE 1. *Cranial Capacities in Different Samples of Neandertals and "Neandertaloids"*[1]

Run 1	Vol. (ml)	Run 2	Run 3	Vol. (ml)
1. Skhūl IV	1,555		Skhūl IV	1,555
2. Skhūl V	1,520		Skhūl V	1,520
3. Skhūl VI	1,585		Skhūl VI	1,585
4. Amud	1,740		Amud	1,740
5. Shanidar I	1,600		Shanidar	1,600
6. Djebel Irhoud I	1,305		Djebel Irhoud I	1,305
7. Qafzeh VI	1,570		Djedel Irhoud II	1,450
8. Neandertal	1,525	Neandertal	Qafzeh VI	1,570
9. Le Moustier	1,565	Le Moustier	Neandertal	1,525
10. La Ferrassie I	1,640	La Ferrassie I	Le Moustier	1,565
11. La Chapelle	1,625	La Chapelle	La Ferrassie I	1,640
12. Spy I	1,553	Spy I	La Chapelle	1,625
13.			Spy I	1,553
14.			Spy II	1,305
15.			Tabūn I	1,270
16.			La Quina	1,350
17.			Krapina B	1,450
18.			Saccopastore I	1,245
19.			Saccopastore II	1,300
20.			Monte Circeo I	1,550
$\overline{X}$ = 1565, S.D. = 101 N = 12		$\overline{X}$ = 1582, S.D. = 49 N = 5	$\overline{X}$ = 1485, S.D. = 142.4 N = 20	

[1]Both Run 1 and Run 2 are for purported males, with Run 2 strictly confined to those "Classic" Western European Neandertals regarded as "cold-adapted." Run 3 includes both males and females, restricted to North Africa, Europe, and the Middle East. Except for Spy I, Spy II, and Djebel Irhoud I, cranial capacities are from Olivier and Tissier (1975). I have only worked on those three endocasts. Run 1 includes Skhūl and Qafzeh fossils which, although not strictly "classic," may nevertheless be associated with more acceptable Near Eastern Neandertals.

task, given that we have never seen one: our cherished hopes of a fully preserved Neandertal corpse imbedded in a block of Siberian ice have not come to light, as yet . . . .

Previous attempts to characterize Neandertal brains as "primitive" have been based on phrenological endocast examinations, which in the "Zeitgeist" of that period (1890–1920) enabled workers to "see" what they believed, or what they wished to believe, particularly with regard to the lunate sulcus or "affenspalte," and the frontal lobes. I will state my conclusions first. I believe the Neandertal brain was fully *Homo*, with no essential differences in its organization compared to our own. It was simply large, and for at least two reasons: 1) it was metabolically efficient within periglacial and/or cold tundra habitats, and 2) it was related to a larger amount of lean body mass than our own, and thus the larger size was, in the main, an allometric scaling effect. Two additional speculations are: 1) Neandertals did have language, and 2) they were highly competent, visuospatially.

BRAIN SIZE AND SAMPLING

It is a vexing problem to decide which discoveries to include in this paper, given the wide geographic distribution of what we commonly recognize as "Neandertal" and the problems of grades and clades, sex determination, and partial crania (e.g., Trinkaus, 1980, 1983; Smith, 1980; Stringer, 1974; Stringer and Trinkaus, 1981; Wolpoff, 1980).

For the purpose of this presentation, I am confining my analysis to the so-called "classic" western European Neandertals, very much along earlier criteria as discussed in Holloway (1981a). There are many statistical analyses that could be performed, depending on one's chosen data base. For example, Table 1 shows three statistical runs; two of these are for purported *males* showing "classic features." Run 1 includes Skhūl IV, V, VI, Djebel Irhoud I, Shanidar I, and Qafzeh II, all from the Middle East or North Africa, showing a considerable range of morphological variations. The Krapina materials, Saccopastore, and Monte Circeo have been deleted, as secure cranial capacities have not been provided and problems of sexing remain (see discussion in Holloway 1981a). In this run, the mean cranial capacity is 1,565 ml, with a standard deviation of 101 ml. If Djebel Irhoud I (1,305 ml) were removed (on the basis of being female), and Amud were also removed on the basis of difficulty of endocast reconstruction, the mean would be 1,574 ml, S.D. = 39 ml, the latter rather low. In run 2, *only* Western European specimens of reasonably certain male identity were used, yielding a mean of 1,582 ml and an S.D. of 49 ml, which is again a low value. Run 3, a "hodge-podge" of males, females, cold-, and supposedly warm-adapted Neandertals (N = 20) still provides a high $\bar{X}$ = 1,485, which is above most averages for northern European populations.

There are, I believe, many weaknesses in these three approaches, and the S.D.s show this. Nevertheless, one must begin somewhere, and I have made the choice fully understanding the vexatious criteria.

A major reason for such constriction really revolves about the choice of modern peoples with possible similar ecological habitats and adaptations, i.e., cold-adapted peoples. Hrdlička's (1942) data on Northern and Eastern Eskimo and on Greenland Eskimo provide male cranial capacities averaging roughly 1,555 ml and these were peoples with high lean body mass compositions, i.e., heavily muscled. Hrdlička's (1942, p. 396) abstract does not provide summary cranial capacities for Northern and Eastern Eskimo females. However, p. 424 shows, for "General Eskimos," male $\bar{x}$ = 1,485 (N = 468) and female $\bar{X}$ = 1,320 (N = 426). The average is thus roughly, 1,402 cm^3. As Hrdlička pointed out, all of these people were of shorter stature than Europeans.

Finally, there is yet another problem with these samples, in that the so-called "archaic" *Homo sapiens sapiens* (e.g., Cro-Magnon, Combe-Capelle, Predmost, Chancelade, etc.) could well have had equal or higher cranial capacities *if* we had accurate volume estimates. Of course, these too were "cold-adapted" populations of the late Pleistocene. The point of these comments and statistical manipulations is the relative certitude that indeed, Neandertals did have large brains, and that for their stature, they were probably somewhat larger than our own.

SULCAL CONFIGURATIONS

In 1911, Boule and Anthony published their detailed study of the endocast of La Chapelle-aux-saints. This was later followed with studies by Kappers (1929) and Anthony (1928), who also published a detailed endocast mapping for the La Quina specimen (Anthony, 1913). Critical commentaries by Symington (1916) provoked a response from Boule and Anthony in 1917, regarding the latter's interpretations that the Sylvian fissure was best compared to a position to be found in a fetal human brain of seven months of age. (The most useful commentary on these publications can be found in Connolly, 1950, pp. 342–348).

In general, Boule and Anthony (1911) "found" several reasons for considering these early Neandertal specimens as primitive, such as a lunate sulcus (left side) in a somewhat anterior position, a parieto-occipital fissure well anterior of lambda, and an overall "simplicity" of convolutional pattern. Their frontal lobes were deficient as well.

On the basis of Boule and Anthony's (1911) drawings, LeMay (1975, 1976) concluded that the left hemisphere showed a lower posterior ramus of the Sylvian fissure than the right (LeMay, 1976, p. 361; 1977). This was accepted as a partial bit of evidence for cerebral asymmetry and thus a possible handedness, which in the case of La Chapelle-aux-Saints would have been *left*.

This author has examined the Neandertal cap, La Quina, La Ferrassie, Spy I and II, La Chapelle-aux-Saints, Djebel Irhoud I, and an endocast of the Amud specimen. I can only say that I have *no* confidence in *unambiguously* identifying *any* convolutional patterns that are suggestive of a "primitive" condition. As I have pointed out elsewhere (Holloway, 1976a,b, 1978, 1981a,b, 1983a,b, 1984), newer methodologies must be developed to increase the probabilities of accurate location of key cerebral landmarks if continuing controversies (e.g., Falk, 1980, 1983; Holloway, 1984) are to be avoided. Until this is done, there is good reason to consider much of hominid paleoneurology as a sort of "paleophrenology," as characterized by Jerison (1975).

Unfortunately, we are presently confined to making "educated" judgments, and in that sense, I can find no reason to assert that Neandertals had smaller or more "primitive" Broca's areas than did modern *Homo*. Moreover, there is no evidence for any critical weakness of organization or mass in what would be the so-called Wernicke's area of superior but caudal temporal lobe, and anterior inferior parietal zones.

ASYMMETRIES

Cerebral asymmetries are clearly present in the Neandertals mentioned above. Except for La Chapelle-aux-Saints, they appear to follow a classic modern *Homo* petalial pattern (left-occipital, right-frontal) as described and elaborated by LeMay (see 1976, 1977; and Trinkaus and LeMay, 1982, for complete references). These observations were published by Holloway (1981a) for Neandertals in particular, and by Holloway and de La Coste-Larymondie (1982) for all of the available hominoid endocasts (see also LeMay et al., 1982), and its significance was speculatively argued in Holloway (1976a,b, 1983a,b) with regard to language behavior and visuospatial integration.

Stereoplotting methods (as per Holloway, 1981b) are still nascent. Neandertal-*sapiens* comparisons are simply very suspect given the small sample sizes for Neandertals (4–6), which reduce the effectiveness of multivariate techniques enormously.

Symap comparisons (unpublished)[2] between Neandertal and *sapiens* endocasts, based on the analysis of *residuals* following allometric correction, do show some significant differences of features in the anterior occipital-posterior parietal zone. Group differences, however, vanish if brain size is used as a covariate to test F-ratios by another method. Perhaps it is a matter of the occipital bun or "chignon" of Neandertal fame, which has recently been reviewed by Trinkaus and LeMay (1982), as an extended growth phase of occipital cortex in Neandertals. This is an interesting possibility, perhaps signaling an increased amount of primary visual striate cortex.[3] Certainly, the frontal lobe is almost indistinguishable from modern *Homo*, and aside from some increased platycephaly in the mid-dorsal region, no significant differences can be found, as the F-ratios are very low between modern *Homo* and Neandertals (see Fig. 1).

CONCLUSIONS AND SPECULATIONS

One cannot help but wonder what modern archaeologists would conclude after studying all Eskimo, Aleut, Australian, Bushman, and tropical rainforest aboriginal material cutlures *if only stone tools remained.* No language? No ritual? No concern for the dead, spirits, etc.? (cf. Solecki, 1971). And if, by chance, *no* archaeological or ethnographic evidence was available, one would have to conclude on the basis of brain size alone (given our obsession with this variable) that Neandertals were more advanced, behaviorally, than living groups whose languages and social customs still defy complete understanding among 20th century anthropologists. Perhaps to someone (or something) from outer space, it would be the brains of the 20th century anthropologists that require study. After all, many of the above aboriginal groups do have rich social culture, and many speak more languages than the average American.

The punch line for all this muted speculation and discussion is that the lean body mass proportion and larger brain size for Neandertals was published before most of us were born. By whom? Eugène Dubois (1921). The suggestion was re-echoed by Trinkaus and Howells (1979), and, using data from Danish brain weights (Pakkenberg and Voigt, 1964), by Holloway (1980, 1981a). Unpublished data from roughly three thousand autopsy cases, multi-ethnic in composition, shows that there is a stronger correlation between brain size and body size (particularly stature) in males than in females, which is particularly strong in age cohorts (by decades) from 30–60 years.[4]

If this relationship is added to the interesting clinical data from Beals et al. (1984), that smaller brain sizes are found in more tropical regions, the large size of Neandertal brains has its explanation in biological interrelationships among size, lean body mass, temperature, and metabolism, and *NOT* complexity (or lack of it) of behavior. One does not have to wait for Neandertal carv-

[2]Unfortunately, prior to 1981, the endocasts of Spy I and II and Debel Irhoud I were not available for inclusion in the pilot study of the dorsal surface, and since that time no further progress has been made in enlarging the sample or replicating the original studies on a more accurate and partially automated stereoplotter device. As no funding support exists for this project, no progress is in sight . . . . As my 1981a paper explains, discriminant analytic techniques must be limited to a number of variables not larger than the sample size of any one relevant group, e.g., Neandertals.

[3]I am aware, through the kindness of Dr. Clive Harper (personal communication) of the Royal Perth Hospital, Western Australia, of work in progress using image-analytic techniques to quantify volumetric distributions of cerebral cortex between Australian aboriginal and Caucasian brains. There appears to be a higher percentage of lateral visual striate cortex among aboriginals, which, if true, would correlate in an interesting way with published cognitive testing of visuospatial abilities between aboriginals and Caucasians, in which the former have higher test scores regardless of cultural training (e.g., Kearins, 1981). As the Symap Figure 1 shows, there are high F-ratios between modern human and Neandertal endocasts in a portion of anterior occipital and posterior parietal cortical zones. These *prove* nothing at present, and thus my suggestion of spatiovisual competency being heightened in Neandertals is purely speculative.

[4]These data, currently being analyzed, are from autopsy cases for Australian aborigines and Caucasians, and Black, Hispanic, and Caucasian samples from New York.

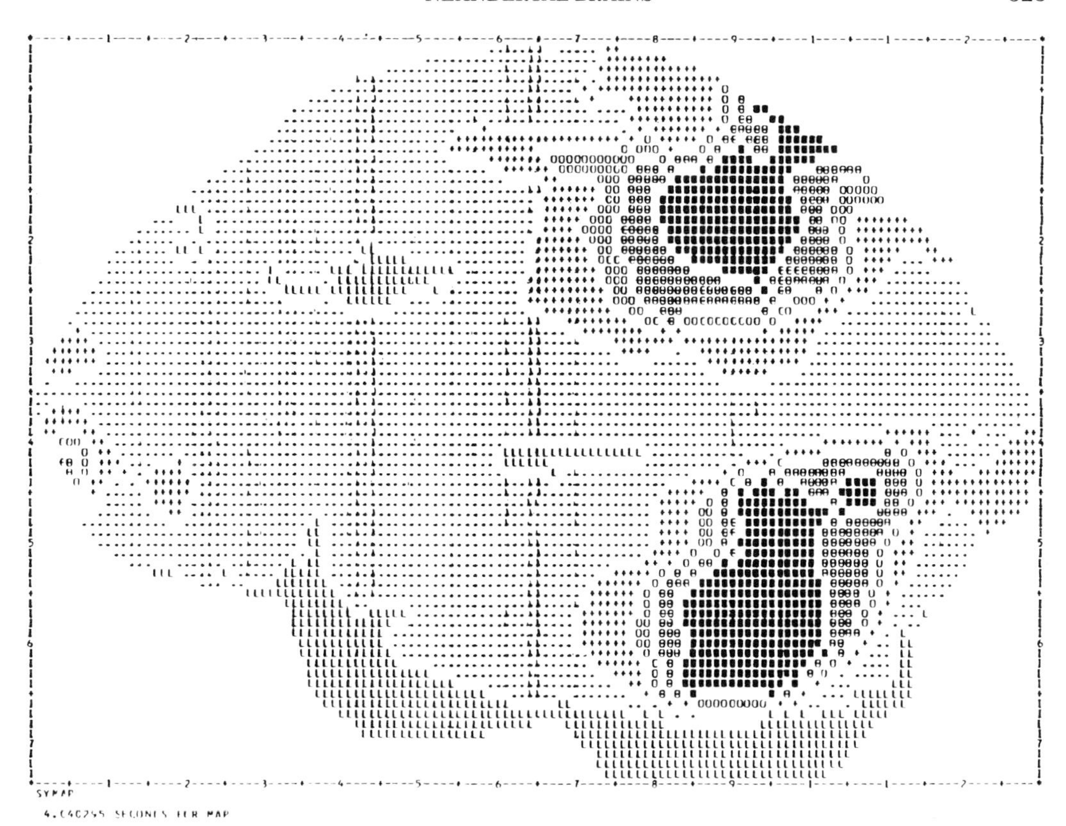

Fig. 1. Symap picture for modern human–Neandertal comparisons of univariate F-ratios and their approximate distribution on the lateral and dorsal endocast surface, *after* allometric corrections have been made (see Holloway, 1981b for details). The "black holes" are those regions with the highest F-ratios for the residuals (after allometric correction), and on this "map" they varied between 16 and 29.6. The two "black holes" are in the following areas (per Brodmann's classification): upper, areas 5 and 7, or superior parietal lobule; lower, areas 22, 37, 40, 39, which are "associative auditory;" and posteriorly, area 19, or "peristriate" cortex. I suspect but cannot demonstrate that the upper "black hole" is an allometric effect produced by a combination of a greater parietal bossing in modern human endocasts and a more pronounced degree of platycephaly in Neandertals. Much more research is needed to clarify the lower "black hole," which may be an artifact of size, or true cerebral organization. In any event, there is no suggestion that this region, so approximate to Wernicke's area, is more primitive in Neandertals.

ings, cave paintings, or sculptures (we anxiously await knowledge about their Venuses . . .) to credit them with communicative skills and complex social behavior. They buried their dead, practiced ritual, and many would be the Upper Paleolithic and Mesolithic populations that could be denied language abilities if we were to rely exclusively on "soft archaeological" remains.

An alternate title for this paper might be "The Tyranny of Brain Size," for its essential message is about this variable, which, when considered alone, is one that is embarrassingly weak when relationships to behavior are to be made, particularly within species. Just as Blacks and Australian aborigines have been the butt of ethnocentric prejudice for their smaller brain sizes, so have Neandertals for "primitive" endocranial features that are simply nonexistent. Logically (in that above sense), Neandertals should be the brightest of all, a proposition I find equally loathesome . . . .

ACKNOWLEDGMENTS

Parts of this research were funded by NSF grant BNS 7911235. I am indebted to Ms. Beverly March for her skill and patience in typing this paper.

LITERATURE CITED

Anthony, R (1913) L'encephale de l'homme fossile de la Quina. Bull. Mem. Soc. Anthrop. Paris. 6e Ser., *4:*117–195.

Anthony, R (1928) Anatomie comparée du cerveau. Paris: Gaston Doin et Cie.

Beals, KL, Smith CL, and Dodd, SM (1984) Brain size, cranial morphology, climate and time machines. Curr. Anthropol. *25:*301–330.

Boule, M, and Anthony, R (1911) L'encephale de l'homme fossile de la Chapelle-aux-Saints. L'Anthropol. *22:*129–196.

Boule, M, and Anthony R (1917) Neopallial morphology of fossil man as studied from endocranial casts. J. Anat. J. Physiol. *51:*95–102.

Connolly, CJ (1950) External Morphology of the Primate Brain. Springfield, Illinois: C.C. Thomas.

Dubois, E (1921) On the significance of the large cranial capacity of *Homo Neanderthalensis.* Proc. Kon. Ned. Akad. Wetenschappen *23:*1271–1288.

Du Brul, EL (1976) Biomechanics of speech sounds. Ann. N.Y. Acad. Sci. *280:*631–642.

Falk, D (1975) Comparative anatomy of the larynx in man and the chimpanzee: Implications for language in Neandertal. Am. J. Phys. Anthropol. *43:*123–132.

Falk, D (1980) A reanalysis of the South African australopithecine natural endocasts. Am. J. Phys. Anthropol. *53:*525–539.

Falk, D (1983) The Taung endocast: A reply to Holloway. Am. J. Phys. Anthropol. *60:*479–490.

Holloway, RL (1976a) Some problems of hominid brain endocast reconstruction, allometry, and neural reorganization. In: Coll. VI, IX Congr. Union Int. Sci. Pre. Protohist., Pretirage. Nice: C.N.R.S., pp. 66–119.

Holloway, RL (1976b) Paleoneurological evidence for language origins. Ann. N.Y. Acad. Sci. *280:*330–348.

Holloway, RL (1978) The relevance of endocasts for studying primate brain evolution. In CR Noback (ed): Sensory Systems in Primates. New York: Plenum, pp. 181–200.

Holloway, RL (1980) Within-species brain–body variability: A re-examination of the Danish data and other primate species. Am. J. Phys. Anthropol. *53:*109–121.

Holloway, RL (1981a) Volumetric and asymmetry determinations on recent hominid endocasts: Spy I and II, Djebel Ihroud I, and the Salé *Homo erectus* specimens, with some notes on Neandertal brain size. Am. J. Phys. Anthropol. *55:*385–393.

Holloway, RL (1981b) Exploring the dorsal surface of hominid brain endocasts by stereoplotter and discriminant analysis. Phil. Trans. Roy. Soc. (Lond.) *B292:* 155–166.

Holloway, RL (1983a) Human brain evolution: A search for units, models, and synthesis. Canad. J. Anthrop. *3*:215–232.

Holloway, RL (1983b) Human paleontological evidence relevant to language behavior. Human Neurobiol *2*:105–114.

Holloway, RL (1984) The Taung endocast and the lunate sulcus: A rejection of the hypothesis of its anterior position. Am. J. Phys. Anthropol. *64*:285–288.

Holloway, RL, and de La Coste-Larymondie, MC (1982) Brain endocast asymmetry in pongids and hominids: Some preliminary findings on the paleontology of cerebral dominance. Am. J. Phys. Anthropol. *58:*101–110.

Hrdlička, A (1942) Catalogue of human crania in the United States National Museum collections: Eskimo in general. Proc. U.S. Nat. Mus. *91:*169–429.

Ivanhoe, F (1970) Was Virchow right about Neandertal? Nature *227:*577–579.

Jaynes, J (1976) The evolution of language in the late Pleistocene. Ann. N.Y. Acad. Sci. *280:*312–325.

Jerison, HJ (1975) Fossil evidence of the evolution of the human brain. Ann. Rev. Anthrop. *4:*27–58.

Kappers, Ariens CU (1929) The Evolution of the Nervous System in Invertebrates, Vertebrates, and Man. Harlem: De Erven F. Bohn.

Kearins, JM (1981) Visual spatial memory in Australian Aboriginal children of desert regions. Cognitive Psychol. *13:*434–460.

LeMay, M (1975) The language capacity of Neandertal Man. Am. J. Phys. Anthropol. *42:*9–14.

LeMay, M (1976) Morphological cerebral asymmetries of modern man, fossil man, and non-human primates. Ann. N.Y. Acad. Sci. *280:*213–215.

LeMay, M (1977) Asymmetries of the skull and handedness. J. Neurol. Sci. *32:*213–225.

LeMay, M, Billig, MS, and Geschwind, N (1982) Asymmetries of the brains and skulls of non-human primates. In E Armstrong and D Falk (eds): Primate Brain Evolution: Methods and Concepts. New York: Plenum, pp. 263–278.

Lieberman, P (1975) On the Origins of Language. New York: MacMillan.

Lieberman, P (1976) Interactive models for evolution: Neural mechanisms, anatomy, and behavior. Ann. N.Y. Acad. Sci. *280:*660–672.

Lieberman, PE, and Crelin, ES (1971) On the speech of Neandertal man. Ling. Inquiry *2:*203–222.

Lieberman, P, Crelin, ES, and Klatt, DH (1972) Phonetic ability and related anatomy of the newborn, adult human, Neandertal man, and the chimpanzee. Am. Anthropol. *74:*287–307.

Marshack, A (1976) Some implications of the Paleolithic symbolic evidence for the origin of language. Ann. N.Y. Acad. Sci. *280:*289–311.

Olivier, G, and Tissier H (1975) Determination of cranial capacity in fossil men. Am. J. Phys. Anthropol. *43:*353–362.

Pakkenberg, H, and Voigt, V (1964) Brain weight of the Danes. Acta. Anat. (Basel) *56:*297–307.

Smith, FH (1980) Sexual differences in European Neanderthal crania with specific reference to the Krapina remains. J. Hum. Evol. *9:*359–375.

Solecki, R (1971) Shanidar: The First Flower People. New York: Knopf.

Stringer, CB (1974) Population relationships of later Pleistocene hominids: A multivariate study of available crania. J. Archaeol. Sci. (London) *1*:317–342.

Stringer, CB, and Trinkaus, E (1981) The Shanidar Neandertal Crania. In Stringer, CB (ed): Aspects of Human Evolution. London: Taylor and Francis Ltd., pp. 129–168.

Symington, F (1916) Endocranial casts and brain form: A criticism of some recent speculations. J. Anat. J. Physiol. *50:*111–130.

Trinkaus, E (1980) Sexual differences in Neandertal limb bones. J. Hum. Evol. *9:*377–397.

Trinkaus, E (1983) The Shanidar Neandertals. New York: Academic Press.

Trinkaus, E and Howells, WW (1979) The neanderthals. Sci. Amer. *241*:118–133.

Trinkaus, E, and LeMay, M (1982) Occipital bunning among later Pleistocene hominids. Am. J. Phys. Anthropol. *57:*27–36.

Wind, J (1976) Phylogeny of the human vocal tract. Ann. N.Y. Acad. Sci. *280:*612–630.

Wolpoff, MH (1980) Paleoanthropology. New York: Knopf.

Ancestors: The Hard Evidence, pages 325–333

The Fate of the Neandertals

Erik Trinkaus and Fred H. Smith
Department of Anthropology, University of New Mexico, Albuquerque, New Mexico 87131 and Department of Anthropology, University of Tennessee, Knoxville, Tennessee 37916

ABSTRACT The paleontological analysis of the Neandertals has been primarily concerned with determining how many, if any, of them may be ancestors of modern humans. Problems in determining the genetic versus stress-induced bases of morphology and in evaluating the genetic foundations of morphological shifts, combined with the multiple mechanisms through which the gene frequencies might change, suggest that the resolution of strictly phylogenetic questions may remain beyond the nature of data provided by the fossil record. Nonetheless, human morphological change during the Late Pleistocene provides insights into human adaptive evolution.

Brain size and sulcal morphology, postural and locomotor anatomy, most upper limb articular morphology, posterior permanent and deciduous dental dimensions, and average body mass remained constant during the Neandertal to anatomically modern human transition. In contrast, midfacial prognathism (and many functionally related features of the facial skeleton), masto-occipital and tympanic morphology, pubic morphology and breadth, and limb segment plus phalangeal length proportions exhibited marked change at the transition. In between with morphological shifts are neurocranial shape, occipital bunning size and frequency, anterior tooth size and morphology, deciduous tooth morphology, thumb carpometacarpal articular morphology, stature, and most aspects of postcranial robusticity. The varying amounts of morphological change in these features, before and during the transition, indicate a mosaic of behavioral shifts that correlate with the cultural evolution indicated by the contemporaneous archeological record.

INTRODUCTION: THE NEANDERTAL PROBLEM

Ever since it was recognized around the turn of the century that there was a group of archaic humans in Europe that immediately preceded the oldest anatomically modern humans and resembled the specimen unearthed in the Neander Valley in 1856, human paleontologists have been arguing about the phylogenetic relationship of the Neandertals to more recent humans (for recent reviews, see: Brace, 1964; Mann and Trinkaus, 1974; Trinkaus and Howells, 1979; Wolpoff, 1980a; Spencer and Smith, 1981; Smith, 1982; Stringer, 1982; Trinkaus, 1982a, 1983a; Spencer, 1984). Originally, most paleoanthropologists saw the Neandertals as our natural predecessors, representing the more archaic stage through which humans must have passed (Mortillet, 1883; Schwalbe, 1904; Sollas, 1907; Keith, 1911; Gorjanović-Kramberger, 1906). However, the discoveries of substantial Neandertal remains at several sites prior to 1910 and their subsequent analysis (esp. Boule, 1911–1913) tipped the scales of opinion in the other direction and began the seemingly interminable argument as to how many, if any, of the Neandertals can be rightfully counted among our ancestors. The basic positions were established by the 1920s, with most paleoanthropologists perceiving the Neandertals as a side-branch in hominid evolution (e.g., Keith, 1915; Osborn, 1918; Boule, 1921; Sollas, 1924) and a persistent minority seeing them as the direct, local ancestors of subsequent human populations (Verneau, 1924; Hrdlička, 1927; Weinert, 1932). Our knowledge of this period of human evolution has increased markedly in recent years, and our models as to what might have happened and how to decipher the prehistoric record have improved in sophistication, but the basic argument regarding the "fate" of the Neandertals remains essentially the same (see comments in Smith (1982) and articles in Smith and Spencer (1984)).

THE NEANDERTALS

The Neandertals, although definable by lists of morphological characters that distinguish them from their geographic and temporal neighbors (e.g., Trinkaus, 1983a; Stringer et al., 1984), are best considered a geographic subspecies (of *H. sapiens*) that occupied the northwestern Old World (Europe and western Asia north to 49° and east to 67°) from the end of the last interglacial to the middle of the last glacial. They represent the final group of archaic humans in the northwestern Old World and have their closest morphological affinities to preceding Middle Pleistocene humans from the same region, differing primarily in a slight reduction of robusticity (cranial and postcranial) and greater average brain size (Hublin, 1978; Wolpoff, 1980b; Cook et al., 1982; Trinkaus, 1982b). The Neandertals differ from their African and perhaps East Asian contemporaries in their greater midfacial prognathism and the expression of certain masto-occipital traits. They contrast with their early anatomically modern human successors in a number of features, which are discussed below.

PHYLOGENETIC ISSUES

Few paleontologists would argue that the morphological differences between the Neandertals and their early modern human successors were trivial, but there is considerable disagreement as to how significant they were. The primary issue remains whether the changes that took place in the middle of the last glacial in Europe and western Asia represent a *major* acceleration in the rate of Late Pleistocene human morphological evolution (and if so, what specific factors brought about this acceleration) or were merely a continuation of previously existing temporal trends with little or no change in the rate of alteration (for recent opinions see: Heim, 1978; Howells, 1978; Trinkaus and Howells, 1979; Wolpoff, 1980a; Smith, 1982; Stringer, 1982; Hublin, 1983; Trinkaus, 1983a; Stringer et al., 1984). A major acceleration in evolutionary tempo would require a marked shift in some combination of selective pressures, gene frequencies, and gene flow, and could be seen as the product of major population replacement. The second perspective would invoke only gradual changes in gene frequencies and associated selective pressures with no alteration of levels of gene flow. Since some continuities of form exist between Neandertals and modern *H. sapiens* in Europe and the Near East (Smith and Ranyard, 1980; Wolpoff, 1980a; Smith, 1982, 1984; Trinkaus, 1983a, 1984a), at least minimal genetic continuity must have occurred, and therefore, total population replacement need not be considered further, even if it occurred in some areas of the northwestern Old World.

One of the persistent problems in sorting out these issues is our inability to determine potential rates of morphological change. Many of the anatomical alterations around the time of this Late Pleistocene transition[1] can be partially accounted for by changes in environmental and biomechanical stress on the anatomy during development, without any change in genotype. However, the relative uniformity of most of the differentiating morphological patterns before and after the transition, and the appearance of most of the diagnostic postcranial and cranial features early in development (Vlček, 1970, 1973; Smith, 1976; Hublin, 1980; Tillier, 1982; Heim, 1982b; Trinkaus, 1983a), argue that there was nonetheless a strong genetic component in the morphological differences. But were these genetic differences due to simple changes at a few loci that had systemic effects (through regulator genes, shifts in the timing and/or levels of endocrine secretion during development, secondary effects of shifts in relative rates of development of ad-

[1]The period during which the morphological pattern of anatomically modern humans replaced that of archaic *H. sapiens* across the Old World will be referred to here as a "transition." The term "transition" is not intended to imply or deny the possibility of direct *in situ* evolution of populations from one group to the other.

jacent structures, and/or pleiotropy), or were they the products of multiple changes at a variety of loci, each affecting one paleontologically perceived portion of the anatomy? We favor the former explanation, but as is also the case with stress-induced versus genetically determined effects, a modest sample of bones is unlikely to be very informative in separating these different, complementary, and not necessarily mutually exclusive processes.

Even if one accepts a large genetic component in the morphological changes evident at the transition, there are a variety of population dynamics that could account for the presumed genetic shift. They could be due to 1) a shift in selective pressure, markedly changing the frequencies of genes within local gene pools; 2) an alteration of selection, shifting the balance between local stabilizing selection and a relatively constant rate of gene flow between neighboring groups (especially in a possible cul-de-sac like Europe); 3) a marked increase in gene flow into the region from neighboring populations; and/or 4) the in-migration of a substantial population from a neighboring region, with little genetic contribution to the subsequent population from the preceding Neandertals.

The decision as to which of these processes was dominant depends upon prior decisions as to the relative genetic versus stress-related components in the morphological changes (and their implications for potential rates of changes) and observations as to how much time was available for the transition across the region occupied by Neandertals. The recent discovery of a Châtelperronian-associated Neandertal at Saint-Césaire (Lévêque and Vandermeersch, 1981) reduces the available time for the transition in Western Europe to, at the most, a few thousand years (Stringer et al., 1984), but available dates for diagnostic human remains from further east in Europe (Smith, 1982, 1984) and in western Asia (Trinkaus, 1983a, 1984a) allow for considerably more time, upwards of 5,000–10,000 years (assuming that the Qafzeh specimens were contemporaneous with the Skhūl sample at ca. 40,000 years BP (Jelinek, 1982; Trinkaus, 1983a); if they were markedly older [per Bar Yosef and Vandermeersch (1981) and Vandermeersch (1981)], there would be no time available in the Levant for such a morphological transition).

Interestingly, the time of the transition, indicated primarily by the earliest unequivocal dates for anatomically modern humans in each area, appears to have taken place first in the Near East (ca. 40,000 years BP) (Trinkaus, 1984a), next in Eastern and Central Europe (ca. 35,000 years BP) (Smith, 1984), and last in Western Europe (30–33,000 years BP) (Stringer et al., 1984). This indicates a sloping horizon, running from east to west through time, and suggests elevated gene flow into the west with the new variants derived from more eastern populations. However, this pattern does not demonstrate to what extent the morphological changes were genetically based or give significant insight into the population dynamics in any one region of Europe or western Asia during the transition. Unfortunately, the earliest appearances of anatomically modern humans in Africa and East Asia are too poorly dated to indicate conclusively whether one or the other of those regions could have been a source for new genetic variation, although the possibility that the transition from archaic *H. sapiens* to modern humans occurred in excess of 50,000 years BP in subsaharan Africa (Beaumont et al., 1978; Singer and Wymer, 1982) makes such an interpretation attractive.

MORPHOLOGICAL CHANGE AT THE TRANSITION

These speculations must come back, eventually, to a consideration of the paleontological record and the nature of change around this Late Pleistocene transition. It should be mentioned that all of the currently available evidence, cranial and postcranial, supports the conclusion that the Neandertals of the early last glacial evolved gradually out of Middle Pleistocene archaic *H. sapiens* during the last interglacial across Europe and western Asia (Hublin, 1978; Wolpoff, 1980a,b; Trinkaus, 1982b, 1983a; Cook et al., 1982). The human morphological changes in the middle of the last glacial, between the Neandertals and early modern appearing humans, were significantly less gradual and more pervasive anatomically. The anatomical complexes for which the fossil record around this transition provides information can be divided into three groups: 1) those that exhibit no significant change, 2) those that underwent a morphological shift in which the ranges of variation on either side of the transition overlapped but the sample means changed markedly, and 3) those that demonstrate a discrete morphological change with little or no overlap between the two samples.

In considerations of morphological change, *all* anatomical complexes for which data are available should be evaluated, regardless of available sample size. There is a tendency for some complexes to be considered "more relevant" to paleontological interpretations, resulting usually in pronounced gnathocentrism. This approach assumes that the researcher knows *a priori* the roles of individual anatomical complexes in human adaptive evolution and can therefore make decisions as to relative importance. In actual practice, this is rarely the case.

A number of important human complexes changed little if at all during this Late Pleistocene transition, and it is primarily these that are responsible for the inclusion of the Neandertals in *H. sapiens*. They include brain size and cerebral sulcal morphology (Vlček, 1969; Kochetkova, 1978; Holloway, 1981; Trinkaus and LeMay, 1982; Trinkaus, 1983a), posture and locomotor anatomy indicative of fully efficient bipedalism (Straus and Cave, 1957; Trinkaus, 1975, 1983a), most upper limb articular morphology (Heim, 1982a; Trinkaus, 1983a), posterior dental dimensions and morphology (Brace, 1979; Smith, 1976; Trinkaus, 1983a), deciduous dental dimensions (P. Smith, 1978; Tillier, 1979; Trinkaus, 1983a), and average body mass (Trinkaus, 1983a, 1984b).

Alterations in the second group of complexes were more ubiquitous. There was a shift in neurocranial shape to higher and more rounded cranial vaults, with an associated reduction in the size and frequency of occipital buns (Stringer, 1978; Smith, 1983; Trinkaus and LeMay, 1982). The frequency of mandibular foramina with the horizontal-oval morphology decreased (F.H. Smith, 1978). Anterior teeth decreased in size, absolutely and relative to posterior teeth, and they became less frequently shovel-shaped (Patte, 1959; Brace, 1979; Smith, 1983; Trinkaus, 1983a). Deciduous teeth, especially molars, decreased in occlusal morphological complexity (P. Smith, 1978). Thumb carpometacarpal articulations became universally double saddle-shaped, rather then frequently lacking a dorso-palmar concavity (Musgrave, 1971; Vlček, 1975; Stoner and Trinkaus, 1981; Trinkaus, 1983a). There was an increase in mean stature (Trinkaus, 1983a), and postcranial robusticity reduced markedly, especially in the Near East (Trinkaus, 1983b, 1984a). This last is reflected in cervical vertebral spine thickness and length, rib thickness, upper limb muscular development, and lower limb diaphyseal proportions and cross-sectional areas; in some of these anatomical regions the change in robusticity was sufficiently pronounced to suggest that they should be included in the third category of pattern of change at the transition.

The third category includes several diverse regions. Midfacial prognathism reduced markedly; the reduction consisted primarily of a general facial shortening without any change in the position of the zygomatic region, and included, as secondary spatial and/or architectural effects, reduction in supraorbital torus thickness and projection (especially laterally), reduced nasal projection, increased zygomatic curvature and associated formation of canine fossae, reduction of retromolar spaces, more anterior positioning of mental foramina, and increased mental protuberance projection (Howells, 1975; Stringer, 1978; Smith and Ranyard, 1980; Wolpoff et al., 1981; Smith, 1983; Trinkaus, 1983a, 1984a). There was a change in masto-occipital morphology and temporal (tympanic) configurations (Vallois, 1969; Smith, 1976; Hublin, 1978; Santa Luca, 1978; Trinkaus, 1983a). Pubic morphology and dimensions shifted, producing less attenuated and shorter pubic bones, and hence smaller pelvic apertures (Trinkaus, 1976, 1984b). Distal/proximal limb segment proportions shifted from the lower to the upper limits of modern human ranges of variation (Trinkaus, 1981). Pollical and hallucial phalanges changed from being subequal in length to the distal one being shorter than its proximal phalanx (Musgrave, 1971; Trinkaus, 1983a,c). The final feature in this group may be meningeal vascular branching patterns, which became more complex in early modern humans (Saban, 1977).

These various changes can also be sorted into 1) those that remained stable throughout the Late Pleistocene; 2) those that evidenced gradual change prior and subsequent to the transition, with no acceleration in the rate of change around the middle of the last glacial; 3) those that changed on either side of the transition but still accelerated their rates of change around the transition; and 4) those that exhibited relative stasis before and after the transition and changed markedly at the time of the transition.

The first are primarily those features that changed little if at all around the transition, namely brain size and sulcal morphology, deciduous dental dimensions, and body mass. The last did decrease slightly during the late last glacial (Frayer, 1981) but changed little during the previous part of the Late Pleistocene.

The features that changed gradually throughout the Late Pleistocene, with little or no acceleration around the transition, are primarily those related to general mastication. These include posterior dental dimensions and reflections of general facial massiveness, such as the relative anterior positioning and rugosity of masticatory muscle attachments, mandibular robusticity, and supraorbital torus thickness and projection. Interestingly, these are features that appear to have been reducing gradually during most of the Middle, as well as the Late Pleistocene.

The third group are primarily those related to paramasticatory use of the anterior dentition but include primarily those aspects relating to levels of stress generated by the activity, rather than patterns of tooth use. They comprise anterior dental dimensions, rates of anterior dental attrition, rugosity of the nuchal muscle attachments, and possibly meningeal vascular branching patterns.

The last group of features is more varied. In it are pubic bone morphology and relative breadth, limb segment proportions, mean stature, thumb phalangeal length proportions and carpo-metacarpal articular morphology, hallucial phalangeal length proportions, most aspects of postcranial robusticity, masto-occipital and tympanic morphology, total facial prognathism, and deciduous dental morphology.

It should be apparent from these lists that the changes in human morphology during the Late Pleistocene in the northwestern Old World were highly mosaic. They varied considerably in degree and constancy of rate of change, and therefore as many as possible should be taken into consideration when using them for formulating phylogenetic or adaptational scenarios.

PHYLOGENETIC IMPLICATIONS OF THE CHANGES

The evaluation of these and other observable morphological traits and their patterns of change through the Late Pleistocene for phylogenetic purposes is dependent upon the assumptions one brings to them. Their genetic and developmental determinants, as discussed above, should be evaluated, if that is possible. Perhaps more important—and more resolvable—are the criteria used to divide up the observed morphology into units of analysis.

For example, does one evaluate the entire dentition as a unit or divide it into anterior and posterior fields? Does one merely measure mid-facial prognathism or consider it to be a secondary reflection of relative positionings of the dentition and the primary masticatory muscles? Are pollical and hallucial phalangeal proportions reflections of manipulative versus locomotor demands or is one merely a pleiotropic effect of demands on the homologous structure in the other limbs? Do the changes in limb segment proportions indicate an improved ability to deal with thermal stress or more efficient long distance locomotion? And is it possible that the changes in neurocranial shape are secondary reflections of relative brain and cranial vault growth rates, determined by levels of perinatal stimulation of brain growth due to changing reproductive patterns, the last of which is best reflected in the pelvis? Or is it possible that shifts in the shape of the posterior cranium (including occipital bunning) relate to a change in the extent and orientation of the nuchal plane, which is in turn associated with the amount of paramasticatory use of the anterior dentition? These and other structurally related questions need to be answered before any attempts are made to identify plesiomorphic, apomorphic, or autapomorphic traits or engage in similar mental exercises. Otherwise, we will merely end up counting traits of uncertain functional and developmental correlation and unequal phylogenetic significance.

In addition, it should be recognized that the evolutionary fate of the Neandertals need not have been the same across the large geographical region occupied by the Neandertals during the early last glacial. In fact, the diversity of habitats and topographies, as well as accessibility to neighboring regions, suggests that patterns of gene flow and population dynamics varied considerably across this region.

The ultimate resolution of whether all, a few, or none of the Neandertals were ancestral to early anatomically modern humans in Europe and western Asia will be achieved when there is general agreement on how one should interpret the paleontological record and/or there is incontrovertible evidence of local contemporaniety of the two morphological patterns across the northwestern Old World. Neither of these conditions are currently met, and it is uncertain whether they will ever be. It may therefore be more profitable to set aside the ultimate phylogenetic question and concentrate efforts on functional evaluations of the discernable morphology and its patterns of change during the Late Pleistocene. Such a concentration of efforts may eventually hold the key to the phylogenetic question, since functional studies should ultimately allow us to evaluate whether, for example, supposed Neandertal autapomorphies were clearly functionally based and likely to change rapidly in response to specific behavioral shifts. We view this approach as the more profitable one in the long run, whether the ultimate goal is the resolution of either hominid phylogeny or the evolution of human adaptive patterns.

BEHAVIORAL IMPLICATIONS OF THE CHANGES

The human morphological changes during the Late Pleistocene indicate a series of behavioral shifts during this time period, most of which were concentrated around the Neandertal to anatomically modern human transition. The implied shifts in human adaptation are reflected in the contemporaneous archeological record.

It should be emphasized that it is possible to make behavioral interpretations irrespective of the actual phylogenetic relationships between the Neandertals and early modern humans. The morphological pattern of the Neandertals was replaced by that of modern humans, so that it is evident that the latter had selective advantages, in the context of a changing cultural adaptive system, *vis-à-vis* the former. The consistent differences between the two morphological patterns can therefore be investigated from a strictly functional perspective and, along with interpretations of the associated archeological record, provide insights into Late Pleistocene human adaptive evolution.

The changes in the masticatory apparatus during this period indicate a general reduction in force and repetitiveness of human chewing (Brace, 1979; F.H. Smith, 1978, 1983; Trinkaus, 1983a). This is reflected in the posterior retreat of the masticatory muscles (which initially produced the mid-facial prognathism of the Neandertals), the reduction of mandibular robusticity and masticatory muscle attachment rugosity, and decreasing tooth size throughout the Late Pleistocene. This was associated with the extensive use of the anterior dentition by the Neandertals for paramasticatory purposes, which maintained their large anterior teeth and anteriorly placed dentitions (Ryan, 1980; Smith, 1983; Trinkaus, 1983a). A marked reduction in non-dietary anterior tooth use apparently occurred at the time of the transition to anatomically modern humans, which resulted in the associated reductions in anterior dental dimensions and rates of attrition and in total facial prognathism. The changes in the morphology of the nuchal region with the advent of modern appearing humans were probably related to the shift in the pattern of use of these muscles once they were no longer counteracting forces applied to the anterior dentition (Wolpoff, 1980a; Smith, 1983).

The shift in pubic morphology produced a decrease in pelvic aperture dimensions relative to adult body size, which indicates a shift from a gestation length close to the 11–12 months expected for humans of the brain and body size of Neandertals and early modern humans to the 9 months characteristic of modern humans (Trinkaus, 1983b, 1984b). It is possible that the earlier exposure of the neonate to environmental stimuli, combined with the generally richer cultural environment of early modern humans, promoted an accelerated rate of brain growth relative to cranial vault ossification. This would produce more anterior and superior brain growth and hence higher and rounder adult neurocrania (Trinkaus and LeMay, 1982; Trinkaus, 1984a).

The marked decrease in postcranial massiveness evident across the transition indicates a shift from habitual use of elevated physical strength and high levels of activity among the Neandertals to a greater reliance on culturally based technology and planning to accomplish regular subsistence activities (Trinkaus, 1983b). When combined with the increased stature, the lengthening of distal limb segments, and the narrowing of the pelvis, all of which would increase locomotor efficiency, it suggests that the early anatomically modern humans were better adapted to covering large distances and hence able to monitor more effectively and exploit more efficiently the available resources with less physical durability and strength. These changes were undoubtedly correlated with the shifts in thumb morphology, all of which indicate a decrease in use of the power grip and associated increased emphasis on the precision grip and fine manipulation (the hallucial phalangeal length shift was probably pleiotropically related to the more important pollical alterations) (Stoner and Trinkaus, 1981; Trinkaus, 1983c).

The shift in limb segment proportions also implies an increased adaptation to thermal stress, since the Neandertal pattern is indicative of cold stress in modern mammals and that of the early modern humans of warmer climates (Trinkaus, 1981). Since both groups were subjected to similar climates, the shift implies a marked increase in heat generating and conserving abilities among early modern humans.

The morphological changes at the time of this Late Pleistocene transition therefore indicate a continuation of previous trends in the reduction of the masticatory apparatus associated with major changes in human reproduction, non-dietary tooth use, habitual activity levels, locomotor efficiency, and thermal adaptation. The contemporaneous archeological record largely confirms these interpretations. It indicates little change in food preparation techniques but a major increase in technology as reflected in artifact assemblages (lithic and osteological) (Sonneville-Bordes, 1963; Klein, 1973; Bordes, 1981; White, 1982; Harrold, 1983; Marks and Volkman, 1983). There are suggestions of a shift toward more efficient exploitation of animal resources and avoidance of carnivores (Binford, 1982, 1984; Straus, 1982; White, 1982). These changes were accompanied by reflections of increased social complexity (larger, more complex, and more variable sites, elaboration of burials, emergence of style zones) (Harrold, 1980; Klein, 1973; Hietala, 1983), major elaboration of information systems ("art") (Marshack, 1972; Conkey, 1983), and improved means of heat production and conservation (pit hearths, structures, sewn clothing) (Movius, 1966; Klein, 1973; Stordeur-Yedid, 1979). All of these could have allowed the habitual birth of altricial infants, which in turn could have provided demographic and energetic advantages.

These adaptive shifts around the Late Pleistocene archaic to anatomically modern human transition undoubtedly involved complex biocultural feedbacks, with changes in each sphere permitting or promoting changes in the other. It is not possible to determine which sphere was the prime mover, but attempts to construct causal directions (e.g., Trinkaus, 1983b) indicate that culture and biology were about equally important in setting the process in motion.

THE FATE OF THE NEANDERTALS

These considerations should make it evident that the available fossil record does not permit us to decide conclusively what the phylogenetic fate of the European and western Asian Neandertals might have been. However, comparative analyses of their functional morphology indicate that their fate was to provide a background for the evolutionary emergence of people behaviorally and anatomically similar to ourselves. It is ironic that it is through contrasts of early modern appearing humans to the much maligned Neandertals that we will have the means to understand the evolutionary processes responsible for the origins of anatomically modern humans.

ACKNOWLEDGMENTS

We would like to express our sincere appreciation to the many individuals in Europe, western Asia, and North America who have allowed us to examine fossil human remains in their care. This research has been supported by grants from the Wenner-Gren Foundation, the National Science Foundation, the National Academy of Sciences, the Alexander von-Humboldt Foundation, Harvard University, the University of Tennessee, and the University of New Mexico.

LITERATURE CITED

Bar Yosef, O, and Vandermeersch, B (1981) Notes concerning the possible age of the Mousterian layers in Qafzeh Cave. In P Sanlaville and J Cauvin (eds): Préhistoire du Levant. Paris: Editions du C.N.R.S., pp. 281–285.

Beaumont, PB, Villiers, H de, and Vogel, JC (1978) Modern man in sub-saharan Africa prior to 49,000 years B.P.: A review and evaluation with particular reference to Border Cave. S. Afr. J. Sci. *74*:409–419.

Binford, LR (1982) Comment on: Rethinking the Middle/Upper Paleolithic transition. Curr. Anthropol. *23*:177–181.

Binford, LR (1984) Faunal Remains From Klasies River Mouth. New York: Academic Press.

Bordes, F (1981) Vingt-cing ans après: Le complexe moustérien revisité. Bull. Soc. Préhist. Franç. *78*: 77–87.

Boule, M (1911–1913) L'homme fossile de La Chapelle-aux-Saints. Ann. Paléontol. *6:*111–172; *7:*21–56, 85–192; *8:*1–70.

Boule, M (1921) Les Hommes Fossiles. Paris: Masson.

Brace, CL (1964) The fate of the "classic " Neanderthals: A consideration of hominid catastrophism. Curr. Anthropol. *5:*3–43.

Brace, CL (1979) Krapina, "Classic" Neanderthals, and the evolution of the European face. J. Hum. Evol. *8:*527–550.

Conkey, MW (1983) On the origins of Paleolithic art: A review and some critical thoughts. In E Trinkaus (ed): The Mousterian Legacy: Human Biocultural Change in the Upper Pleistocene. Brit. Archaeol. Rep. *S164:*201–227.

Cook, J, Stringer, CB, Currant, AP, Schwarcz, HP, and Wintle, AG (1982) A review of the chronology of the European Middle Pleistocene Hominid record. Yrbk. Phys. Anthropol. *25:*19–65.

Frayer, DW (1981) Body size, weapon use, and natural selection in the European Upper Paleolithic and Mesolithic. Am. Anthropol. *83:*57–73.

Gorjanović-Kramberger, D (1906) Der diluviale Mensch von Krapina in Kroatien. Wiesbaden: Kriedels Verlag.

Harrold, FB (1980) A comparative analysis of Eurasian Palaeolithic burials. World Archaeol. *12:*195–211.

Harrold, FB (1983) The Châtelperronian and the Middle Upper Paleolithic transition. In E Trinkaus (ed): The Mousterian Legacy: Human Biocultural Change in the Upper Pleistocene. Brit. Archaeol. Rep. *S164:*123–140.

Heim, JL (1978) Contribution du massif facial à la morphogenèse du crâne néanderthalien. In J Piveteau (ed): Les Origines Humaines et les Époques de l'Intelligence. Paris: Masson, pp. 183–215.

Heim, JL (1982a) Les hommes fossiles de La Ferrassie II. Arch. Inst. Paléontol. Hum. *38:*1–272.

Heim, JL (1982b) Les Enfants Néandertaliens de La Ferrassie. Paris: Masson.

Hietala, H (1983) Boker Tachtit: Intralevel and interlevel spatial analysis. In AE Marks (ed): Prehistory and Paleoenvironments in the Central Negev, Israel III. Dallas: Dept. of Anthropology, Southern Methodist University, pp. 217–282.

Holloway, RL (1981) Volumetric and asymmetry determinations on recent hominid endocasts: Spy I and II, Djebel Irhoud I, and the Salé *Homo erectus* specimens, with some notes on Neandertal brain size. Am. J. Phys. Anthropol. *55:*385–393.

Howells, WW (1975) Neanderthal man: Facts and figures. In RH Tuttle (ed): Paleoanthropology: Morphology and Paleoecology. The Hague: Mouton, pp. 389–407.

Howells, WW (1978) Position phylétique de l'homme de Néanderthal. In J Piveteau (ed): Les Origines Humaines et les Époques de l'Intelligence. Paris: Masson, pp. 217–235.

Hrdlička, A (1927) The Neanderthal phase of man. J. Roy. Anthropol. Inst. *57:*249–274.

Hublin, JJ (1978) Le Torus Occipital Transverse et les Structures Associées: Évolution dans le Genre *Homo.* Thèse de Troisieme Cycle, Univ. de Paris.

Hublin, JJ (1980) La Chaise Suard, Engis 2 et La Quina H 18: Développement de la morphologie occipitale externe chez l'enfant prénéandertalien et néandertalien. C.R. Acad. Sci. Paris *291D:*669–672.

Hublin, JJ (1983) Les origines de l'homme de type moderne en Europe. Pour la Science *64:*62–71.

Jelinek, AJ (1982) The Tabun Cave and paleolithic man in the Levant. Science *216:*1369–1375.

Keith, A (1911) Ancient Types of Man. London: Harper.

Keith, A (1915) The Antiquity of Man. London: Williams and Norgate.

Klein, RG (1973) Ice-Age Hunters of the Ukraine. Chicago: University of Chicago Press.

Kochetkova, VI (1978) Paleoneurology. Washington: Winston and Sons.

Lévêque, F, and Vandermeersch, B (1981) Le néandertalien de Saint-Césaire. La Recherche *12:*242–244.

Mann, AE, and Trinkaus, E (1974) Neandertal and Neandertal-like fossils from the Upper Pleistocene. Yrbk. Phys. Anthropol. *17:*169–193.

Marks, AE, and Volkman, PW (1983) Changing core reduction strategies: A technological shift from the Middle to Upper Paleolithic in the southern Levant. In E Trinkaus (ed): The Mousterian Legacy: Human Biocultural Change in the Upper Pleistocene. Brit. Archaeol. Rep. *S164:*13–34.

Marshack, A (1972) Cognitive aspects of Upper Paleolithic engraving. Curr. Anthropol. *13:*445–477.

Mortillet, G de (1883) Le Préhistorique, origine et antiquité de l'homme. Paris: Reinwald.

Movius, HL, Jr (1966) The hearths of the Upper Perigordian and Aurignacian horizons at the Abri Pataud, Les Eyzies (Dordogne) and their possible significance. Am. Anthropol. *68:*296–325.

Musgrave, JH (1971) How dextrous was Neanderthal man? Nature *233:*538–541.

Osborn, HF (1918) Men of the Old Stone Age, 3rd ed. New York: Scribners.

Patte, E (1959) La dentition des Néanderthaliens. Ann. Paléontol. *45:*221–305.

Ryan, AS (1980) Anterior Dental Microwear in Hominid Evolution: Comparisons With Humans and Nonhuman Primates. Ph.D. Thesis, University of Michigan. Ann Arbor: University Microfilms.

Saban, R (1977) Les impressions vasculaires pariétales endocrâniennes dans la lignée des Hominidés. C.R. Acad. Sci. Paris *284D:*803–806.

Santa Luca, AP (1978) A re-examination of presumed Neandertal-like fossils. J. Hum. Evol. *7:*619–636.

Schwalbe, G (1904) Die Vorgeschichte des Menschen. Braunschweig: Vieweg und Sohn.

Singer, R, and Wymer, J (1982) The Middle Stone Age at Klasies River Mouth in South Africa. Chicago: University of Chicago Press.

Smith, FH (1976) The Neandertal Remains from Krapina: A Descriptive and Comparative Study. Univ. Tenn. Dept. Anthropol. Rep. Invest. *15:*1–359.

Smith, FH (1978) Evolutionary significance of the mandibular foramen area in Neandertals. Am. J. Phys. Anthropol. *48:*523–532.

Smith, FH (1982) Upper Pleistocene hominid evolution in South-Central Europe: A review of the evidence and analysis of trends. Curr. Anthropol. *23:*667–703.

Smith, FH (1983) A behavioral interpretation of changes in craniofacial morphology across the Archaic/Modern *Homo sapiens* transition. In E Trinkaus (ed): The Mousterian Legacy: Human Biocultural Change in the Upper Pleistocene. Brit. Archaeol. Rep. *S164:* 141–164.

Smith, FH (1984) Fossil hominids from the Upper Pleistocene of Central Europe and the Origin of Modern Europeans. In FH Smith and F Spencer (eds): The Origins of Modern Humans. New York: Alan R. Liss, Inc., pp. 137–209.

Smith, FH, and Ranyard, GC (1980) Evolution of the supraorbital region in Upper Pleistocene fossil hominids from South-Central Europe. Am. J. Phys. Anthropol. *53:*589–610.

Smith, FH, and Spencer, F (eds) (1984) The Origins of Modern Humans. New York: Alan R. Liss, Inc.

Smith, P (1978) Evolutionary changes in the deciduous dentition of Near Eastern Populations. J. Hum. Evol. *7:*401–408.

Sollas, WJ (1907) On the cranial and facial characters of the Neandertal race. Phil. Trans. Roy. Soc. London, ser. B *199:*281–339.

Sollas, WJ (1924) Ancient Hunters, 3rd ed. New York: Macmillan.

Sonneville-Bordes, D de (1963) Upper Paleolithic cultures in western Europe. Science *142:*347–360.

Spencer, F (1984) The Neandertals and their evolutionary significance: A brief historical survey. In FH Smith and F Spencer (eds): The Origins of Modern Humans. New York: Alan R. Liss, Inc., pp. 1–49.

Spencer, F, and Smith, FH (1981) The significance of Aleš Hrdlička's "Neanderthal Phase of Man:" A historical and current assessment. Am. J. Phys. Anthropol. *56:*435–459.

Stoner, BP, and Trinkaus, E (1981) Getting a grip on the Neandertals: Were they all thumbs? Am. J. Phys. Anthropol. *54:*281–282 (Abstract).

Stordeur-Yedid, D (1979) Les aiguilles à chas au Paléolithique. Gallia Préhist. Suppl. *11:*1–215.

Straus, LG (1982) Carnivores and cave sites in Cantabrian Spain. J. Anthropol. Res. *38:*75–96.

Straus, WL, Jr, and Cave, AJE (1957) Pathology and the posture of Neanderthal man. Quart. Rev. Biol. *32:*348–363.

Stringer, CB (1978) Some problems in Middle and Upper Pleistocene Hominid relationships. In DJ Chivers and KA Joysey (eds): Recent Advances in Primatology, Vol. 3. London: Academic Press, pp. 395–418.

Stringer, CB (1982) Towards a solution to the Neanderthal problem. J. Hum. Evol. *11:*431–438.

Stringer, CB, Hublin, JJ, and Vandermeersch, B (1984) The origin of anatomically modern humans in Western Europe. In FH Smith and F Spencer (eds): The Origins of Modern Humans. New York: Alan R. Liss, Inc., pp. 51–135.

Tillier, AM (1979) La dentition de l'enfant moustérien Chateauneuf 2 découverte à l'Abri de Hauteroche (Charente). L'Anthropol. *83:*417–438.

Tillier, AM (1982) Les enfants néanderthaliens de Devil's Tower (Gibraltar). Z. Morphol. Anthropol. *73:* 125–148.

Trinkaus, E (1975) A Functional Analysis of the Neandertal Foot. Ph.D. Thesis, University of Pennsylvania. Ann Arbor: University Microfilms.

Trinkaus, E (1976) The morphology of European and Southwest Asian Neandertal pubic bones. Am. J. Phys. Anthropol. *44:*95–104.

Trinkaus, E (1981) Neanderthal limb proportions and cold adaptation. In CB Stringer (ed): Aspects of Human Evolution. London: Taylor and Francis, pp. 187–224.

Trinkaus, E (1982a) A history of *Homo erectus* and *Homo sapiens* paleontology in America. In F Spencer (ed): A History of American Physical Anthropology, 1930–1980. New York: Academic Press, pp. 261–280.

Trinkaus, E (1982b) Evolutionary continuity among archaic *Homo sapiens.* In A Ronen (ed): The Transition from Lower to Middle Palaeolithic and the Origin of Modern Man. Brit. Archaeol. Rep. *S151:*301–314.

Trinkaus, E (1983a) The Shanidar Neandertals. New York: Academic Press.

Trinkaus, E (1983b) Neandertal postcrania and the adaptive shift to modern humans. In E Trinkaus (ed): The Mousterian Legacy: Human Biocultural Change in the Upper Pleistocene. Brit. Archaeol. Rep. *S164:*165–200.

Trinkaus, E (1983c) Functional aspects of Neandertal pedal remains. Foot Ankle *3:*377–390.

Trinkaus, E (1984a) Western Asia. In FH Smith and F Spencer (eds): The Origins of Modern Humans. New York: Alan R. Liss, Inc., 251–293.

Trinkaus, E (1984b) Neandertal pubic morphology and gestation length. Curr. Anthropol. *25:*509–514.

Trinkaus, E, and Howells, WW (1979) The Neanderthals. Sci. Amer. *241(6):*118–133.

Trinkaus, E, and LeMay, M (1982) Occipital bunning among later Pleistocene hominids. Am. J. Phys. Anthropol. *57:*27–35.

Vallois, HV (1969) Le temporal néandertalien H-27 de La Quina: Étude anthropologique. L'Anthropol. *73:*365–400, 525–544.

Vandermeersch, B (1981) Les Hommes Fossiles de Qafzeh (Israël). Paris: Éditions du C.N.R.S.

Verneau, R (1924) La race de Neanderthal et la race de Grimaldi; leurs rôles dans l'humanité. J. Roy. Anthropol. Inst. *54:*211–230.

Vlček, E (1969) Neandertaler der Tschechoslowakei. Prague: Tschechoslowakische Akademie der Wissenschaften.

Vlček, E (1970) Étude comparative onto-phylogénétique de l'enfant du Pech-de-l'Azé par rapport à d'autres enfants néandertaliens. Arch. Inst. Paléontol. Hum. *33:*149–178.

Vlček, E (1973) Postcranial skeleton of a Neandertal child from Kiik-Koba, U.S.S.R. J. Hum. Evol. *2:* 537–555.

Vlček, E (1975) Morphology of the first metacarpal of Neanderthal individuals from the Crimea. Bull. Mém. Soc. Anthropol. Paris, sér. 13 *2:*257–276.

Weinert, H (1932) Ursprung der Menschheit. Stuttgart: Enke.

White, R (1982) Rethinking the Middle/Upper Paleolithic transition. Curr. Anthropol. *23:*169–192.

Wolpoff, MH (1980a) Paleoanthropology. New York: Knopf.

Wolpoff, MH (1980b) Cranial remains of Middle Pleistocene European hominids. J. Hum. Evol. *9:*339–358.

Wolpoff, MH, Smith, FH, Malez, M, Radovčić, J, and Rukavina, D (1981) Upper Pleistocene human remains from Vindija Cave, Croatia, Yugoslavia. Am. J. Phys. Anthropol. *54:*499–545.

Ancestors: The Hard Evidence, pages 334–338

Early *Homo* From Narmada Valley, India

Arun Sonakia

Regional Palaeontological Laboratory, Geological Survey of India, Central Region, Nagpur—440 010, India

ABSTRACT This paper records the long-awaited discovery of ancient man in India. Part of a cranium of *Homo* sp. from the basal gravel-conglomerate bed of the Middle Pleistocene Narmada Valley Alluvium was collected at Hathnora, near Hoshangabad. The cranium was found associated with Acheulian artefacts and a contemporary mammalian fauna. The diagnostic features of the cranium are reviewed and compared with the known fossil hominids. The Narmada Valley cranium, informally named Narmada Man, is similar to the "anteneanderthals" of Europe, but in parietal and dorsal-outline shape, low (bimastoid) position of greatest breadth, and presence of a sagittal keel, it appears more comparable to *Homo erectus*.

INTRODUCTION

The Pleistocene alluvial deposits of the Narmada Valley have been of global interest to geologists and archaeologists alike for more than a century because of the rich faunal and cultural material recovered there. The rich mammalian fauna of the Narmada Valley Alluvium of Central India has been discussed by Princep (1833), Splisbury (1833, 1837, 1844), Lydekker (1880, 1882, 1884), De Terra and De Chardin (1936), De Terra and Patterson (1939), Hooijer (1963), Khatri (1966), Badam (1979), Dassarma (1979), and Biswas and Dassarma (1981), among others. Theobald (1881) noted that part of a human cranium, which was collected from beds similar to the younger Narmada gravels, was lost from the Museum of the Asiatic Society of Bengal, Calcutta. However, no unequivocal evidence of the presence of fossil man from these alluvial deposits has been recorded by previous workers on the Narmada Valley Pleistocene fauna.

The author's discovery of an "Early *Homo*" cranium in the Narmada Valley came as a sequel to a long-term research project of the Regional Palaeontological Laboratory, Central Region, Geological Survey of India, Nagpur, to elucidate the Quaternary biostratigraphy of the Narmada Valley based on mammalian fauna. Field work commenced in the Hoshangabad area of Madhya Pradesh in the last week of November, 1982, and a hominid skull cap was discovered on December 5, 1982 (Anonymous, 1983). The skull cap was found well embedded in a fossiliferous gravel-conglomerate bed that forms the basal part of the middle to upper Pleistocene deposits of the Narmada Valley, near the village of Hathnora, 40 km east of Hoshangabad in Madhya Pradesh. The bed is exposed close to the river channel as a 50 m-wide platform for over 300 m all along the river course west of Hathnora. The thickness of these deposits at the site of the *Homo* find is 18 m, reaching a maximum of over 200 m in the central part of the Narmada Basin. These deposits are believed to be of lacustrine and fluviatile origin. In the Hathnora area four stratigraphic units of Pleistocene age, capped by black cotton soil, are exposed (Fig. 1). The basal unit is composed of cemented gravel conglomerate interlayered with thin bands of sand. Although

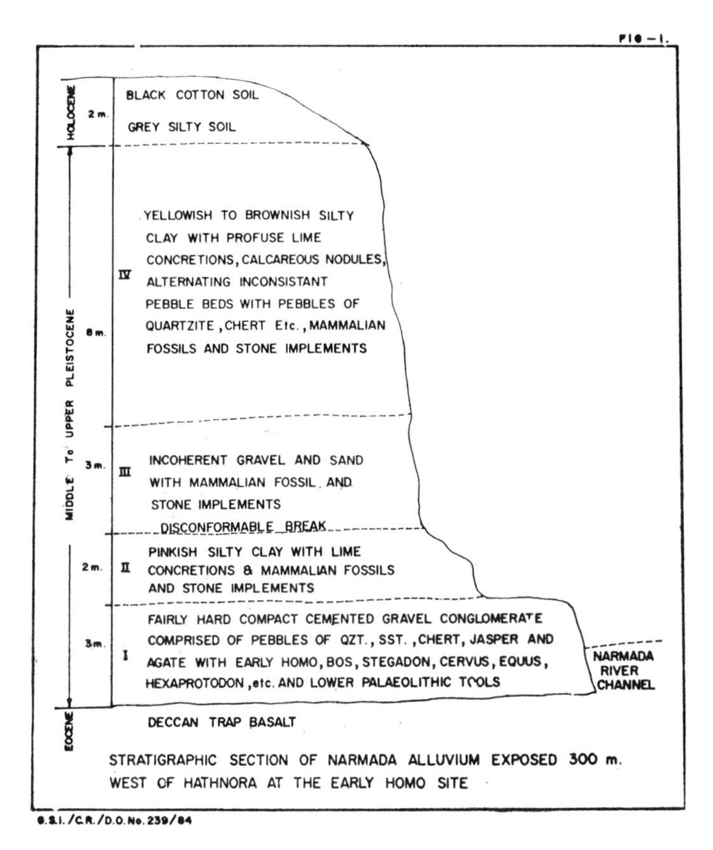

Fig. 1. Stratigraphic section of Narmada Alluvium exposed 300 m west of Hathnora (near Hoshangabad, India) at the early *Homo* site.

Badam (1979) and Dassarma (1979) consider this unit as of probable early Late Pleistocene age (or possibly latest Middle Pleistocene), the author considers it to be fully Middle Pleistocene.

The fossil skullcap was found embedded in the upper part of this gravel-conglomerate bed. The *Homo* cranium was found associated with a complete skull of a bovid, *Bos namadicus*, with horncores measuring 80 cm in length, and large skeletal remains of *Stegodon* sp. Besides these, other mammalian remains found associated with the *Homo* skullcap included *Bubalus palaeindicus*, *Leptobos* sp., *Cervus duvaucelli*, *Hexaprotodon namadicus*, *Stegodon ganesa*, *Stegodon insignis*, *Elephas namadicus*, and *Gazella* sp.

MORPHOLOGICAL DETAILS OF THE SKULL CAP

The *Homo* fossil consists of the right half of the skull cap with the posterior part of the left parietal attached to it. The face, most of

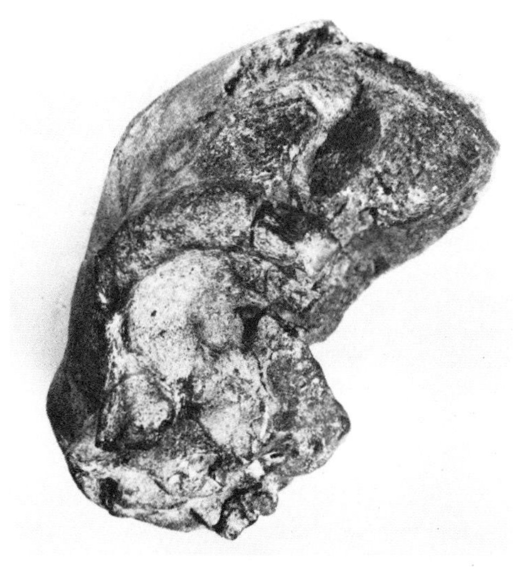

Fig. 2. Frontal view of Narmada skullcap. × 0.48.

Fig. 3. Lateral view of Narmada skullcap. × 0.50.

the left parietal, the occipital, and the basal bones are lacking. The sagittal, coronal, and lambdoid sutures are completely closed suggesting that the skull belonged to a fairly old individual. However, the obliterated traces of the sutures provide clues to the identification of bregma, lambda, and asterion.

In frontal view (Fig. 2), the cranium presents a forwardly projecting, arched supraorbital torus. The thickness of bone at the torus margin is 13 mm. There is a slight post-orbital constriction. The malar bone is not prominent.

In lateral view (Fig. 3), the forehead recedes smoothly. The greatest length of the cranium (glabella to opisthocranion, minimally reconstructed) is 198 mm, and its greatest breadth (bimastoid, minimally reconstructed) is 142 mm. Based on these measurements, the cranial index is 72.4. The thickness of bone in the bregma region is 10 mm. From the region of glabella the profile of the vault rises smoothly to bregma. The vertex appears to coincide with bregma when the skull is oriented in the Frankfurt horizontal plane. This profile is almost flat in the sagittal arc region and then slopes gently to a posteriorly bulging occipital protuberance passing over a very shallow supratoral depression. Opisthocranion appears to coincide with inion. The occipital bone does not roll over the nuchal plate. There is a flat, large nuchal plane. The impressions of the nuchal muscles are prominent and may be confused with a few bony islands that are seen on the right parieto-occipital region. There is a laterally bulging, very conspicuous supramastoid crest. The mastoid process must have been small, as evidenced by its broken base. The supramastoid crest is continuous with a heavy zygomatic base. The supraorbital torus passes laterally to a thick zygomatic process and also gives rise to a prominent temporal line. The nearest approach of the temporal line to the sagittal suture is 60 mm.

In dorsal view (Fig. 4), the skull is markedly ovoid, and dolichocephalic. The broken glabella exposes the right frontal sinus. There is a marked raising of the parietals and a linear depression on the sagittal arc at the junction of the parietals. In the basal region the mandibular fossa is shallow. The position of the foramen magnum could be identified by its right lateral margin. The provisional estimate of the cranial capacity, based on the outer dimensions, is a little over 1200 cm^3.

ARTEFACTS

The stone implements found associated with the skullcap, in the same bed, include heavy hand axes, cleavers and scrapers made of quartzite, and smaller tools made of flint and chert, probably of pre-Soan type. Khatri

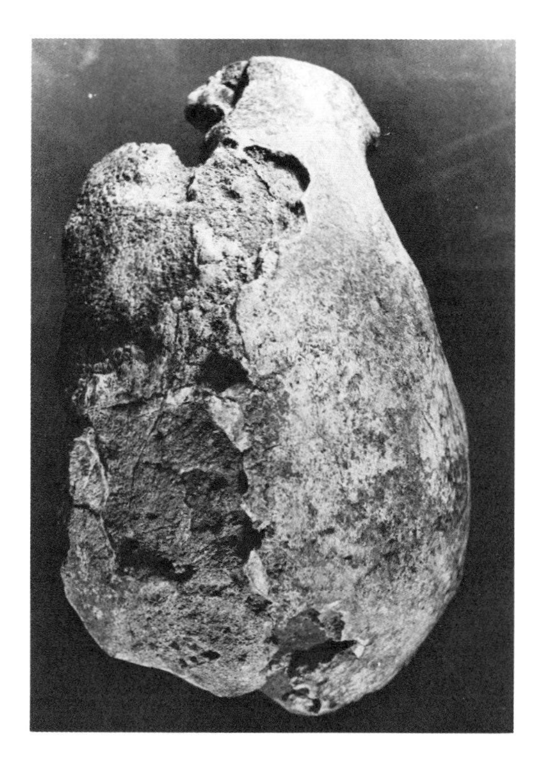

Fig. 4. Dorsal view of Narmada skullcap. × 0.50.

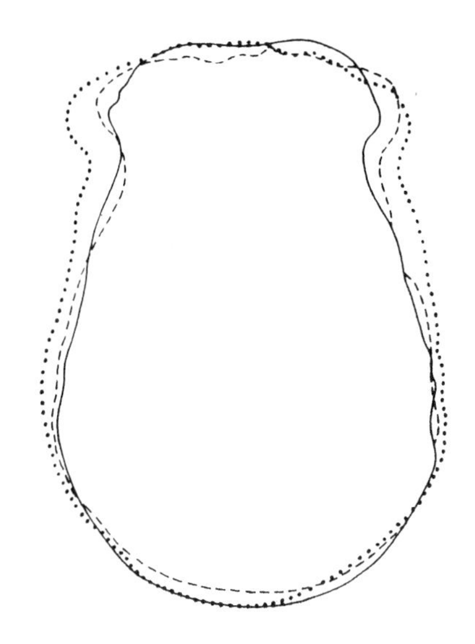

Fig. 5. Superimposition of norma superior outlines of the Narmada specimen (——), *Homo erectus* (Java) (- - -), and "classic" Neanderthal Man (· · · ·).

(1966) describes pear-shaped, ovate hand-axes and "U"-shaped cleavers of Acheulian type from the same basal gravel conglomerate horizon, in other parts of the Narmada Valley.

DISCUSSIONS

The precise assessment of the taxonomic position of Narmada Man will be possible after detailed anatomical studies of the cranium are carried out. Geological, palaeontological, and archaeological correlation must also be taken into account. In having a forwardly projecting, thick, arched supraorbital torus, receding forehead, flat vault dorsum, rapidly sloping parietals, thick cranial walls, and depressed bulging occipital, the Narmada skull appears to resemble the European "anteneanderthals" such as Steinheim or Arago. However, it differs markedly from a Neanderthal or "anteneanderthal" skull in certain features of the occipital and parietal regions. In section, the occipital of Narmada Man is not rounded, but forms a sort of trapezoid, as evidenced by the first rapidly sloping and then nearly vertical parietals. There is also some development of a sagittal keel, and the greatest cranial breadth is in the mastoid region. These characteristics of the mastoid, parietal, and squamosal regions resemble more closely the *Homo erectus* of Java; China, Tanzania, and Kenya. In the outline of its norma superior it also comes closer to Java Man than to a Neanderthal (Fig. 5).

On the basis of the comparative studies carried out so far, it could be tentatively concluded that Narmada Man can be considered as an advanced form of *Homo erectus*. In fact, Sonakia (1984) named it *H. e. narmadensis* and provided additional data. The greatest importance of the present find is that it is the first tangible record of the skeletal remains of Pleistocene man from India.

ACKNOWLEDGMENTS

The author is grateful to the Director General, Geological Survey of India, for very kindly recommending his name to the Government of India for participation in the symposium. He is also thankful to the organisors of the symposium, particularly Dr. Eric Delson, for permitting him to present his finding from India at the Symposium. He also wishes to thank those who have kindly helped with their comments, suggestions, and materials.

LITERATURE CITED

Anonymous (1983) Discovery of fossilised human skull in Narmada Valley Alluvium, Hoshangabad Area. M.P.G.S.I.,C.R., News, Vol. 4(2), July 1983.

Badam, GL (1979) Pleistocene Fauna of India. Deccan Pune: College.

Biswas, S, and Dassarma, DC (1981) New species of *Stegodon* from Narmada Valley. Proc. Field Conference on Neogene/Quaternary boundary, India (IGCP 41) G.S.I. Calcutta.

Dassarma, DC (1979) Some observations on the Quaternary stratigraphy and mammal assemblages of India. G.S.I. Misc. Publ. *45:*279–287.

De Terra H, and De Chardin, T (1936) Observation on the Upper Siwalik formation and later Pleistocene deposits in India. Proc. Am. Phil. Soc. *76:*791–822.

De Terra, H and Paterson, TT (1939) Studies on ice age in India and associated Human Culture. Washington, D.C.: Carn. Inst. Publ. 493.

Hooijer, DA (1963) Preliminary identification of some fossil mammals from Narmada Valley, India. Rev. Sci. Preist. *18:*20.

Khatri, AP (1966) The Pleistocene mammalian fossils of the Narmada river valley and their horizons. Asian Perspec. *9:*113–133.

Lydekker, R (1880) Siwalik and Narbada Equidae. Pal. Ind. Ser X, *1:*162–292.

Lydekker, R (1882) Siwalik and Narbada Proboscidea. Pal. Ind. Ser. X, *2:*63–92.

Lydekker, R (1884) Siwalik and Narbada Bunodont Suidae, Pal. Ind. Ser X, *3:*35–104.

Princep, J (1833) Note on the fossil bones discovered near Jabalpur. J. Asiat. Soc. Bengal *2:*583–588.

Sonakia, A (1984) The skull-cap of early man and associated mammalian fauna from Narmada Valley Alluvium Hoshangabad Area, Madhya Pradesh (India). Rec. Geol. Surv. India *113:*159–172.

Splisbury, CG (1833) Account of the fossil bones discovered in the bed of the Omar Nadi near Narsinghpur at Bedarwara in the Valley of Nerbudda. J. Asiat. Soc. Bengal *2:*388–395.

Splisbury, CG (1837) Notice of new sites of fossil deposits in Nerbudda Valley. J. Asiat. Soc. Bengal *6:*487–489.

Splisbury, CG (1844) Notes on Narbudda fossils. J. Asiat. Soc. Bengal *13:*765–766.

Theobald W (1881) The Siwalik Group of the Sub-Himalayan Region. Rec. Geol. Surv. India *14:*66–125.

Ancestors: The Hard Evidence, pages 339–345

Problems of Paleoanthropological Research in Pakistan

S.M. Ibrahim Shah and S. Mahmood Raza
Geological Survey of Pakistan, Quetta, Pakistan

ABSTRACT Paleoanthropological research in the Indo-Pakistan subcontinent has always been undertaken as a subsidiary part of studies aimed at understanding the total mammalian assemblages of Neogene Siwalik Group rocks. The historical review summarised here clearly indicates that these researches were mostly carried out by European and American scientists with negligible scientific input from local scientists. It is realised that pure and academic scientific pursuits are contingent upon socioeconomic aspects of the country. From this perspective, developing countries have very little appreciation of the scientific importance of such achievements. In addition to this, developing countries also suffer from the lack of appropriate technical manpower, comparative collections, and laboratory facilities. However, in a few countries, government patronage and liberal foreign assistance have helped establish paleoanthropological (and other related paleobiological) research on a firm foundation. The Geological Survey of Pakistan (GSP) has, for the last ten years, engaged in geological studies of the Neogene Siwalik Group sediments in collaboration with the Yale (and later Harvard) Peabody Museum and with Howard University. Although the GSP–Harvard Peabody project is primarily designed around hominoid origins research, the collaborative project's activities encompass all important aspects of vertebrate paleontology and fluvial sedimentology. Such multi-disciplinary approaches, the scientific results from which promise better understanding of paleoenvironmental and paleogeographic interpretations, have better appeal for obtaining government favour and assistance in developing countries. Based on the scientific progress made through these collaborative projects, the Geological Survey of Pakistan has approached the government to establish a modern, well-equipped laboratory to continue researches in vertebrate paleontology, paleoanthropology, and fluvial sedimentology. Minor finances involved in provisioning for equipment, reference specimen, and library facilities will be raised through requests to international aid-giving agencies. We will endeavour to provide through this laboratory a centre where academic as well as applied research can be carried out in complete harmony by scientists from all over the world, on an equal and reciprocal basis.

INTRODUCTION

Continental sedimentary rocks of Pakistan have long been known for their richness in vertebrate fossils, including hominoid primates. Sedimentary rocks make up more than 70% of the total rock units exposed in Pakistan. Of the total sedimentary deposits, fossil producing Neogene continental rocks constitute about 20%. Since the Early Miocene, when land conditions were established south of the main Himalayas, an almost unintrerrupted sequence of non-marine sediments more than 7,000 metres thick has been deposited, forming the sub-Himalayan ranges of the Indo-Pakistan subcontinent. Enclosing numerous well preserved fossils with occasional layers of bentonite, the rocks provide a unique opportunity for both taxonomic study of the fossils and geochronological control for world wide correlation. With more than 150 years of collection in easily accessible areas, it has been demonstrated that the continental deposits of Pakistan contain the best vertebrate faunal succession in the world. Although the previous workers have collected a plentiful variety of vertebrate fossils and carried out diverse geological studies, the sediments still have enormous potential to provide animated geological interest and unlimited fossil reserves. As a matter of fact, large areas of Neogene sediments are still unexplored, which may lead to important and rich fossil discoveries of immense paleontological interest.

The Neogene of Pakistan, therefore, is capable of yielding material crucial to the solution of critical questions in the evolutionary history of mammals, including hominoids. Major advances certainly have been made in paleoanthropological research elsewhere in the world, yet major questions about the origin, initial evolution, and differentiation of the human family remain unanswered. It may not be out of context to say that answers to many of these questions have or presumably will come from rocks exposed in those parts of the developing world, where people are struggling for the basic amenities of life. On the other hand, the interested scientists and research workers belong to affluent societies. What are the prospects for carrying out paleoanthropological research in such paradoxical circumstances? In an attempt to answer this question, we shall examine the subject in the light of our own experiences. We will proceed to discuss the subject matter in light of the historical background of vertebrate paleontological and paleoanthropological research in the Indo-Pakistan subcontinent, followed by the present status of this research in Pakistan, and finally the future prospects of continuing paleoanthropological enquiries in the developing countries.

HISTORICAL BACKGROUND

The history of vertebrate paleontological investigations began long before the formal establishment of the Geological Survey of India in 1851. Perhaps the first collection of bones for study, from the Siwalik hills of British India, was picked up by Captain (Sir) Proby Cautley in the early thirties of the last century. Vertebrate fossils from the eastern extension of these continental deposits were, however, collected much earlier, probably in the early twenties. By the middle of the last century, appreciable advances had been made in various paleontological attempts to determine the pattern of Late Tertiary mammalian evolution in the subcontinent. The discovery of hominoid fossils had been almost simultaneous with other mammalian fauna from Siwalik rocks. Due to their possible bearing on the phylogeny of man, their taxonomic affiliation had been a question of great interest. In fact, like today, the antiquity of man was one of the most eagerly investigated questions of that time. Dr. Hugh Falconer, one of the earliest workers on Siwalik fauna, wrote, "Among the four or five species of Siwalik quadrumana . . . one was inferred by Sir Proby Cautley and myself, in 1837, to have been a large ape exceeding the size of the Orang-Outang but of unknown immediate affinity" (see Oldham, 1868). In a later contribution of 1842, Dr. Falconer instituted a close comparison between the fossil canine and the corresponding tooth in recent skulls of orangutan and confirmed it to have been of a large ape allied to "*Pithecus satyrus.*" In the same paper, Dr. Falconer continued, "Captain Cautley and myself were constantly on the look-out for the turning up, in some shape or other of evidences of man out of the strata of the Siwalik hills, partly from considerations of a different order" (for detail see Oldham, 1868). He presumed that modern forms of life had originated during the Siwalik period; he hoped that rigorous search in these continental deposits might provide the vestiges of man's earliest abode.

In the later part of the last century, however, enthusiasm in hominoid research declined. Then major field and research activities with bearing on paleoanthropological thinking were re-activated between 1915 and 1935, when the Geological Survey of India, under the leadership of Dr. Guy Pilgrim carried out extensive fossil collection with stratigraphic control in the Siwaliks of Pakistan and adjoining areas of India. Important discoveries of *Sivapithecus* and various species then called *Dryopithecus* were made between 1910 and 1927 (see Pilgrim, 1910, 1915, 1927). The American Museum of Natural History, as well, entered into this research when, in 1922, Dr. Barnum Brown undertook extensive collection from the Potwar Plateau and nearby areas of India. Brown's collections led to important contributions on mammalian systematics and phylogeny, which also revised the biostratigraphy of the Siwalik rocks (Colbert, 1935). The primates were described by Brown et al. (1924) and later monographed by Gregory and Hellman (1926). The 1932–1934 "Yale University–North India expedition" led by G. Edward Lewis made good collections of fossils from the Potwar Plateau of Pakistan as well as from the adjoining Ramnagar and Haritalyangar areas of India. Lewis's, work resulted in the recognition of one of the most important hominoid fossils, *Ramapithecus*, which was long considered the most likely candidate for an ancestor of hominids (Lewis, 1934). Since then, *Sivapithecus* and *Ramapithecus* have been considered the most crucial link between the early Miocene hominoids and Pliocene hominids.

During these thirty years of extensive faunal collections, Siwalik mammalian systematics, evolutionary trends, and correlation with the European and American faunal zonations were established. Wadia and Aiyengar (1938) recorded that 82 primate specimens belonging to ten species and four genera had been described from the Siwalik sediments. A slack period followed again for a long time, with hardly any concerted effort to find hominoid fossils from the Siwaliks. Even after the partition of India and Pakistan, the status quo was maintained for a considerable time (but see the summaries of some work by Dehm, 1983 and von Koenigswald, 1983).

The purpose of presenting this brief history of paleoanthropological research in the Indo-Pakistan subcontinent is to emphasise that all the scientific work was carried out by European and American scientists. One often finds acknowledgment to Indian geologists for discovering some important fossils, but they were generally field assistants to the foreign scientists, rather than being equal peers, or had peripheral interests in paleontology. Evidently, a scholarly paleoanthropological group could not be organised in the subcontinent during the British domination or even for a long time after the establishment of the two independent nations of India and Pakistan. However, India's position in producing native paleoanthropologists was comparatively better than Pakistan. In terms of the establishment of reference collection and library facilities in India and Pakistan, the situation remained in favour of India. The famous Siwalik gallery, built as a result of 100 years of collection and research work from the entire subcontinent, remained intact at the headquarters of the Geological Survey of India at Calcutta. Thus, at the time of independence, India inherited a well-established Museum including the Siwalik gallery at Calcutta, while Pakistan was in the process of establishing the nucleus of a geological survey in the country.

DEVELOPMENTS IN PAKISTAN

Beginning in 1947, when Pakistan came into being, almost everything had to be started from scratch. Five geologists from the Geological Survey of India who opted to serve Pakistan initiated the Geological Survey of Pakistan (GSP) at Quetta (northwest Pakistan). Existing universities at the time of independence had no facilities for teaching geology and paleobiology. In view of pressing demands for better economic conditions in the country, all efforts of the newly established GSP were directed to tapping natural resources. By the late fifties, the Geological Survey of Pakistan began to make steady progress when liberal technical guidance and support from the U.S. Geological Survey became available. And in the meantime, some universities in Pakistan also started teaching geology.

As in any other developing country with poor economic resources, the policy of the Pakistani government has been to acquire a better standard of living for the people. As such, all technical activities of the GSP have been focused towards the developmental side, with hardly any consideration given to taking up pure research. Academic institutions also emphasized applied aspects and taught such basic courses as petrology, mineralogy, stratigraphy, and invertebrate and micro-paleontology. Vertebrate paleontology in general, and paleoanthropology in particular, did not get any attention, and when taught in some of the universities remained of very poor quality and low importance. No qualified vertebrate paleontologist is in the faculty of any academic institution in Pakistan save one member of the Zoology Department of Punjab University, Lahore, who has established a modest fossil collection with a small research group. The teaching of paleoanthropology has always remained beyond the scope of the curriculum of zoology, anthropology, or geology departments. The Archeology Department of the government of Pakistan and the few archeology institutes in the universities currently do not concern themselves beyond the pre-history of the last few thousand years.

CURRENT RESEARCH IN VERTEBRATE PALEONTOLOGY

With the brief survey outlined above, it is evident that Pakistan provided previous workers with an abundance of vertebrate fossils, including hominoids. Most of the published material on various investigations is available in Pakistan, but the fossils documented by previous workers now serve as a part of the reference collections and exhibits of many different museums around the world, while practically no worthwhile collection of vertebrate fossils exists in the country itself. Moreover, some collections available at the GSP and other academic institutions are devoid of proper documentation with respect to their modern taxonomy, stratigraphic position, and, in some cases, even geographic locations.

This deplorable situation led the Geological Survey of Pakistan to initiate projects aimed at developing its own collection, so that researches in vertebrate paleobiology could be made viable. Therefore, beginning in 1973, two independent projects, undertaken in collaboration with the Yale Peabody Museum and Howard University, began work on the Neogene rocks exposed in various parts of Pakistan. The aim of both collaborative projects is a broader understanding of mammalian evolution, paleoecology, and relationships with Neogene faunas from other parts of the Old World. The Yale–GSP (since 1981 the Harvard Peabody Museum–GSP) project emphasized comprehensive understanding of the Neogene faunas and sediments, with special stress on paleoanthropological studies in the Potwar Plateau.

The advantages of working in the Siwalik Group rocks of the Potwar Plateau are considerable. They offer a virtually continuous record of about the last 18 or 20 million years (m.y.) of Neogene time and contain one of the best known mammalian paleofaunas in the world. These factors make possible the precise definition of various biostratigraphic zonations, reconstruction of faunal community structures and habitats at a particular time, and documentation of changes in communities and habitats through time, within a sequence that can be independently calibrated by fission-track and paleomagnetic techniques (see details in Pilbeam et al., 1979). Disadvantages are that the predominantly fluvial sediments preserve relatively incomplete vertebrate remains, and that the ambitious scope of the project involves a long term commitment of time, effort, and adequate funds.

Substantial progress has been made in our GSP–Harvard Peabody Museum collaborative project. Over 30,000 fossils have been collected from more than 600 localities, while over 150 papers on various subjects have issued from this multidisciplinary project (Badgley, 1981; Shah and Pilbeam, 1981). Besides, field and laboratory training have been imparted to many local workers. A large percentage of the collected fossils have well-documented stratigraphic records and are under study by a team of international scientists. All specimens are being stored in the "Fossil Reference Collection" of the newly established National Laboratory in the Geological Survey of Pakistan, where important and relatively complete specimens will also be on display.

In terms of paleoanthropology, early workers' hopes to find a "Siwalik Man" in one form or another were brought close to reality when Lewis's 1932 collection of primate fossils generated the concept that the Siwalik anthropoids may include the vestiges of man's earliest ancestor. Infrequent but meaningful field attempts made by foreign teams collected critical material from the Siwaliks. In light of these earlier efforts, the materials recovered from Pakistan during the last decade produced substantial changes in the interpretation of hominoid evolution.

During the current investigations, 140 hominoid specimens have been collected from 24 localities. A large number of hominoid specimens comes from closely spaced localities in the Khaur area (northern Potwar Plateau), which are dated to between 10 and 7 m.y. BP (Pilbeam et al., 1980). However, within this sequence representing three million years' duration most of the hominoid fossils come from a section about 200 m thick dated to around 8 m.y. BP

(Badgley, 1982). A few scattered specimens have also been collected from the Chinji Formation in the southern Potwar, which represent a time period between 13 and 11 m.y. BP (Raza, 1983; Raza et al., 1983).

The hominoids collected by the GSP–Harvard collaborative team represent at least three species: *Ramapithecus punjabicus, Sivapithecus indicus,* and *Gigantopithecus* sp. These and other Miocene species have been variously classified at one time or the other in the Pongidae or the Hominidae. A large number of samples recovered from Pakistan helped the team to reclassify these Neogene hominoids (Pilbeam et al., 1977).

Material from 13 to 8 m.y. BP collected in Pakistan greatly expanded our knowledge of these hominoids; previously unknown parts are now represented, in particular, postcranial remains, which are so important in locomotor anatomical studies and even for precise taxonomic placement of these species and a face that showed links to *Pongo*. New finds of hominids from other parts of the world have greatly enhanced our understanding of the stages of hominid evolution since about 5 m.y. BP but an enormous gap in our knowledge still exists for the period between 8 and 5 m.y. BP. Fortunately, a large area in Pakistan is occupied by the continental sediments of this critical period. Provided the said stock did not become extinct after 8 m.y. in Pakistan, it is believed that a thorough search in the Siwaliks, where so many reasons combine to indicate greater probability of success, will hopefully yield material crucial to the understanding of further critical phases in hominoid evolution.

During the last ten years of extensive research on the Miocene sediments and faunas of the Siwalik Group rocks in the Potwar Plateau, notable progress has been made in all aspects of mammalian paleobiology, fluvial stratigraphy, and hominoid phylogeny. The findings of our collaborative projects, however, have left many questions unanswered, and at the same time pointed out avenues in need of more rigorous investigation. We need more and better faunal samples from the 11 to 8 m.y. sediments from other parts of the Potwar Plateau for assessment and comparison of community composition and faunal assemblages with the better-known fauna of the Khaur area. Intensive collections of hominoids and other fossils are required from the 14 to 10 m.y. BP strata. Furthermore, older and less fossiliferous units of the Siwalik Group should be thoroughly scanned. Higher in the section, we are concerned with younger rocks of less than 8 m.y. to record the link between the extinct hominoids and hominids.

Our research activities are also aimed at establishing the origin of the Siwalik faunas, improving temporal control on appearance/disappearance of various mammalian taxa, and assessing the extents and episodes of mixing of the immigrant groups. The first appearance of hominoids and their extinction, if any, in the Siwaliks are eagerly investigated questions. Concomitant with the paleontologic activities, sedimentologic and taphonomic studies of the fossiliferous strata are being carried out. Together, these analyses will provide a comprehensive picture of the climatic, geographic, and habitat changes that happened during the last 18–20 m.y. All these efforts will result in providing the "contextual" information that will enhance our interpretations of hominoid paleobiology.

FUTURE PROSPECTS OF PALEOANTHROPOLOGICAL RESEARCH

In the foregoing pages, we have reviewed important aspects of the history and present status of paleoanthropological research in the Indo-Pakistan subcontinent. The purpose of presenting this synthesis of our experience in initiating vertebrate paleontological (including paleoanthropological) research in Pakistan is the hope that similar efforts could secure the future of this discipline in other countries as well. It may be emphasized that prerequisite to the development of a scientific institution (especially for paleoanthropology, as in our case here) in any country are general literacy, a strong infrastructure of education in science, and the encouragement of scholars. Public awareness of problems and liberal government financial aid are certainly crucial to success.

We are aware of the fact that research centres just do not suddenly emerge in a country. However, both the educational and to some extent financial capabilities of the developing country, as well as meaningful foreign collaboration from developed countries, can accelerate the process. This is true where the host country and visiting scientists can benefit on a "symbiotic" basis. In our case, the GSP has developed a small team with a nucleus of laboratory facilities that can successfully undertake paleoanthropological research with foreign collaboration. To make them more meaningful, we believe that our collaborative projects should be based on the principles of equality, mutual respect, and fair play in scientific as well as in personal matters. Furthermore, our colleagues' acceptance of Pakistan's proprietary rights to the fossils collected during joint field expeditions and their readiness to assist in developing laboratory and library facilities at the GSP are some of the factors ensuring the effective continuance of the collaborative projects.

The authorities at the GSP being fully convinced of the scientific importance of vertebrate paleontology and paleoanthropology research, are actively undertaking to establish a well-equipped laboratory so that this important scientific pursuit can be carried out on a permanent basis. Future collaboration will provide technical and financial assistance for the GSP National Paleontological and Sedimentological Laboratory. It has been realised that, in addition to continuing active field expeditions, efforts should be concentrated on establishing a fossil reference collection and a library. The GSP plans to include more geologists so that all phases of research in vertebrate paleontology, paleoanthropology, and continental stratigraphy can be undertaken. Needless to say, an effective training program for the GSP personnel is eagerly desired. All fossil specimens to be collected during our field expeditions will permanently reside in Pakistan and will constitute the Fossil Reference Collection of our proposed Laboratory. On a voluntary basis, we hope visitors will help in our efforts to acquire comparative fossil and recent materials from abroad. The main emphasis of collaborative projects with foreign countries will be to acquaint GSP participants with modern field and research techniques in vertebrate paleontology and fluvial sedimentology.

The GSP will in future shift emphasis from just conducting field work to developing the proposed National Laboratory to provide modern facilities, especially for fossil preparation, systematic studies, and sedimentological analyses. It is hoped that gradually the GSP personnel would be able to undertake independent field and laboratory studies on Neogene vertebrate faunas and sediments. It must, however, be emphasised that foreign participation will always be welcome.

In a developing country like Pakistan, pure scientific research can only survive if its achievements have some application in developing (or have an attached sister project to tap) economic natural resources of the country. That is why the technical activities of the proposed National Laboratory will also emphasise regional sedimentological studies to understand the genesis, paleoenvironment, and depositional mode of the continental sediments, which in turn may assist in delineating various prospective areas of secondary mineral deposits. This scheme, together with researches in vertebrate paleontology and paleoanthropology, will ensure the continuation of these studies in the developing countries.

ACKNOWLEDGMENTS

We thank the organizers of the "Ancestors" symposium, particularly Dr. Eric Delson, for their invitation and excellent arrangements. Mr. Asrarullah, Director General, Geological Survey of Pakistan, is thanked for his review of the manuscript. The Government of Pakistan kindly allowed our participation at the symposium.

LITERATURE CITED

Badgley, C (1981) Bibliography of the Geological Survey of Pakistan–Yale Peabody Museum Research Project on the Miocene sediments, faunas and hominoids of Potwar Plateau Pakistan 1973–Present. Mem. Geol. Surv. Pakistan *11*:19–42.

Badgley, CE (1982) Community reconstruction of a Siwalik Mammalian assemblage. Unpublished Ph.D dissertation, Yale University.

Brown, B, Gregory, WK, and Hellman, M (1924) On three incomplete anthropoid jaws from the Siwaliks, India. Am. Mus. Novitates *130*:1–9.

Colbert, EH (1935) Siwalik mammals in the American Museum of Natural History. Trans. Am. Phil. Soc. *27*:1–401.

Dehm, R (1983) Miocene hominoid primate dental remains from the Siwaliks of Pakistan. In RL Ciochan and RS Corruccini (eds): New Interpretations of Ape and Human Ancestry. New York: Plenum, pp 527–537.

Gregory, WK, and Hellman, M (1926) The dentition of *Dryopithecus* and the origin of man. Anthrop. Papers Am. Mus. Nat. Hist. *28*:1–123.

Lewis, GE (1934) Preliminary notices of new man-like apes from India. Am. J. Sci. *27*:161–179.

Oldham, T (1868) On the agate-flake found by Mr. Wynne, in the Pleiocene (?) deposits of the Upper Godavery. Rec. Geol. Surv. India, 1(*3*):65–69.

Pilbeam, D, Rose, MD, Badgley, C, and Lipschutz, B (1980) Miocene hominoids from Pakistan. Postilla *181*:1–94.

Pilbeam, D, Behrensmeyer, AK, Barry, JC, and Shah, SMI (eds) (1979)Miocene sediments and faunas of Pakistan. Postilla *179*:1–45.

Pilbeam, D, Meyer, GE, Badgley, C, Rose, MD, Pickford, MHL, Behrensmeyer, AK, and Shah, SMI (1977) New hominoid primates from the Siwaliks of Pakistan and their bearing on hominoid evolution. Nature *270*:689–695.

Pilgrim, GE (1910) Notices of new mammalian genera and species from the Tertiaries of India. Rec. Geol. Surv. India *40*:63–71.

Pilgrim, GE (1915) New Siwalik primates and their bearing on the question of the evolution of man and the Anthropoidea. Rec. Geol. Surv. India *45*:1–74.

Pilgrim, GE (1927) A *Sivapithecus* palate and other primate fossils from India. Geol. Surv. India, Palaeont. Indica, N.S. *14*:1–24.

Raza, SM (1983) Taphonomy and paleoecology of Middle Miocene vertebrate assemblages, Potwar Plateau, Pakistan. Unpublished Ph.D. dissertation, Yale University.

Raza, SM, Barry, JC, Pilbeam, D, Rose, MD, Shah, SMI, and Ward, S (1983) New hominoid primates from the Middle Miocene Chinji Formation, Potwar Plateau, Pakistan. Nature *305*:52–54.

Shah, SMI, and Pilbeam, D (eds) (1981) A contribution to the geology of Siwaliks of Pakistan. Mem. Geol. Surv. Pakistan *11*:1–77.

von Koenigswald, GHR (1983) The significance of hitherto undescribed Miocene hominoids from the Siwaliks of Pakistan. In RL Ciochan and RS Corruccini (eds): New Interpretations of Ape and Human Ancestry. New York: Plenum, pp. 517–526.

Wadia, DN, and Aiyengar, NKN (1938) Fossil Anthropoids of India: A list of the fossil material hitherto discovered from the Tertiary deposits of India. Rec. Geol. Surv. India *72*:467–494.

Ancestors: The Hard Evidence, pages 346–351

Ancestors for Us All: Towards Broadening International Participation in Paleoanthropological Research

Glynn Ll. Isaac
Department of Anthropology, Harvard University, Cambridge, Massachusetts 02138

ABSTRACT It is highly desirable that enquiry into human origins becomes a pursuit in which scientists from all nations of the world are actively engaged. Current imbalances can best be redressed if scientists and institutions that already have developed research programs: 1) form partnerships with colleagues in countries where such programs are not yet as well established; 2) help secure training for students and technical staff; 3) help develop laboratory and museum facilities; and 4) help to educate the public and to arouse the interest of governments in the fossils and antiquities of their own countries.

INTRODUCTION

Fossilised remains of human ancestors and archaeological vestiges of their handicrafts are being found in all the inhabited continents. Clearly, the narrative of human origins spans the globe, and the pursuit of knowledge concerning the events and processes of human evolution needs to be a cooperative venture shared among peoples of all nations. However, in the world as it is today, there are some important imbalances. To overcome these, thoughtful consideration by the scientific community is needed. After thought and discussion, deliberate action will be essential.

The discussion of the topic offered here is inspired and informed by the session that concluded the "Ancestors" symposium in New York. The session was entitled "Problems of Sponsoring Paleoanthropology Faced by Institutions in Developing Countries." The issues were addressed by an international panel of scientists consisting of: P.K. Basu (India), J.D. Clark (U.S.A.), M. Day (Britain), H. de Lumley (France), A. Nkini (Tanzania), L. Osmundsen (U.S.A.), J. Radovčić (Yugoslavia), S.M. Raza (Pakistan), Wu Rukang (China), M. Sakka (France), S.M.I. Shah (Pakistan), E. Vrba (South Africa), and G. Isaac (U.S.A.), Moderator.

This brief report and commentary cannot pretend to represent fully the views of any one member of the panel, let alone the ensemble, but an attempt is made to convey the shared sense of the major issues. This report should be read in conjunction with the preceding paper by Drs. Ibrahim Shah and Mahmood Raza, who eloquently summarise the past experiences and the aspirations of one country—Pakistan.

THE SITUATION

A situation of imbalance exists as a consequence of several intersecting historical circumstances. Notably, 1) the early ancestors of humankind lived only in the Old World tropics. Consequently, while later fossils are being recovered and studied all over the globe, fossils of the earliest stages can only be found in Africa and the southern portions of Eurasia. However, 2) over the last two centuries, the great international movement that we call "modern science" has had its location of maximum growth and intensity in an almost exactly complementary sector of the globe—namely, first in Western Europe and then later also in North America, the Soviet Union, China, and Japan. Consequently, the great majority of the world's trained scientists and the great majority of well-equipped laboratories and museums are situated in regions where the early fossils do not occur. This, in turn, has meant that over the past century, as scientific curiosity about human evolutionary origins rose in intensity, expeditions and colonial organisations reached out into the scientifically less developed countries of the Old World tropics and collected specimens that were often taken back to the industrial centers from which the expeditions emanated. Participation of the citizens of the fossil-homeland countries has often been restricted to nonscientific roles.

There has been a kind of inevitability to this pattern until recently, but now, with the emergence of independent, self-governing nation states over most of Africa and Asia, and with the steadily increasing numbers of university educated young scientists in these countries, such a situation is no longer appropriate or tolerable. Equally, with the rise of sovereign nation states around the world, the peoples of each country take pride in the heritage of the human past that is to be found within their borders. The fossils and artifacts from each country are legitimately expected to be appropriately housed and preserved within the country, and there is mounting concern that, while knowledge should be shared, members of each nation's own scientific community should be fully involved in the discovery, study and interpretation of evidence derived from the country itself.

The challenge confronting the international scientific community is clear. What can be done to broaden participation in the pursuit of knowledge about human origins? How can nations with large numbers of trained scientists and well-equipped institutions help with the development of proportionally equivalent levels of activity in those countries that have hitherto had lower intensities of involvement?

The panel at the "Ancestors" symposium considered several lines of possible action: 1) recruitment and training of young scientists, 2) museum and research facilities, and 3) educating the public and informing governments. It was clear to all participating in the discussion in New York that these lines of action are highly interconnected. For instance, recruitment and training cannot proceed successfully unless the public and the government of the country in question are aware of, and interested in, the paleoanthropological materials of the country. Similarly, training would be futile if no facilities are developed where research can be pursued and where specimens and records can be safely cared for. There are thus no priorities among these headings. All are vital from the outset.

RECRUITMENT AND TRAINING

There exist numbers of countries in which strata preserve important paleoanthropological evidences that are already known and which are being found and studied largely by expeditions emanating from developed nations. For these nations, there was a strong sense that the first step is for the expedition scientists to form partnerships with colleagues in the most appropriate existing institutions of the host country—even if there are not perfectly matched counterparts. This commonly means partnerships with scientists in museums, geological surveys, and universities. The second step is to encourage university students to participate in the field researches. For those students who show aptitude and enthusiasm, the visitors have an obligation to help arrange for further undergraduate—or more usually, graduate—education. Often this involves the securing of scholarship funds that will allow the student to go abroad to a well-established institution so as to secure a full and in-depth training. It is increasingly common for expeditions to participate in friendly reciprocation of this kind, but it does also depend on the sympathetic help of funding agencies in their countries of origin.

The training of technical personnel is also important. This too can begin with the apprenticeship of appropriate young men and women in the work of visiting expeditions. Further technical training can then proceed by having experienced personnel from developed nations come and work in the host country and simultaneously design and establish facilities and train apprentices in their operation. Alternatively, prospective technical personnel can go abroad for training and experience. Both can occur, which is what happened in the case of the excellent casting laboratory attached to the National Museums of Kenya, a laboratory that supplies the world demand for high quality casts of the numerous important fossils that have been discovered in Kenya and in Tanzania. Regional centers of expertise such as this can then help other neighbouring countries develop their capabilities. For instance, the Nairobi casting lab has helped train technical staff for the Malawi Antiquities Service.

Modes of recruitment are an important aspect of securing a vigorous, well-trained paleoanthropological work force in developing countries. Paleoanthropology, with its arduous fieldwork and its elusive fragmentary evidences, is a pursuit to which, regardless of ability, not all temperaments are equally well suited. How are countries to find the youngsters that have the particular flair and drive? One possibility that was discussed was that the organisers of research—both visitors and hosts—look for opportunities to run field schools as a part of their programs. Such schools will have the combined effect of broadening well-informed public awareness of what is involved in paleoanthropology and hopefully of intensifying educated enthusiasm for the subject. Many more students will participate in the schools than can become life-long professional paleoanthropologists. Some among them are liable to become schoolteachers of biology, geography, history, etc. Some will become agricultural officers, geological surveyors, etc. Many benefits for paleoanthropological discovery flow from having such members of professional communities be aware of and interested in prehistory. Last, but not least, from among the participants in field schools from time to time students with the aptitude and passion for archaeology, paleontology, and paleoecology will emerge and go on to take further training.

At the panel discussion, Dr. Mahmood Raza gave expression to several other important points:

> What are the prospects that the well-trained, new returnee will be able to carry out research? It's quite common that when well-qualified people return, they are involved in all kinds of office duties. They wear ties and suits, and do not get much time for research. Given the circumstances that they do manage to get away from these exercises, then there's not adequate comparative material, and often only very poor library facilities, that they can use to keep in touch with what is happening in other parts of the world. One of the possible solutions, in addition to requesting their own institution or government for funds, would be to request funds from various international agencies to develop a modest library and raise a comparative collection. That could be done when people are about to finish their studies and want to go home. One could apply for that kind of support and could prepare a library which eventually will go to one's future institutional base.
>
> The last point I would like to make concerns exchange among scientists. I personally feel that would be beneficial if scientists of developing countries could visit each other's institutions and participate in one another's field research projects. This would provide an opportunity to supplement each other's facilities and also help in broadening understanding of regional problems.

LABORATORY, MUSEUM AND RESEARCH FACILITIES

It is now widely agreed and recognised that fossils and antiquities recovered in any country are part of the heritage of that country and that such materials should, wherever possible, reside in their country of origin. Sometimes, where facilities do not yet exist for full-scale study in the source country, temporary export is appropriate, but for many reasons this should be avoided whenever possible. Prompt return is always desirable. However, this kind of policy entails the establishment of suitable, safe storage and study facilities such as are characteristic of well-designed modern museums.

Expeditions working as visitors in a country should be increasingly aware of the need and obligation to provide facilities, where these are lacking, for the safe housing and study of their collections. Clearly, it is irresponsible to collect irreplaceable antiquities from the field and then not be able to preserve them properly in the country to which they belong. To solve the problem by long term export is equally unacceptable nowadays. This principle has had clear expression in the action, for instance, of the French and the American funding agencies in helping to develop storage and study facilities in Ethiopia and Kenya. Similar moves are underway to help Pakistan and other countries.

In addition to storage, research facilities and apparatus are essential—both to facilitate research activity by young, newly-qualified scientists and to allow visiting scientists to work on materials that they are no longer permitted to take away for study.

The facilities that are needed minimally include such things as cameras, microscopes, X-ray machines, cast-making labs, and libraries. Eventually, such things as scanning electron microscopes, dating laboratories, and chemical laboratories will also be needed.

Quite as important as buildings and apparatus are generous terms of employment that enable young scientists to pursue research. Such terms need to include appropriate support staff (secretarial and technical), vehicles, and fieldwork opportunities; funds and encouragement to attend appropriate conferences; periodic study leave so as to keep up to date; libraries and reprints, plus computer facilities. All this means substantial amounts of money, of which initially the host country can commonly supply only a part. The international paleoanthropological community will need to work hard in securing the support of its sponsors to help develop such facilities where they are needed and where they will be well used.

Clearly, the cooperative development of a network of research facilities in countries where important paleoanthropological evidence is found presupposes a spirit of common enterprise that simultaneously recognises legitimate national pride along with the sharing of enquiry and new knowledge. It will be essential that protocols governing access to field sites and to collections be developed. These will presumably involve the normal etiquette of the reservation of access to recently-recovered unpublished materials and the setting up of research clearance procedures to which would-be visitors can openly apply.

The interest of the specimens is also important. It is predictable that in ensuing decades the number of scientists and students interested in human evolution will steadily increase. If every member of this growing community handles and measures every fossil, the specimens will be worn out before the end of the century. Members of the panel reported on steps that curators are taking to keep access open when it is needed and yet preserve the fossils. Providing excellent casts, making available good, standard photographs, and issuing lists of standard measurements were all possible procedures that were mentioned.

The question of curation and access leads to the issue of the production, quality, and distribution of *casts* and *replicas*. The symposium in New York had the benefit of a report and commentary from Mrs. Lita Osmundsen, Director of Research for the Wenner-Gren Foundation, which organization for many years pioneered in the production and circulation of high quality casts. Mrs. Osmundsen stated:

> If techniques of mold- and cast-making are thought of in the light of needs for wide distribution rather than in terms of occasional, individual needs and interests, new demands are placed on the production process. The leap from producing casts in a hit-and-miss way to one of developing a network of facilities for a maturing field of study is not a simple one. It requires not only finances and ongoing support but mutual collaboration on a global basis and the commitment of scholars, technicians, and administrators at the institutional, provincial, and governmental levels. We must also take into account the fact that despite the crying need this community feels for access to these materials, in mass-production terms the market is very limited, and the process, if well-designed for the control of quality, can be more costly than most people foresee. Furthermore, added to that cost is the factor of upkeep and update, which very few people think about . . . .
>
> The existence of casts of high and consistent quality can contribute to the protection, preservation, and interpretation of originals and the dissemination of knowledge. Much depends, however, on acquisition and on the integrity of the scholarly community at large to honor the standards of quality and respect the inherent rights of ownership. Copyrights attest to those rights and to the authenticity of product, but piracy does exist. It is the profession itself that must monitor these acts and prevent them primarily by moral pressure. Recipients of known pirated materials are equally responsible for undermining the total enterprise and rewarding such practices.
>
> Issues that need to be further discussed include access, conceptualization, funding, appropriate methods of quality control, limits to fidelity and accuracy, and respect for the rights of ownership. These begin to set the agenda for a conference of the best and most responsible minds in the profession, who can develop the guidelines for managing a constructive and useful enterprise for disseminating reliable materials. The developing countries must play a major role. Developing such an enterprise offers a great opportunity.

PUBLIC EDUCATION

We confront here a classic dilemma—a which-comes-first-the-chicken-or-the-egg situation! Public knowledge, interest, and pride are important as eggs to justify and secure support for enquiry into prehistory, but first one has to have a chicken in the form of trained personnel and effective museums! Dr. Desmond Clark, as a member of the panel, put the matter well:

> I was for 23 years a museum director in an African country (Zambia), and I fully appreciate the problems faced more recently by scientists starting up in independent nations. I had the same problems. The first is involved in educating the public, and that is bound up with educating the government to give you some money! This is absolutely essential. It took us about 15 years to make my government (in colonial Northern Rhodesia) realise that a curator did not simply sit at the museum door and wait to show people around.
>
> But after a lot of pushing and so on, once they realized the worthwhileness of what we were doing, then they gave us as much as they could afford, not only for conservation and display, but also for research . . . so I recommend first of all to educate the government for some funding, then also to encourage scientists from outside to come and collaborate so as to help provide collections and help provide the basic research that goes to understanding the collections. All this must be done in close collaboration. No longer is there any room for the hit-and-run expeditions of the past, where people descended from the sky, dug big holes, scooped up everything, took it away, and that was that. The people in the countries like to see the originals. They don't want to see casts in their national museums; they want to see originals and take pride in what it is that comes from their country.

Starting from scratch to disseminate knowledge and arouse interest clearly is not easy, but clearly it is important that scientific guests in developing countries put effort into helping their host scientists with this task, by giving classes and lectures, providing newspaper stories, and helping with exhibits. As already discussed, one possibility is the operation of field schools that cater to as wide a range of participants as possible—certainly more than just a handful of trainees.

Paleoanthropology is mainly important for the answers that it provides about origins and history, but its researches also do bring practical benefits, and the scientific community needs to help developing countries realise and make use of those benefits. Among others, these include the links between comparative anatomical research and medical school anatomy training; or links between paleoanthropology, stratigraphic geology, and the mapping of economic rock and mineral resources—to witness, diamond-bearing deposits in Angola, diatomites in Kenya, and gravels everywhere.

The "Ancestors" exhibit has given the public of New York an extraordinary opportunity to become more interested and more informed. Should such exhibits of original specimens now go to capital cities all over the world in order to promote interest and support? Many scientists would join with Dr. Mary Leakey, who argued at the symposium that New York's good fortune should not become a universal expectation, since if it were to do so, the irreplaceable fossils would be subject to repeated wear and risk of damage or loss. However, independent plans are already underway for carefully designed travelling exhibits involving casts. Such exhibits can surely go to cities all over the world and perhaps, if combined with the exhibition of originals from the host country, can serve to arouse interest and to symbolise the international characteristics of human origins research.

THE FUTURE

The panel adopted no resolutions and made no formal recommendations. However, some aspects of what should be done stood out particularly clearly. Most important is the development of a sense of partnership linking scientists in countries with well-established traditions of research and those where local involvement is just beginning. Whenever possible, research should be a joint collaborative effort. Happily, this is rapidly becoming common practice.

Members of such partnership enterprises then need to work together to foster interest and support in the developing nation, even if the support is modest in financial terms. The partners also need to work to *educate the sponsors of paleoanthropological research* to recognise that it is no longer sufficient for them just to buy tents, jeeps, and airplane tickets. Funds must be included to facilitate the participation of colleagues and students from the host country and to cover the development of crucial facilities for the storage and study of the specimens that are found. Helping to secure scholarship funds for young trainees is also a crucial reciprocal responsibility. For example, the Baldwin Fellowships of the L.S.B. Leakey Foundation is a pioneer program providing training for African nationals. All these kinds of changes for the better are already well underway, but they need all the help they can get.

The "Ancestors" symposium as a whole and the session on problems for developing countries in particular gave ample evidence of the existence of a grand common *esprit* that links scientists from around the world as they share the pursuit of enquiry into human origins. Given this *esprit*, there are surely grounds for optimism about redressing current imbalances—optimism, yes, but complacency, no! Perhaps, the "Ancestors" symposium can help us to start thinking about a formal international research and education drive modeled on the IBP—a decade of human origins research that, among other things, will organise funds to train personnel and develop facilities where they are most needed.

Index